Woven Fabric Engineering

edited by

Prof. Dr. Polona Dobnik Dubrovski

Woven Fabric Engineering
Edited by Prof. Dr. Polona Dobnik Dubrovski

Published by InTech
Janeza Trdine 9, 51000 Rijeka, Croatia

Publishing Process Manager Jelena Marusic

Technical Editor Teodora Smiljanic

Cover Designer InTech Design Team

Additional hard copies can be obtained from orders@intechopen.com

Woven Fabric Engineering, Edited by Prof. Dr. Polona Dobnik Dubrovski
p. cm.
ISBN 978-953-307-194-7

For EU product safety concerns:
IN TECH d.o.o., Prolaz Marije Krucifikse Kozulić 3, 51000 Rijeka, Croatia,
info@intechopen.com or visit our website at intechopen.com.

Contents

Preface

Woven Fabrics are flexible, porous materials used for clothing, interior and technical applications. Regarding their construction they posses different properties which are achieved to satisfy project demands for specific end-use. If woven fabrics are to be engineered to fit desired properties with minimum production costs, then the relationship between their constructional parameters and their properties must be first quantitatively established. So a great attention should be focused on woven fabric engineering, which is an important phase by a new fabric development predominantly based on the research work and also experiences. For the fabric producer's competitiveness fabric engineering is the important key for success or at least better market position.

Nowadays, a great attention is focused on the fastest growing sector of textile industry, e.g. technical textiles, which are manufactured primary for their technical performance and functional properties rather than their aesthetic or decorative characteristics. Technical woven fabrics are used in a large number of diverse applications such as protective clothing, in agriculture, horticulture, finishing, building and construction, filtration, belting, hygiene, automobiles, packaging, etc. Woven technical fabrics are also the reinforcement component in engineering material no.1, e.g. composites, which offer significant opportunities for new applications of textile materials in the area of aerospace, defence, construction and power generation, land transportation, marine.

The main goal in preparing this book was to publish contemporary concepts, new discoveries and innovative ideas in the field of woven fabric engineering, predominantly for the technical applications, as well as in the field of production engineering and to stress some problems connected with the use of woven fabrics in composites.

The book is organized in five main topics and 20 chapters. First topic deals with the Mechanical Properties Engineering. For technical applications the mechanical properties of woven fabrics are one of the most important properties. Many attempts have been made to develop predictive models for mechanical properties of woven fabrics using different modelling tools and to define the influence of woven fabric structure on some mechanical properties. This topic includes six chapters dealing with: prediction of woven fabric tensile strength using design experiment, artificial neural network and multiple regression (chapter 1), prediction of plain fabric elastic properties using finite element method (chapter 2), woven fabrics tensile properties modelling and measuring (chapter 3), the influence of biodegradable yarns (bamboo, polylactic acid, soybean protein) on mechanical properties of woven fabrics (chapter 4), experimentally verified theory of identification of stages of cotton fabric by static tearing process (chapter 5), and the data base system of the fabric structure design and mechanical properties (chapter 6).

The second topic is focused on Porous Properties Engineering. Woven fabrics are porous materials which allow the transmission of energy (electromagnetic radiations: UV, IR, light,etc.) and substances (liquid, gas, particle) and are, therefore, interesting materials for different applications. The chapters involved within this topics cover: the theory of flat textiles porosity and the description of a new method for porosity parameters assessment based on the air flow through the flat fabrics (chapter 7), the modelling of air permeability behaviour of woven fabrics using artificial neural network method (chapter 8), and the theory of the UV protective properties of woven fabrics with the emphasis on the influence of woven fabric geometry on ultraviolet protection factor (chapter 9).

The woven fabric surface unevenness prediction and evaluation (chapter 10), detection of woven fabric faults by morphological image processing (chapter 11), and the research dealing with the surface properties of PES fabrics on the basis of chromatic aberration and dynamic wetting measurements (chapter 12) are discussed within the third topic Surface Properties Engineering.

The forth topic of the book deals with the Textile Production Engineering, where the use of microwaves by finishing processes (chapter 13), basic properties and advantages of coated fabrics with woven component as substrate (chapter 14), and the importance of the cloth-fell position (chapter 15) are discussed.

The last topic Textile Composites Engineering involves the contributions dealing with mostly woven fabrics as reinforcement phase in polymer composites. It comprehends the characteristics of natural fabrics in composite ranging from mat to woven fabrics (chapter 16), crashworthiness investigation of polyamid composite crash boxes with glass woven fabric reinforcement (chapter 17), the influence of the immersion time in different environments (water, natural seawater, detergent/water liquid) on some mechanical properties of E-glass woven fabric reinforced polymer composites (chapter 18), the investigation of weight loss of composites with glass woven fabric reinforcement as well as with aramid fibres reinforcement (chapter 19), and simulations of composite reinforcement forming (chapter 20).

The advantage of book Woven Fabric Engineering is its open access fully searchable by anyone anywhere, and in this way it provides the forum for dissemination and exchange of the latest scientific information on theoretical as well as applied areas of knowledge in the field of woven fabric engineering. It is strongly recommended for all those who are connected with woven fabrics, for industrial engineers, researches and graduate students.

Editor

P D Dubrovski
University of Maribor,
Slovenia

a. Mechanical Properties Engineering

1

Anisotropy in Woven Fabric Stress and Elongation at Break

Radko Kovar
Technical University of Liberec
Czech Republic

1. Introduction

Anisotropy is a characteristic of most fabrics, especially woven; the impact of the direction of loading on tensile properties can be enormous and is frequently examined, for example in (Dai & Zhang, 2003; Hu, 2004; Kilby, 1963; Kovar & Dolatabadi, 2009; Kovar, 2003; Lo & Hu, 2002; Pan & Yoon, 1996; Postle et al., 1988 etc.). Anisotropy of properties comes out of anisotropy of the structure, based on longitudinal fibers. For woven fabric there are two principal directions - warp and weft (fill), in which yarns and majority of fibres are oriented. Load in principal directions results in minimum breaking elongation and maximum initial modulus. For arbitrary load direction the values of tensile properties change and fabric deformation becomes more complex, often incorporating fabric shear and bend deformation.

Although weave anisotropy is well known, tensile properties are usually theoretically and experimentally investigated namely for principal directions; the main reason is probably complexity of deformation and stress distribution when the load is put at non-principal direction. In this section we shall try to make a step to describe and perhaps to overcome some of these problems.

In practical use, the fabrics are often imposed load in arbitrary direction, bi-axial load or complex load composed of elongation, bend, shear and lateral compression. To predict tensile properties becomes more and more important with development of technical textiles.

Now only main difficulties, connected with the topic of this section, will be outlined:

a. At diagonal load great lateral contraction occurs. It causes complex distribution of stresses. It results in stress concentration at jaws when experiment in accordance with EN ISO 13934-1 is used.
b. There are yarns cut ends in the sample where tensile stress starts from zero.
c. Shear deformation causes jamming of yarns, what can change yarn properties. Strength of the yarn in the fabric can be higher than the strength of free yarn.

There are not available many publications, based on real fabric structure and solving the problem of woven fabric tensile properties in different directions. The reason is mentioned long range of problems and difficulties. Monographs (Hearle et al., 1969 and Postle et al., 1988) are involved in problems of bias fabric load only marginally. (Hu, 2004) is oriented on influence of direction on properties such as tensile work, tensile extension, tensile linearity etc. and uses another approach. Fabric shear at bias extension is investigated in (Du & Yu, 2008). Model of all stress-strain curve of fabric, imposed bias load, is introduced for example in (King, M. J. et al., 2005) with the respect to boundary conditions (stress concentration at

jaws). In (Peng & Cao, 2004) is area of fabric sample separated into 3 zones with different characteristics of bias deformation. Experimental models of woven fabric deformation in different directions are presented by (Zouari et al., 2008). Often the mechanics of continuum approach, coming out of prediction of Hook's law validity, is used, for example, in (Du & Yu, 2008; Hu, 2004; Peng & Cao, 2004 and Zheng et al., 2008). A new method of anisotropy measuring is proposed by (Zheng, 2008) etc.
This section is oriented first of all on anisotropy of rupture properties of weaves, imposed uniaxial load in different directions. The main goal is to develop algorithm for calculation of plain weave fabric breaking strain and stress under conditions of simulated idealized experiment. There are two ways of ideal uni-axial woven fabric loading (details are in section 2): (a) Keeping stable lateral (i.e. perpendicular to direction of load) dimension, (b) Keeping lateral tension on zero (i.e. allowing free lateral contraction).
In this chapter rupture properties will be analyzed for plane weave structure.

2. Nomenclature

β_0, β – angle of warp yarns orientation to the load direction before and after load [rad].
γ - shear angle [rad].
ε - relative elongation or strain [1].
μ - yarn packing density, a share of volume of fibrous material and volume of yarn [1].
ν - Poisson's ratio [1]
b - width of the fabric sample [m].
c - yarn crimp [1]
d - diameter of yarn [m].
F - force [N].
h_0, h - length of the fabric, taken for calculation, before and after fabric elongation [m].
l_0, l - length of the yarn in a crimp wave [m].
L_0, L - projection of the length of the yarn in fabric plane before and after load [m].
s_0, s - component of yarn length L into direction perpendicular to load [m].
p - spacing of yarns (pitch) [m].
S - fabric sett (yarn density) [m^{-1}].
t - fabric thickness [m^{-1}].
T - yarn linear density [Mtex].
Main subscripts: y – yarn, f – fabric, 1 – warp yarn or direction of warp yarns, 2 – weft yarn or direction of weft yarns, 1,2 – warp or weft yarns, 0 – status before load (relaxed fabric), b – status at break, d – diagonal direction (45 °), n – not-broken yarn, h – horizontal or weft direction, v – vertical or warp direction.

3. Models of woven fabrics rupture properties

Modelling always means simplification of reality and, in our case, idealizing the form of the load. When we wish to simulate experimental investigation of similar property, we should start with brief description of standard fabric rupture properties measuring with the use of EN ISO 13934-1 (strip test) standard. Fast jaws keep the sample in original width (width before load) what results in tension concentration at these jaws. Break usually occurs near the sample grip sooner then real fabric strength is reached. In Fig. 1 a, b are these critical points of the sample marked by circles.

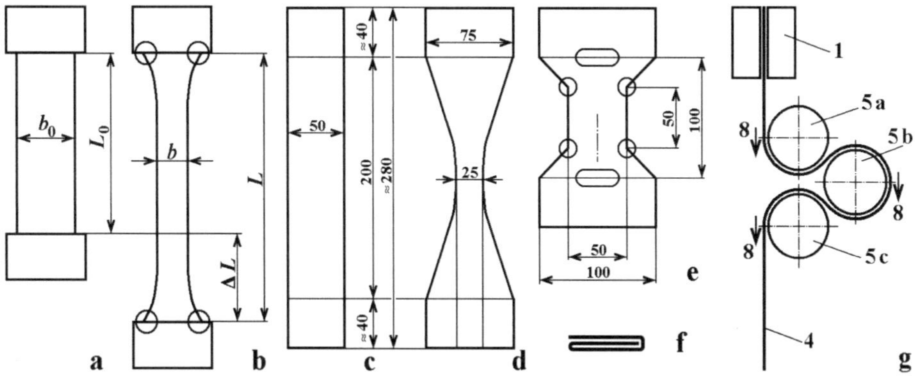

Fig. 1. Tension concentration at jaws and methods of its elimination

There are two main ways how to avoid the problem of tension concentration that occurs when using standard method; sample dimensions are in Fig. 1 c. First is reduction of fabric tension at jaws by narrowing the sample in central part, Fig. 1 d (Zborilova & Kovar, 2004) and 1 e (CSN standard 80 0810). This solution improves the results but some places with tension concentration stay. Till now the best results provides a new method (Kovar & Dolatabadi, 2010), Fig. 1 g; details of the method will be described in section 4.1.

When we wish to model fabric rupture properties and to avoid the problems with uneven tension distribution, we need to analyze a part of the fabric imposed constant load. Virtually there are two idealized situations:

a. To prevent sample from lateral contraction, i.e. to keep the fabric at original width b_0. This can simulate an experiment with infinite fabric width, when the influence of the sample margins becomes negligible. Restriction of lateral contraction must be, in practical experiments, connected with biaxial load, because some complementary load arises in the direction perpendicular to the direction of main load.
b. To allow free lateral contraction of the fabric. This model can simulate an experiment with flexible jaws that change the width simultaneously with fabric lateral contraction, or partly infinite fabric length, where the effect of fast jaws will not change sample relative elongation at break. This model is more complicated owing to fabric jamming and cut ends of yarns under load. Tensile stress in yarn at the cut end is zero and increases gradually due to yarn-to-yarn friction.

Yarn parameters and properties

From parameters and properties of yarn are, for fabric tensile properties investigation, most important: (a) Yarn cross-section as variable parameter. For simplification we can use yarn diameter d. For rough estimation of d can be used well known formula (1), where ρ is density of fibrous material and μ is average yarn packing density, the most problematic parameter. Its average value in free yarn used to be around $\mu_0 \approx 0.5$. This could be used for fabric with low packing density (lose fabric). At tight fabric yarn cross-section becomes flat and packing density increases. Here can be used effective yarn diameter d_{ef}. It is variable parameter, described as distance of yarns neutral axes in cross-over elements. In tight fabric can packing density reach, near warp and weft yarn contact, approximately $\mu_{ef} \approx 0.8$. In fabric near the break, mainly at diagonal load, yarn packing density reaches maximum possible value $\mu_b \approx 0.9$. (b) Yarn stress-strain curve, which can be for some purposes replaced by yarn breaking stress F_{yb} (strength) and strain ε_{yb}. Due to yarn jamming breaking stress can increase and strain decrease. (c) Unevenness of yarn geometry and other properties. In this section this will be neglected.

$$d = \sqrt{\frac{4T}{\pi \cdot \rho \cdot \mu}}, \quad d_{\mathrm{ef}} = \sqrt{\frac{4T}{\pi \cdot \rho \cdot \mu_{\mathrm{ef}}}} \qquad (1)$$

3.1 Model for infinite sample width

Restriction of lateral contraction makes the models relatively simple. This model and its experimental verification have been described in (Kovar & Gupta 2009). A conception of this theory is the test with infinite sample width that does not allow fabric lateral contraction. Experimental verification was based on keeping the tubular sample in original width by two fast wires (see Fig. 19). The yarns in model fabric are shown in Fig. 2; 1 is upper jaw, 2 bottom jaw before and 3 after elongation or at break.

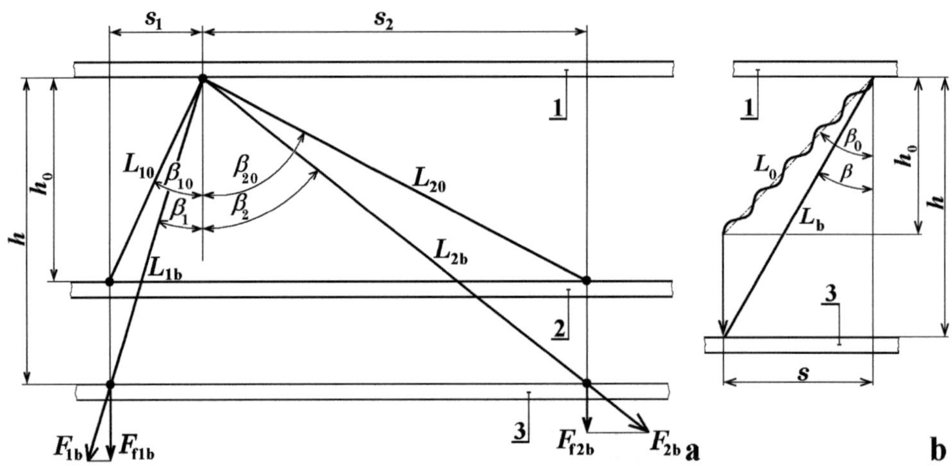

Fig. 2. Warp and weft yarns dimensions before load and at break

Relative fabric elongation, ε_f, and warp and weft yarns elongations, $\varepsilon_{1,2}$, are defined as

$$\varepsilon_{\mathrm{f}} = \frac{h - h_0}{h_0} = \frac{h}{h_0} - 1 \text{ and } \varepsilon_{1,2} = \frac{L_{1,2\mathrm{b}} - L_{1,20}}{L_{1,20}} = \frac{L_{1,2\mathrm{b}}}{L_{1,20}} - 1 \qquad (2)$$

Original lengths of warp L_{10} and weft L_{20} yarns before elongation between the jaws are

$$L_{10} = \frac{h_0}{\cos\beta_{10}} \text{ and } L_{20} = \frac{h_0}{\cos\beta_{20}} = \frac{h_0}{\sin\beta_{10}} \qquad (3)$$

where β_0 is the angle between direction of warp or weft yarns and direction of load. After fabric elongation from h_0 on h yarn angles decrease from $\beta_{1,20}$ to $\beta_{1,2}$.

Note: subscripts 1, 2 denotes validity of expression either for warp or for weft yarns, valid are either the first or the second subscripts, so often one expression contains two equations.

The final lengths of yarns segments will be L_{1b} and L_{2b}. As lateral contraction is restricted, horizontal projections of yarn segments will not change and so $s_{1,2} = s_{1,20}$. Fig. 2 b describes two main means of each yarn elongation, i.e. decrimping and yarn axial elongation.

With the exception of one particular load angle β_0 only one system of yarns reaches breaking elongation; in so called square fabric, when all the parameters are for warp and weft directions the same, first break yarns with $\beta_0 < 45\,^{\circ}$. For these broken yarns their lengths at fabric break will be $L_{1,2b} = L_{1,20} \cdot (1+\varepsilon_{1,2b}) \cdot (1+c_{1,20})$, where $\varepsilon_{1,2b}$ is yarn relative elongation at break and $c_{1,20}$ is crimp of the yarn, see Fig. 6 and Equation (7). As in this case it is s = const. we shall obtain, using Pythagorean Theorem, the length of the fabric at break $h = \sqrt{L_{1,2b}^2 - s_{1,2}^2}$ and fabric breaking elongation ε_{fb} will be

$$\varepsilon_{fb} = \frac{h}{h_0} - 1 = \frac{\sqrt{\left(L_{1,20} \cdot (1+\varepsilon_{1,2b}) \cdot (1+c_{1,20})\right)^2 - s_{1,2}^2}}{h_0} - 1 \tag{4}$$

where $s_{1,2} = h_0 \cdot \tan \beta_{1,20}$.

Characteristics of this formula is shown, for one value of warp yarns extensibility ε_{1b} = 0.2 and five values of weft yarns extensibility ε_{2b} = 0.1; 0.15; 0.2; 0.25 and 0.3, in Figure 3. As it was mentioned, with the exception of one critical angle only yarns of one system (warp ore weft) will break. The critical angle can be found as crossing points of the curves for warp and weft yarns.

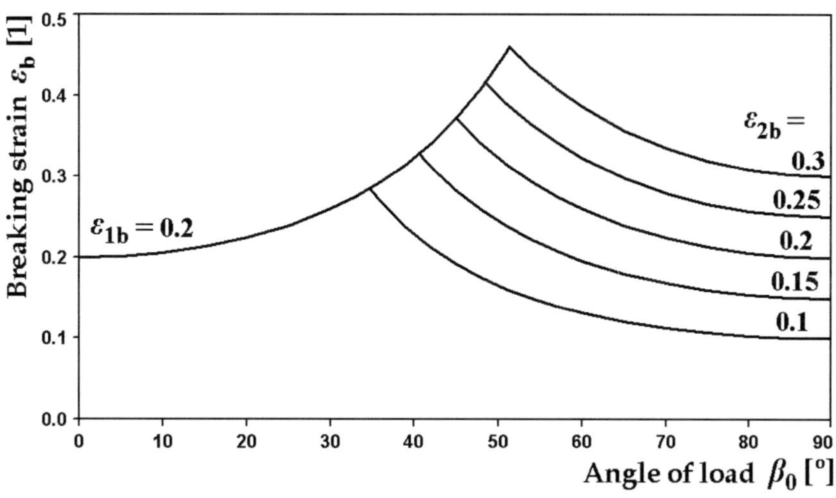

Fig. 3. Influence of angle load on breaking elongation for different yarn

Breaking stress, i.e. force necessary for damage of 1 m width of the fabric, can be calculated as the sum of the stresses from all yarns in 1 m of fabric width. Number of yarns in fabric width unit is $n = n_1 + n_2 = S_1 \cdot \cos \beta_{10} + S_2 \cdot \cos \beta_{20}$, where $S_{1,2}$ are fabric setts. The component of yarns axial stresses to the direction of external load from one broken yarn is $F_{f1,2b} = F_{1,2b} \cdot \cos \beta_{1,2}$ (see Fig. 2), $F_{1,2b}$ is strength of one yarn (axial force at break). The force from one non-broken yarn $F_{1,2n}$ depends on this yarn elongation at fabric break $\varepsilon_{1.2n}$ and on yarn stress-strain curve. For simplification we shall assume linear yarn deformation and so $F_{1,2n} = F_{1,2b} \cdot \frac{\varepsilon_{1,2n}}{\varepsilon_{1,2b}}$. The final result, the force necessary for breakage of 1 m fabric width is

$$F_{fb} = F_{f1,2b} + F_{f2,1n} = F_{1,2b} \cdot \cos\beta_{1,2} \cdot S_{1,2} \cdot \cos\beta_{1,20} + F_{2,1b} \cdot \cos\beta_{2,1} \cdot S_{2,1} \cdot \cos\beta_{2,10} \cdot \frac{\varepsilon_{2,1n}}{\varepsilon_{2,1b}} \quad (5)$$

where $F_{f1,2b}$ is the force in 1 m fabric width from all broken yarns, $F_{f2,1n}$ is the same from non-broken yarns. If change of angle of yarn incline during the elongation is neglected, i.e. if $\beta_{1,2} = \beta_{1,20}$, equation (5) would be simplified as

$$F_{fb} = F_{1,2b} \cdot \cos^2\beta_{1,20} \cdot S_{1,2} + F_{2,1b} \cdot \cos^2\beta_{2,10} \cdot S_{2,1} \cdot \frac{\varepsilon_{2,1n}}{\varepsilon_{2,1b}} \quad (6)$$

Examples of results of calculation are shown in the following charts. In Fig. 4 is set warp and weft yarns extensibility at break, that includes de-crimping and yarns breaking elongation, on $\varepsilon_{1b} = \varepsilon_{2b} = 0.2$, warp yarn strength $F_{1b} = 8$ and weft yarn strengths are variable. Warp and weft yarns set is $S_1 = S_2 = 1000$ m^{-1}. Similarly in Fig. 5 is shown influence of yarn extensibility on fabric breaking stress for $\varepsilon_{1b} = 0.2$, variable ε_{2b}, $F_{1b} = 8$ N and $F_{1b} = 6$ N. In Figs. 4 and 5 thick lines represent calculation according to equation (5) and thin lines follows equation (6).

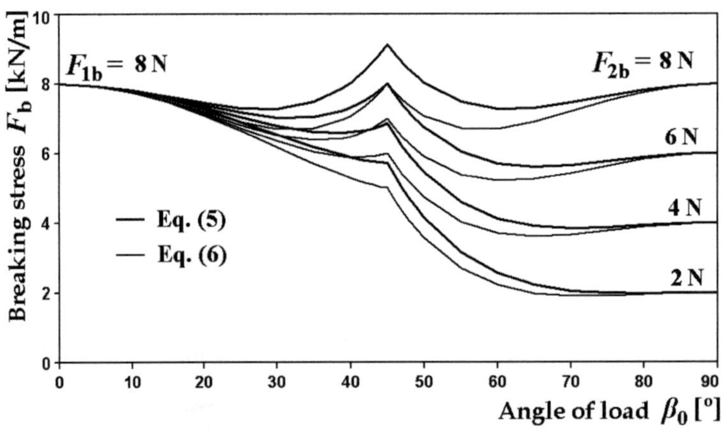

Fig. 4. Influence of angle β_0 on breaking stress for $F_{1b} = 8$ N, different F_{2b}, $\varepsilon_{1b} = \varepsilon_{2b} = 0.2$

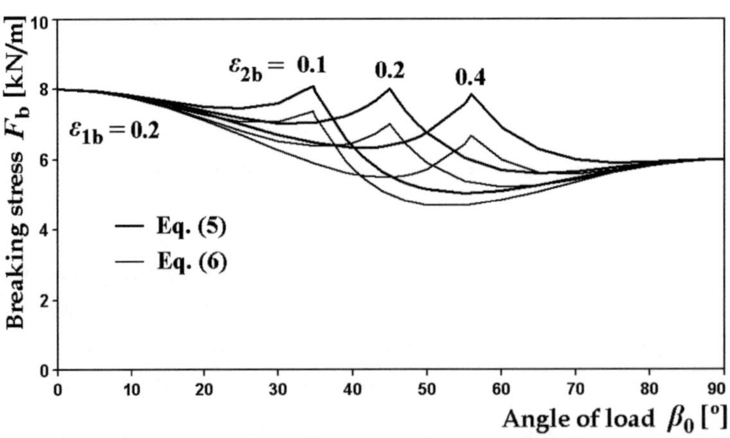

Fig. 5. Influence of angle β_0 on breaking stress for $\varepsilon_{1b} = 0.2$, different ε_{2b}, $F_{1b} = 8$ and $F_{2b} = 6$ N

Figures 3, 4 and 5 show significant peeks at critical angle where lines for warp yarns and weft yarns meet; this is not recognized in experimental results. This disagreement is caused by simplified model approach that assumes ideally even yarn properties. In real fabric, variability in yarn breaking strain and in other parameters causes, near mentioned critical point, break of some warp and some weft yarns together.

3.2 Model for infinite sample length

Models of ideally uni-axial woven fabric load in variable directions are rather complicated, first of all from the next reasons: (a) Substantial change in yarns incline (angle β) toward load direction during fabric elongation results in combination of tensile and shear deformation. The change of angle β allows itself certain fabric elongation and so its diagonal extensibility and lateral contraction is greater. (b) Reduction of yarns crimp in bias directions is limited, whereas at load in principal directions crimp of the yarns, imposed load, could be practically zero. (c) Change in fabric properties caused by jamming of yarns at diagonal load is great. The jamming could improve utilization of fibers strength and so the strength of the yarn could become better than that at load in principal directions. (d) There are cut ends of the yarns, bearing fabric load, what changes the results when the fabric width is limited. In the next steps will be modeled relaxed fabric, fabric at load in principal and in different directions.

3.2.1 Relaxed fabric

Investigation of fabric tensile properties starts at definition of relaxed state. It is described in (Lomov et all, 2007) etc. Simple model of plain weave balanced fabric in shown in Figure 6. Wavelength λ_1 of warp is defined by weft pitch p_2 and vice versa. Fabric thickness is t; average p value corresponds with reciprocal value of fabric sett S of the opposite yarn system and so $\lambda_{1,20} = 2p_{2,10} = \dfrac{2}{S_{2,10}}$. Main parameters of crimp wave are: wavelength λ, wave amplitude a and length of the yarn axis l. Wave amplitudes a are dependent on yarn diameters and in non-square fabric (i.e. $S_1 \neq S_2$, $d_{10} \neq d_{20}$, $l_{10} \neq l_{20}$ etc.) as well on fabrics setts, yarns diameters, imposed load (contemporary or in fabric history) and so on.

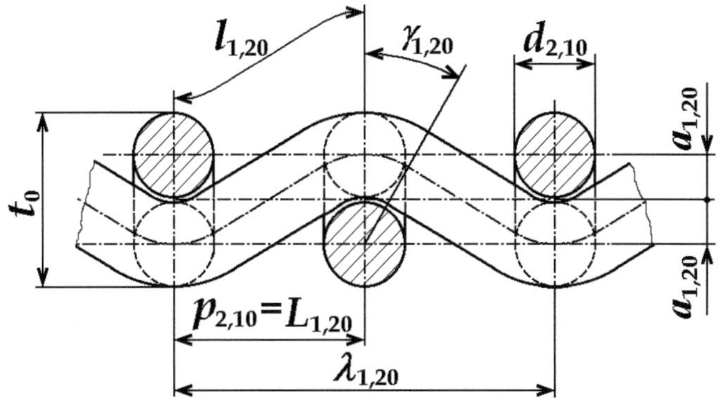

Fig. 6. Definition of yarn crimp in a woven fabric.

Crimp of the yarns in woven fabric as numeric parameter c is defined by equation (7); wavelength of warp λ_1 corresponds with pitch of weft p_2 and vice versa.

$$c_{1,20} = \frac{l_{1,20}}{p_{2,10}} - 1 \quad \text{or} \quad l_{1,20} = \left(c_{1,20} + 1\right) \cdot p_{2,10} \tag{7}$$

Lengths of the yarn in a crimp wave l will be counted with the help of equation (8). This formula approximates crimp c of loose fabrics (low packing density) using sinusoid crimp wave model and of tight fabrics (high packing density) using Pierce's model.

$$c_{1,20} = 2.52 \cdot \left(\frac{a_{1,20}}{p_{2,10}}\right)^2 \quad \text{or} \quad c_{1,2} = 2.52 \cdot \left(\frac{a_{1,2}}{p_{2,1}}\right)^2 \tag{8}$$

Relative crimp wave amplitude a/p can reach maximum value of 0.57735 at so called fabric limit packing density (maximal available fabric sett) for square fabric construction. In Figure 7 is curve, following equation 8, compared with calculation of crimp using sinusoid and Peirce crimp wave models (Kovar & Dolatabadi 2008).

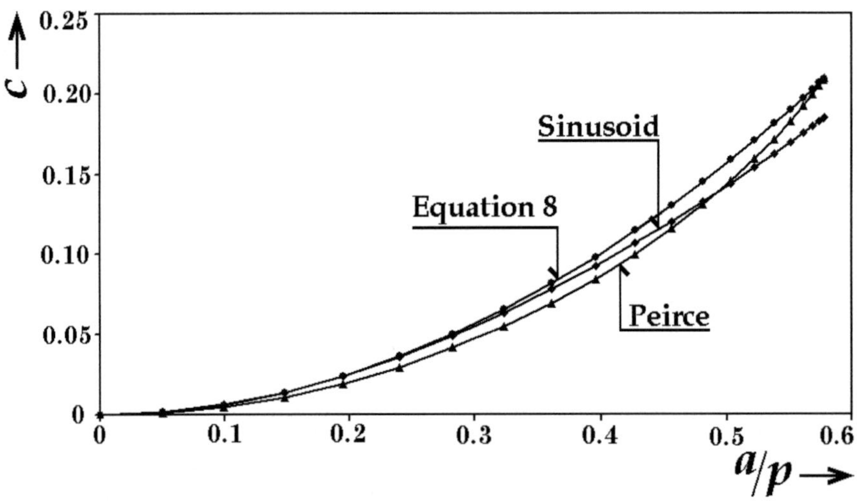

Fig. 7. Dependance of crimp c on relative crimp amplitude a/p.

Using equations 7 and 8 one get for length of the yarn in a crimp wave l:

$$l_{1,20} = \left(2.52 \cdot \left(\frac{a_{1,20}}{p_{2,10}}\right)^2 + 1\right) \cdot p_{2,10} = 2.52 \cdot a_{1,20}^2 + p_{2,10}^2 \quad \text{or} \quad l_{1,2} = 2.52 \cdot a_{1,2}^2 + p_{2,1}^2 \tag{9}$$

Note: equations 7, 8 and 9 are valid both for fabric status before (with subscript 0) and after load. Parameters p_0 are known (reciprocal values of setts S) and height of the crimp wave can be, for square fabric, estimated as $a_{10} = a_{20} = 0.5\ d$.

3.2.2 Load in principal directions

These directions are analyzed in different publications (Hearle et al., 1969; Hu, 2004; Pan, 1996 etc.) sufficiently and so this paragraph will describe only few important parameters.

a. **Fabric breaking force** $F_{f1,2}$ (subscripts 1, 2 specify warp or weft direction of imposed load). For its calculation simple equation (10) can be used (Kovar, 2003), in which $F_{1,2b}$

are breaking forces of one warp or weft yarn and $C_{1,2u}$ are coefficients of utilization of these forces at fabric break. We shall assume, that in principal directions it will be $C_{1,2u} \approx 1$, as there are two opposite tendencies; yarn and fabric unevenness results in decreasing of $C_{1,2u}$ and fabric jamming can, on the contrary, this parameter increase.

$$F_{f1,2} = S_{1,2} \cdot F_{1,2b} \cdot C_{1,2u} \tag{10}$$

b. **Fabric breaking strain,** $\varepsilon_{f1,2}$. There are two main resources of fabric elongation (Kovar & Gupta, 2009): yarn straightening (de-crimp) and yarn axial elongation. For principal direction it will be assumed that all yarns at break are straight and so in equation (8) $a_{1,2} = 0$. For yarn axial elongation experimental results of yarn at breaking strain, $\varepsilon_{1,2b}$, can be used.

$$\varepsilon_{f1,2} = \left(1 + \varepsilon_{1,2b}\right) \cdot \left(1 + c_{1,20}\right) - 1 \tag{11}$$

Explanation: equation (11) can be derived from general definition of relative elongation with the use of (7) and Fig. 6:

$$\varepsilon_{f1,2} = \frac{l_{1,2} - p_{2,10}}{p_{2,10}}, \quad \text{in which} \quad l_{1,2} = p_{2,10} \cdot \left(\varepsilon_{1,2b} + 1\right) \cdot \left(c_{1,20} + 1\right),$$

where $l_{1,2}$ are lengths of the yarn in a crimp wave after straightening and elongation.

c. **Fabric width**. Fabric elongation in principal directions is attached with straightening (de-crimping) of the yarns imposed load, whereas opposite yarns crimp amplitude increases and fabric contracts. We shall assume that lateral contraction is similar as elongation in lateral direction. There are two opposite tendencies again: quicker increase of yarn crimp at greater a/p, Fig. 7, and yarn cross-section deformation (flattening). Original width of the sample b_0 will be changed into $b_{b1,2}$:

$$b_{b1,2} = \frac{b_0}{1 + \varepsilon_{fb1,2}} \text{ and } b_0 = b_{b1,2} \cdot \left(1 + \varepsilon_{fb1,2}\right) \tag{12}$$

d. **Lateral contraction**. Fabric Poisson's ratio ν can be counted using

$$\nu_{1,2} = \frac{b_0 - b_{b1,2}}{b_0} = \frac{b_{b1,2} \cdot \left(1 + \varepsilon_{fb1,2}\right) - b_{b1,2}}{b_{b1,2} \cdot \left(1 + \varepsilon_{fb1,2}\right)} = \frac{\varepsilon_{fb1,2}}{1 + \varepsilon_{fb1,2}} \tag{13}$$

3.2.3 Load in diagonal direction (45 °) for structural unit

Load at diagonal directions is connected with shear deformation and lateral contraction (Sun & Pan, 2005 a, b). This analysis helps with recognition of yarns spacing p_d and angle of yarns incline β_d at fabric break, Fig. 8. Elongation of woven fabric in principal directions is restricted by the yarn system that lays in direction of imposed load, whereas load in angle of 45 ° with free lateral contraction enables greater breaking strain thanks to shear deformation. For description of fabric geometry at break it is necessary to describe jamming in the fabric; break can't occur sooner than maximum packing density is reached.

There are two opposite trends for originally circular yarn cross-section change: (a) Fabric lateral contraction is connected with increase of compressive tension between neighboring yarns. This tension causes tendency to increase the fabric thickness. (b) Crimp of the yarns could not be near zero as it was at loading in warp or weft directions, because now both yarn systems are imposed load. Axial stress in all yarns leads to tendency of de-crimping and so to reduction of fabric thickness.

The situation for initial value of warp and weft yarns decline $\beta_{1,20} = \beta_{d0} = 45°$ is shown in Figure 8 (a before, b after uniaxial elongation), where $p_{1,2} = p_d$ describes yarn perpendicular spacing, p_h and p_v are projections of these parameters in horizontal and in vertical direction, respectively.

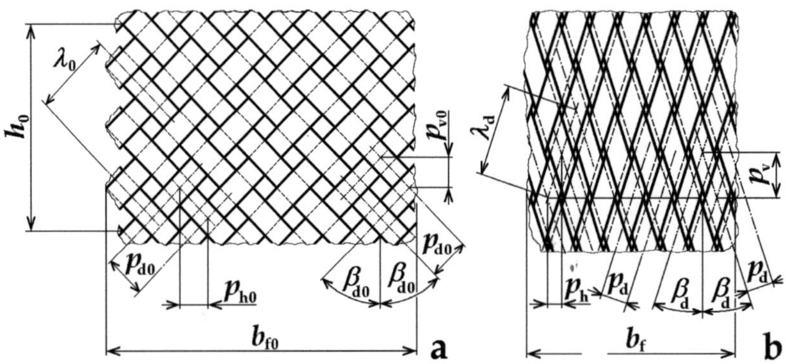

Fig. 8. Diagonal fabric deformation at ideally uniaxial load for $\beta_0 = 45°$ for square fabric.

Spacing of yarns, p_0, is independent parameter corresponding with reciprocal value of fabric sett ($p_0 = 1/S$). During diagonal deformation p decreases and at fabric break it reaches minimum value that restricts fabric lateral contraction and breaking elongation. A hypothesis of even packing density μ distribution in all fabric thickness at break is accepted and so profiles of yarns in crossing points can be as shown in Fig. 9 a for the same material in warp and in weft or in Fig. 8 b for different diameters in warp and weft. Parameter δ, defined as $\delta = \frac{p_d}{t_d}$, can be variable, or another hypothesis of maximum area $p_d \cdot t_d$ can be incorporated and then $\delta = \frac{p_d}{t_d} = 1$ (Figs 9 a and b; Fig. c is for $\delta \neq 1$).

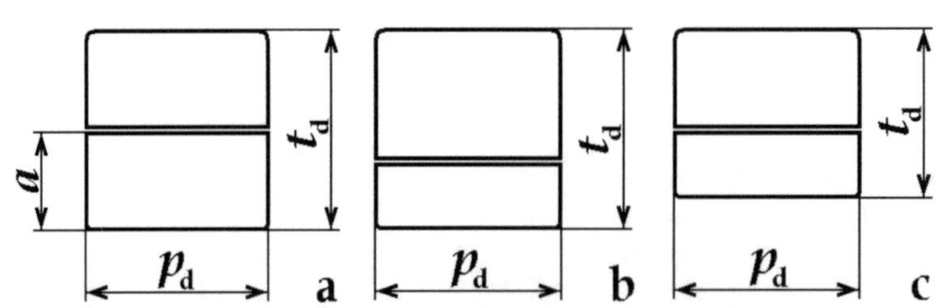

Fig. 9. Model profiles of warp and weft yarns in crossing elements.

If A_0 is original area of warp and weft yarns cross-section at packing density μ_0 ($A_0 = \frac{\pi \cdot \left(d_{10}^2 + d_{20}^2\right)}{4}$, approximately $\mu_0 = 0.5$), final area $A = p_d \cdot t_d$ after yarn compression (at packing density $\mu = 0.8$) will be $A = A_0 \cdot \mu_0/\mu$. Then yarn spacing will be, for $p_d = t_d$:

$$p_\mathrm{d} = \frac{A}{p_\mathrm{d}} = \frac{A_0 \cdot \frac{\mu_0}{\mu}}{p_\mathrm{d}} = \frac{\frac{\pi \cdot \left(d_{10}^2 + d_{20}^2\right)}{4} \cdot \frac{\mu_0}{\mu}}{p_\mathrm{d}} \text{ and so } p_\mathrm{d} = \sqrt{\frac{\pi \cdot \left(d_{10}^2 + d_{20}^2\right)}{4} \cdot \frac{\mu_0}{\mu}} \tag{14}$$

Note: experiments show, that $p_1 = p_2 = p_\mathrm{d}$ also for fabric in which $S_1 \neq S_2$. When yarn diameters are different ($d_1 \neq d_2$), then p_1 and p_2, β_1 and β_2 will be different, but these changes will be small.

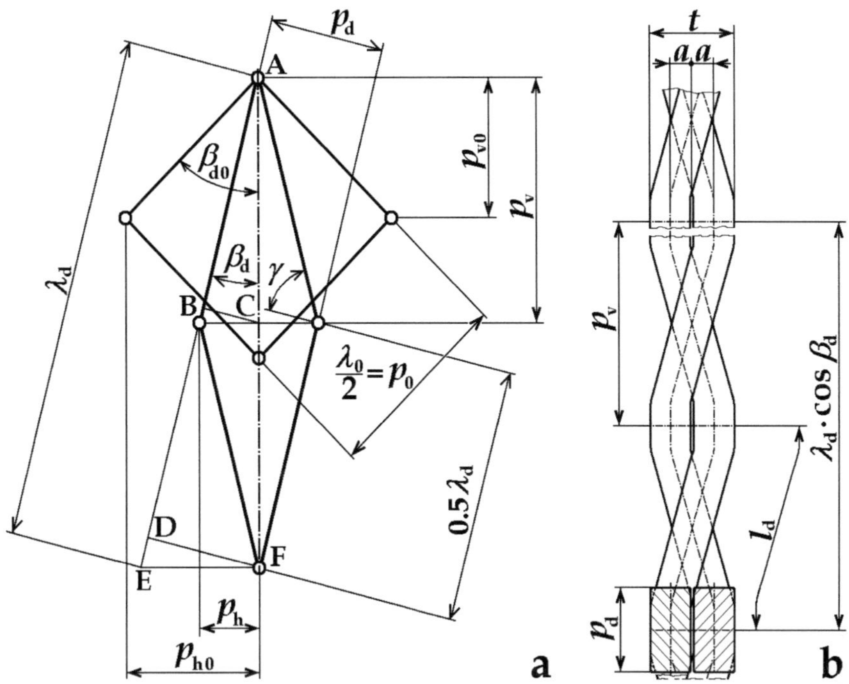

Fig. 10. Geometrical analysis of structural unit deformation at $\beta_0 = 45\,^\circ$ for square fabric.

Geometrical changes, connected with shear fabric deformation at diagonal load of one fabric structural element (this is a square connecting four adjacent crossing points), are described using Figure 10 a, where β_{d0}, β_d are angles of yarns incline before deformation and at break ($\beta_{d0} = 45\,^\circ$), γ is shear angle ($\gamma = \frac{\pi}{2} - 2\beta_\mathrm{d}$, $\gamma_0 = 0$), p_0, p_d is spacing (pitch) of yarns before and after deformation. Parameter p_0 corresponds with fabric sett and with ½ of wavelength of crimped yarns λ_0 (Figs. 6 and 8). λ_d is wavelength of crimped yarns at break. Elongation of λ_0 is enabled both by yarns straightening and by yarn axial elongation, residual crimp of warp and weft yarns can be counted using parameter $a = 0.5\ t_\mathrm{d} = 0.5\ p_\mathrm{d}$ (Figures 9, 10 b) or neglected, because for $a/p < 0.1$ is c < 0.006, Fig. 7. p_{h0} and p_h are horizontal projections of yarn spacing p_0 and p_d before and after deformation and similarly p_{v0} and p_v are vertical projections of p_0 and p_d before and after deformation; these parameters are necessary for calculation of fabric breaking elongation and of maximum shear angle γ, see as well Fig. 10 b (view in direction perpendicular to the load).

Length of the yarn in a crimp wave, l_d, increases in the course of elongation, but due to jamming not so much as in free yarn. We shall assume that yarn axial elongation at break is reduced on axial elongation at break of fibers ε_{fib} and so it will be, see Fig. 6 and equation (9):

$$l_{1,2\mathrm{d}} = l_{1,20} \cdot \left(1+\varepsilon_{\mathrm{fib}}\right) = \left(2.52 \cdot a_{1,20}^2 + p_{2,10}^2\right) \cdot \left(1+\varepsilon_{\mathrm{fib}}\right) \tag{15}$$

Experimentally such yarn breaking elongation can be measured with a very short test length (shorter than is the length of the fibers) to simulate status of fibers and yarn in the fabric. Getting wavelength of crimped yarn at break, λ_d, needs to know crimp wave amplitude a (Figs. 9, 10 b); it is assumed, that $a = \frac{t_d}{4} = \frac{p_d}{4}$. Using crimp definition $c_{1,2} = \frac{l_{1,2}}{p_{2,1}} - 1$ and replacing $p_{2,1}$ for $0.5 \cdot \lambda_{1,2}$ we get $0.5\lambda_{1,2\mathrm{d}} = \frac{l_{1,2\mathrm{d}}}{c_{1,2\mathrm{d}} + 1}$. Parameter $c_{1,2\mathrm{d}}$ can be counted with the help of equation (8): $c_{1,2\mathrm{d}} = 2.52\left(\frac{a}{0.5 \cdot \lambda_{1,2\mathrm{d}}}\right)^2$. After connection and conversion we get quadratic equation $0.5\lambda_{1,2\mathrm{d}}^2 - l_{1,2\mathrm{d}} \cdot 0.5\lambda_{1,2\mathrm{d}} + 2.52 \cdot a^2 = 0$ that leads to the result

$$0.5 \cdot \lambda_{1,2\mathrm{d}} = \frac{l_{1,2\mathrm{d}} + \sqrt{l_{1,2\mathrm{d}}^2 - 4 \cdot 2.52 \cdot a^2}}{2} \tag{16}$$

Using Fig. 10 one can calculate horizontal and vertical projection of the yarn spacing before deformation, p_{h0} and p_{v0}, and after deformation (triangle ABC), p_h and p_v:

$$p_{\mathrm{h0}} = 0.5 \cdot \lambda_0 \sin\frac{\pi}{4} = 0.5 \cdot p_0 \cdot \sqrt{2} \quad \text{and} \quad p_{\mathrm{h}} = 0.5 \cdot \lambda_{\mathrm{d}} \cdot \sin\beta_{\mathrm{d}} \tag{17}$$

$$p_{\mathrm{v0}} = 0.5 \cdot \lambda_0 \cos\frac{\pi}{4} = 0.5 \cdot p_0 \cdot \sqrt{2} \quad \text{and} \quad p_{\mathrm{v}} = 0.5 \cdot \lambda_{\mathrm{d}} \cdot \cos\beta_{\mathrm{d}} \tag{18}$$

Minimum yarn spacing, p_d, has already been known from equation (14). It can help with calculation of p_v and p_h that give fabric breaking elongation and lateral contraction. From Fig. 10, triangles AEF and DEF, it will be $\sin\beta_{\mathrm{d}} = \frac{2p_{\mathrm{h}}}{\lambda_{\mathrm{d}}}$ and $\sin\beta_{\mathrm{d}} = \frac{p_{\mathrm{d}}}{2p_{\mathrm{h}}}$ and hence $\frac{2p_{\mathrm{h}}}{\lambda_{\mathrm{d}}} = \frac{p_{\mathrm{d}}}{2p_{\mathrm{h}}}$, $p_{\mathrm{h}} = \frac{\sqrt{p_{\mathrm{d}}\lambda_{\mathrm{d}}}}{2}$, what results in angle of yarn incline at break β_d:

$$\sin\beta_{\mathrm{d}} = \frac{\sqrt{p_{\mathrm{d}} \cdot \lambda_{\mathrm{d}}}}{\lambda_{\mathrm{d}}} \quad \text{and} \quad \beta_{\mathrm{d}} = \arcsin\frac{\sqrt{p_{\mathrm{d}} \cdot \lambda_{\mathrm{d}}}}{\lambda_{\mathrm{d}}} \tag{19}$$

Fabric elongation at diagonal break ε_{db} is, using eq. (18):

$$\varepsilon_{\mathrm{db}} = \frac{p_{\mathrm{v}} - p_{\mathrm{v0}}}{p_{\mathrm{v0}}} = \frac{\lambda_{\mathrm{d}} \cdot \cos\beta_{\mathrm{d}} - p_0 \cdot \sqrt{2}}{p_0 \cdot \sqrt{2}} \tag{20}$$

Maximum value of shear angle γ_d will be $\gamma_{\mathrm{d}} = \frac{\pi}{2} - 2 \cdot \cos\beta_{\mathrm{d}}$

3.2.4 Load in diagonal direction (45 °) for fabric strip

Strip of the tested fabric of original width b_0 at angle of warp and weft yarns decline to vertical load direction $\beta_{10} = \beta_{20} = 45$ ° is shown in Figure 11 (a before load, b after load for square fabric, c after load for non-square fabric with $S_1 \neq S_2$ or/and $\varepsilon_{1b} \neq \varepsilon_{2b}$).

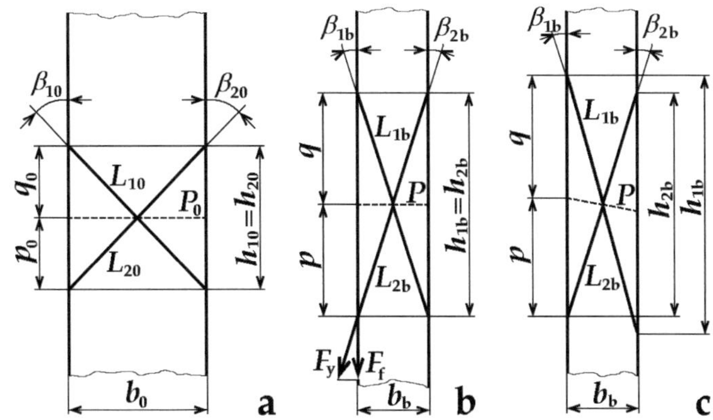

Fig. 11. Geometrical analysis of sample deformation at $\beta_0 = 45$ °.

Lengths of the weft and the warp yarns in fabric plane before elongation, L_{10} and L_{20}, are $L_{1,20} = \dfrac{b_0}{\sin\beta_{1,20}}$, where for $\beta_{10} = \beta_{20} = \beta_0 = 45$ ° is $\sin\beta_{1,20} = \dfrac{\sqrt{2}}{2}$. The width of the sample at break, b_b, and fabric lateral contraction, ν, can be counted using equation (17), in which p_h corresponds with b ($\dfrac{p_h}{p_{h0}} = \dfrac{b_b}{b_0}$):

$$b_b = b_0 \cdot \frac{p_h}{p_{h0}} = b_0 \cdot \frac{0.5 \cdot \lambda_d \cdot \sin\beta_d}{p_0 \cdot \sqrt{2}} \quad \text{and} \quad \nu_b = \frac{b_0 - b_b}{b_0} \tag{21}$$

Note: equation (21) gives, for non-square fabric (Figure 11 c), different results for weft and warp yarns ($b_{1b} \neq b_{2b}$). In this case we shall assume that minimum value of b and maximum value of ν is valid; the reason is that maximum load is only in one yarn system, the opposite systems of yarns with lower or negative load cannot keep fabric strip wider as the yarns are able to bear only negligible compressive load.

Now we shall count fabric breaking elongation, using knowledge of sample width b_b, equation (21), and determination of projections of lengths of the yarns axes l_1 and l_2 into fabric plane L_1 and L_2 (Figs. 6, 11):

$$l_{1,20} = L_{1,20}\left(1 + c_{1,20}\right) \quad \text{and} \quad l_{1,2b} = l_{1,20}\left(1 + \varepsilon_{fib}\right) = L_{1,20} \cdot \left(1 + c_{1,20}\right) \cdot \left(1 + \varepsilon_{fib}\right) \tag{22}$$

Relation between lengths of yarns axes calculation before load, l_{10} and l_{20}, and at break, l_{1b} and l_{2b}, is described by equation (9). The lengths of yarns axes projection into fabric plane, L_{1db} and L_{2db}, will be from (22)

$$L_{1,2b} = \frac{l_{1,2b}}{1 + c_{1,2b}} = \frac{L_{1,20} \cdot \left(1 + c_{1,20}\right) \cdot \left(1 + \varepsilon_{fib}\right)}{1 + c_{1,2b}} \tag{23}$$

Yarns elongation at break is reduced on the value, corresponding with elongation at break of fibers. Crimp of yarns at break is taken from equation (8), where p is replaced with 0.5 λ:

$$c_{1,2\mathrm{db}} = 2.52 \cdot \left(\frac{a}{0.5 \cdot \lambda_{1,2}} \right)^2 \tag{24}$$

Horizontal projections of lengths L_{1b} and L_{2b} are $h_{1,2} = \sqrt{L_{1,2\mathrm{b}}^2 - b_{\mathrm{db}}^2}$.
Finally fabric sample breaking elongation $\varepsilon_{\mathrm{fbd}}$ is (Fig. 11) for $h_{1b} = h_{2b} = h_b$ and $h_0 = b_0$:

$$\varepsilon_{\mathrm{fbd}} = \frac{h_{\mathrm{b}} - h_0}{h_0} = \frac{h_{\mathrm{b}} - b_0}{b_0} \tag{25}$$

More complex is breaking elongation calculation for non-square fabric (Fig. 11 c) when $L_1 \neq L_2$. It leads to skewed fabric in which originally horizontal line P_0 will get some another angle to load direction. In this case it could be assumed, that thanks to possibility of reaching new equilibrium, fabric breaking elongation will be average of the values, get from equation (24) for weft and warp yarns; this assumption is in good agreement with experimental results.
Breaking stress of the fabric, F_{fb}, equation (26), can be got by modification of equation (10). The role of both warp and weft yarns strength is for one critical angle β_{0c} identical (for square fabric $\beta_{0c} = 45\ ^{\circ}$, for other fabrics it is near 45 $^{\circ}$). Incline of yarns is as well considered; only component of yarn axial stress F_y into direction of fabric load F_f supports fabric strength (it is shown in Fig. 11 b). Prediction of coefficients of yarn strength utilization C_u is in this case difficult, because due to jamming strength of the yarn can be higher than the strength of free yarn. On the contrary, cut yarn ends causes gradual tensile stress increase from zero at sample edge to maximum value (it will be described later).

$$F_{\mathrm{fb}} = S_1 \cdot F_{1\mathrm{b}} \cdot C_{1\mathrm{u}} \cdot \cos\beta_{1\mathrm{b}} + S_2 \cdot F_{2\mathrm{b}} \cdot C_{2\mathrm{u}} \cdot \cos\beta_{2\mathrm{b}} \tag{26}$$

3.2.5 Load in variable directions – breaking strain

Example for angle of load $\beta_{10} = 30\ ^{\circ}$ is described in Fig. 12. The lengths of warp and weft yarns projections into fabric plane within the sample width b_0 before elongation, L_{10} and L_{20}, can be counted similarly as in previous paragraph.
For further calculations it is necessary to know sample width b_b at break as variable, dependent on angle β_0. We shall assume, in accordance with experimental results, that the change of b is slower near the critical angle β_{0c} ($\beta_{0c} \approx 45\ ^{\circ}$, value of b is here at minimum) and quicker near $\beta_0 = 0\ ^{\circ}$ and 90 $^{\circ}$. This is the reason, why parabolic approximation in equation (27) is used. In this, $b_{1,2b}$ is sample width at break for angle of load 0 $^{\circ}$ or 90 $^{\circ}$ and b_{db} is the same for angle of load 45 $^{\circ}$.

$$b_{\mathrm{b}}(\beta_0) = (b_{1,2\mathrm{b}} - b_{\mathrm{db}}) \cdot \left(1 - \frac{4}{\pi} \cdot \beta_0\right)^2 + b_{\mathrm{db}} \tag{27}$$

Sample strain at break is, with the exception of critical angle β_{0c}, restricted namely by yarns of one system (warp or weft), usually by that with smaller value of β_0. Elongation at break near β_{0c} is influenced first of all by: (a) Crimp interchange. At mentioned β_{0c} both warp and weft yarns are crimped and so an important crimp interchange occurs; yarns of one system get straight whereas crimp of the opposite yarns grows. This mechanism enables, in some range of angles β_0, to utilize strength of warp and weft yarns simultaneously. Out of this range, stress in the yarn of the system with greater β_0 becomes negligible or even negative (compression). (b) Effect of yarn jamming, that increases yarn breaking stress, is maximum at β_{0c} and minimum for $\beta_0 = 0\,^{\circ}$ or $90\,^{\circ}$.

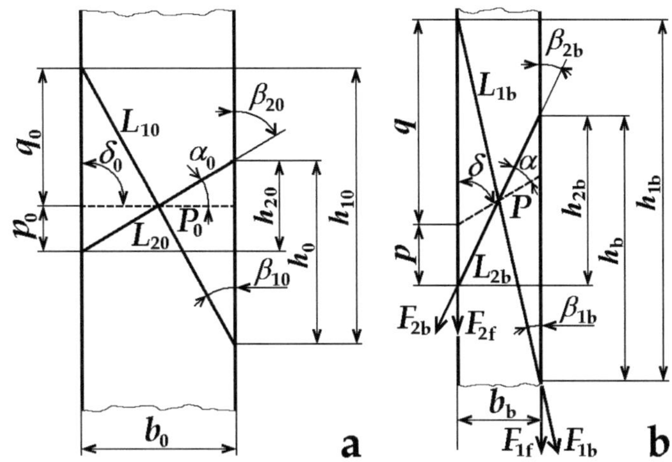

Fig. 12. Deformation of a fabric strip before deformation (a) and at break (b).

One of the results of different stresses in warp and weft yarns is that originally horizontal line P_0 changes after load its direction δ_0 for δ, see Fig. 12 (dash lines) and (Zouari, Amar & Dogui, 2008). Lengths of the yarn axis projections at break, $L_{1,2b}$, corresponding with the lengths $L_{1,20}$ before load, can be counted similarly as in equation (23). Crimps $c_{1,2b}$ are variable near critical angle β_{0c}, near principal directions are crimps of broken yarns neglected. Yarn elongations at break $\varepsilon_{1,2b}$ are as well variable; near angle β_{0c} it is $\varepsilon_{1,2b} = \varepsilon_{fib}$.

$$L_{1,2b} = \frac{L_{1,20} \cdot (1 + c_{1,20}) \cdot (1 + \varepsilon_{1,2b})}{1 + c_{1,2b}} = \frac{b_0 \cdot (1 + c_{1,20}) \cdot (1 + \varepsilon_{1,2b})}{\sin \beta_{1,20} \cdot (1 + c_{1,2b})} \tag{28}$$

For fabric breaking strain it is necessary to know vertical projections of $L_{1,2b}$, parameters $h_{1,2b}$ (Fig. 12). Using Pythagorean Theorem and equations (27) and (28) it will be

$$h_{1,2b} = \sqrt{L_{1,2b}^2 - b_b^2} \tag{29}$$

The same parameters before load, $h_{1,20}$, are

$$h_{1,20} = L_{1,20} \cdot \cos \beta_{1,20} \tag{30}$$

Now we can count, separately for warp and weft yarns, fabric breaking elongation, $\varepsilon_{1,2b}$. Smaller of the results will be valid (break yarns of only one system).

$$\varepsilon_{1,2b} = \frac{h_{1,2b} - h_{1,20}}{h_{1,2b}} \tag{31}$$

In Fig. 13 is shown an example of results for calculated and measured breaking elongation. Fabric was plain weave, cotton yarn linear density 35 tex (warp and weft), warp sett 2600 ends/m, weft sett 2500 ends/m, finished fabric. Experiment 1 was carried on in accordance with the standard EN ISO 13934-1, experiment 2 is in accordance with (Kovar & Dolatabadi, 2010).

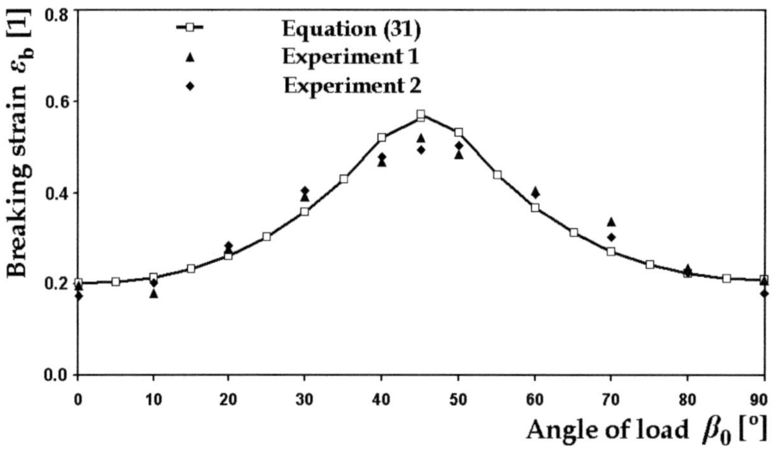

Fig. 13. Example of calculated and measured fabric strain at break.

3.2.6 Load in variable directions – breaking stress

Force from one broken yarn. Basic algorithms for breaking stress calculation were explained sooner, equations (10) and (26). For variable β_0 it is necessary to consider that only component of yarn axial stress F_1 and F_2 aims in direction of load, Fig. 12. Components of axial tensile force of one yarn at fabric break, $F_{1,2b}$, into fabric load direction, $F_{1,2f}$, are $F_{1,2f} = F_{1,2b} \cdot \cos\beta_{1,2b}$, where angles of yarns incline at break $\beta_{1,2b}$ can be counted using (27) and (29):

$$\tan\beta_{1,2b} = \frac{b_b}{h_{1,2b}} \text{ and } \beta_{1,2b} = \arctan\frac{b_b}{h_{1,2b}} \tag{32}$$

Force from all broken yarns. Number of broken yarns in fabric width at break corresponds with fabric sett S, width b and angle of load β_0 and so comprehensive force from all warp or weft yarns in fabric strip, $F_{a1,2b}$, is

$$F_{a1,2b} = S_{1,2} \cdot b_0 \cdot \cos\beta_{1,20} \cdot F_{1,2f} = S_{1,2} \cdot b_0 \cdot \cos^2\beta_{1,20} \cdot F_{1,2b} \tag{33}$$

where $n_{1,2} = S_{1,2} \cdot b_0 \cdot \cos\beta_{1,20}$ is number of yarns in sample width.

Correction for fabric jamming

At diagonal load, utilization of fibers strength can be better then at load in principal directions. It will be described by coefficient C_{fu}, equation (10). In this chapter only rough estimation will be presented. Fibers strength utilization, by other words a share of broken fibers to all fibers in yarn cross-section, can be: (a) In free spun yarn around value of C_{fuy} =

0.5; it depends on fibers length (staple), friction coefficients, yarn twist etc. (b) In fabric at break in principal directions, C_{fup} is similar or slightly higher; it depends on fabric packing density and on other parameters. (c) In fabric at break in diagonal directions, C_{fud} is maximal due to jamming. Extremely it can be near to 1. From these reasons, final parameters $C_{fu1,2}$ as a function of β_0, will be predicted as parabolic relation (without derivation):

$$C_{fu1,2}(\alpha_0) = (C_{fud} - C_{fup}) \cdot \left(\frac{\beta_0}{\frac{\pi}{4}} \right)^2 + C_{fup} \tag{34}$$

Fabric strip strength from broken yarns with implementation of jamming, $F_{j1,2b}$, is from (33) and (34):

$$F_{j1,2b} = C_{fu1,2} \cdot F_{a1,2b} = C_{fu1,2} \cdot S_{1,2} \cdot b_0 \cdot \cos^2 \beta_{1,20} \cdot F_{y1,2b} \tag{35}$$

Correction for cut yarn ends

With the exception of $\beta_0 = 0$ and $\beta_0 = 90\,°$ there are yarns, bearing fabric load, having one or two free ends (Kovar & Dolatabadi, 2007). Near cut yarn end axial stress is zero and gradually increases (linear increase is assumed) due to friction till it reaches yarn strength in length l, see Fig. 14 a. In this area fabric jamming is not as important as in sample inner parts and shear angle is smaller. This length l is hardly predictable and depends on many parameters (setts, yarn properties including frictional, fabric finishing, shear deformation, angle of load, jamming etc.). It can be evaluated experimentally by testing yarn pullout force from the fabric (Pan & Yoon, 1993) or testing the samples of variable widths. By this effect, some width on each side of fabric $b_{in} = l \cdot \sin \beta_{1,20}$ is inefficient; this is important mainly for broken yarns. This strip b_{in} can bear only about 50 % of full load. It results in reduction of original sample width to effective one $b_{ef} = b_{b0} - l \cdot \sin \beta_{1,20}$.

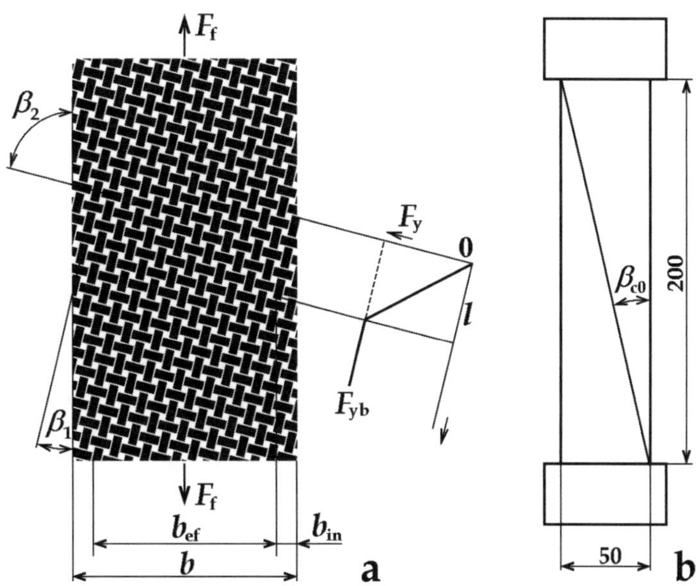

Fig. 14. Free ends of yarns in fabric at bias load.

Total effective force from broken yarns with reduced fabric width, $F_{f1,2b}$, is then from (35)

$$F_{f1,2b} = C_{fu1,2} \cdot S_{1,2} \cdot b_{ef} \cdot \cos^2 \beta_{1,20} \cdot F_{y1,2b} = C_{fu1,2} \cdot S_{1,2} \cdot \left(b_{b0} - l \cdot \sin \beta_{1,20}\right) \cdot \cos^2 \beta_{1,20} \cdot F_{y1,2b} \quad (36)$$

Correction for critical angles

In our theory, unlimited sample length is assumed and the effect of critical angles is neglected. Nevertheless for comparison with real experiments it should be mentioned; tension concentration at jaws reaches high value for critical angles, at which only 1 yarn is kept simultaneously by both pair of jaws and all others yarns have 1 end free. For critical angle β_{c0} it will be: $\tan \beta_{c0} = 50 : 200$, see Fig. 14 b (sample width is 50 mm, test length 200 mm). Near this angle an important drop in tested fabric strength is observed.
Example of results for plain weave fabric, warp and weft yarns are polypropylene/cotton 35/65 %, linear density T = 29.5 tex, warp sett S_1 = 2360 ends/m, weft sett S_2 = 1920 (lines 1 and 3) and S_2 = 1380 ends/m (lines 2 and 4) is shown in Fig. 15. Lines 3, 4 describes standard experiment (EN ISO 13934-1), lines 1, 2 results of the new method (Kovar & Dolatabadi, 2010) with the same size of samples. Drop in the sample strength near critical angles is evident.

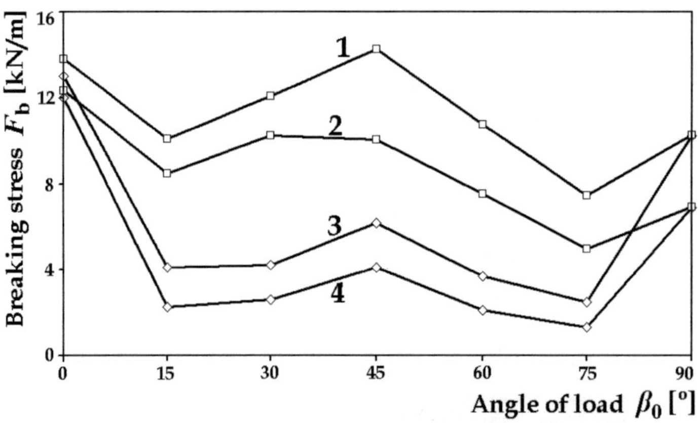

Fig. 15. Influence of critical angles on fabric breaking stress.
Note: linear connection of measured points only assembles these points together; in any case it is does not mean approximation of the results.

Force from unbroken yarns at fabric break

These yarns are, for fabric strength, important only near critical angle β_{0c} (near 45 °). At other load angles, tensile stress in these yarns is low or negative. We shall assume, that maximum force corresponds with maximum length $L_{1,2b}(\beta_0)$, Fig. 12, and that it can be calculated using formula (33) on condition of similar tensile properties of warp and weft yarns.
Vertical projection of unbroken yarn length at fabric break, $h_{u1,2}$, depends on this parameter before load ($h_{1,20}$) and on sample elongation at break (elongation of sample is proportional), identified by broken yarns of the opposite system: $h_{u1,2} = h_{1,20} \cdot \left(1 + \varepsilon_{2,1b}\right)$. Length of unbroken yarns in fabric width before load is $L_{1,20} = \dfrac{b_0}{\sin \beta_{1,20}}$, corresponding length of unbroken yarns at fabric break (Fig. 12), $L_{u1,2}$, is using (29), $L_{u1,2} = \sqrt{b_b^2 + h_{1,2b}^2}$.

Relative elongation of unbroken yarns is then

$$\varepsilon_{u1,2} = \frac{\left(L_{u1,2} - L_{1,20}\right)}{L_{1,20}} \tag{37}$$

and hence force, by which unbroken yarns contribute to sample strength, will be:

$$F_{u1,2b} = F_{a1,2d} \cdot \frac{L_{u1,2}}{L_{u1,20}} \tag{38}$$

where Fa1,2d is breaking load, calculated in accordance with (33) for $\beta 0 = 45\,^{\circ}$.

Final results

Force $F_{12,b}$ is the sum of the forces from broken and unbroken yarns, equations (36) and (38):

$$F_{1,2b} = F_{f1,2b} + F_{u1,2b} \tag{39}$$

In Fig. 16 is an example of results, carried on the same fabric and with the same experimental methods as shown in Fig. 13. Agreement is not excellent; it is caused by simplifications in calculation and as well by imperfection of known experimental methods. Results of patented method (experiment 2, Kovar & Dolatabadi, 2010) shows, with exception of principal directions, higher breaking stress than does standard method (experiment 1, EN ISO 13934-1). Important drop is observed near previously mentioned critical angles β_0 14 and 76 °. Slower decrease of breaking stress near angle β_0 = 45 ° is due to interactions between warp and weft yarns that were not implemented into calculation yet.

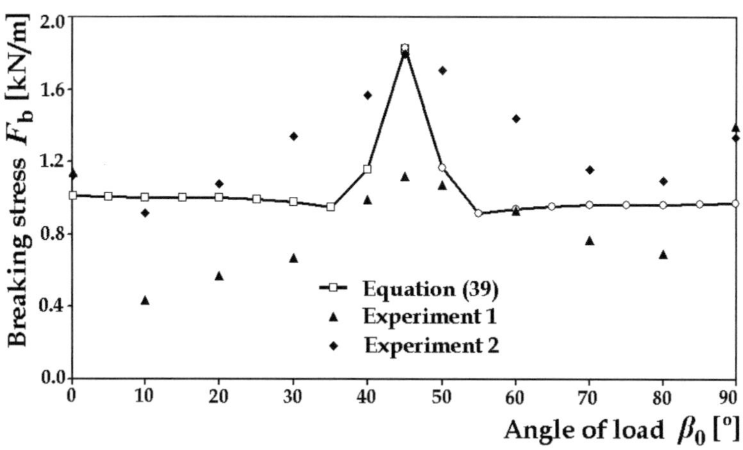

Fig. 16. Example of calculated and measured fabric stress at break.

4. Measuring of rupture properties

Experiments always mean some scale of unification and simplification in comparison with fabric real loading at the use. To simulate real practical situations is not possible – it would result in too many different experimental methods. In general, the load put on textile fabric, can be (a) tensile uniaxial, (b) tensile biaxial or (c) complex as combination of different form of the load (elongation, bend, shear etc.). Nevertheless uniaxial and biaxial stresses are the most important forms of load for investigation of textile fabrics rupture properties. Other forms of deformation (bending, shear, lateral pressure etc.) seldom result in fabric break.

4.1 Uniaxial stress

The problems, connected with breaking test of woven fabrics due to great lateral contraction that accompanies load in diagonal directions, have already been described in section 2 (Fig. 1). The principle of a new method (Kovar & Dolatabadi, 2010) is sample tension reduction by fabric capstan friction, Fig. 17 (scheme and photographs at three stages of sample elongation). A set of fast cylinders 5, 6 is connected with each pair of dynamometer jaws 1, 2. At sample elongation fabric slips towards central fabric part 4 in directions 8, what results in tension reduction due to capstan friction; however, fabric lateral contraction on cylinders is enabled. Total angle of contact is on each sample side is approximately 8.03 π (460 °) and for friction coefficient f = 0.17 (this is low value of f, valid for fabric to smooth steel surface friction at high load near break of the sample) decrease of sample tension will be $\frac{F_c}{F_j} = e^{\alpha \cdot f} \doteq 3.9$ (390 %). In Fig. 17 right is example of tested sample before elongation (a), at elongation of 40 % (b) and 90 % near the break point (c).

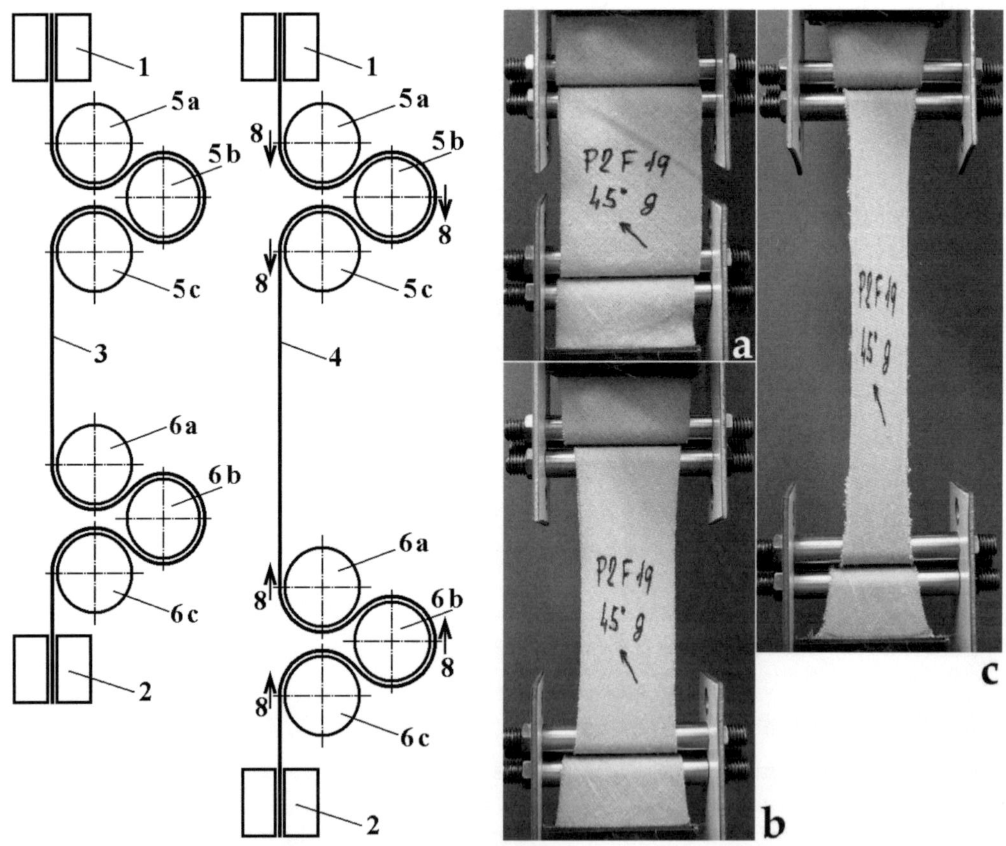

Fig. 17. Patented method for fabric tensile properties measuring

4.2 Biaxial stress

Measuring of fabric tensile properties at biaxial stress is more complicated task, described for example in (Bassett, Postle & Pan, 1999). If fast jaws 1 are used, Fig. 18 a, fabric would soon break at sample corners as relative elongation of L_2 is many times greater than that of sample length and width L_1. MA is measured area of the sample. Two of solutions are shown. In Fig. b are fast jaws replaced with sets of individual narrow free grippers and in Fig. c is measured sample MA connected with four auxiliary fabrics cut into strips that enable 2-D sample elongation, although jaws 1 are fast. Two mentioned methods are suitable for measuring fabric anisotropy, nevertheless they need special equipment and much of labor. It is not easy to investigate rupture properties by these methods. As the load in two directions can be different, it would be useful to reduce number of tested samples by election of only some variants such as: (a) uniaxial load (but different than at standard methods, lateral contraction is now enabled), (b) restriction of lateral contraction similarly with chapter 2.2, (c) the same load (absolutely or recounted per one yarn in the sample width) or tension in two directions, (d) the same elongation in two directions.

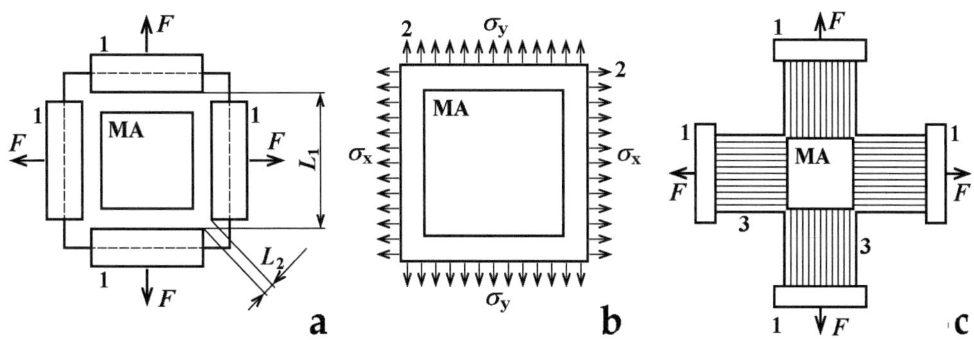

Fig. 18. Principles of tensile properties measuring at biaxial load

The principle of measuring tensile properties when fabric lateral contraction is restricted (simulation of sample infinite width, section 2.1) is shown in Fig. 19. The sample 1 is sewn by several individual stitches into tubular form and by wires 3, placed beside jaws 2, is kept in original width.

5. Discussion, current trends and future challenges in investigated problems

The problems of anisotropy of woven fabric rupture properties are very complex and till now not in the gravity centre of researches. This section could make only a short step in bringing new knowledge on this field. Partly another approach to similar problem solution is used in (Dolatabadi et al., 2009; Dolatabadi & Kovar, 2009). Anisotropy of different fabric properties is often investigated for textile based composites, where rupture properties are very important, for example in (Hofstee & van Keulen, 2000).

There are lots of possibilities how to go on in research on this topic, for example:

a. Investigation of influence of sample width on tensile properties with the goal to specify better impact of cut yarn ends (Fig. 14).
b. Research on biaxial and combined fabric load, the aim could be, for example, better description of fabric behaviour at practical usage.

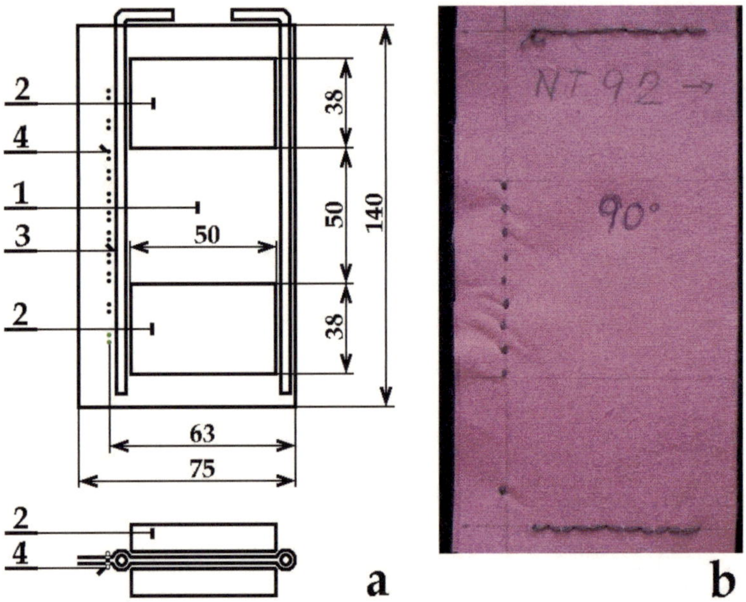

Fig. 19. Measuring of tensile properties at restricted lateral contraction (scheme, sample)

c. Development of suitable experimental methods and its standardization; till now there is no standard method for measuring rupture properties of fabrics with great lateral contraction.
d. Implementation of other variable parameters into calculation, such as variability in yarns properties, unevenness of fabric structure etc.
e. Research of another weaves (twill, sateen…), influence of structure on utilization of strength of used fibres.
f. Developing of suitable methods for simulation of fabric tension distribution at particular load with the stress to be put on a great and variable Poison's ratio of fabrics etc.

There are other important anisotropic forms of fabric deformation, which are not described in this chapter, such as bend (Cassidy & Lomov, 1998) and shear. Lateral contraction is as well very important.

6. Acknowledgement

This work was supported by the research project No. 106/09/1916 of GACR (Grant Agency of Czech Republic).

7. References

Bassett, R. J.; Postle, R. & Pan, N. (1999). Grip Point Spacing Along the Edges of an Anisotropic Fabric Sheet in a Biaxial Tensile Test. *Polymer composites*, Vol. 20, No. 2

Cassidy, C. & Lomov, S. V. (1998). Anisotropy of fabrics and fusible interlinings. *International Journal of Clothing Science and Technology*, Vol. 10 No. 5, pp. 379-390

Dai, X.; Li, Y. & Zhang, X. (2003). Simulating Anisotropic Woven Fabric Deformation with a New Particle Model, *Textile Res. J.* 73 (12), 1091-1099

Dolatabadi, K. M.; Kovar, R. & Linka, A. (2009). Geometry of plain weave fabric under shear deformation. Part I: measurement of exterior positions of yarns. *J. Text. Inst.*, 100 (4), 368-380

Dolatabadi, K. M. & Kovar, R. (2009). Geometry of plain weave fabric under shear deformation. Part II: 3D model of plain weave fabric before deformation and III: 3D model of plain weave fabric under shear deformation. *J. Text. Inst.*, 100 (5), 381-300

Du, Z., & Yu, W. (2008). Analysis of shearing properties of woven fabrics based on bias extension, *J. Text. Inst.*, 99, 385-392

Hearle, J. W. S.; Grosberg, P. & Backer, S. (1969). *Structural Mechanics of Fibres, Yarns and Fabrics*. Vol. 1. New York, Sydney, Toronto

Hofstee, J. &van Keulen, F. (2000). Elastic stiffness analysis of a thermo-formed plain-weave fabric composite. Part II: analytical models. *Composites Science and Technology*, 60, 1249-1261

Hu, J. (2004). *Structure and mechanics of woven fabrics*. Woodhead Publishing Ltd. P 102, ISBN 0-8493-2826-8

Kilby, W. F. (1963). Planar stress-strain relationships in woven fabrics. *J. Text. Inst.*, 54, T 9-27

King, M. J.; Jearanaisilawong, P. & Socrate, S. (2005). A continuum constitutive model for the mechanical behavior of woven fabrics. *International Journal of Solids and Structures* 42, 3867–3896

Kovar, R. & Gupta, B. S. (2009). Study of the Anisotropic Nature of the Rupture Properties of a Woven Fabric. *Textile Research Journal* Vol 79(6), pp. 506-506

Kovar, R. & Dolatabadi, M. K. (2010). The way of measuring of textile fabric deformation and relevant equipment. Czech patent No. 301 314

Kovar, R. & Dolatabadi, M. K. (2008). Crimp of Woven Fabric Measuring. Conference Strutex 2008, TU of Liberec 2008, ISBN 978-80-7372-418-4

Kovar, R. & Dolatabadi, M. K. (2007). Impact of yarn cut ends on narrow woven fabric samples strength. Strutex, TU Liberec, ISBN 978-80-7372-271-5

Kovar, R. (2003). *Structure and properties of flat textiles* (in Czech). TU of Liberec, ISBN 80-7083-676-8, Liberec, CZ, 142 pages

Lo, M. W. & Hu, J. L. (2002). Shear Properties of Woven Fabrics in Various Directions, *Textile Res. J.* 72 (5), 383-390

Lomov, S. V. et all, (2007) Model of internal geometry of textile fabrics: Data structure and virtual reality implementation. *J. Text. Inst.*, Vol. 98, No. 1 pp. 1–13

Pan, N. & Yoon, M. Y. (1996). Structural Anisotropy, Failure Criterion, and Shear Strength of Woven Fabrics. *Textile Res. J.* 66 (4), 238-244

Pan, N. & Yoon, M. Y. (1993). Behavior of Yarn Pullout from Woven Fabrics: Theoretical and Experimental. *Textile Res. J.* 63 (1), 629-637

Pan, N. (1996 b). Analysis of Woven Fabric Strength: Prediction of Fabric Strength Under Uniaxial and Biaxial Extension, *Composites Scence and Technology* 56 311-327

Peng, X. Q. and Cao, J. (2004). A continuum mechanics-based non-orthogonal constitutive model for woven composite fabrics. *Composites: Part A* 36 (2005) 859–874

Postle, R.; Carnaby, G. A. & de Jong, S. (1988). The Mechanics of Wool Structures. Ellis Horwood Limited Publishers, Chichester. ISBN 0-7458-0322-9

Sun, H. & Pan, N. (2005 a). Shear deformation analysis for woven fabrics. *Composite Structures* 67, 317–322

Sun, H. & Pan, N. (2005 b). On the Poisson's ratios of a woven fabric. University of California Postprints, Paper 662

Zborilova, J. & Kovar, R. (2004). Uniaxial Woven Fabric Deformation. Conference STRUTEX, TU of Liberec, pp. 89-92, ISBN 80-7083-891-4

Zheng, J. et all (2008). Measuring technology of the Anisotropy Tensile Properties of Woven Fabrics. *Textile Res. J.*, 78, (12), pp. 1116-1123

Zouari, R., Amar, S. B. & Dogui, A. (2008). Experimental and numerical analyses of fabric off-axes tensile test. *JOTI*, Vol. 99, iFirst 2008, 1–11

European standard EN ISO 13934-1. Determination of maximum force and elongation at maximum force using the strip method

CSN standard 80 0810 Zistovanie trznej sily a taznosti pletenin (Recognition of breaking stress and strain of knitted fabrics)

2

Mechanical Properties of Fabrics from Cotton and Biodegradable Yarns Bamboo, SPF, PLA in Weft

Živa Zupin and Krste Dimitrovski
University of Ljubljana, Faculty of Natural Sciences and Engineering, Department of Textiles
Slovenia

1. Introduction

Life standard is nowadays getting higher. The demands of people in all areas are increasing, as well as the requirements regarding new textile materials with new or improved properties which are important for the required higher comfort or industrial use. The environmental requirements when developing new fibres are nowadays higher than before and the classical petroleum-based synthetic fibres do not meet the criteria, since they are ecologically unfriendly. Even petroleum as the primary resource material is not in abundance. The classical artificial fibres, e.g. polypropylene, polyacrylic, polyester etc, are hazardous to the environment. The main problems with synthetic polymers are that they are non-degradable and non-renewable. Since their invention, the use of these synthetic fibres has increased oil consumption significantly, and continues even today. It is evidenced that polyester is nowadays most frequently used among all fibres, taking over from cotton. Oil and petroleum are non-renewable (non-sustainable) resources and at the current rate of consumption, these fossil fuels are only expected to last for another 50–60 years; the current petroleum consumption rate is estimated to be 100,000 times the natural generation rate (Blackburn, 2005).

Environmental trends are more inclined to the development of biodegradable fibres, which are environment-friendly. A material is defined as biodegradable if it can be broken into simpler substances (elements and compounds) by naturally occurring decomposers - essentially, anything that can be ingested by an organism without harming the organism. It is also necessary that it is non-toxic and decomposable in a relatively shot period on a human time scale (Blackburn, 2005). The biodegradability of fibres also depends on their chemical structure, molecular weight and super-molecular structure.

Biodegradable polymers can be classified into three main groups, i.e.:

- natural polysaccharides and biopolymers (cellulose, alginates, wool, silk, chitin, soya bean protein),
- synthetic polymers, esp. aliphatic polyesters (poly (lactic acid), poly (ε-caprolactone)), and
- polyesters produced by microorganisms (poly (hydroxyalkanoate)s) (Blackburn, 2005).

All known natural fibres are biodegradable; however, they have some disadvantages in the growing up and production processes. At growing cotton and other vegetable fibres, large amounts of pesticides are used which has a negative influence on the environment.

In the research, three biodegradable fibres, i.e. bamboo fibres, fibres form polylactic acid (PLA) and soybean protein fibres (SPF) were used for which the industrial procedures already exist. At the same time, there are enough natural resources for the latter and they are environment-friendly. The physical-mechanical properties of fabrics with biodegradable yarns in weft and cotton yarns in warp were researched. We would like to determine whether and to what extent physical and mechanical properties change and whether they are acceptable in terms of today's criteria.

The researchers have been investigating and researching the production of biodegradable fibres and their properties. This research focuses on the mechanical properties of yarns made prom biodegradable fibres and first of all, on the mechanical properties of woven fabrics made from biodegradable yarns in weft and cotton yarns in warp. The latter is the most common way of producing woven fabrics, since the warp threads do not need to be changed.

2. Properties of bamboo, PLA and SPF fibres

New trends are being sought for naturally renewable resources in order to protect the nature. With the help of chemical processes, new biodegradable materials can be produced. Such materials can successfully replace or improve the existing artificial or natural materials. Many different sources can be used to produce biodegradable materials. Fibres from naturally renewable resources are made chemically as fibres from polylactic acid (PLA fibres) or as a secondary product of other technologies. Such products are soybean fibres, which are made from soy proteins after the extraction of oil from soybean. New, natural resources are also used for fibre-making purposes, e.g. bamboo tree for bamboo fibres. These are by far not the only existing fibres from renewable resources; nevertheless, in our research, these three types of yarns are used. All presented fibres have compatible properties with classical natural fibres and some additional properties with a good influence on the comfort of clothing to the human body.

2.1 Bamboo fibres

Bamboo is considered by many to be the ultimate green material (Netravali, 2005). Since it is a fast growing plant, it can be harvested in as little as six weeks, although more typically in three to five years. Bamboo reproduces through its extensive system of rhizomes. As such, there is a continuous supply of bamboo, which meets the definition of a renewable resource. And, of course, it is also a sustainable material, capable of sustaining itself with minimal impact to the environment.

Bamboo can thrive naturally without using any pesticide. It is seldom eaten by pests or infected by pathogen.

The bamboo fibre is a kind of regenerated cellulose fibre, which is produced from raw materials of bamboo pulp refined from bamboo through the process of hydrolysis-alkalization and multi-phase bleaching, then processed and pulp is turned into bamboo fibres.

The properties of bamboo fibre are:

- strong durability, stability and tenacity,
- thinness and whiteness degree similar to the classically bleached viscose,
- antibacterial and deodorizing in nature (even after being washed fifty times),
- incredibly hydroscopic (absorbing more water than other conventional fibres, e.g. cotton),

- fabric garments make people feel extremely cool and comfortable in hot conditions,
- fabric is exceptionally soft and light, almost silky in feel, and
- fabric has a high level of breathability, for the cross-section of bamboo fibres is filled with various micro-gaps and -holes (Das, 2010).

2.2 Polylactide fibres (PLA)

Polylactic acid is a natural, biodegradable organic substance, which is harboured in the bodies of animals, plants and microbes. The polylactic acid as such cannot be found in the nature but needs to be industrially prepared with the lactic acid polymerisation.
The lactic acid used for the synthesis of polylactic acid is derived from genetically altered corn grains (Rijavec, Bukošek, 2009).
Unlike other synthetic fibre materials with vegetable resources (e.g. cellulose), PLA is well suited for melt spinning into fibres. Compared to the solvent-spinning process required for the synthetic cellulose fibres, melt spinning allows PLA fibres to be made with both lower financial and environmental cost, and enables the production of fibres with a wider range of properties (Dugan, 2000). The polymerisation occurs with the condensation of acid with alcohol, forming polyester. The misguidance in this observation is to assume that since PLA is polyester, it will behave in many ways similarly to PES or PA 6 fibres (Rekha et al., 2004).
The fundamental polymer chemistry of PLA allows control of certain fibre properties and makes the fibre suitable for a wide range of technical textile applications and special apparel (Farrington et al, 2005).
The properties of PLA fibre are:

- low moisture absorption,
- good natural regulation of the body temperature through moister absorption,
- low flammability,
- high resistance to UV and a low index of refraction, and
- excellent mechanical properties and module of elasticity (Lou et al., 2008).

2.3 Soybean fibres

Soy protein fibre (SPF) is the only plant protein fibre (Rijavec, Bukošek, 2009). It is a liquefied soy protein that is extruded from soybean after the extraction of oil, and processed mechanically to produce fibres by using new bioengineering technology. Fibres are produced by wet spinning, stabilized by acetylating, and finally cut into short staples after curling and thermoforming.
A soybean protein fibre has not only the superiorities of natural fibres but also the physical properties of synthetic ones.
The properties of SPF fibres are:

- noble appearance and similar look as silk fibres, however, they are considerably cheaper (Yi-you, 2004),
- very comfortable to wear, soft, smooth, with soft handle,
- fabric has the same moisture absorption as cotton fibres (Brooks, 2005),
- better moisture transmission than a cotton fabric, which makes it comfortable and sanitary,
- higher tensile strength than wool, cotton, and silk, however, lower than polyester fibres,
- does not shrink when washed in boiling water,
- outstanding anti-crease, easy-wash and fast-dry properties,

- antibacterial properties, and
- high UV resistance.

In the table below, the physical and mechanical properties of fibres, e.g. length, fineness, dry tenacity, wet tenacity, dry breaking extension and physical density are shown.

Properties	Bamboo	Cotton	Viscose	PLA	PES	PA	SPF	Silk	Wool
Length (mm)	38–76	25–45	30–180		32–150		38–76	$3.5 \cdot 10^6$–$9 \cdot 10^6$	50–200
Fineness (dtex)	1.3–5.6	1.2–2.8	1.3–25		1.3–22		0.9–3	1–3.5	4–20
Dry tenacity (cN/dtex)	2.33	1.9-3.1	1.5–3.0	3.2–5.5	3–7	3–6.8	3.8–4.0	2.4–5.1	1.1–1.4
Wet tenacity (cN/dtex)	1.37	2.2–3.1	0.7–1.11		2.4–7	2.5–6.1	2.5–3.0	1.9–2.5	1.0
Dry breaking extension (%)	23.8	7-10	8–24	20–35	20–50	26–40	18–21	10–25	20–40
Moisture regan (%)	13.3	8.5	12.5–13.5	0.4–0.6	0.4	4.5	8.6	11.0	14.5
Density (g/cm³)	0.8–1.32	1.5–1.54	1.46–1.54	1.25–1.27	1.36–1.41	1.15–1.20	1.29–1.31	1.34–1.38	1.32

Table 1. Comparison of physical and mechanical properties of bamboo fibres, PLA, SPF, cotton, viscose, wool and PES

3. Mechanical properties of woven fabrics

With mechanical properties, the phenomenon on textile material is described which is a result of the material resistance on the activity of external forces causing the change of shape. The response of the textile material depends on the material properties, the way of load and its tension. With regard to the direction of the applied force, deformations at stretch and compression are known. To the mechanical properties of fabrics uniaxial or biaxial tensile properties, compression, shearing properties, bending rigidity, bursting and tear resistance can be listed.

Numerous parameters influence the mechanical properties of woven fabrics. Firstly, there are fibre properties, and their molecular properties and structure. The mechanical properties of fibres depend on their molecular structure, where macromolecules can be arranged in crystalline (unique arrangements of molecules) or amorphous (coincidental arrangements of molecules) structure. The macromolecules are orientated mostly along the fibre axis and are connected to each other with intermolecular bonds. When a force is applied, the supramolecular structure starts changing (Geršak, 2006).

The fibre properties and the type of spinning influence the yarn properties, while the fabric properties are also influenced by warp and weft density of the woven fabrics, and weave. The mechanical properties are also influenced by the weaving conditions, e.g. speed of weaving, warp insertion rate, weft beat-up force, the way of shed opening, warp preparation for weaving, warp and weft tension, number of threads in reed dent etc.

The properties of raw fabrics consequently depend on the construction and technological parameters. For the final use, raw fabrics have to be post-treated to add different functional properties. In most cases, these post-treatments worsen some mechanical properties, while again some other mechanical properties improve. In Figure 1, the procedure from fibres to the end of woven-fabric production is presented.

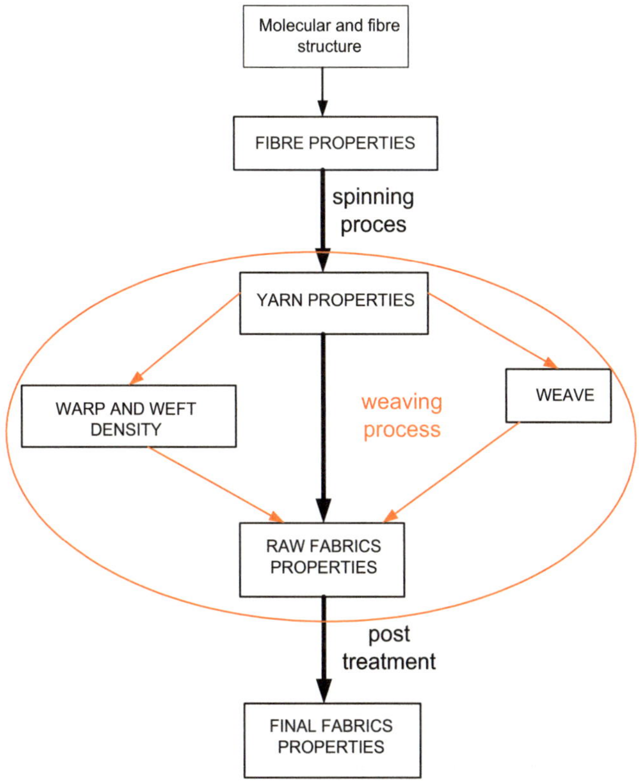

Fig. 1. Interrelation of fibre, yarn and fabric structure and properties

A lot of researches have been investigating the mechanical and tensile properties of fabrics. The approaches to the problem have included geometric, mechanical, energy and statistical models (Realff et al, 1997). The first geometric model of fabrics was presented by Pierce (Pierce, 1937), who presumed that yarn has an ideal circular cross section, which is rigid and inextensible. His work was continued by Womersley (Womersley, 1937), who presented a mathematical model of deformation of fabrics if exposed to a load. Similarly, other researchers have taken Pierce's work as a fundament. Kemp (Kemp, 1958) improved Pierce's model with the introduction of ecliptic shape of yarn. With the help of Pierce's and Kemp's geometry, Olofsson (Olofsson, 1965) presented a mechanical model of fabrics under uniaxial loading. His work was continued by Grosberg with co-authors (Grosberg et al, 1966), who were investigating tensile, bending, bulking and shearing properties, and fabrics and forces acting at counted properties on a fabric and yarn in the fabric. Kawabata approached the geometry of the interlacing point. He set the interlacing point in space, presented it as a space curve, and researched how the fabric behaves when forces act upon it and what deformations occur (Kawabata, 1989). Apart from the geometric and mechanical models, the researchers have also developed energy, statistical and numerical models of woven fabrics.

In more recently, many researches are still based on the already known models, trying to improve or reform the already existed models. A lot of researchers have performed work based on real woven fabrics, studying their physical and mechanical properties. They have been investigating the influence of differently used yarn (material or different technique of spinning), the influence of different density of warp and weft threads, and weave.

Our research is also based on the investigation of the physical and mechanical properties of woven fabrics with different yarns used in weft.

3.1 Tensile properties of fabrics

For designing apparel as well as for other uses, the knowledge about the tensile properties of woven fabrics is important. Strength and elongation are the most important performance properties of fabrics governing the fabric performance in use. Their study involves many difficulties due to a great degree of bulkiness in the fabric structure and strain variation during deformation. Each woven fabric consists of a large amount of constituent fibres and yarns and hence, any slight deformation of the fabric will subsequently give rise to a chain of complex movements of the latter. This is very complicated, since both fibres and yarns behave in a non-Hooken way during deformation (Hu, 2004)

At the beginning of loading, extension occurs in amorphous parts, where primary and secondary bonds are extending and are shear loaded. If in this stage, an external force stops acting, most of the achieved extension will recover and the material shows elastic properties. If the loading continuous, a plastic deformation of the material occurs. Long chains of molecules are reciprocally re-arranged as a consequence of the disconnection of secondary bonds. The re-arrangements of the reciprocal position of molecules give material better possibility to resist additional loading. If the loading continuous, a final break will occur (Saville, 2002).

The stress-strain curve has three parts as it is shown in Figure 2. A higher initial module at a tensile test occurs, due to the resistance against friction and bending of fabrics. In the tested direction, in the direction of force, crimp yarns are straightened. When the yarns are straightened, the force in the fabrics increases quickly and fibres and yarns begin to extend, as it is shown in Figure 2b. The tensile properties of fabrics mostly depend on the tensile properties of yarns (Grosberg, 1969)

In the region of elasticity, where Hook's law exists, tenacity (σ) is given with Equation 1.

$$\sigma = E \cdot \varepsilon \tag{1}$$

Where:

σ – tenacity (N/mm^2),
E – elastic or Young's module (N/mm^2),
ε – extension – deformation (%).

A major difference between the shapes of the curves above occurs in the first part of the curve, i.e. in the Hook's zone (I – zone). This is influenced by a crimp of warp or weft yarns, when they begin to straighten. The elongation of the fabric is already increasing under a low force (still before the zone in which Young's modulus is calculated). Here, the crimp is interchanged between the threads of the two systems. The crimp decreases in the direction investigated, however, it increases in the perpendicular direction. Consequently, the tension of the threads of the system, which is perpendicular to the direction investigated, increases. When a tensile force acts on the threads of one system, the threads of both systems undergo extension. Due to the crimp interchange, the maximum possible elongation of perpendicular threads depends on the fabric geometry (Saville, 2002, Gabrijelčič et al, 2008).

The elastic or Young's module provides resistance against the deformation of the material (fabric). Lower the value of Young's module, the more deformable (extensible) is material. The Young's module in the diagram stress-strain represents the tangents of the inclination angle α. The more resistant the material, the higher the angle of inclination α.

$$E = tg\alpha = \frac{\sigma}{\varepsilon} \tag{2}$$

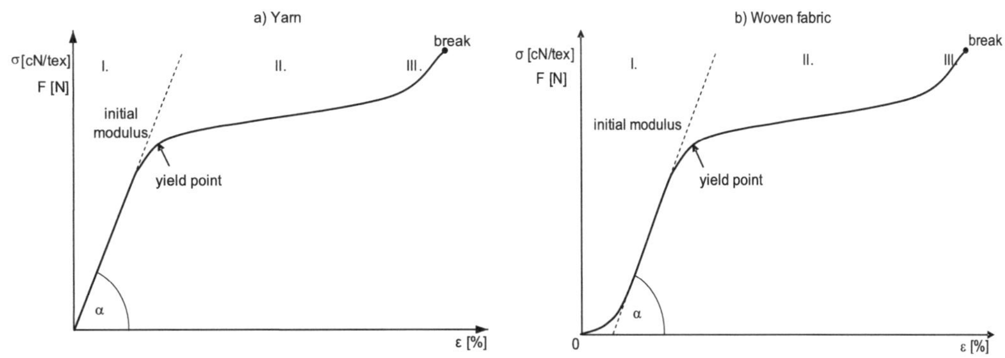

Fig. 2. Stress-strain curve of yarn and fabrics

As it can be seen in Figure 2, the load-extension curve is divided into three zones:

- the zone of elastic deformation or Hook's zone (zone I) of both yarn and fabric: If the extension occurs inside the Hook's zone, the material recovers to its initial length after the relaxation. This zone is also called the zone of linear proportionality or linear elasticity.
- the zone of viscoelastic deformation (zone II): After the loading, the material recovers to its initial length after a certain time of relaxation. The relationship between the stress and deformation is not linear. The limit between the elastic and plastic deformation is the yield point, on the stress-strain curve seen as a turn of curve.
- the zone of permanent deformations (zone III): The material does not recover after the relaxation (Geršak, 2006, Reallf et al, 1991)

3.2 Mechanical properties measured with KES evaluation system

Measuring other physical and mechanical properties and not only tensile properties is of great help in controlling and in the quality processes during the manufacture and post-treatment of textiles. Many researchers have been trying to develop a system for measuring the mechanical properties of textiles. The Kawabata Evaluation System (KES) is the first system for testing fabric mechanical properties. And it is also the system which evaluates fabric handle. This system has four different machines, and 16 parameters in warp and weft direction can be obtained, covering almost all aspects of physical properties of fabrics measured at small load. Tensile, bending, shearing, compressional and surface properties can be measured. From these measurements, properties such as stiffness, softness, extensibility, flexibility, smoothness and roughness can be inferred.

Tensile property

The tensile behavior of fabrics is closely related to the inter-fiber friction effect, the ease of crimp removal and load-extension properties of the yarn themselves as it was discussed before. Four tensile parameters can be determined through the KES instruments LT, WT, RT and EMT. LT represents the linearity of the stress-strain curve. A higher value of LT is supposed to be better. EMT reflects fabric extensibility, a measure of fabric ability to be stretched under a tensile load. The larger the EMT, the more extensible is the fabric (Chan et al, 2006). A proper amount of extensibility is desirable, while both excessive and insufficient extensibility will cause problems for the production. LT represents the linearity of the stress–strain curve. A higher value of LT is supposed to be better. WT denotes the tensile energy per unit area, taking care of the effect of both EM and LT. Thus, the conclusion about WT can be deduced from the comparison of EM and LT. RT (tensile resiliency) measures the recovery from tensile deformation. A tight fabric structure contributes to a better recovery.

Property	Symbol	Parameter measured	Unit
Tensile	LT	Linearity of load extension curve	/
	WT	Tensile energy	cN cm/cm^2
	RT	Tensile resilience	%
	EMT	Extensibility, strain at 500 cN/cm	%
Shear	G	Shear rigidity	cN/cm degree
	2HG	Hysteresis of shear force at 0.5°	cN/cm
	2HG5	Hysteresis of shear force at 5°	cN/cm
Bending	B	Bending rigidity	cN cm^2/cm
	2HB	Bending hysteresis	cN cm/cm
Compression	LC	Linearity of compression thickness curve	/
	WC	Compressional energy	cN cm/cm^2
	RC	Compressional resilience	%
Surface	MIU	Coefficient of fabric surface friction	/
	MMD	Mean deviation of MIU	/
	SMD	Geometric roughness	µm
Thickness	T	Fabric thickness at 50 N/m^2	mm
Weight	W	Fabric weight per unit area	mg/cm^2

Table 2. Parameters measured on KES system

Shear property

Whenever bending occurs in more than one direction, so that the fabric is subjected to double curvature, shear deformations of the fabric are involved. As revealed by its definition, shear property is highly related to the fabric bending property. The shear property in conjunction with the bending property is thus a good indicator of the ability of a fabric to drape. A shear deformation is very common during the wearing process, since the fabric needs to be stretched or sheared to conform to the new gesture of a body movement. During the making-up of a garment, the shear deformation is also indispensable for an intended garment shape. Shear rigidity *G* provides a measure of the resistance to the rotational movement of the warp and weft threads within a fabric when subjected to low levels of shear deformation. The lower the value of *G*, the more readily the fabric will conform to three-dimensional curvatures. If the shear rigidity is not enough, a fabric distortion will easily occur. So does the skewing or bowing during handling, laying up, and sewing. On the other hand, too high shear rigidity might also present a problem to form, mould, or shape, especially at the sleeve head. 2HG and 2HG5, the hysteresis of shear force at 0.5° and 5°, are two other measures of the shear property of a fabric. Like 2HB, the lower the 2HG and 2HG5, the better the recovery from shear deformation.

Bending property

The fabric bending property is apparently a function of the bending property of its constituent yarns. Two parameters can be used to measure this property, i.e. *B* and 2HB. *B* is bending rigidity, a measure of a fabric ability to resist to a bending deformation. In other words, it reflects the difficulty with which a fabric can be deformed by bending. This parameter is particularly critical in the tailoring of lightweight fabrics. The higher the bending rigidity, the higher the fabric ability to resist when it is bent by an external force, i.e. during fabric manipulation in spreading and sewing. Apart from for the bending rigidity of the constituent yarns and fibers, the mobility of the warp/weft within the fabric also comes into play in this aspect. In addition, the effect of density and fabric thickness are also very profound for this property. 2HB represents the hysteresis of the bending moment. It is a measure of recovery from bending deformations. A lower value of 2HB is supposed to be better.

Compression

Fabric compression is one of the most important factors when assessing fabric mechanical properties, since it is highly related to the fabric handle, i.e. fabric softness and fullness, and fabric surface smoothness. Especially, this property might even influence the thermal property of a fabric. For example, when a fabric is compressed, a subsequent drop in its thermal insulation will be found as well due to the loss of still air entrapped in the fabric. The compressional property can be influenced in many ways. Generally speaking, this property can reflect the integrated effect of a fabric structure like yarn crimp level and thickness, the constituent fiber and/or yarn surface property, and lateral compressional property. LC, the linearity of compression-thickness curve, WC, the compressional energy per unit area, and the last one RC, the compressional resilience, reflect the ability of a fabric to recover from a compressional deformation.

Surface property

Apparently, the fabric handle bears a close relationship with the surface property of a fabric. Three parameters are used as indices of fabric surface property, i.e. MIU, the coefficient of friction, MMD, a measure of the variation of the MIU, and SMD, a measure of geometric roughness. MIU is mainly governed by the contact area and type of weave. The greater the contact area, the higher the MIU value. Generally, a plain weave exhibits a higher geometric roughness in comparison with twill weave due to its shorter floats. [5, 6]

4. Experimental

The research was focused on the mechanical properties of fabrics with cotton warp and biodegradable yarns (bamboo, PLA and SPF) as well as cotton in weft. Pure cotton fabrics were made for the comparison with other fabrics with biodegradable yarns in weft.

Fabrics were made in four most commonly used weaves (i.e. plain weave, basket weave, twill 1/3 and twill 2/2). All fabrics were made on the same loom with the same density for all fabrics, 30 threads/cm in warp and 28 threads/cm in weft. Fabrics were washed after desizing.

For all fabrics, the physical characteristic warp and weft crimp, mass per square meter, thickness of fabrics, as well as tensile properties of used yarns and tensile properties of fabrics in warp and weft direction were measured in compliance with the SIST EN ISO 13934 standard. For a better comparison between the fabrics with different materials in weft,

breaking tenacity was calculated as well and presents how much force can yarn hold per linear density.
Moreover, other mechanical properties were measured on the KES system, e.g. bending, tensile properties at small load, shearing and compression. The measurements were statistically estimated and analyzed with multivariate statistical methods.

	Weave	Warp	Weft	Warp crimp	Weft crimp	Mass	Thickness
		Tt_1 (tex)	Tt_2 (tex)	C_1(%)	C_2(%)	(g/m²)	(mm)
1	Plain	Cotton 28 tex	Bamboo 21 tex	9.24	13.32	170.83	0.163
2	Basket			2.94	5.44	164.30	0.241
3	1/3 Twill			2.72	15.08	168.21	0.266
4	2/2 Twill			3.28	13.06	167.44	0.247
5	Plain		PLA 20 tex	8.86	17.74	174.42	0.203
6	Basket			3.58	19.44	168.66	0.264
7	1/3 Twill			4.16	20.06	169.09	0.279
8	2/2 Twill			3.94	21.52	169.35	0.269
9	Plain		SPF 15 tex	8.04	23.32	156.93	0.162
10	Basket			2.54	21.80	153.97	0.234
11	1/3 Twill			2.84	25.62	159.32	0.247
12	2/2 Twill			3.16	22.98	152.35	0.244
13	Plain		Cotton 19 tex	11.06	14.86	164.40	0.201
14	Basket			2.34	13.46	158.58	0.281
15	1/3 Twill			3.36	14.70	161.61	0.283
16	2/2 Twill			3.76	15.86	161.80	0.278

Table 3. Construction parameters of fabrics and measured physical parameters of fabrics

Plain weave (PL) Basket weave (BW) Twill 1/3 (T 1/3) Twill 2/2 (T 2/2)

Fig. 3. Used weaves in fabrics

5. Results

5.1 Tensile properties of yarns

As said before, the strength of a fabric depends not only on the strength of the constituent yarn, but also on the yarn structure, yarn bending behaviour, fabric geometry, thus tensile properties (i.e. tensile force and tensile elongation) of all used yarns were measured and for a better comparison, breaking tenacity of yarns was calculated. It was established that SPF yarn is the strongest and can withstand the most stress per linear density. Warp and weft cotton yarns have almost the same breaking tenacity (i.e. around 16 cN/tex). The breaking tenacity of PLA yarn is around 12.5 cN/tex and the lowest is for bamboo yarns.
Furthermore, the tenacity-extension curves were elaborated for each yarn, where the stress-strain behaviour of the used materials can be observed. In Figure 2, it can be seen that

biodegradable yarns differ from cotton yarns especially at tensile elongation, which is approximately two times (bamboo yarn), three times (SPF) and five times (PLA) higher than at cotton weft yarns. On the other hand, the tensile strength of weft cotton yarns is comparable with the tensile strength of bamboo and PLA, while the SPF yarn has a considerably higher tenacity.

	COTTON - WARP	BAMBOO	PLA	SPF	COTTON - WEFT
F (cN)	**444.38**	**218.84**	**249.77**	**287.22**	**258.49**
CV	8.44	12.32	8.03	8.39	9.21
E (%)	**4.18**	**8.52**	**27.52**	**13.72**	**4.45**
CV	9.39	12.34	8.27	6.41	11.24
σ (cN/tex)	16.35	10.42	12.49	19.17	16.88

Table 4. Tensile properties (breaking force, breaking elongation and breaking tenacity) of yarns used in fabrics

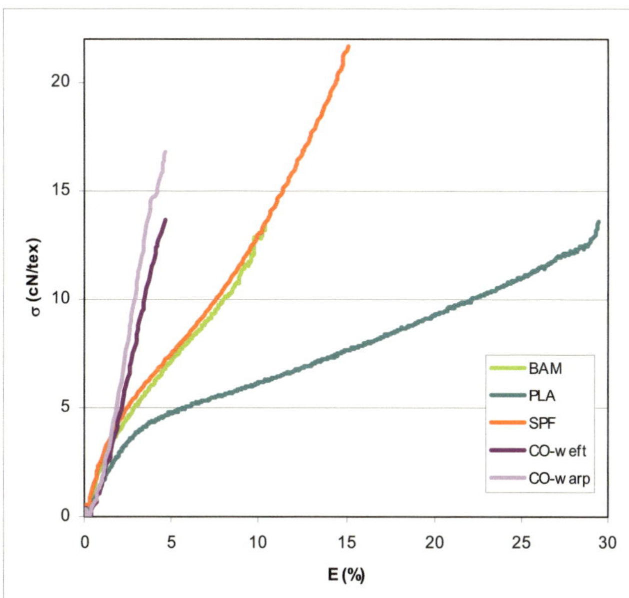

Fig. 4. Tenacity-extension curve for bamboo, PLA, SPF and cotton yarns

5.2 Tensile properties of fabrics

Tensile properties of all 16 fabrics were measured. The results of all measurements (breaking force and breaking elongation) are shown in Table 4. Moreover, the breaking tenacity of one yarn in weft direction of the fabric was calculated for a better comparison of yarns with different linear densities.

Firstly, it was established that the type of weave has a greater influence on the breaking force of fabrics in warp direction than different types of yarns. With a multivariate statistical analysis, it was proved that weave is a 5-time more important factor than different material in weft. The highest tensile force is, as it was expected, in plain weave, due to the maximum number of interlacing points resulting in higher friction between yarns and consequently also higher tensile strength in warp direction. Twill 2/2, twill 1/3 and basket weaves follow with lower values, which are presented in Figure 5. The differences in the same weave depend considerably on the material used in weft. It was found out that the extensibility of yarns in weft direction influences the breaking force in warp direction. The highest breaking force in warp direction is shown at fabrics with PLA and SPF yarn in weft and the lowest tensile force at pure cotton fabrics, since cotton yarn has the lowest elongation.

No			WARP				WEFT				
			F (N)	CV	E (%)	CV	F (N)	CV	σ (cN/tex)	E (%)	CV
1	PL	Bamboo	903.75	1.24	13.95	2.06	322.13	3.3	54.78	19.58	4.26
2	BW		779.27	2.39	7.48	2.53	334.68	4.31	55.34	19.18	4.4
3	T1/3		815.44	3.18	7.98	3.44	326.05	3.63	54.29	21.69	3.12
4	T2/2		812.6	4.43	7.83	2.68	350.98	2.42	59.69	21.99	2.74
5	PL	PLA	907.77	3.44	15.82	2.51	376.91	3.91	63.67	44.14	4.36
6	BW		796.93	3.13	7.58	2.79	370.18	1.44	65.17	44.29	2.92
7	T1/3		806.68	2.92	7.93	1.7	360.12	1.46	64.31	46.6	2.04
8	T2/2		845.43	4.97	8.08	2.62	352.82	4.02	61.68	48.75	2.53
9	PL	SPF	965.67	3.99	14.06	3.99	351.04	2.68	82.11	32.34	2.53
10	BW		779.76	1.74	6.22	3.43	347.26	7.24	80.38	30.87	6.1
11	T1/3		788.65	3.53	7.23	2.91	332.87	2.74	78.14	30.53	1.24
12	T2/2		817.92	1.8	6.98	2.97	340.38	4.88	78.79	33.19	4.93
13	PL	Cotton	857.76	5.39	15.16	2.52	474.14	2.64	85.76	16.16	1.77
14	BW		766.48	4.11	7.13	4.04	424.42	5.75	79.49	16.82	16.86
15	T1/3		730.2	2.86	7.93	3.64	433.9	1.42	81.56	15.62	3.5
16	T2/2		771.94	4.8	7.58	2.77	451.16	2.1	84.20	19.18	12.73

Table 5. Tensile properties (breaking force, breaking elongation in warp direction and breaking force, breaking elongation and breaking tenacity in weft direction) of fabrics

Among the fabrics woven in plain weave, the fabric with SPF yarn in weft is distinguished with the highest breaking force (965.67 N), followed by the fabrics with PLA (907.77) and bamboo (903.75) yarn in weft. The lowest tensile strength belongs to the fabric with cotton yarn also in weft (857.76 N). In the case of basket weave, the difference between the highest value (fabric with PLA yarn in weft - 796.93 N) and the lowest value (fabric with cotton yarn in weft - 766.48 N) is small. Among the fabrics woven in twill 1/3, the highest breaking force is observed in the fabric with bamboo yarn in weft (815.44 N), and the lowest tensile strength again in the fabric with cotton also in warp (730.2 N). Among the fabrics woven in 2/2 twill, the fabric with PLA yarn in weft has the highest value of tensile strength (845.43 N); the lowest tensile strength is observed, as in previous weaves, in the fabric with cotton weft (771.94 N).

The tensile force in weft direction influences mostly the material used in weft, while the weave has practically no influence. For a better comparison and understanding how different weft yarns influence tensile properties, esp. breaking force, the breaking tenacity of fabrics in weft direction was calculated and is presented in Figure 7.

From Figure 6, it can be seen that the highest breaking force characterises pure cotton fabrics, since weft cotton yarn has also high breaking tenacity of yarns, however, not the highest one. SPF yarns have the highest breaking tenacity (19.17 cN/tex); nevertheless, the fabrics have lower breaking force and also lower breaking tenacity calculated on one thread than pure cotton fabrics (83 cN/tex). The breaking tenacity of SPF yarns in fabrics is approximately 80 cN/tex. The reason could be that SPF yarns are much smoother than cotton yarns and less friction occurs between warp and weft yarns. The second highest breaking force in weft direction is typical of fabrics with PLA yarn in weft, although the breaking tenacity of PLA yarn (12.49 cN/tex) and the breaking tenacity calculated on one thread in fabrics are lower than for cotton and SPF. The average breaking tenacity of all fabrics with PLA in weft is approximately 64 cN/tex. It can be expected that the SPF fabrics with the same linear density of weft yarn will have higher tensile strength than the fabrics with PLA yarn in weft. The lowest breaking force in weft direction characterises the fabrics with bamboo yarn in weft. Bamboo yarn has the lowest tenacity (10.42 cN/tex) and in fabrics, the breaking tenacity is approximately 56 cN/tex.

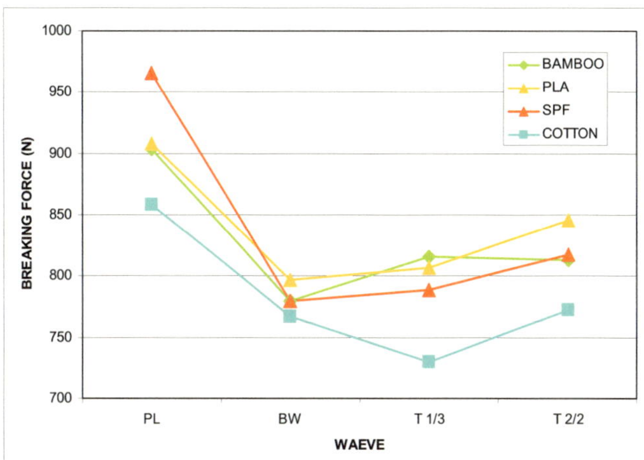

Fig. 5. Breaking force of fabrics in warp direction

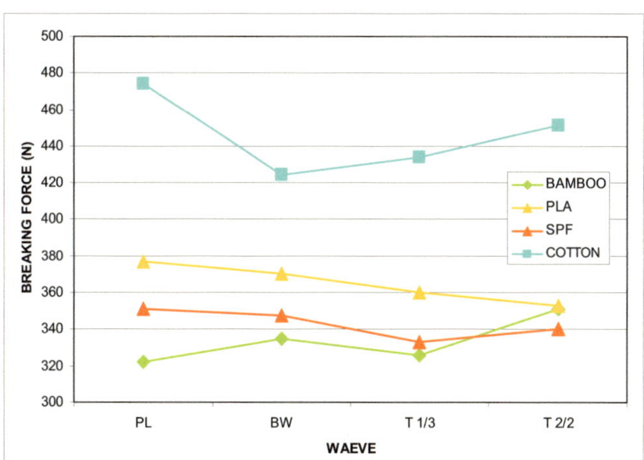

Fig. 6. Breaking force of fabrics in weft direction

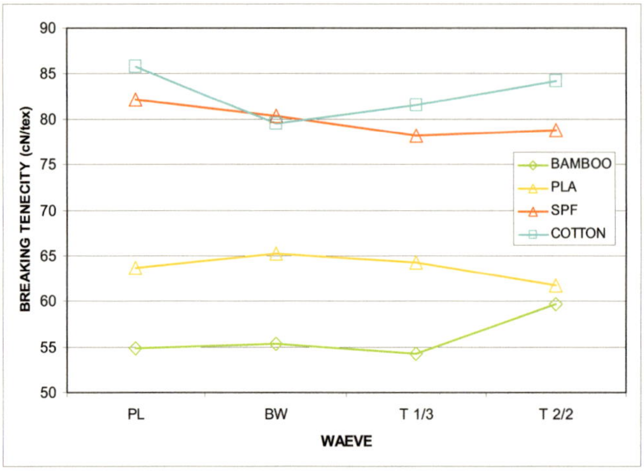

Fig. 7. Breaking tenacity of fabrics in weft direction

It was also found out that different properties of yarns have almost no influence on the tensile elongation of fabrics in warp direction, while mostly the weave type influences the tensile elongation in warp direction. The weave type is statistically 50 times more important than different materials used in weft. Figure 8 shows that fabrics in plain weave have the highest tensile elongation, which is approximately 15%, whereas the tensile elongation of all other fabrics, as can be seen in the diagram, is about 7%. Plain weave has the maximum number of interlacing points, which is twice as high as that of other weaves and, as a result, tensile elongation is higher. Also, warp crimp is the highest in plain weave, which influences tensile elongation as it was said before. The lowest tensile elongation is typical of the fabrics woven in basket weave.

Both the weave type and the material in weft influence the tensile elongation in weft direction, but the material used in weft is statistically 90 times more important. The highest tensile elongation is at fabrics with PLA yarns in weft, which are also the most extensible yarns. Then there are fabrics with SPF yarn, followed by fabrics with bamboo yarn and the lowest tensile elongation is at pure cotton fabrics, since cotton yarns have the lowest extensibility.

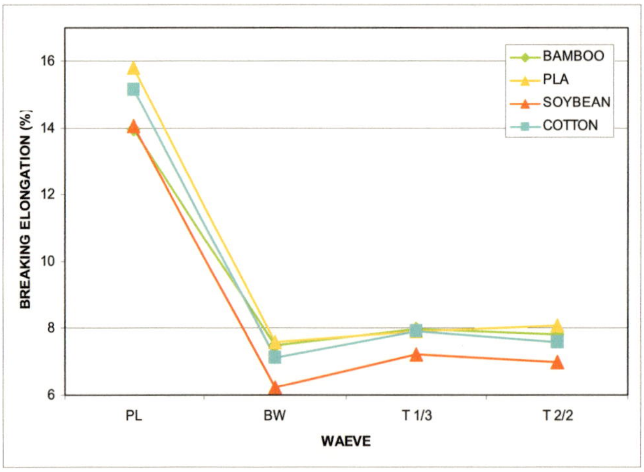

Fig. 8. Breaking elongation of fabrics in warp direction

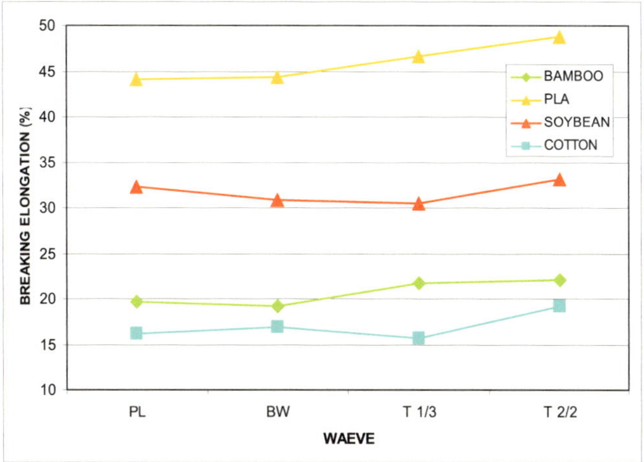

Fig. 9. Breaking elongation of fabrics in weft direction

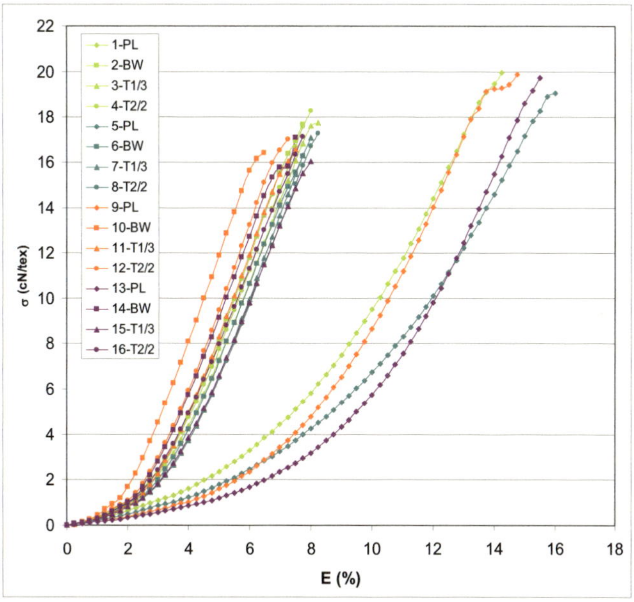

Fig. 10. Tenacity - extension curves for fabrics in warp direction

The fabrics with PLA yarn have the highest tensile elongation, for PLA yarn itself already has the highest tensile elongation (27.52%) and is the most extensible yarn. All these fabrics have tensile elongation about 45%. It is evident in Figure 9 that the fabrics in twill 1/3 and twill 2/2 have higher tensile elongation than plain and basket weave. The fabrics with SPF yarn in weft come in the second place. The tensile elongation of SPF yarn (13.72%) is ranked second. Figure 8 also shows that tensile elongation of these fabrics is ca 32%. The highest tensile elongation belongs to plain weave and twill 2/2 weave. Next to them, there are the fabrics with bamboo yarn with mean tensile elongation at about 20%. The tensile elongation of bamboo yarn is 8.52%. It is noticed again that twill 1/3 and twill 2/2 weaves have higher tensile elongation than the plain and basket weave. Cotton fabrics have the lowest tensile elongation, since the cotton yarn itself also has the lowest tensile elongation of only 4.45%. The tensile elongation of cotton fabrics is 16%. The highest elongation is observed in the fabric in twill 2/2 weave.

For all fabrics in warp and weft direction, tenacity-extension curves were made to compare different behaviour at the tensile test.

The tenacity-extension curves in Figure 10 show that it is the weave, which has the highest influence on the shape of curves in warp direction. The curves of plain weave have almost the same shape, whereas the shapes of other weaves have very similar shapes. All curves for each group of materials are arranged in a defined order, i.e. twill 1/3, twill 2/2 and basket. Plain weave has a completely different shape of the curve due to a more frequent interlacing of threads in the weave, which results in a higher shrinkage of the fabric and, consequently, higher elongation.

The shapes of the curves for weft show that it is solely the material, which influences the shape of the curve. The weave has practically no influence, which has already been proved by previous results. Each group of materials has its own specific shape of the curve. The fabrics with cotton weft have the most vertical shape of the curve, for they have the lowest tensile elongation. The fabrics with PLA yarn in weft have a very specific shape of the curve. If the tenacity-extension of yarns (cf. Figure 5) is compared with the tenacity-extension curve of fabrics, some similarities can be detected. However, if the curves are compared with the shapes of the curves of standard materials, i.e. cotton, cellulose, PES or PA, and silk, it can be stated that the curve with bamboo yarn in weft has the same shape of the curve as cellulose fibres. The shape of the curve with SPF yarn in weft is similar to the shape of the curve of silk. Cotton fabrics have the same shape of the curves as cotton fibres.

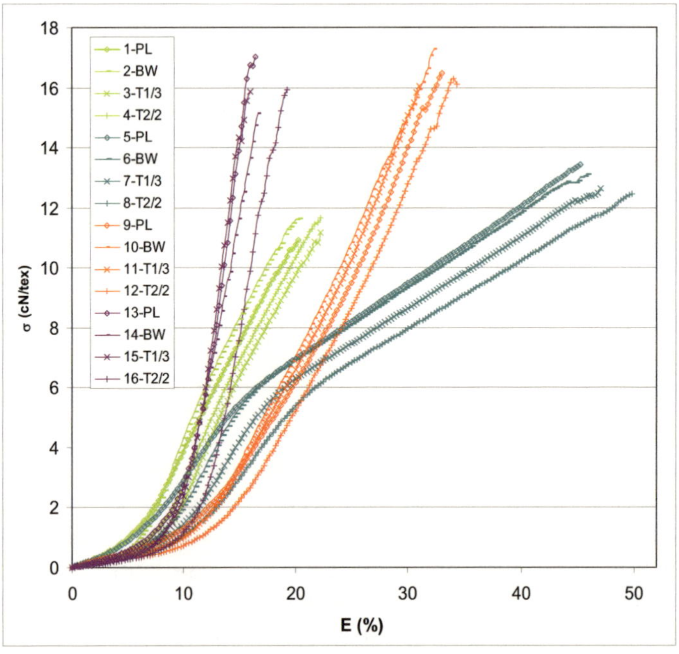

Fig. 11. Tenacity - extension curves for fabrics in weft direction

5.3 Other mechanical properties measured with KES evaluation system

The measurements on the KES system show which of the investigated fabrics is the most suitable for the clothing industry and what kind of behaviour can be expected. From the results, it can be seen that fabrics with SPF yarn in weft are very extensible and flexible. Cotton fabrics are the softest and fabrics with bamboo weft have very similar properties as cotton fabrics.

As it was discussed above, the measurements of tensile properties on the KES system confirmed as well that the tensile behavior of fabrics is closely related to the inter-fiber friction effect, the ease of crimp removal and load-extension properties of the yarn themselves. The measurements of extensibility (EMT) of fabrics and tensile work (WT) show that as at tensile test, the weave mostly influences the tensile properties in warp, and the material used in weft the tensile properties in weft. At EMT, it can be seen that the highest extensibility characterizes the fabric with the SPF yarn in weft direction; however, the SPF yarn is not the most extensible material (13.72%) but the fabric with the highest weft crimp, which has the highest influence on EMT. The fabrics in plain weave usually demonstrate a higher tensile work (WT), as it is also seen at our fabrics. It is also demonstrated that the fabrics wit SPF yarn in weft have the highest WT.

Shear rigidity *G* provides a measure for the resistance to the rotational movement of warp and weft threads within a fabric when subjected to low levels of shear deformation. The lower the value of *G*, the more readily the fabric will conform to three-dimensional curvatures. If the shear rigidity is not enough, a fabric distortion will easily occur. Shear properties are most commonly influenced by weave, while the material used in weft has practically no influence.

The KES system tensile properties influence both the type of weave and the material used in weft. It was also established that some properties measured on the KES system have very good correlation with each other (e.g. thickness, and compressional properties, bending and shearing properties) and some properties inversely proportional (e.g. tensile energy and tensile resilience, bending and shearing properties, and compressional properties). If it is known which properties correlate with each other, it is easier to predict what kind of properties the fabric will fabrics.

	LT	WT	RT	EMT	G	2HG	2HG5	B	2HB	LC	WC	RC	TO	TM	THIC
LT	1														
WT	0.588	1													
RT	-0.696	**-0.819**	1												
EMT	0.202	**0.906**	-0.623	1											
G	**0.745**	0.223	-0.449	-0.147	1										
2HG	**0.702**	0.159	-0.367	-0.201	**0.992**	1									
2HG5	**0.770**	0.181	-0.411	-0.204	**0.983**	**0.974**	1								
B	0.600	0.315	-0.298	0.060	0.648	0.631	0.691	1							
2HB	0.652	0.114	-0.213	-0.211	**0.842**	**0.855**	**0.868**	**0.893**	1						
LC	0.302	0.018	-0.084	-0.146	0.527	0.531	0.528	0.069	0.207	1					
WC	-0.128	-0.287	0.314	-0.311	0.272	0.325	0.199	-0.247	0.027	0.422	1				
RC	-0.180	0.003	0.136	0.127	-0.363	-0.368	-0.299	-0.167	-0.301	-0.064	-0.402	1			
TO	-0.683	-0.312	0.543	0.007	**-0.820**	**-0.794**	**-0.821**	-0.684	**-0.742**	-0.476	0.189	0.266	1		
TM	-0.670	-0.268	0.572	0.072	**-0.799**	**-0.753**	**-0.784**	-0.541	-0.658	-0.232	-0.019	0.582	**0.801**	1	
THIC	**-0.794**	-0.415	0.640	-0.057	**-0.805**	**-0.756**	**-0.811**	**-0.701**	**-0.729**	-0.238	0.132	0.470	**0.840**	**0.942**	1

Table 6. Correlation table of measured data on KES system

The principal components analysis (PCA) enables the visualization of linear correlations between the measured data on the KES system. PCA transforms multivariate data into a

lower-dimensional space where trends and patterns in the data can be detected. It extracts a set of new variables - the so-called principal components (PCs) - that are linear combinations of the original variables. The number of PCs equals the number of variables, however, generally, the first few (two or three) PCs already account for most of the variability in the data. By plotting two PCs, the relationships among the samples as well as among the studied variables can be visualized (Zupin et al, 2009).
From the correlation Table 6 and Figure 12 below, it can be seen that between the mechanical properties measured on the KES system, positive and negative correlations exist. The surface properties (MIU, MMD and SMD) and mass per square meter are not included into the PCA analysis. The surface properties are mostly influenced by weave, which makes the structure and surface of fabrics. Mass per square metre is almost the same at all fabrics, since the set density is the same at all fabrics.
From the correlation Table 6 and Figure 12 where close proximity, clustering, of points suggests that a high degree of positive correlation exists among the linearity of tensile load-extension curve (LT) and shear properties of fabrics (G, 2GH, 2GH5); tensile energy per unit area (WT) and extensibility (EMT); shear rigidity (G) and shear hysteresis (2HG, 2HG5) and bending properties (B, 2HB). This result was expected, for the shear and bending properties are highly related, both influenced by construction parameters of fabrics, esp. yarn properties and weave. Bending and shear properties are the most important properties for explaining the fabric stiffness, hardness and tailorability (Lam, Postle, 2006). It is also known that tensile work and extensibility correlate with each other.

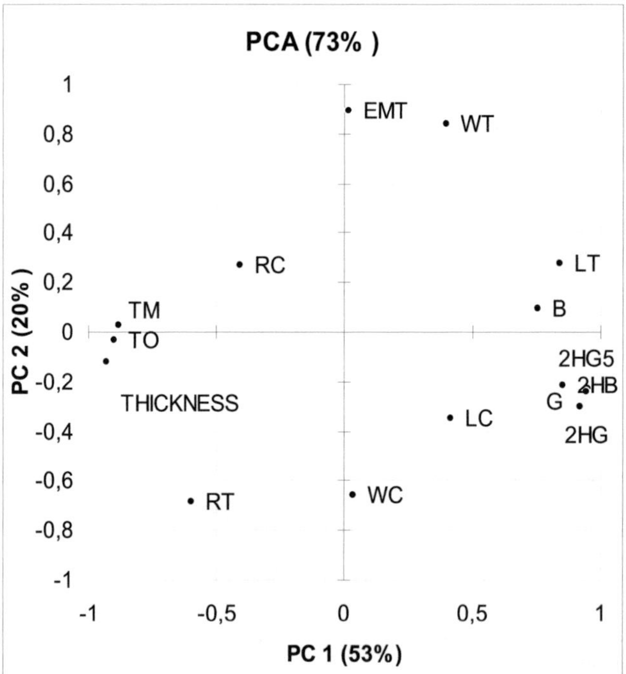

Fig. 12. Plot of original variables in PC1-PC2 coordinate system made with multivariate statistical method PCA, which presents correlations between mechanical properties measured on KES evaluation system

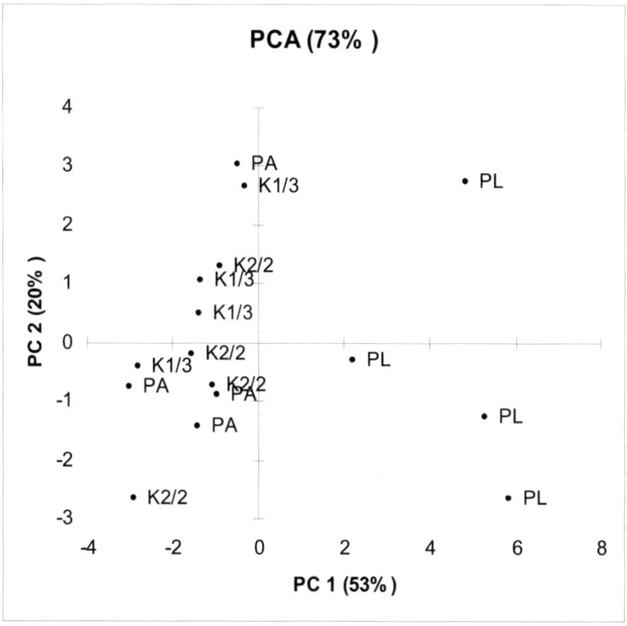

Fig. 13. Plot of 16 fabrics samples in PC1-PC2 coordinate system, each data label representing particular weave type

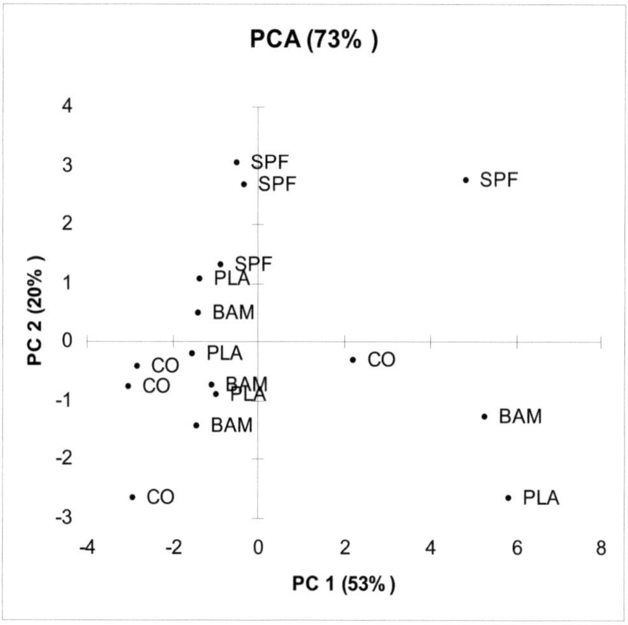

Fig. 14. Plot of 16 fabrics samples in PC1-PC2 coordinate system, each data label representing particular material used in weft

A strong negative correlation is observed where the points are opposite each other. A high degree of negative correlation exists among the linearity of tensile load-extension curve (LT) and thickness of fabrics; tensile energy per unit area (WT) and tensile resilience (RT); thickness of fabrics and shear (G, 2GH, 2GH5) and bending (B, 2HB) properties. These findings are expected, since the higher the fabric thickness, the harder it is to bend and shear the material. A negative correlation is also expected between WT and RT, for the more the fabric extends, the more energy is required to extend the fabrics to the final extension percentage.
In Figure 13, it can be seen that the fabrics made in plain wave (PL) have different properties than all other fabrics, as has already been discussed. And Figure 14 shows that fabrics are clustered according to different material used in weft.

6. Conclusion

From the research results, it can be seen how different materials used in weft influence the mechanical properties of fabrics. The highest breaking force in warp direction is exhibited in the fabrics with the highest tensile elongation: SPF, PLA and bamboo yarn, followed by the fabrics with cotton yarn in weft. The material used in weft has almost no influence on the breaking elongation in warp direction; it is influenced almost only by weave. In weft direction, cotton fabrics have the highest tensile force, whereas lower tensile strength is observed in the fabrics with PLA, SPF and bamboo wefts. The materials also have the highest influence on the tensile elongation of fabrics in weft direction. Each yarn selected in weft has its own specific properties, which is reflected in the fabric.
It was established that all fabrics in weft direction have elongation from 20 to 45%, which is completely suitable to use as a fabric with increased elasticity.
It can also be seen that different weave types also influence fabric properties; however, their influence is lower.
The research of other mechanical properties measured with the KES system shows that fabrics with SPF yarn in weft are the most extensible and flexible, while the fabrics with cotton and bamboo in weft have higher tensile resilience than the fabrics with PLA and SPF in weft. Fabrics with PLA in weft are the stiffest and the most rigid, while the cotton fabrics are the softest.
In general, it can be claimed that fabrics with PLA, SPF and bamboo yarn in weft have a lower tensile force in comparison with pure cotton fabrics but higher tensile elongation in comparison with pure cotton fabrics. The tensile elongation or extensibility of fabrics is an important property for comfortable feeling when wearing clothes.
With the help of multivariate statistical analysis, the correlation between the measured mechanical properties on the KES system was established. Good positive correlation can be seen between shear and bending properties, and between tensile work and extensibility, while good negative correlation is seen between thickness and shear, and bending properties.
In the future researches, apart from the mechanical properties, we would like to research permeability properties of fabrics made with biodegradable material in weft and cotton in warp, *inter alia* air permeability, water vapour permeability, thermal resistance of fabrics and permeability of UV rays.
Moreover, we would like to test the mechanical and permeability properties of woven fabrics made from 100% biodegradable fibres (bamboo, PLA or SPF) and compare the properties with presented samples. Nevertheless, for a successful weaving of biodegradable

yarns in warp, it would be necessary to determine the behaviour of warp yarns in the weaving process, since, as it can be seen from the research, these yarns have high extensibility, i.e. from 8.5 to 27.5%, and not very high strength in comparison with cotton yarns.

It would also be interesting to test how the biodegradable materials act in other application as knittings and nonwovens.

For a more frequent use of biodegradable materials, various constructions, woven and knitted fabrics, and nonwovens should be tested and fabric properties and behaviour after the wearing, usages and after caring, e.g. washing and ironing, should be determined.

7. Acknowledgements

I would like to thank Prof. Dr. Manuela Neves from the University of Minho, Portugal, for providing us with the yarns and woven samples used in this research.

8. References

Rijavec, T., Bukošek, V. (2009). Novel Fibres for the 21st Century. *Tekstilec,* Vol. 52, No.10-12/2009, 312-327, ISSN 0351-3386

Blackburn, R. S. (2005). *Biodegradable and sustainable fibres,* Woodhead Publishing, ISBN 1-85573-916-x, Cambrige England

Brooks, M. M. (2005). Soya bean fibres - past, present and future, In: *Biodegradable and sustainable fibres,* Blackburn, R. S. (Ed.), 389-440, Woodhead Publishing, ISBN 1-85573-916-x, Cambrige England

Farrington, D. W., Lunt, J., Davies, S., Blackburn, R. S. (2005). Poly(lactic acid) fibres, In: *Biodegradable and sustainable fibres,* Blackburn, R. S. (Ed.), 389-440, Woodhead Publishing, ISBN 1-85573-916-x, Cambrige England

Dugan, J.S. (2000). Novel properties of PLA fibres Vice President. Research Fiber Innovation Technology. Inc. Available on-line: http://www.fitfibers.com/files/PLA%20Fibers.doc

Das, S. (2010). Properties of bamboo fibres. [citated 25.3.2010]. Available on -line: http://www.fibre2fashion.com/industry-article/textile-industry-articles/properties-of-bamboo-fibre/properties-of-bamboo-fibre1.asp

Netravali, A.N. (2005). Biodegradable natural fiber composites, In: *Biodegradable and sustainable fibres,* Blackburn, R. S. (Ed.), 389-440, Woodhead Publishing, ISBN 1-85573-916-x, Cambrige England

Rekha, R., Purnima, D. C. (2004). Biodegradable Polylacitic Acid Polymers for Textile Applications. *Man-made Textiles in India.* February 2004, 42-44, ISSN:0377-7535

Lou, C.W. et al (2008). Manufactering and Properties of PLA Absorbable Surgical Suture. *Textile Research Journal,* Vol. 78, No.11, 958-965, ISSN 0040-5175

Yi-you, L. (2004) The soybean protein fibre - A healthy & comfortable fibre for 21st century. *Fibres & Textile in Eastern Europe.* Vol. 12, No. 2 (46), 8-9, ISSN 1230-3666

Realff, M.L., Boyce, M.C, in Backer, S. (1997) A Micromechanical Model of the Tensile Behavior of Woven Fabric. *Textile Research Journal,* Vol. 67, No. 6, 445-459, ISSN 0040-5175

Grosberg, P. (1969). The tensile properties of woven fabrics. In: *Structural Mechanics of Fibres, Yarns and Fabrics. Volumne 1.* ,John Wiley&Sons, 339-354, ISBN 471-36669-2

Saville, B. (2002). *Physical testing of textiles*. Woodhead Publishing and CRC Press, ISBN 1-85573-367-6, Cambridge

Gabrijelčič, H., Černoša, E. in Dimitrovski, K. (2008). Influence of weave and weft characteristics on tensile properties of fabrics. *Fibres & textiles in Eastern Europe,* Vol. 16, No. 2 (67), 45-51, ISSN 1230-3666

Geršak, J. (2006). *Mehanske in fizikalne lastnosti tekstilnih materialov*. Fakulteta za strojništvo, Laboratorij za oblačilno inžinerstvo ter fiziologijo in konstrukcijo, ISBN 86-435-0754-7, Maribor

Hu, J. (2004). *Structure and mechanics of woven fabrics*. Woodhead Pub., ISBN 0-8493-2826-8, Cambridge

Realff, M. L., Seo, M. H., Boyce, M. C., Schwartz, P., and Backer, S. (1991). Mechanical properties of fabric woven from yarns produced by different spinning technologies : yarn failure as a function of gauge length. *Textile Research Journal,* Vol. 61, No. 9, 517-530, ISSN 0040-5175

Brooks, M. M. (2005). Soya bean fibres - past, present and future, In: *Biodegradable and sustainable fibres,* Blackburn, R. S. (Ed.), 389-440, Woodhead Publishing, ISBN 1-85573-916-x, Cambrige England

Kawabata, S. (1989). Nonlinear mechanics of woven and knitted materials, *Textile Structural Composites,* Chou, T. W. and Ko, F. K. (Ed.), Elsevier Science Publisher, 67-116, ISBN 0-444-42992-1, Amsterdam

Zupin, Ž., Dimitrovski, K. (2008). Tensile properties of fabrics made from new biodegradable materials (PLA, soybean, bamboo fibres). *Proceedings of Magic world of textiles,* pp. 185-190, ISBN 978-953-7105-26-6, Dubrovnik, Croatia, October, 2008, Faculty of Textile Technology, University of Zagreb, Zagreb

Pierce, F.T. (1937). The geometry of cloth structure. *The Journal of the Textile Institute,* Vol. 28, No. 3, T45-T96, ISSN 0368-4474

Womersley, J.R. (1937). The application of differential geometry to the study of the deformation of cloth under stress. *The Journal of the Textile Institute,* Vol. 28, No. 3, T97-T112, ISSN 0368-4474

Zupin, Ž., Dimitrovski, K., Hladnik, A. (2009). Determination of woven fabrics influential parameters using multivariate statistics. *Proceedings of the 9th Autex Conference,* pp. 585-590, ISBN 978-975-483-787-2, Izmir, Turky, May, 2009, Ege University, Engineering Faculty, Department of Textile Engineering, Izmir

Lam, J.K.C., Postle, R. (2006). Multivariate Studies of Mechanical Properties for Wool and Cotton Fabrics. *Textile Research Journal,* Vol. 76, No. 5, 414-425, ISSN 0040-5175

Chan, C.K. et al (2006). Evaluation of mechanical properties of uniform fabrics in garment manufacturing. *Journal of Materials Processing Technology,* Vol. 174, 183-189, ISSN 0924-0136

Kemp, A. (1958). An extension of Peirce's cloth geometry to the treatment of non-circular theads. *The Journal of the Textile Institute,* Vol. 49, No. 1, T44-T48, ISSN 0368-4474

Olofsson, B. (1965). A general model of a fabrics as a geometric-mechanical structure. *The Journal of the Textile Institute,* Vol. 55, No. 11, T541-T557, ISSN 0368-4474

3

Wing Tear: Identification of Stages of Static Process

Beata Witkowska[1] and Iwona Frydrych[2]
[1]*Textile Research Institute,*
[2]*Technical University of Lodz,*
Poland

1. Introduction

In 1945 Krook and Fox (Krook & Fox 1945) analyzing photos of torn tongue shape samples distinguished and described the fabric tearing zone, i.e., the zone, which appears between both thread systems of torn fabric sample. The fabric tearing zone is limited by two threads of stretched thread system arising from the cut strips of torn sample, and the torn thread system being in the position "just before the breakage". Krook's and Fox's research was an inspiration for the next researchers, who were interested in the fabric tearing. They have very often based their considerations on Krook's and Fox's conclusions.

The research was continued by Teixeira et al. in 1955 (Teixeira et al., 1955). They developed Krook's and Fox's (Krook & Fox, 1945) observations, and also formulated the theoretical equation, which described the tearing phenomenon.

In 1959 Taylor (Taylor, 1959) elaborated the mathematical relationship, which on the basis of fabric and thread structure parameters enables to predict the tear strength. He analyzed the geometry of tearing zone in the aspect of friction forces of stretched thread system (not torn) along the threads of perpendicular system (torn). Taylor didn't analyze the phenomena taking place in perpendicular (torn) threads.

Research on the tearing process of trapezoidal samples was carried out by Hager, Gagliardi and Steele (Hager et al., 1947). They (as Taylor) did not take into consideration the influence of the perpendicular thread (torn) system and occurring friction forces in the torn system. Moreover, they assumed the linear relationship between the load and strain of stretched thread system.

Hamkins and Backer (Hamkins & Backer, 1980) presented their results concerning the tearing mechanics on the basis of two quite different in their structure fabrics: made from a glass yarn of „loose" weave and a big possibility of threads to be moved as well as from an elastomeric yarn of small possibility of threads to be moved. The application of previous models was not fully satisfactory for different variants of fabric structure.

Then the problem was analyzed by Seo (Scelzo et al., 1994a), who broaden the Taylor's (Taylor, 1959); (Letters to the Editor, 1974) model by the proposed by him parameters of the tearing zone geometry. The common feature of both models was an existence of such parameters as: the friction coefficient (thread by thread in a fabric), mean thread breaking force, distance between the axes of neighbor threads.

Scelzo, Backer and Boyce (Scelzo et al., 1994a); (Scelzo et al., 1994b) proposed the description of phenomenon taking place in the cotton fabric during the tearing basing on the tearing zone shape analysis and adapting the rheological model consisted of spring arrangement for a description of tearing process.
Now, the resistance on the static tearing is one of the most important assessment criteria of strength parameters of fabrics destined for the work and protective clothing, high-tech textiles, technical and interior textiles, upholstery and daily used clothing. In experiments the different specimen shapes are used, a choice of which can be found in the appropriate standards (Witkowska & Frydrych, 2004) (Witkowska & Frydrych, 2005).
In the chapter, there are presented the theory verified experimentally concerning the stages of the cotton fabric static tearing with taking into consideration the fabric tearing zone assessment, i.e., its length, depth and the thread number in the tearing zone. The analysis was carried out for the wing shape specimen described in the standard PN-EN ISO 13937-3:2002. An observation of tearing stages required the application of appropriate measurement system, which enabled the registration of changes taking place during the tearing process; and next, an analysis of obtained images. Such an opportunity was guaranteed by a specially elaborated for this purpose software working in the Windows environment. The whole software enables the observation of the video image on the computer screen, an image registration on CD with the given speed and image analysis. The special stand for the on-line tearing phenomenon registration was designed and built. Experiment was done for the model cotton fabrics of such a structure, which enables the observation of changes of tearing zone parameters dependably on the fabric weave at the assumption that the parameters of torn and stretched thread systems are unchanged.

1.1 Methods used for determination of static tear strength

In the whole period of fabric static tear resistance research, which has been estimated for over 95 years (Harrison, 1960), i.e., from 1915 up till now, there were proposed ca. ten different specimen shapes. Dependably on the assumed specimen shape, the particular investigators proposed their own specimen sizes and a measurement methodology, and also their own way of assessment of fabric tearing strength as well as own expression of results. The fabric tear strength (resistance) is a property determining the fabric strength on the static tearing action (static tearing), kinetic energy (dynamic tearing) and tearing on the "nail" of appropriate prepared specimen.
Different ways of tearing were reflected in the measurement methodology. The methods were diversified by the shape and size of specimen, a length of torn fabric distance, and also a way of tear force determination. The most popular methods were standardized. Now, in the all methods the parameter, which characterizes the fabric tear strength, is the tear force. In the static tearing methods as well as in the dynamic one the tearing process is a continuation of tearing started by an appropriate cutting the specimen before the measurement.
In Fig. 1, there are presented the specimen shapes, which are now used in the laboratory measurements of static tear strength; whereas in Table 1, there are presented the important data concerning the applied specimen shapes and dealt with them a measurement methodology.
Similarly as shapes and sizes of specimens also a way of tear force calculation has been changed since 95 years. This process was finished by a standardization, which unified a way of calculation of the static tear strength result. The result of static tearing can be read:

- directly from the measurement device;
- from the tearing chart dependably on the assumed measurement methodology.

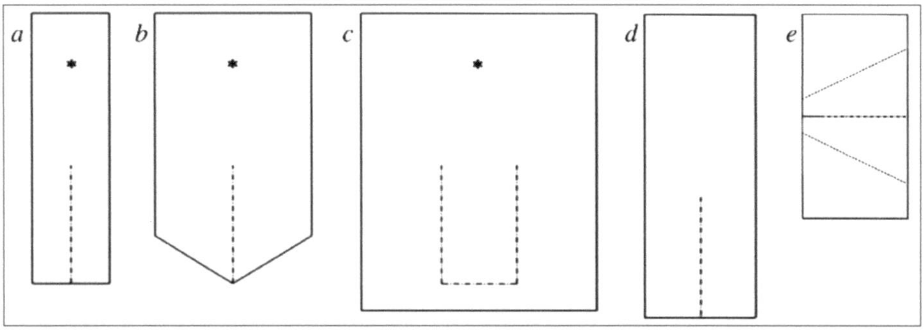

Fig. 1. The actual used shapes of specimens; a) trousers according to PN-EN ISO 13937-2 and PN-EN-ISO 4674-1-method B (for rubber or plastic-coated fabrics); b) wing according to PN-EN ISO 13937-3; c) tongue with double tearing according to PN-EN ISO 13937-4 and PN-EN ISO 4674-1-method A (for rubber or plastic-coated fabrics); d) tongue with single tearing according to ISO 4674:1977-method A1; e) trapezoidal shape according to PN-EN ISO 9073-4 (for nonwoven) and PN-EN 1875-3 (for rubber or plastic-coated fabrics)
A source: Own on the basis of present-day standards concerning the static tearing

Fig. 1a single tearing	Fig. 1b single tearing	Fig. 1c double tearing	Fig. 1d single tearing	Fig. 1e single tearing
Tearing direction to the acting force				
∥	⊥	∥	∥	⊥
Tearing distance, mm				
75	75	75	145	210
Measurement rate, mm/min				
100	100	100	100	100
Distance between jaws, mm				
100	100	100	75	25
Specimen dimension,mm; length, depth				
200; 50	200; 100	220, 150	225; 75	150; 75
Length of cut, mm				
100	100 ; angle 55°	100	80	15

Table 1. The description of static tearing methods
A source: Own on the basis of present-day standards concerning the static tearing

Now reading the tear forces from the tearing chart is possible for all used measurement methods of static tearing, i.e., for specimens of tongue shape with a single (trousers) and double tearing, for the wing and trapezoidal shape. As the tearing chart the curve, which occurs as a result of sample tearing on the tearing distance, is assumed. The initial point of tearing curve is a peak registered in the moment of breakage of the first thread (or thread group) being on the tearing distance, and the end of tearing curve is a breakage of the last thread (or thread group) on the given distance.
According to the standardized measurement procedure the following methods are used now:

- the method described in the standard series PN-EN ISO 13937 part 2 (trousers), part 3 (wind) and part 4: tongue – double tearing (Witkowska & Frydrych, 2004). The tearing graph is divided onto four equal parts starting from the first and finishing on the last peak on the tearing distance. The first part of graph is neglected in calculations. From the rest three parts of graph six the highest and lowest peaks are chosen (by the manual method) or all the peaks on 3/4 of tearing distance are calculated (by the electronic method). As a result an arithmetic mean of tear forces is given.
- the method A1 described in ISO 4674:1997, which is in agreement with the *American Federal Specifications* (Harrison, 1960), was elaborated in 1951. It relies on a median determination from five tear force values represented by maximum peaks for the medium graph distance creating 50% of tearing distance (Witkowska & Frydrych, 2004).
- the method described in PN-EN ISO 9073-4 for nonwovens and according to the PN-EN 1875-3 for rubber and plastic-coated fabrics - relies on the calculation of arithmetic mean from the registered maximum peaks on the assumed tearing distance.

1.2 Tear force as the criterion of fabric assessment

Analysis of fabric strength parameters showed that the tear strength is one of the important criteria of fabric assessment destined for the protective and work clothing; high-tech and technical textiles, interior textiles, upholstery and daily clothing.

There are many possibilities of fabric destruction by tearing. To the most characteristic cases of tearing belong:

- fabric tearing by nail or the other sharp tool;
- tearing in the places, where textile elements are joined together (in the utility or decorative purpose) by the elements of the high strength, for example rivets. In these places there can be often observed the characteristic tearing of its element from the fabric and the hole appearance;
- the hole occurence (random) in the fabric by, for example, cigarette or spark. If this place is stressed, for example, during the utility, the local area decrement causes the increase of stresses on the hole circumference and the fabric is torn;
- fabric cutting on the same distance and next, its stretching. Such a situation is observed during the utility process, for example, cutting on the pocket can cause the clothing destruction, if these places are not protected at the ends.

The diversity of fabric tearing processes as well as the existence of many measurement methods often cause the problems dealt with a choice of the appropriate measurement method for the given fabric assortment. The selection of tearing strength measurement method for the given fabric should follow the analysis of criteria of such fabric assessment. These criteria are described in:

- harmonized standards (concerning the protective clothing);
- other standards (national, European, international);
- in so-called a list of technological and utilization indices (old Polish national standard introduced before 2000 year);
- contracts between textile producers and their customers.

Below, in Table 2, there is presented a division of static tear methods dependably on the chosen fabric assortment.

The other aspect is an application possibility of the admitted tearing method for the given fabric structure. It often happens that, for example, for a high tear strength fabric, i.e., of tear force above 100 N or for fabric destined for the protective and work clothing (often cotton or cotton similar) of diversified tear strength dependably on the thread direction (warp-weft) and for fabric of weaves with long floating threads there is possible the application (undependably on the mentioned above rules) of only one tearing method, i.e., according to PN-EN ISO 13937 Part 3 (Fig. 1c). It is caused by the fact that the sample shape in this method and a way of mounting it in the tensile tester jaws enables the obtaining the better its clamping in the tensile tester jaws. Thanks to it the sample is not broken in the jaws and the measurement can be performed.

Textiles - rubber or plastic-coated fabric	
PN-EN 1875	Technical textiles
ISO 4674:1977	Method A1: protective clothing (high-visibility warning for professional use; protection against rain), Method A2: textiles for tarpaulin
PN-EN ISO 4647-1	Method A: Protective clothing (protection against cold), Method B: protective clothing (for firefighters for firefighting)
Textiles - non-coated	
PN-EN ISO 13937-2	protective clothing (for firefighters in action), work clothing (overalls, shirts, trousers), mattress – woven, daily textiles, textiles for flags, banners
PN-EN ISO 13937-3	upholstery (furniture) textiles, bed clothing, textiles for beach chairs, technical textiles (roller bind)
PN-EN ISO 13937-4	work clothing (changeable with PN-EN ISO 13937-2)

Table 2. The division of static tear methods dependably on the chosen fabric assortment
A source: Own on the basis of present-day standards concerning the static tearing

Summing up, the significance of static tear force measurement increased in the last years. Laboratory practice showed that this parameter has the same importance for the fabric assessment as a tensile strength. The main reason of such situation is the growing significance of safe fabric utility, first of all destined for the protective clothing.

It should be pointed out that the fabric producers taking into consideration the significance of strength parameters in the complex fabric assessment apply modern raw materials of better quality, for example, PES, PA, PI, AR and their blends with natural fibers, which guarantee the obtaining the required level of these parameters.

The fabric tear strength is the complex problem, character of which is still not fully explained. The big number of tearing methods and not useful models with parameters, which are not available in the laboratory practice make difficult the prognosing the tear strength; therefore, the experimental research in this field has been necessary.

2. Analysis of stages of cotton fabric static tearing process for the wing shape specimen

The tearing process of cotton fabric sample of the wing shape (according to PN-EN ISO 13937-3) started by loading the specimen by the tensile force was divided into three stages, which were presented schematically in Fig. 2.

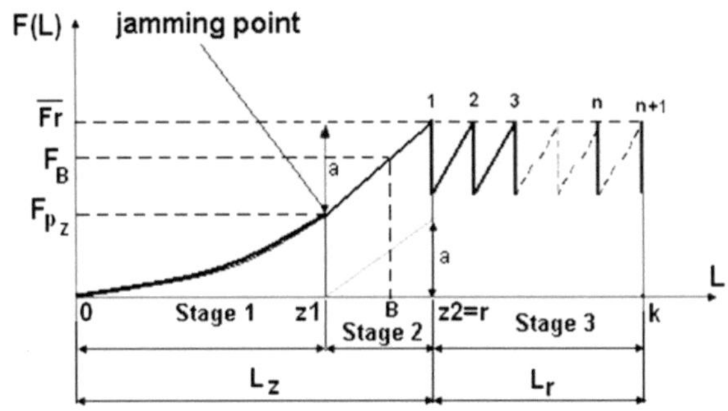

Fig. 2. The graph of tear force of specimen in the function of tensile tester clamp displacement, i.e., the tearing process graph. Stages of tearing process of cotton fabric for the wing shape specimen
Source: own

In Fig. 2 the following designations were assumed:
point 0 - start of sample tearing process, i.e., start of movement of tensile tester clamp. *Point 0* means also the beginning of stage of both thread system displacement;
point z1 - the end of stage of thread displacement and the beginning of stretching the torn thread system;
point z2 - the end of stretching stage and the beginning of thread breakage - *point r;*
point k - the end of specimen tearing process, i.e., the end of measurement;
point B - any point on the distance *z1-z2;*
distance a - the value of breaking force, i.e., the value, which is „added" to the value of displacement in the moment of achieving the jamming point;
L - the way of movement of the tensile tester clamp;
L_z - the way of the movement of the tensile tester clamp up to the first thread breakage on the distance L_r;
L_r - tearing distance, i.e., the distance of displacement of tensile tester clamp since the moment of first thread breakage up to the breakage of the last thread on the marked tearing distance;
F(L) - the stretching force acting on the torn sample on the distance of displacement of tensile tester clamp;
$\overline{F_r}$ - the mean value of tearing force calculated as an arithmetic mean of local tear forces represented by peaks 1, 2, 3 … n, n+1 on the tearing distance L_r, (for ideal conditions $F_{r1}=F_{r2}=F_{r3}=F_{rn}=F_{rn+1}$);
F_B - the value of tensile force in any point B;
Line z1 - the end of distance a – the relationship between the breaking force and strain for the single thread, i.e., $W_z = f(\varepsilon_z)$;
Curve 0 - jamming point - the relationship between the distance, on which the clamp of tensile tester is moving and the force causing the displacement of both thread systems of torn specimen up to achieving the thread jamming point;
Curve 0 - 1 - the relationship between the distance, on which the clamp of tensile tester is moving and the stretching force up to the first thread breakage on the tearing distance. *Curve 0-1* on the distance *z1-z2* is the value of *line z1 - the end of a distance* - moved about a displacement force value in the jamming point.

The analysis of tearing stages is presented assuming that the process of forming the fabric tearing zone on the assumed tearing distance has been started since the moment that the tensile tester clamp has started to move. In the tearing zone dependably on the stage of tearing the following areas can be distinguished: displacement, stretching and breaking (Witkowska, 2008).

Stage 1. Mutual displacement of both specimen system threads. The appearance of the *displacement area in the tearing zone*. The phenomena taking place in this stage are initiated in the moment of starting the movement of tensile tester clamp. The clamp movement on the *distance 0-z1* (Fig. 2) causes the displacement of both thread systems of torn fabric specimen, i.e., threads of stretching system - mounted in the clamps and threads of torn system - perpendicular to the thread system mounted in the clamps. It was assumed that in this stage the threads of torn system are not deformed.

Stage 2. Stretching the threads of torn system. It is done due to the further increase of loading the threads of stretched system, but without the mutual displacement of both thread systems of torn fabric sample. In this stage there are *two areas of tearing zone: displacement and stretching*. Continued in this stage the movement of tensile tester clamp on the *distance z1-z2* (Fig. 2) causes that due to a lack of possibility of further mutual displacement of both fabric system threads the first thread of torn system (being in the displacement area) goes into the stretching area and started to elongate up to the achievement of critical value of elongation, i.e., the value of elongation at the given thread breaking force. Therefore, it was assumed that in the successive tearing process moments in the stretching area there was only one thread of torn system, for which the relationship between the load and strain was linear.

Stage 3. The breakage of torn system thread on the assumed tearing distance. In this stage of tearing process the tearing zone is *built from three areas: displacement, stretching and breaking*. The further movement of tensile tester clamp on the *distance r-k* (Fig. 2) causes the breakage of successive threads of torn system on the tearing distance up to finishing the tearing process (*point k*, Fig. 2).

Between stages 1 and 2 there is so-called the *jamming point* (Fig. 2), i.e., the point, in which fabric parameters and values of friction force between both system threads make impossible the further mutual displacement of both thread systems. Therefore, the achievement of jamming point finishes the stage 1, and the breakage of the first thread on the given tearing distance finishes the stage 2.

Since the moment of the first thread breakage on the tearing distance, the phenomena described in *stages 1, 2* and *3* have been taking place simultaneously up to the moment of breakage of the last thread of torn system on the tearing distance.

The characteristics of tearing process stages have some features similar to the description of this phenomenon for the wing shape specimen presented by the previous researchers of tearing process, i.e.:

1. Distinguishing in the torn fabric sample two thread systems, i.e., the stretched thread system - mounted in the tensile tester clamps; and the torn thread system, which is perpendicular to the stretched one (Krook & Fox, 1945); (Scelzo et al., 1994a); (Scelzo et al., 1994b); (Teixeira et al., 1955); (Taylor, 1959). Both systems can be called also un-torn and torn.
2. Distinguishing the fabric tearing zone (Krook & Fox, 1945); (Scelzo et al., 1994a); (Scelzo et al., 1994b); (Teixeira et al., 1955); (Taylor, 1959) in the torn wing shape specimen.
3. Stating that in the torn fabric sample there are simultaneously the displacement and stretching of both system threads, displacement of stretched system of threads (Taylor, 1959), displacement of both thread systems (Teixeira et al., 1955).

4. Limiting the fabric tearing process to three components represented by threads being in the tearing zone (Scelzo et al., 1994a) (Scelzo et al., 1994b) (Teixeira et al., 1955): the first component - torn system thread in the position "just before the breakage", second and third components - "at the border of tearing zone" threads of stretched system (threads on the inner edge of cut sample elements).

To the most important differences between the descriptions of fabric tearing process presented in this paper and done by the previous Authors belong:

1. Division of fabric tearing zone onto the following areas: the displacement, stretching and breaking.
2. Distinguishing the jamming point of both thread systems of torn sample.
3. Stating that the both thread system displacement (*stage* 1) phenomenon and stretching (*stage* 2) of torn thread system are not taking place at the same time. This statement is true at the assumption that it is possible to find such a point, in which the first thread of torn system being in the displacement area cannot be further displaced. This thread goes into the stretching area and starts to elongate up to achieving the critical value of elongation and the thread breakage.
4. Stating that the tear force is a sum of vector forces, i.e., the force, which causes the displacement without both system thread deformation up to achieving the so-called jamming point; and the force, which causes the elongation of torn system thread up to achieving the critical value of elongation and thread breakage.

Consideration that assumptions for elaboration of the theoretical model of cotton fabric tearing for the wing shape sample are proceeded by the separation of three tearing stages in the fabric tearing zone, i.e., displacement of both system threads to achieve the thread jamming point, stretching and breaking the threads of torn system. Next, there was stated that from the moment of the first thread breakage on the tearing distance the phenomena described in the mentioned above stages have a place simultaneously up to the breakage of the last thread of torn system. The force put to the torn specimen in the function of displacement of tensile tester clamp, i.e., $F=f(L)$ was described as follows:

$$F=f(L)=F_p(L)+Fw_z(L) \tag{1}$$

where:

$F_p(L)$ - force F in the function of displacement of tensile tester moving clamp during the displacement of both system threads of torn sample;

$Fw_z(L)$ - force F in the function of displacement of tensile tester moving clamp during the stretching one thread of torn system, which is in the tearing zone at this moment. It was assumed that the relationship $Fw_z(L)$ is described by the Hookean low.

According to the proposed stages of tearing process the relationship (1) takes the form:

$$\text{stage 1: } F=f(L)=F_p(L) \tag{2}$$

$$\text{stage 2 and stage 3: } F=f(L)=F_p(L)+Fw_z(L) \tag{3}$$

and the thread breakage in the tearing zone:

$$F=f(L)=F_r \tag{4}$$

where:

F_r - local value of breaking force.

The value of fabric breaking force in the moment of the first thread breakage on the given tearing distance on the border of stretching and breaking of tearing zone is described by the relationship (5) (general model of fabric tearing):

$$F_r=F_p(z_1) + Fw_z(r) = F_{pz1} + Fw \quad (5)$$

where:

r - the end of stages of stretching of torn thread system and the beginning of thread breaking stage;

F_{pz1} - the value of displacement force in the jamming point of both thread systems of torn fabric sample;

Fw_z - the value of breaking force of torn thread system.

3. Experiment

3.1 The stand for image analysis

The stand for image analysis was built from the following elements: the tensile tester Zwick model 1120 (1), jaws of tensile tester (2), computer with the software *test-Xpert (3)* for textile static tearing, color camera TV (5) (zoom 3.5 to 8.0 mm), computer with the software *Microstudio* (4) or image analysis, tripod camera (6) (Witkowska & Frydrych, 2008); (Witkowska, 2008). In order to observe the cotton fabric tearing stages and tearing zone the video system was used (Fig. 3 and Fig. 4).

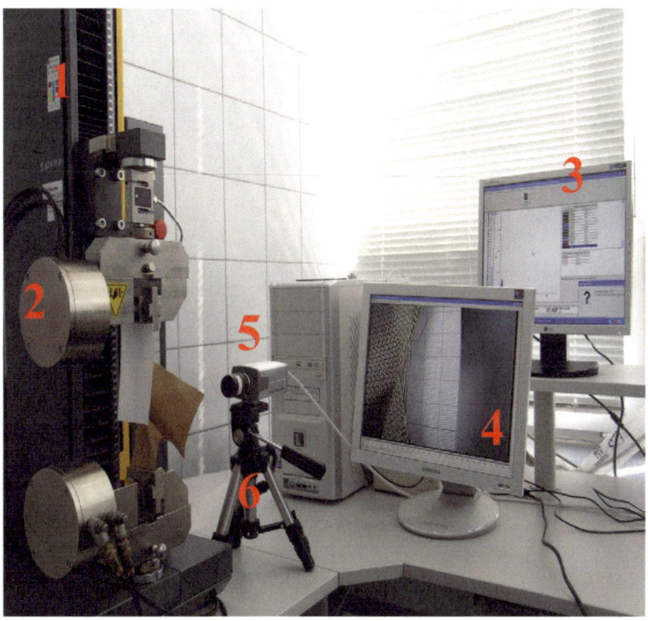

Fig. 3. Stand for the registration of tearing process and computer analysis of images of tearing zone
Source: own

The system enables also to archive and analyze the static images. To obtain a dynamic image the stand assured the registration of 25 images per second. For the image acquisition there was used the digital camera CCD CP-720 with a special optical system enabling the analysis of moving images, choosing the observed area and obtaining the clear (sharp) images of analyzed objects. In order to archive and analyze the static images the *MicroStudio Video* software was applied. From the video images showing the fabric tearing there were chosen the static images representing the fabric in the exactly the same measurement points. On the basis of static images there were determined the length of tearing zone (l_{Δ}), depth of tearing zone (d_{Δ}), and the thread number in the tearing zone $(L_{n-\Delta})$ for each marked measurement point. An analysis of static images representing tearing zones enables to state that the tearing zone in many cases is not symmetrical according to *x-x* axis going through the point determining the depth of tearing zone. Therefore, the length of distance from the *x-x* axis to the end of marked tearing zone is $(l_{\Delta\uparrow})$, the length (l_{Δ}) and depth (d_{Δ}) of tearing zone were assumed in the agreement with Teixeira's (Teixeira et al., 1955) model of tearing zone. From the static images the following values were read: the total tearing zone length (l_{Δ}), length $(l_{\Delta\uparrow})$ and depth (d_{Δ}) of tearing zone in pixels; and next, these values were recalculated into millimeters.

Fig. 4. The way of sample mounting in the tensile machine jaws for the tearing zone registration
Source: own

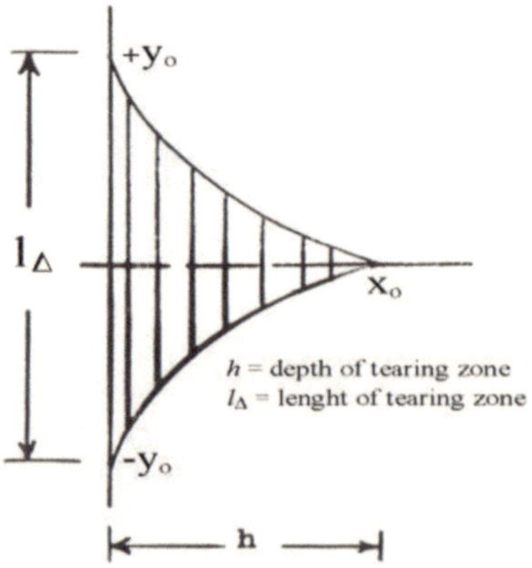

Fig. 5. Teixeira's model of fabric tearing zone (Teixeira et al., 1955)

Successive operations at registration of tearing process of cotton fabrics:

1. Mounting the marked specimen of wing shape in the tensile machine jaws.
2. Setting the video camera in such a position to have on the screen of computer with *MicroStudio Video* software the specimen with the marked measurement points and appropriate scale (millimeter paper).
3. Simultaneous starting of tensile machine and software *MicroStudio Video* registering the video camera image.
4. Finishing the movement registration and the specimen tearing process after the third measurement point.
5. Repeating the points 1 to 4 for each prepared specimen.

3.2 Testing materials

100% cotton woven fabrics of characteristics presented below from woven color threads were design and manufactured for the need of experiment. An application of color threads facilitated the observation of video images of the fabric tearing process.
Fabric characteristics (Witkowska, 2008):
- weaves: plain, twill 3/1 Z, satin 7/1 (5), and broken twill 2/2 V4;
- warp and weft thread linear density: 25 tex x 2;
- the mean number of threads per 1 dm: warp - 188, weft - 174.

From the each fabric three specimens in each direction (warp/weft) were cut. On the right side of fabric starting from the cut edge three points were marked every 5 mm representing three measurement distances, i.e., 5 mm, 10 mm and 15 mm (Fig. 6). The speed of testing was 75 mm/min.

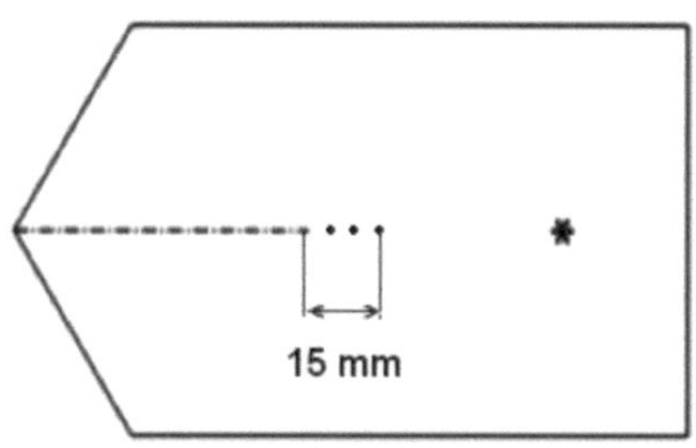

Fig. 6. The way of sample marking for the measurement
Source: own

The static tear resistance of analyzed fabrics was determined according to PN-EN ISO 13937-3:2002. From each fabric 10 specimens in the warp and weft directions were measured; and next, the tear force values using the tensile tester Zwick 1120 controlled by the software *test-Xpert version 4.12* were determined. The measurement was carried out with the speed 100 mm/min. From charts the tear force (F_r) for the whole tearing distance, i.e., from the first to the last peak registered on the assumed tearing distance were read. The obtained values of force of 10 specimens were statistically assessed; the mean value and variation coefficient were determined.

3.3 Results

The set of mean values of cotton fabric tear forces dependably on the weave of analyzed fabrics are presented in Table 3. The data of obtained parameters of tearing zone dependably on the fabric weave, length of tearing distance are presented in Tables 4÷7; whereas the chosen static images of cotton fabrics tearing zones of registered in the measurements point 15 mm (specimen 1) for the wing shape specimens in plain and satin weave variants are presented in Fig. 7 and 8.

Parameter/fabric	Fabric weave							
	plain		twill 3/1 Z		satin 7/1 (5)		broken twill 2/2 V4	
	warp	weft	warp	weft	warp	weft	warp	weft
Mean tear force, N	19.7	19.8	23.9	23.0	63.2	52.2	26.5	21.0
Variation coefficient, %	3.6	3.7	3.8	3.7	4.4	5.4	1.4	5.1

Table 3. The set of tear force results for cotton fabrics dependably on the weave

Specimen	Plain warp - measurement point											
	5 mm				10 mm				15 mm			
	l_{Δ}	$l_{\Delta\uparrow}$	d_{Δ}	$L_{n-\Delta}$	l_{Δ}	$l_{\Delta\uparrow}$	d_{Δ}	$L_{n-\Delta}$	l_{Δ}	$l_{\Delta\uparrow}$	d_{Δ}	$L_{n-\Delta}$
1	7.0	3.5	0.7	1	4.5	2.1	0.7	2	5.7	2.8	0.7	2
2	7.2	3.6	0.7	2	4.8	2.3	0.8	1	6.3	3.1	0.7	2
3	7.7	3.9	1.0	2	5.0	2.4	0.8	2	6.5	3.3	1.0	2
Specimen	Plain weft - measurement point											
	5 mm				10 mm				15 mm			
	l_{Δ}	$l_{\Delta\uparrow}$	d_{Δ}	$L_{n-\Delta}$	l_{Δ}	$l_{\Delta\uparrow}$	d_{Δ}	$L_{n-\Delta}$	l_{Δ}	$l_{\Delta\uparrow}$	d_{Δ}	$L_{n-\Delta}$
1	5.2	2.6	0.5	1	4.6	2.3	1.0	1	5.2	2.6	1.0	1
2	5.5	2.7	1.0	1	4.7	2.3	0.9	2	5.1	2.5	0.8	1
3	5.1	2.5	0.8	1	5.6	2.8	1.0	1	5.2	2.6	0.9	1

Table 4. Parameters of tearing zone and the thread number in the tearing zone for the model plain cotton fabrics

Specimen	Twill 3/1(Z) warp - measurement point											
	5 mm				10 mm				15 mm			
	l_{Δ}	$l_{\Delta\uparrow}$	d_{Δ}	$L_{n-\Delta}$	l_{Δ}	$l_{\Delta\uparrow}$	d_{Δ}	$L_{n-\Delta}$	l_{Δ}	$l_{\Delta\uparrow}$	d_{Δ}	$L_{n-\Delta}$
1	18.7	9.3	2.5	5	24.7	12.3	3.6	5	31.9	15.4	4.8	6
2	18.7	9.4	2.6	4	28.3	14.2	4.1	5	36.5	18.6	5.3	7
3	18.7	9.4	2.6	5	26.1	13.0	3.2	5	35.0	15.6	5.0	7
Specimen	Twill 3/1(Z) weft - measurement point											
	5 mm				10 mm				15 mm			
	l_{Δ}	$l_{\Delta\uparrow}$	d_{Δ}	$L_{n-\Delta}$	l_{Δ}	$l_{\Delta\uparrow}$	d_{Δ}	$L_{n-\Delta}$	l_{Δ}	$l_{\Delta\uparrow}$	d_{Δ}	$L_{n-\Delta}$
1	13.1	6.3	2.1	5	25.9	13.0	3.6	6	32.9	14.7	5.2	8
2	12.2	5.9	2.1	4	25.8	12.8	3.8	5	31.1	14.9	5.4	8
3	13.1	6.6	2.4	3	24.3	12.2	4.1	5	34.3	16.4	5.2	7

Table 5. Parameters of tearing zone and the thread number in the tearing zone for the model twill 3/1 Z cotton fabrics

Specimen	Satin 7/1 (5) warp – measurement point											
	5 mm				10 mm				15 mm			
	l_Δ	$l_{\Delta\uparrow}$	d_Δ	$L_{n-\Delta}$	l_Δ	$l_{\Delta\uparrow}$	d_Δ	$L_{n-\Delta}$	l_Δ	$l_{\Delta\uparrow}$	d_Δ	$L_{n-\Delta}$
1	12.1	6.0	2.9	7	22.3	11.1	4.1	10	41.5	21.2	5.6	14
2	13.7	7.7	2.9	7	24.9	12.2	3.5	9	39.8	18.2	5.7	12
3	13.6	6.9	3.0	5	24.4	14.2	3.6	8	40.3	24.0	5.2	11
Specimen	Satin 7/1 (5) weft – measurement point											
	5 mm				10 mm				15 mm			
	l_Δ	$l_{\Delta\uparrow}$	d_Δ	$L_{-n\Delta}$	l_Δ	$l_{\Delta\uparrow}$	d_Δ	$L_{n-\Delta}$	l_Δ	$l_{\Delta\uparrow}$	d_Δ	$L_{n-\Delta}$
1	13.1	7.1	2.5	6	24.5	12.1	3.1	9	34.5	13.5	4.6	14
2	14.0	7.0	2.4	7	25.0	12.4	3.8	10	34.4	15.1	4.6	15
3	12.7	6.2	2.3	6	26.1	13.1	3.8	13	34.0	17.2	4.9	17

Table 6. Parameters of tearing zone and the thread number in the tearing zone for the model satin 7/1(5) cotton fabrics

Specimen	Broken twill 2/2 V4 warp - measurement point											
	5 mm				10 mm				15 mm			
	l_Δ	$l_{\Delta\uparrow}$	d_Δ	$L_{n-\Delta}$	l_Δ	$l_{\Delta\uparrow}$	d_Δ	$L_{n-\Delta}$	l_Δ	$l_{\Delta\uparrow}$	d_Δ	$L_{n-\Delta}$
1	14.5	7.2	2.3	5	23.9	12.0	4.2	7	34.3	15.9	5.1	10
2	15.7	7.8	2.2	5	24.1	9.0	3.6	6	35.5	12.7	5.5	9
3	16.0	7.0	2.4	4	25.4	12.6	4.0	7	38.5	18.6	5.6	10
Specimen	Broken twill 2/2 V4 weft - measurement point											
	5 mm				10 mm				15 mm			
	l_Δ	$l_{\Delta\uparrow}$	d_Δ	$L_{n-\Delta}$	l_Δ	$l_{\Delta\uparrow}$	d_Δ	$L_{n-\Delta}$	l_Δ	$l_{\Delta\uparrow}$	d_Δ	$L_{n-\Delta}$
1	14.8	7.5	2.0	4	22.8	11.5	3.6	7	34.3	17.2	4.3	10
2	14.3	7.1	1.6	4	23.9	12.0	3.7	9	35.2	16.7	5.0	13
3	13.4	6.8	2.0	4	24.9	12.5	3.3	7	34.4	16.7	4.9	11

Table 7. Parameters of tearing zone and the thread number in the tearing zone for the model broken twill 2/2 V4 cotton fabrics

Plain weave

Warp direction: specimen 1, point - 15 mm

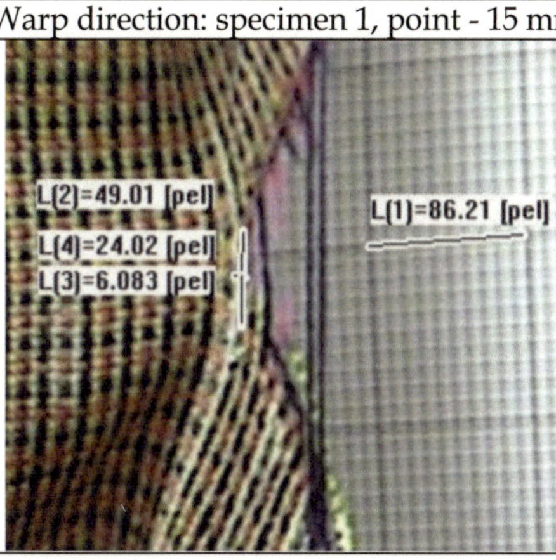

Weft direction: specimen 1, point - 15 mm

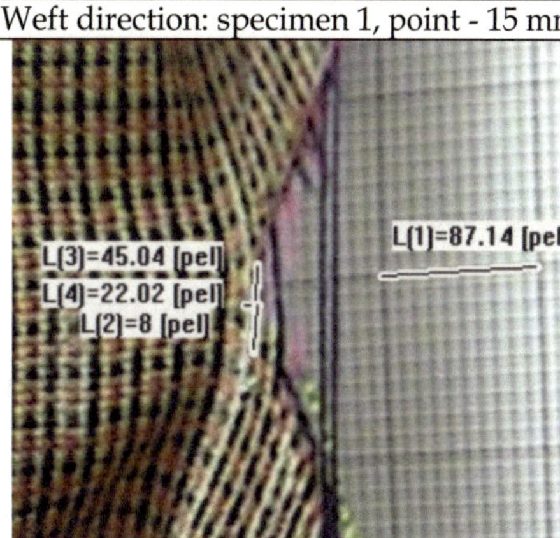

Fig. 7. The fabric images of tearing zone registered in the measurement point 15 mm for the wing shape specimen for plain weave

Satin 7/1(5) weave	
Warp direction: specimen 1, point - 15 mm	Weft direction: specimen 1, point - 15 mm

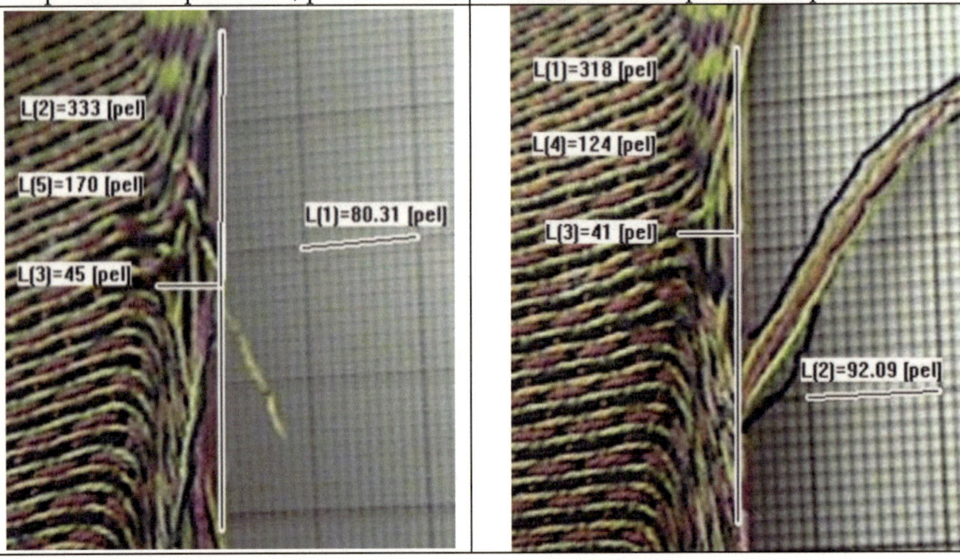

Fig. 8. The fabric images of tearing zone registered in the measurement point 15 mm for the wing shape specimen for satin 7/1(5) weave

4. Analysis and modelling the fabric tearing zone geometry

4.1 Analysis of fabric tearing zone geometry parameters

The graphic presentation of obtained results of the thread number in the tearing zone as well as the mean values of length and depth of zone dependably on the fabric weave, the torn arrangement (warp/weft) and tearing distance are shown in Fig. 9, 10 and 11.

The analysis of presented photos allows drawing out the following conclusions:

1. The smallest changes and at the same time the lowest values of length and depth of tearing zone as well as the thread number in the tearing zone dependably on the tearing distance were observed for zones registered for plain fabrics. The torn thread system (warp/weft) does not influence significantly the changes of mentioned parameters. An analysis of movies presenting the plain fabric tearing showed that in the assumed in measurements the maximum specimen length distance 15 mm there are many zones, in which there are the most often one thread. When it is broken, the next tearing zone is started to create. The threads in the plain fabric are displaced insignificantly. It is worth to remind (Tab. 3) that the obtained mean values of tear forces for plain cotton fabrics are on the lowest level in the comparison to the rest fabrics of different weaves.

The obtained values of plain cotton fabric tear resistance and carried out observations of tearing zones confirmed the Teixeira's et al. (Teixeira et al., 1955), Hamkins's and Backer's (Hamkins & Backer, 1980) also Scelzo, Backer's, Boyce's (Scelzo et al., 1994b) conclusions that for fabrics of higher number of interlacements (on the assumed tearing distance) the value of tear force and also of tearing zone parameters are the lowest ones.

In Fig. 12, there are shown the torn (on the distance of 15 mm) plain cotton fabric specimens. In the place of tearing the sample deformation is not observed, the successive threads on the tearing distance are broken.

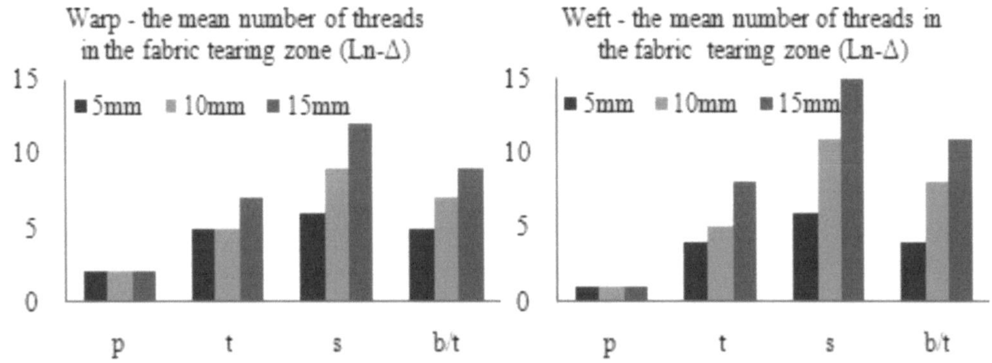

Fig. 9. The mean number of threads in the cotton fabric tearing zone ($L_{n\Delta}$) dependably on the weave, torn thread system (warp/weft) and the tearing distance

where: p, t, s, b/t are symbols of cotton fabric: plain, twill 3/1 (Z), satin 7/1 (5), broken twill 2/2 V4;
$L_{n\text{-}\Delta}$ is the mean thread number in the fabric tearing zone;
5, 10, 15 are symbols of specimen measurement points successively on 5-th, 10-th and 15-th millimeter of tearing distance.

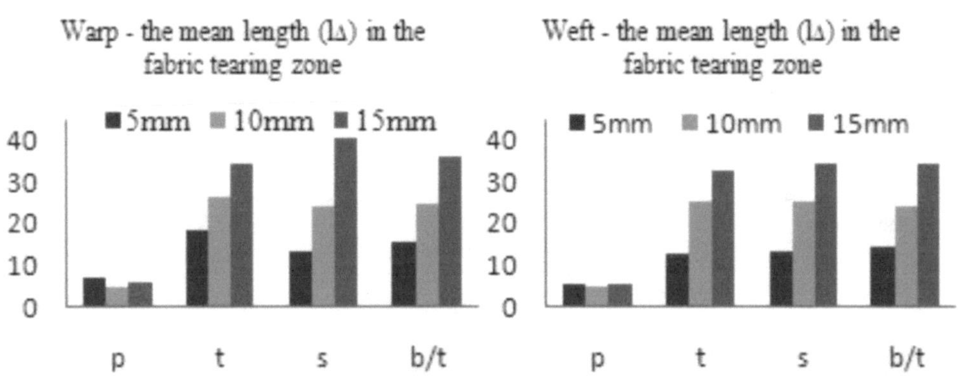

Fig. 10. The mean length (l_Δ) of tearing zone dependably on the fabric weave, torn thread system (warp/weft) and the length of tearing distance

where: ***p, t, s, b/t*** are symbols of cotton fabric: plain, twill 3/1 (Z), satin 7/1 (5), broken twill 2/2 V4,
l_Δ is the mean length of tearing zone, in mm,
5, 10, 15 are symbols of specimen measurement points successively on 5-th, 10-th and 15-th millimeter of tearing distance.

2. Mean values of length and depth of tearing zones registered for twill 3/1 Z, satin 7/1 (5) and broken twill 2/2 V4 fabrics increase in the torn thread system with the increase of tearing distance. But in the area of given measurement points (5, 10, 15 mm) these lengths are smaller dependably on the weave change and these changes concern the mean tearing zone length. The observation of images of specimen tearing allow stating that the increase of length and depth of tearing zone of mentioned weaves is caused by the graduated process of zone formation, i.e., in the last measurement point (15 mm). Such a way of tearing zone formation results from the bigger possibility of thread displacement in fabrics of weaves with the smaller number of interlacements (Fig. 10 and 11) than the plain weave has.

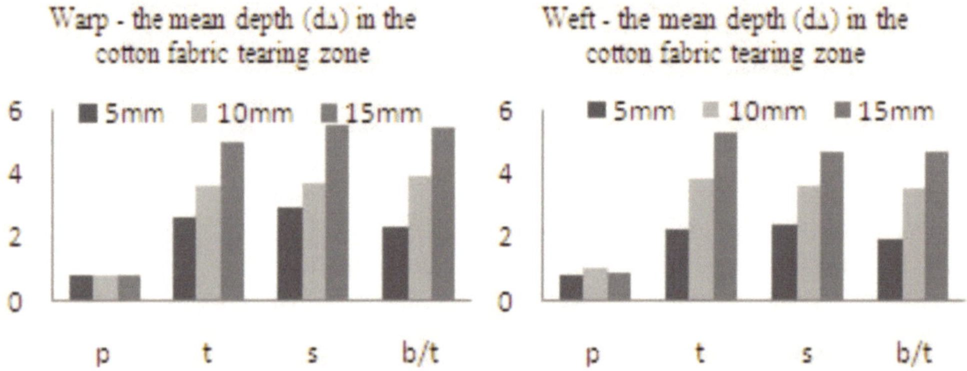

Fig. 11. The mean depth (d_Δ) of tearing zone dependably on the fabric weave, torn thread system (warp/weft)) and the length of tearing distance

where: ***p, t, s, b/t*** are symbols of cotton fabric: plain, twill 3/1 (Z), satin 7/1 (5), broken twill 2/2 V4,
d_Δ is the mean of depth of tearing zone, in mm,
5, 10, 15 are symbols of specimen measurement points successively on 5-th, 10-th and 15-th millimeter of tearing distance.

Plain weave

warp direction	weft direction

Fig. 12. The wing shape specimens of plain fabrics torn on the measurement distance 15 mm
Source: own

3. The analysis of results of mean tear force of fabrics of proposed weaves showed that in the warp as well as weft direction the highest mean values of tear resistance were obtained for fabric of satin 7/1 (5) weave. But the mean values of length and depth of tearing zone for fabric samples of this weave in the given measurement point only in two cases, i.e., warp system – measurement point 15 mm, and weft system – measurement point 10 mm (Fig. 10

and 11) are a little higher than analogous values for twill and broken twill samples. It can be explain in the following way:
3.1 Observation of graphs $F_r=f(L_r)$ (Fig. 13).

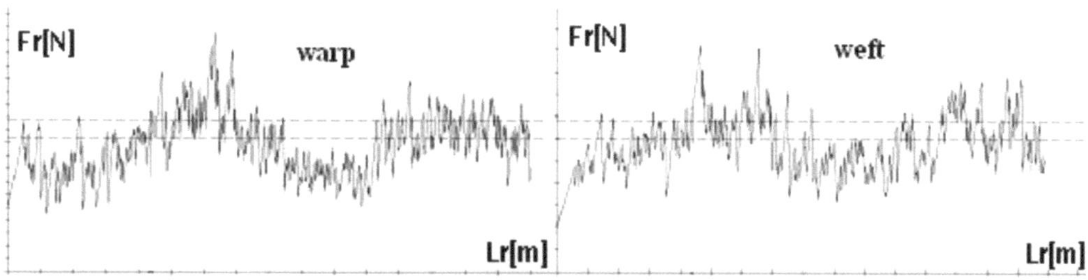

Fig. 13. The example of cotton fabric of satin 7/1 (5) weave tearing process graphs
Source: own

The local values of tear forces (F_r) of satin 7/1 (5) fabric registered on the tear distance (L_r), are characterized by a big variability. Moreover, the highest values of local tear forces are observed the most often from the tearing distance about ¼ of chart, i.e., about 40 mm. Therefore, the highest local values of tear forces, and at the same time - the highest values of length and depth of tearing zone are observed in the assumed observation points (5, 10, 15 mm).
3.2 Observation of torn fabric samples of the following weaves: twill 3/1 Z, satin 7/1 (5) and broken twill 2/2 V4 (Fig. 14).
Analyzing the obtained images it can be noticed a strong deformation of torn fabric specimens of mentioned weaves – the bigger than in the case of deformation of plain fabric sample. Particularly visible it is for satin fabric samples 7/1(5). The successive threads on the tearing distance are not broken, they are only „pulled out" from the fabric sample (from many of them totally). It is caused by a fact that threads of fabric of this weave have a big possibility of displacement and it is difficult to achieve the so called „jamming point" (the point, in which fabric parameters and values of friction forces between threads of both systems enable the further mutual thread displacement in the torn samples. The force, which is necessary to pull out the single thread and in the further part of specimen the group of threads, is higher than the force necessary for breaking threads. From one side it explains very high tear resistance values for satin fabrics; and from the second one – it explains the variability of local tear forces on the tearing distance. The variability of local values of tear forces results from the variable number of pulled out threads from the fabric on the tearing distance. At the beginning of tearing process the number of pulled out threads increases, what causes the growth of local values of tear forces. If the tearing is continued, the first thread or a few first threads being on the tearing distance are pulled out from the fabric and the local values of tear forces will be lower. Next, on the tearing distance there are appeared the successive threads, which gradually are pulled out from the fabric and the local values of tear forces again start to rise. This process is repeated up to the end of tearing distance.
The described sample deformation during the tearing process is also observed, but in the lower extent in the twill and broken twill fabric samples. In the torn samples of mentioned weaves on the tearing distance there are visible the broken as well as „pulled out" from the fabric threads.

4. The highest number of threads in the tearing zone in the each measurement point as well as in the torn thread system (warp/weft) were observed for satin 7/1 (5) (Fig. 9).

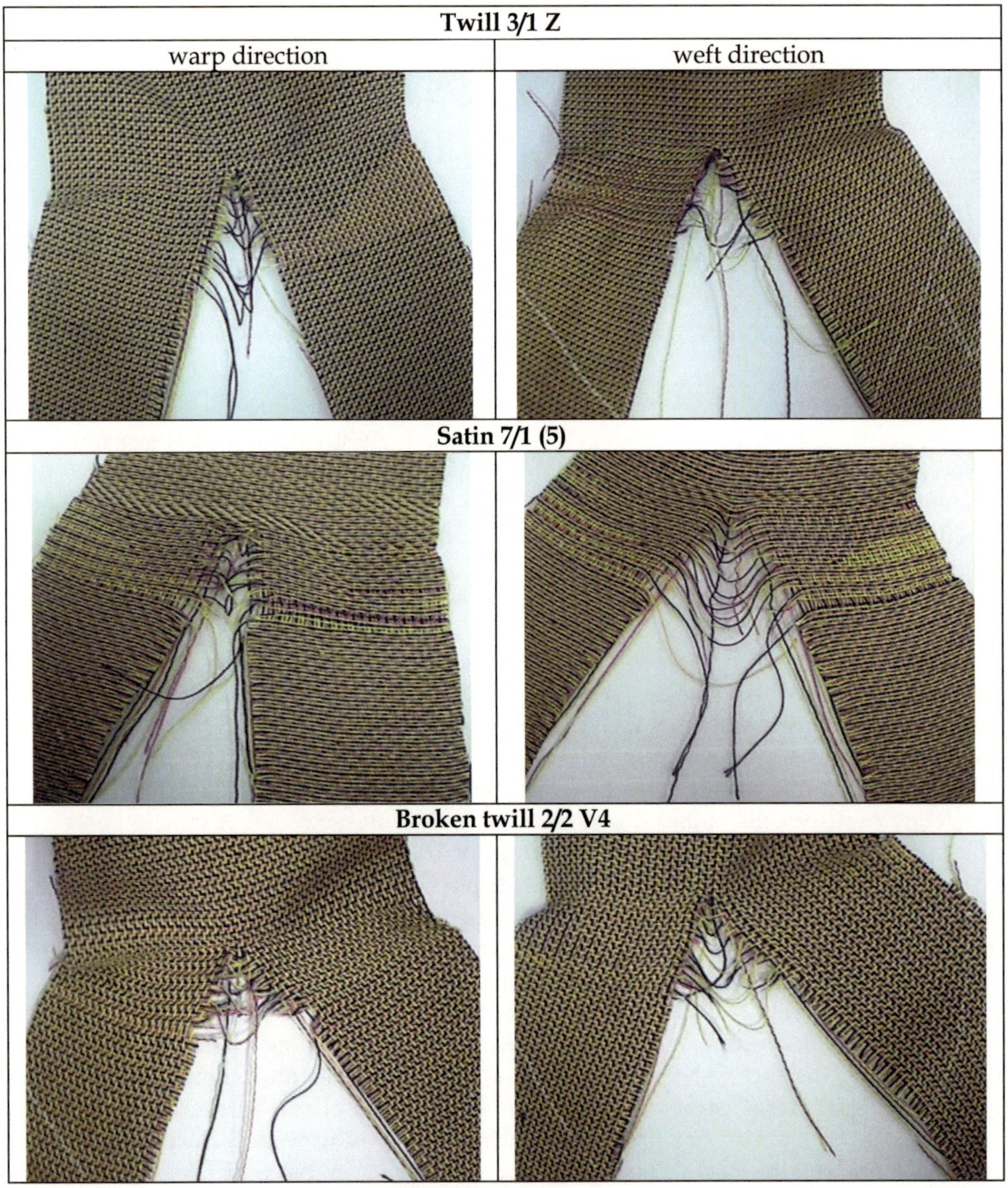

Fig. 14. The wing shape specimens of twill 3/1 Z, satin 7/1(5) and broken twill 2/2 V4 fabrics torn on the measurement distance 15 mm
Source: own

4.2 Modelling the shape of „arms" of tearing zone

In this chapter the process of modelling the shape of tearing zone was divided into two stages. In the first one in order to make a better visualization of sizes and to compare the tearing zones of samples of four weaves in Fig. 15, there were set in the same scale the obtained values of parameter of fabric tearing zone for all the registered zones. The marked on the axes points, i.e., the depth of tearing zone d_{Δ}, length of tearing zone $l_{\Delta\uparrow}$ [length $l_{\Delta\uparrow}$ is the length measured from the abscissa x-x (axis going through the point determining the depth of tearing zone) up to the end of the marked length of tearing zone l_{Δ}.] and the length difference l_{Δ}-$l_{\Delta\uparrow}$ were connected by the straight lines.

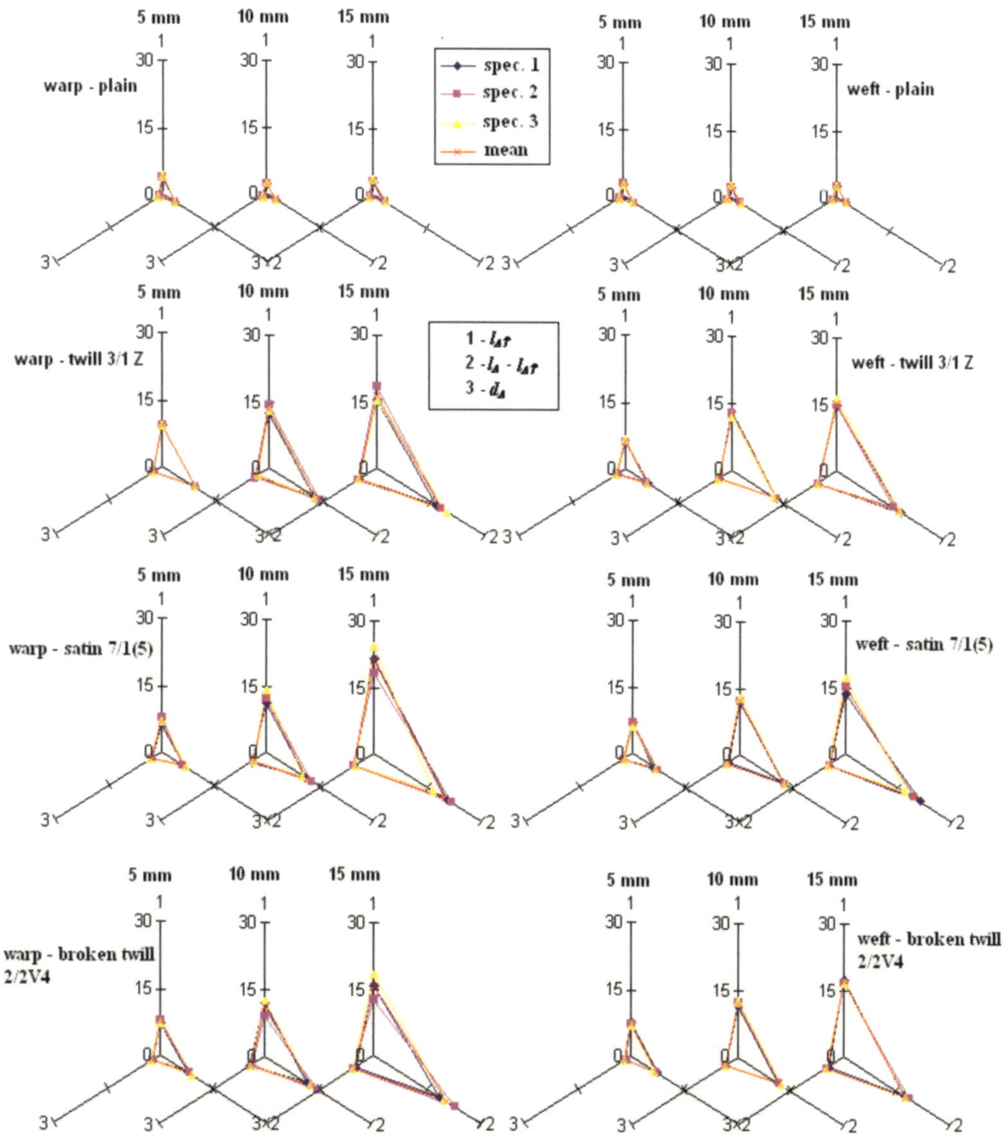

Fig. 15. The size of fabric tearing zone dependably on the fabric weave, torn thread system (warp/weft) and measurement point

The schematic presentation of fabric tearing zones enabled to notice the following observation:

- disproportions between the size of tearing zones dependably on the torn fabric weave;
- insignificant differences between the sizes of tearing zones in the given weave dependably on the given thread system (warp/weft);
- small variability of obtained length and depth for three specimens measured in the thread system (warp/weft);
- symmetry of tearing zone (l_Δ-$l_{\Delta\uparrow}$ = ½ l_Δ) according to abscissa *x-x* going through the point determining the zone depth for torn fabric specimens of the following weaves:
  - plain in the all measurement points for the thread system (warp/weft);
  - twill 3/1 Z for the measurement points (5 and 10 mm) for the thread system (warp/weft);
  - broken twill 2/2 V4 for the measurement point 5 mm for the weft direction;
- a lack of symmetry of tearing zone (l_Δ-$l_{\Delta\uparrow} \neq$ ½l_Δ) in the relation to the abscissa *x-x* for the rest variants of torn specimens. The biggest differences between the values $l_{\Delta\uparrow}$ and ½l_Δ, i.e., differences ≥ 1 mm were calculated for the tearing zones of fabrics of the following weave:
  - satin 7/1 (5): warp thread system - measurement point (10 and 15 mm), weft thread system - measurement point 15 mm;
  - broken twill 2/2 V4: warp thread system - measurement point (10 and 15 mm).

In the second stage of modelling process of the shape of fabric tearing zone, speaking more precisely the shape of zone "arms" there was assumed the following procedure:

1. In the theoretical considerations carried out (Witkowska, 2008) it was shown that the shape of tearing zone, which was formed in the both system threads the jamming point can be described by an exponential function. Therefore, the shape of "arms" of tearing zone observed on the static images presenting the tearing zones in three measurements points were also described by the exponential function:

$$y = A\exp(-Bx)+C \quad (6)$$

where: *A*, *B* and C are the constant of exponential function;
y is a value of half length of fabric tearing zone;
x is a value of depth of fabric tearing zone.

2. Modelling the shape of tearing zone „arms" was carried out for the described weaves for zones registered in the measurement point 15 mm and for the given thread system (warp/weft) for specimen No. 3. On the basis of analysis of static images of tearing zones it was assumed that the constant *C=0*.

3. In each weave variant and torn thread system (warp/weft) for calculating the constant values A and B two pairs of points (x, y) were used, i.e.,:

$$(x_1, y_1) = G(0,0, l_{\Delta\uparrow}) \quad (7)$$

$$(x_2, y_2) = E(½d_\Delta, l_{\Delta\uparrow e}) \quad (8)$$

where:
$l_{\Delta\uparrow}$, d_Δ, $l_{\Delta\uparrow e}$, are the values of tearing zone parameters read out from the static images shown in the Fig. 16.

4. The appropriate modelled curves describing the tearing zone „arms" should go through the point:

$$(x_3, y_3) = D(d_\Delta, 0,0) \quad (9)$$

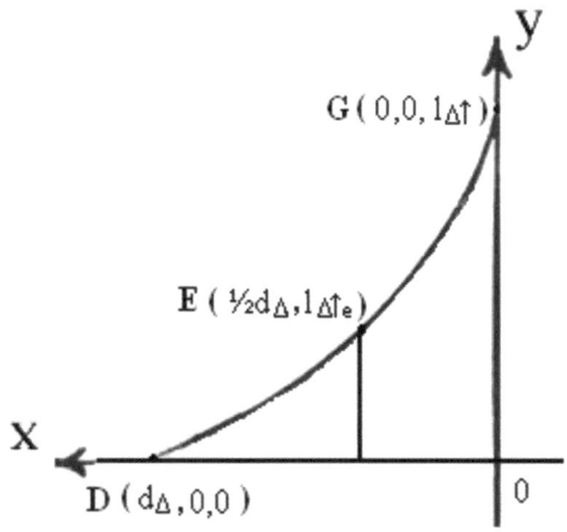

Fig. 16. The way of reading out the measurement points for modelling the fabric tearing zone „arms”
Source: own

Example:
The tearing zone “arm” shape was described for sample No. 3 made of twill 3/1Z for the warp thread system (the measurement point 15 mm).
Data: $l_{\Delta} = 6.5$ mm, $l_{\Delta\uparrow} = 3.3$ mm, $d_{\Delta} = 1.0$ mm.
$(x_1, y_1) = G(0.0, 3.3)$; $(x_2, y_2) = E(0.5, 0.2)$
Introducing the values x_1, y_1 into the equation 6 the value of constant A was calculated:

$$3.3 = A\exp(-B0) \rightarrow A = 3.3 \quad (10)$$

Introducing the values x_2, y_2 and the value of constant A into the equation 6 the value of constant B was calculated:

$$0.5 = 3.3\exp(-B0.2) \rightarrow B = -5.61 \quad (11)$$

Introducing the values of constant A and B into the equation 6 the exponential equation for the following form was obtained:

$$y = 3.3\exp(-5.61x) \quad (12)$$

Equation 12 describes the shape of tearing zone “arms” (in the first quarter of coordinate system x-y) for plain fabric in the case, when the warp was the torn thread system, and the observation point was on 15-th mm of tearing distance. The curve described by the equation 12 goes through the point $(x_3, y_3) = D(1{,}0, 0{,}0)$ determining the depth of tearing zone (Fig. 17). Applying such a procedure the A and B constants of exponential functions describing the tearing zone „arm” shape for the rest fabric weaves were determined. In Table 8, there are set the values of point coordinates (x_1, y_1, and x_2, y_2,) necessary for determination A and B constants of exponential functions and coordinate values (x_3, y_3) determining the depth of tearing zone. In Fig. 17, there are set the chosen curves describing the “arm” shape of tearing zone of examined fabrics and presented equations of the exponential functions describing these curves.

Weave / torn system / measurement point 15 mm, specimen No. 3								
Coordinates	plain		twill 3/1 Z		satin 7/1 (5)		broken twill 2/2 V4	
	warp	weft	warp	weft	warp	weft	warp	weft
(x_1, y_1)	(0.0, 3.3)	(0.0, 2.6)	(0.0, 16.5)	(0.0, 16.4)	(0.0, 21.2)	(0.0, 17.2)	(0.0, 18.5)	(0.0, 16.7)
(x_2, y_2)	(0.5, 0.2)	(0.5, 0.2)	(2.5, 1.4)	(2.6, 1.2)	(2.8, 1.8)	(2.4, 1.6)	(2.8, 1.6)	(2.5, 1.5)
(x3, y3)	(1.0, 0.0)	(0.9, 0.0)	(5.0, 0,0)	(5.2, 0.0)	(5.5, 0.0)	(4.7, 0.0)	(5.6, 0.0)	(4.9, 0.0)

Table 8. The set of values of coordinates points necessary for calculation of A and B constants of exponential functions describing the shape of „arms" of fabric tearing zone

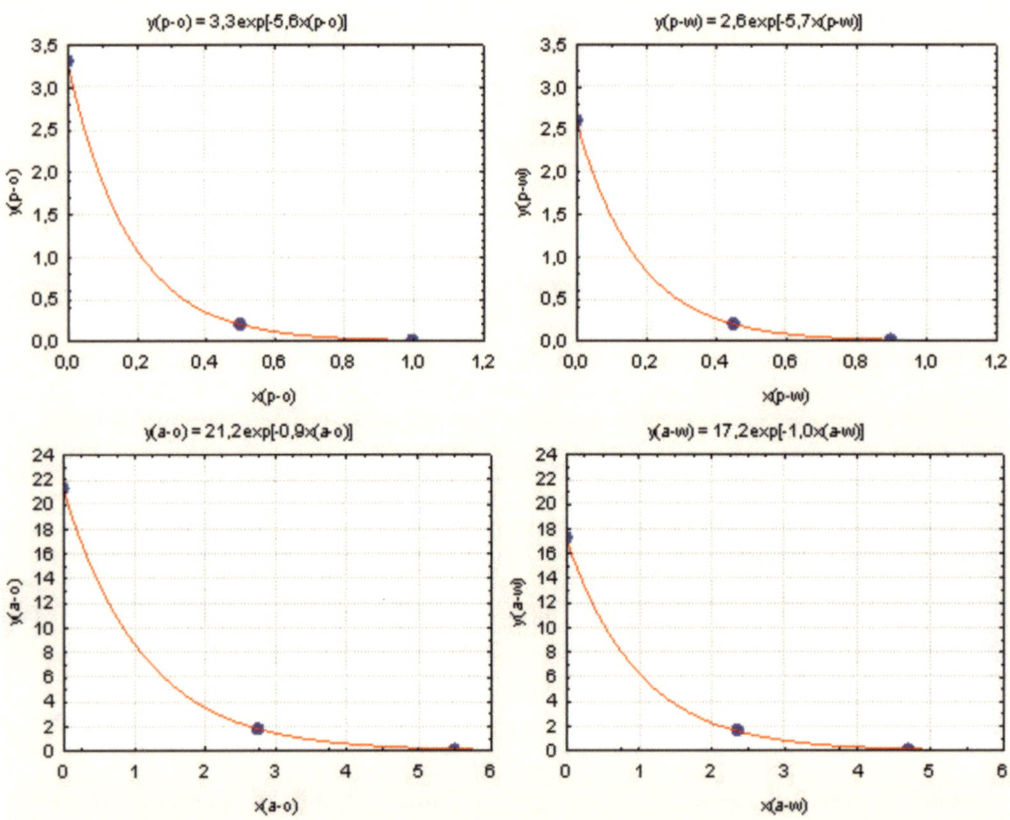

Fig. 17. Set of „arms" of fabric tearing zones of plain and satin 7/1 (5) weaves dependably on the torn thread system (warp/weft) for the measurement point 15 mm (specimen No. 3)

where: *p-o* - warp system of plain fabric,
p-w - weft system of plain fabric,
a-o - warp system of satin 7/1(5) fabric,
a-w - weft system of satin 7/1(5) fabric,
y - ½ of length of tearing zone (symbol l_Δ, Fig. 5),
x - depth of tearing zone (symbol d_Δ, Fig. 5),
exp i.e., an exponential function of basis *e*, of natural logarithm.

5. Summing up

5.1 The theoretical considerations on the static tearing process stages were confirmed by an experiment. It was enabled by the digital analysis of video images of fabric sample tearing. The image analysis was a very powerful measurement tool used for an identification purpose as well as for measurement of changes in the fabric structure in time due to the force acting.

5.2 The designed stand of computer image analysis, with the use of which the cotton fabric tearing was registered, aided the process of fabric phenomenon identification. The elaborated methodology enables the measurement of changes in the fabric structure in time. The additional advantage is its universality, i.e., the possibility of adapting its elements (video camera and *MicroStudio Video* software) for registration and analysis of the other phenomena. Additionally, *MicroStudio Video* software is a good tool for the result presentation. Moreover, the advantage of presented system for the computer image analysis is its multi-functionality. It can be also used for the reconstruction and measurements of changes during the other tensile destructive measurements like strength measurements and for the observation, for example, the flammability or wetting processes.

5.3 The registration of fabric tearing process and analysis of obtained static images helped to describe the parameters of fabric tearing zone such as: the length and depth of tearing zone and the number of threads in this zone.

5.4 The significant influence of fabric weave on the size of fabric tearing zone is observed. With the drop of interlacement number in the fabric, the increase of cotton fabric tear strength is observed. The lower number of interlacements causes the lower number of so called "pseudo-jammings" between the threads and in the same way, it enables the thread displacement and deformation. Simultaneously, the deformation of torn fabric specimens of weaves with the small number of interlacements is observed. Modelling the shape of "arms" of cotton fabric tearing zones showed that the "arm" shape can be described by an exponential function.

6. Acknowledgements

This work has been supported by: European Social Fund and Polish State in the frame of "Mechanism WIDDOK" programme (contract number Z/2.10/II/2.6/04/05/U/2/06); and the Polish Committee for Scientific Research, project no 3T08A 056 29.

7. References

Harrison, P. (1960). The tearing strength of fabrics. Part I: A review of the literature, Journal of Textile Institute 51, T91-T131.

Hamkins, Ch. P. & Backer, S. (1980). On the Mechanisms of Tearing in Woven Fabrics, Textile Research Institute, May, 323-327.

Hager, O. B.; Galiardi, D. D. & Walker, H. B. (1947). Analysis of Tear Strength; Textile Research Journal; July, 376-381.

Krook, C. M. & Fox, K. R. (1945). Study of the tongue tear test, Textile Research Journal, No. 11, 389-396.

Scelzo, W. A.; Backer, S. & Boyce, C. (1994a). Mechanistic role of yarn and fabric structure in determining tear resistance of woven cloth - Part I: Understanding tongue tear, Textile Research Journal, Vol. 64, No. 5, 291-304.

Scelzo, W. A.; Backer, S. & Boyce, C. (1994b). Mechanistic role of yarn and fabric structure in determining tear resistance of woven cloth. Part II: Modelling tongue tear, Textile Research Journal, Vol. 64, No. 6, 321-329.

Teixeira, N. A.; Platt, M. M. & Hamburger, W. J. (1955). Mechanics of elastic performance of textile materials: Part XII: Relation of certain geometric factors to the tear strength of woven fabrics, Textile Research Journal, No. 10, 838-861.

Taylor, H. M. (1959). Tensile and tearing strength of cotton cloths, Journal Textile Research, Vol. 50, T151-T181.

Letters to the Editor (1974). Modified Tearing Model, Journal of the Textile Institute 65, No 10, 559-561.

Witkowska, B. & Frydrych, I. (2004). A Comparative analysis of tear strength methods, Fibres & Textiles in Eastern Europe, Vol. 12, No 2(46), 42-47.

Witkowska, B. & Frydrych, I. (2005). Protective clothing - test methods and criteria of tear resistance assessment, International Journal of Clothing Science and Technology (IJCST), Vol. 17, No 3/4, 242-252.

Witkowska, B. & Frydrych, I. (2008). Static tearing. Part II: Analysis of stages of static tearing in cotton fabrics for wing - shaped test specimens, Textile Research Journal, 78, 977 - 987.

Witkowska, B. (2008). Modelling of the cotton fabric static tearing, Doctoral dissertation Technical University of Lodz, Poland.

4

Effects of Topographic Structure on Wettability of Woven Fabrics

Alfredo Calvimontes[1], M.M. Badrul Hasan[2] and Victoria Dutschk[3]

[1]*Leibniz Institute of Polymer Research Dresden*

[2]*Institute of Textile Machinery and High Performance Material Technology, Dresden*

[3]*EFSM group, Faculty for Engineering Technology, University of Twente, Enschede*

[1,2]*Germany*

[3]*The Netherlands*

1. Introduction

Each surface is characterised by its chemical composition, a certain surface geometry and roughness. The interaction of liquids with textile fabrics may involve one or several physical phenomena such as fibre wettability, depending on the intermolecular interaction between the liquid and fibre surface, their surface geometry, the capillary geometry of the fibrous assembly (Hasan et al., 2008), the amount and chemical nature of the liquid as well as on external forces. A rough textile surface possesses pores, crevices, capillaries or other typical structures with their own characteristic wetting and penetration properties. As a consequence, the apparent contact angle on these surfaces will be affected by thermodynamics and kinetics associated with such intrinsic structures.

Fabric texture affects the porosity and strongly influences the textile characteristics such as fabric mass, thickness, draping ability, stress-strain behaviour, or air permeability (Potluri et al., 2006; Milašius et al., 2003; Kumpikaitė, 2007). The surface topography of fabrics is responsible for their functionality - appearance and handle, wettability, soiling behavior and cleanability (Calvimontes et al., 2005), abrasion resistance and wear (Dutschk et al., 2007). However, there are very few systematic investigations of quantitative relations between construction parameters, topography of fabrics and their wettability.

One technique that has been extensively used for studying the wetting properties of solid surfaces, is dynamic wetting measurements. Calvimontes et al. (2005), demonstrated its utility in characterizing both textile materials and interactions between them and aqueous solutions of soil release polymers. Differences between the soiling behaviour and cleanability of three polyester textile materials with various topographical structures were determined despite the similarity of their chemical nature. In other studies (Calvimontes et al., 2005; Hasan et al., 2009), the usefulness of the application of a relatively new imaging technique based on the principle of chromatic aberration was shown in characterizing the surface of textile materials before and after impregnation with soil release polymers.

2. Textile topography

In almost all the studies cited above, a large number of roughness and waviness parameters were obtained that did not take into account the scales-morphologic periodicity of each surface studied and its influence on the whole topography. All textile materials having periodic surfaces show some horizontal and vertical repeating units; therefore, different length scales have to be taken into account for a proper interpretation of the topographic data measured (Hasan et al., 2009; Calvimontes et al., 2009).

2.1 Topography measurements

Depending on fabric characteristics, and the structure and size of repeating units, several non-contact measurement methods such as chromatic with-light sensor (CWL) - also called chromatic confocal imaging-, high-resolution scandisk confocal microscopy (SDCM), scanning electron microscopy (SEM), confocal laser canning microscopy (CSOM), conoscopic holography (CSL), etc. can be used.

In Calvimontes et al. (2009) CWL was used and recommended for the optical topographic analysis of textile materials. This instrument allows a lateral and a vertical measure range up to 100 mm and 380 µm, respectively, and a lateral and vertical resolution up to 1 µm and 3 nm, respectively. In Lukesch (2009) and Calvimontes (2009), the use of CWL to measure topography of textile surfaces was compared with the use of SDCM. According to these studies, wider cut-off lengths and larger z-ranges make CWL more appropriate than SDCM to measure topographic characteristics of polyester and cotton fabrics.

It is important to note, that the selection of a method due to its high resolution could be inadequate if the cut-off length available or z-range is too small. On the other hand, the use of a very high resolution and larger cut-off lengths (scan areas) results in data whose excessive size could demand extremely long calculation times and special or non-existent hardware and software.

2.2 Optimal sampling conditions

Cut-off length (L_m), defined as the length of one side of the square sampling area, and resolution (distance between measured points Δ_x, assuming that $\Delta_x = \Delta_y$) are the most important sampling parameters, which apart from particular instrumental dependent parameters, such as light intensity, measuring frequency, etc., have to be optimally defined before characterising topography.

Tsukada & Sasajima (1982) and Yim & Kim (1991), discussed the problem of an optimum sampling interval (L_m) by checking the variance of the root mean square roughness (R_q) for a surface under different sampling intervals. According to Stout et al. (1993), recommendation mentioned above for the choice of sampling interval is doubtful because of the fact that the optimum L_m seems to influence the amplitude parameters (wave height W_t and waviness W_z).

The use of tables that relate foreseen the mean rough height (R_z), root mean square roughness (R_q) and arithmetic mean roughness (R_a) with L_m is frequently recommended to set the optimal value of L_m for periodic as well as non-periodic surfaces. As optimal sampling conditions are strongly dependent on the type of material to be characterized, researcher experience is usually required. A systematic procedure to define optimal cut-off length and resolution values was proposed and probed in Calvimontes (2009).

2.3 Topographic characterization using different length scales concept

The use of a scale concept to characterize and study textile surfaces is a new skill that helps to correlate textile parameters, topography and topographical changes with interface phenomena such as spreading, wetting, capillary penetration, and soil release (Calvimontes et al., 2007).

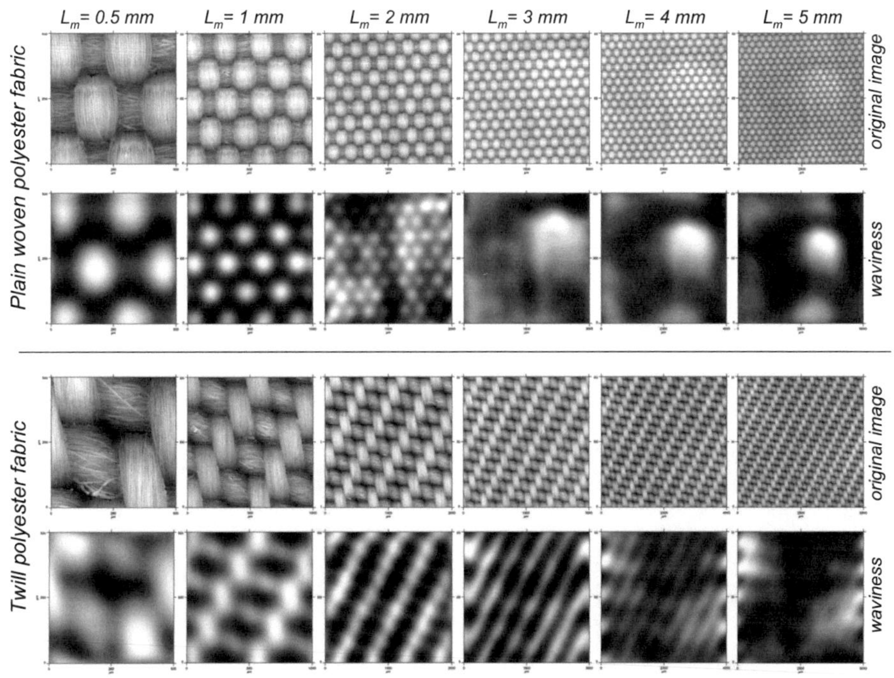

Fig. 1. Original and waviness images of polyester fabric surfaces as a function of L_m

As shown in Figure 1, waviness images obtained by Fast Fourier Transformation-filtering (FFT) (Raja & Radhakrishnan, 1979) provide different type of information depending on cut-off length used. Due to the structural diversity of textile materials, their classification by unit size and morphology on different length scales is necessary. A suggestion to find general range values of L_m in order to identify different measurable length scales is not reasonable. However, the specification of at least three different length scales (macro-, meso- and microscale) is absolutely necessary to describe morphologically homogeneous textile groups. From a conceptual point of view, each one of the length scales proposed for a textile structure has to provide specific information about the surface morphology and topometry of the materials.

Macro-morphological irregularities of textile surfaces such as folds and wrinkles can be studied using FFT-filtering of topographical data measured by large values of L_m. A cut-off length value larger than 3 mm was suggested by Calvimontes et al. (2009) to quantify plane irregularities (waves and wrinkles) of polyester fabrics.

Dimensional changes (relaxation/shrinkage) of fabrics at macro scale influence their meso- and micro-topography due to the modification of repeating unit dimensions and, therefore, distances between yarns, filaments and fibres.

In Figure 2 macro-waviness diagram alone for woven plain does not show any morphological influence of repetitive units (r.u.) morphology, in this case r.u. > 13^2. For macro-topographical characterization of twill and Panama types of weave, optimal L_m values have to be larger than 5 mm.

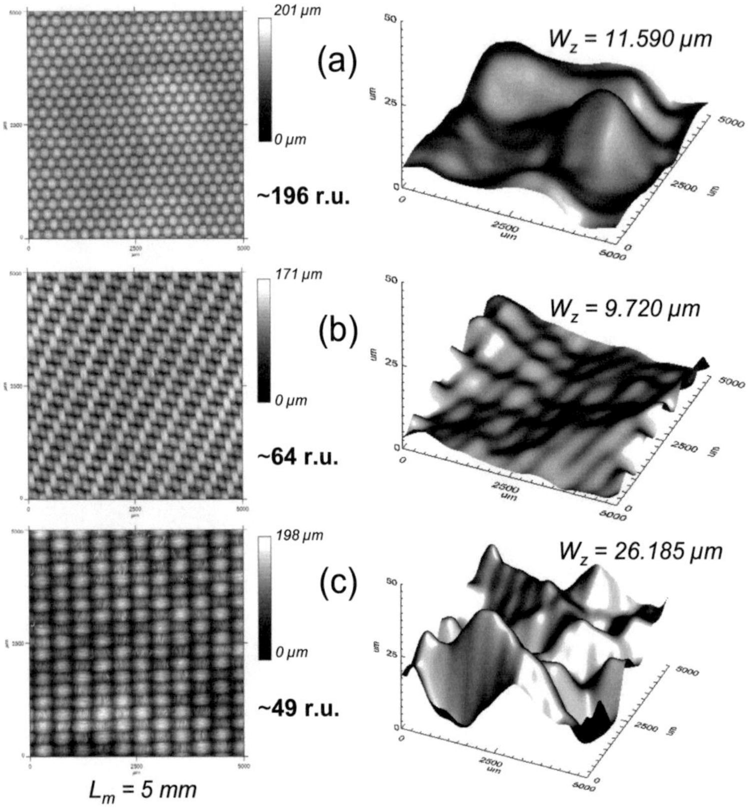

Fig. 2. 2D images (left) and 3D waviness diagrams (right) with W_z values for: (a) woven plain; (b) twill and (c) panama polyester fabrics.

Meso-scale of textile materials should be used to describe the surface topography produced by the type of weave and yarn used, without attending previous defined macro-topographic irregularities and details corresponding to fibres or filaments. A study of fabric surface topography at a mesoscopic scale using FFT-filtering starts with the selection of a new optimal L_m value, which basically depends on the size of fabric repeating unit. From a large amount of experimental data for polyester fabrics studied, it was revealed that a sample area (L_m x L_m) has to cover about 8 repetitive units (Calvimontes et al. 2009).

Another way to construct meso-topographic diagrams is the use of digital surface filtering, which calculates the arithmetical mean of each data point with its neighbourhood (Stout et al., 1993). Filter density used depends on fabric characteristics and has to be able to produce a surface without topographical details of fibres or filaments. Figure 3 shows the construction of meso-topographic surfaces by using FFT-filtering and Smooth filtering. In order to compare morphologies and W_z values obtained, L_m and filtering method used should be remained the same during the characterisation process.

An application of the study on meso- scales claims to know relative z-distances between warps and wefts for woven fabrics and the amplitude of their wave (sinoidal) trace. As shown in Figure 3, wefts describe almost a linear trace (their amplitudes are small). As a consequence, the first contact of any solid with the fabric surface takes place by the warps ("hills") and the final penetration of fluids into the fabric surface takes place principally on wefts ("valleys"). This finding plays a crucial role in understanding the wetting behaviour of textile materials (Calvimontes et al. 2007).

Unlike macro- and meso-scales, characterisation at a micro-length scale reveals the influence of filaments and fibres characteristics on the resulting topography. Profile, fineness, as well as natural or machined texture of these elements or distances between them are only some of possible characteristics which define the resulting morphology and topometry at this length scale.

The selection of an optimal cut-off length in this case no longer depends on some statistical or mathematical criteria as seen at macro and meso length scales, rather on the size and location of the set of filaments or fibres by type and orientation.

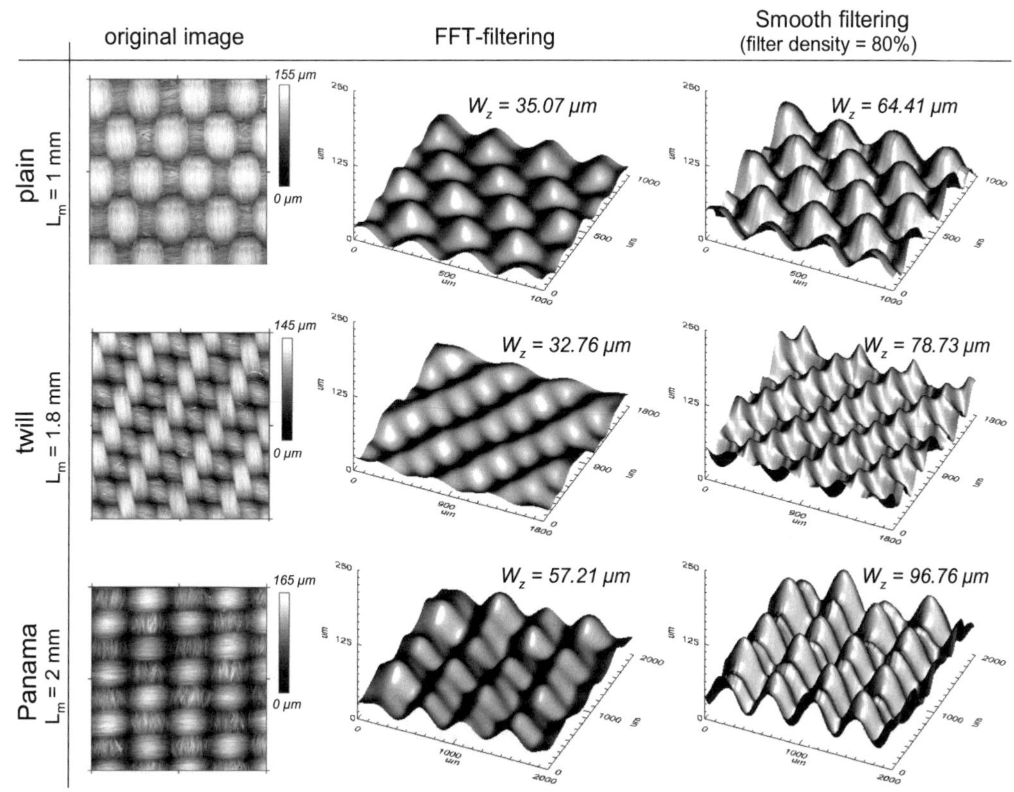

Fig. 3. Meso-topography of different polyester fabrics by FFT filtering and smooth filtering

To study the micro-topography of woven plain fabrics, warps and wefts should be zoomed separately. Optimal L_m values of warps and wefts depend on the type of weave and construction parameters such as yarn types, their diameters, warp densities, weft densities, etc. Depending on the textile structure, more than one L_m value could be necessary for a complete micro-topographical characterisation, as shown in Figure 4.

The number of sub-areas to be isolated depends on topographical parameters studied and on standard deviations of their mean values. Usually, five different zooms should be enough to characterise polyester monofilament fabrics. Depending on the characterization criteria, the elimination of micro-waviness, a consequence of yarn profile and fabric meso-topography, is possible by FFT-filtering, as shown in Figure 5.

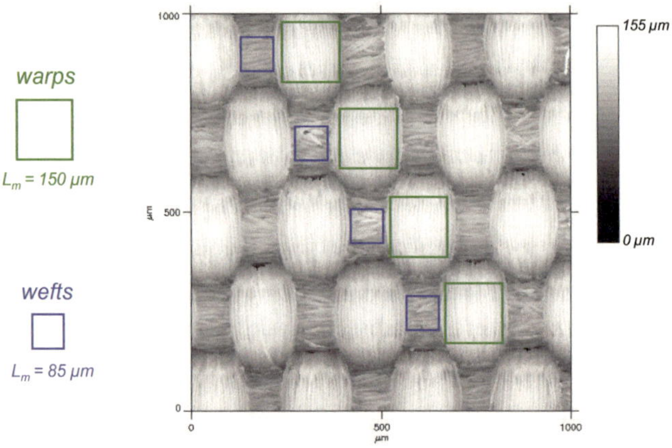

Fig. 4. Optimal L_m values for the characterisation of warps and wefts micro-topography separately at the surface of woven plain polyester fabric.

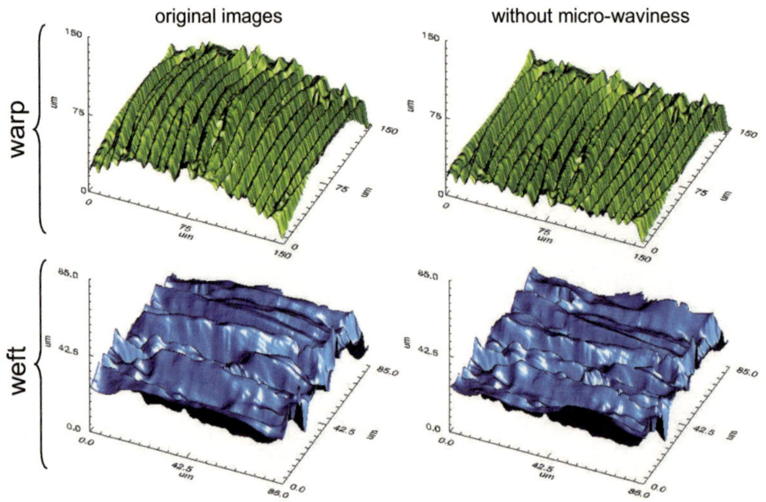

Fig. 5. Micro-topographical images of a warp and a weft. The elimination of micro-waviness was possible by FFT-filtering

Using the new topographical data generated, it is possible to calculate any micro-topographical parameter by profiling or by using the whole surface. The volumetrical characterization is a good tool to measure textile surfaces through evaluating of porosity or filling quantities at different deep heights. Fractal dimension or Wenzel roughness factor could be of interest to characterize micro-topographical modifications of natural fibres, e.g. changes caused by plasma or enzymatic action.

3. Influence of topography on wettability of textile structures

One technique that has found extensive use in studying the wetting properties of solid surfaces is dynamic wetting measurements. Seven different cases of connection between topography and wetting of textiles will be described in following pages. In all of them, dynamic wetting measurements were carried out by a FibroDAT 1122 HS dynamic contact angle tester (Fibro System, Sweden). Some advantages of this equipment over other contact angle measuring systems as well as the measuring procedure have been detailed by Dutschk et al. (2003).

3.1 Type of weave and construction parameters

Hasan et al. (2008) studied the influence of the type of weave on topography and wettability of polyester fabrics produced using multi-microfilament yarn with different filament structures with filament diameter being between 6 and 7.5 µm. The characteristics of yarns used are summarised in Table 1. Two basic types of weaves - plain (1/1) and twill (2/2 Z) - were produced by variation yarn combinations and the weft density without changing the warp density. After the manufacturing, the fabrics were desized under laboratory conditions raising the primary fabric density. The composition of each sample are detailed in previous studies (Calvimontes et al., 2006).

Yarn	Number of filaments	Structure	Filament fineness, dtex[1]	Filament diameter, µm	Yarn fineness, dtex[1]
A	128	flat	0.78	6.0	9.9
B	128	textured, tangled	0.92	7.5[2]	11.5
C	256	false-twist textured	0.78	6.5[2]	16.3
D	384	textured, tangled	0.67	6.5[2]	24.6

[1] Measured according to DIN EN ISO 1973:1996; [2] before texturing.

Table 1. Characteristics of the multi-microfilament yarns used by Hasan et al. (2008)

Relevant topographic characteristics obtained from the fabrics are given in Table 2. Comparing the macroscopic roughness parameters for different types of weave, it was ascertained that a decrease occurs for the plain weave and an increase for the twill weave if the weft density increases. Fabric density calculated according to Walz & Luibrand (1947), is inversely proportional to air permeability (for the same weft yarn) measured, as expected, and the results are summarized in Table 3. This nearly linear relationship is independent of the fabric texture, contrary to the macroscopic roughness parameters as shown in Figure 6. An increase in weft density smoothes the surface waviness independently of the type of weave, as illustrated in Figure 7. The values of waviness are higher for the twill weave than for the plain weave with the same weft density.

Type of weave		Yarn variations: A–B, weft/cm*		Yarn variations: A–C, weft/cm*		Yarn variations: A–D, weft/cm*	
Plain		39 / 43	40 / 46	32 / 36	36 / 40	26 / 30	29 / 32
Twill		50 / 56	54 / 57	42 / 45	47 / 53	35 / 39	38 / 44

* Weft density is given for each sample before (above) and after desizing (below), respectively.

Table 2. Composition of polyester fabrics woven varying construction parameters (68 warps (cm)

Type of weave	Yarn composition/ weft density	Sample specification	Root mean square roughness, R_q, µm	Mean peak to valley height R_z, µm	Porosity, µm³/µm²	Waviness, µm
	A–B / 39	pB39	9.7	95.5	1.189	48.7
	A–B / 40	pB40	9.5	90.1	0.942	47.5
Plain	A–C / 32	pC32	10.9	115.8	2.528	52.8
1/1	A–C / 36	pC36	9.7	87.3	1.325	50.3
	A–D / 26	pD26	11.1	93.7	1.574	67.9
	A–D / 29	pD29	10.6	83.7	0.760	60.2
	A–B / 50	tB50	9.4	48.8	0.552	47.5
	A–B / 54	tB54	9.3	88.5	0.916	40.5
Twill	A–C / 42	tC42	9.4	70.9	0.583	59.8
2/2 Z	A–C / 47	tC47	9.8	79.6	0.628	48.9
	A–D / 35	tD35	11.2	111.6	1.221	76.8
	A–D / 38	tD38	11.6	114.9	1.438	65.3

Table 3. Topographic characteristics obtained by means of CWL (chromatic white light sensor

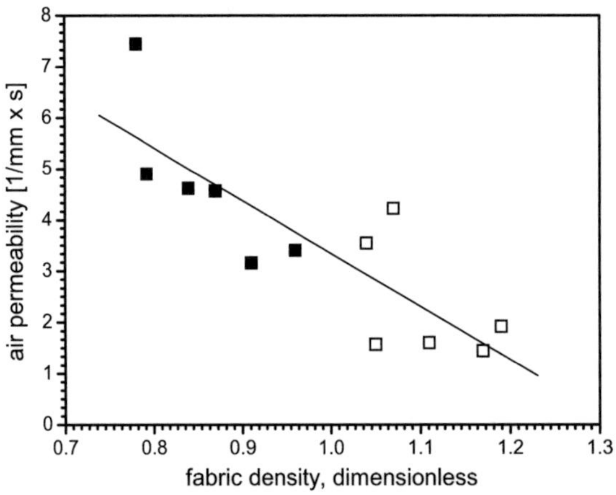

Fig. 6. Nearly linear relationship between fabric density calculated and air permeability measured: (□) plain and (■) twill weave

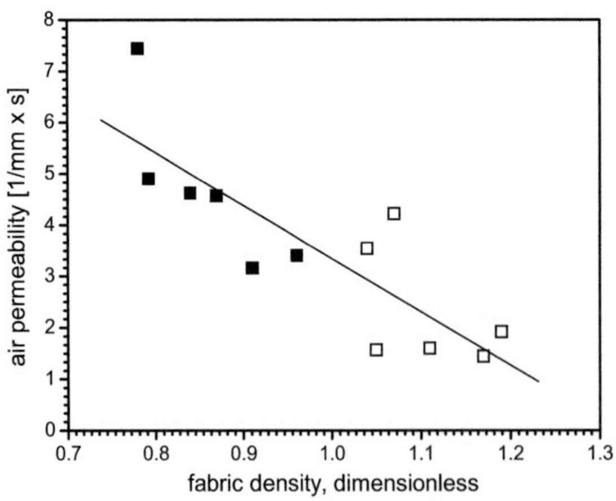

Fig. 7. Dependency of the waviness on the weft density for (□) plain and (■) twill weave

Variations in interlacing are also reflected in the fabric wettability considered in terms of the spreading rate as shown earlier (Calvimontes et al., 2006). The spreading rate decreased with increasing waviness for the plain weave, whereas it increased in the case of twill. It was concluded that the fabric wettability could be adjusted (in certain limits) by variation of density and interlacing, keeping in mind the same chemical nature of microfilaments.

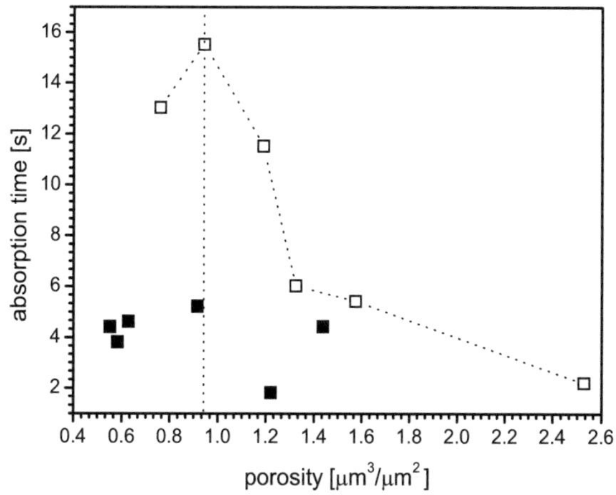

Fig. 8. Absorption time vs. surface porosity calculated as the ratio between the real pore volume and the corresponding geometric surface for (□) plain and (■) twill structures

Noticeable differences in the wetting behaviour of water were seen between the two types of weave, if changes in porosity are considered. In the case of the plain weave, higher weft density leads to lower porosity and, as a consequence, to higher water absorption time, as shown in Figure 8. Moreover, water penetration into the plain texture is slightly slowed down with increasing porosity, reaches a maximum value of the absorption time (about 16 s) in the porosity range of approximately 1 $\mu m^3/\mu m^2$, and then accelerates towards the higher porosity values. It can be speculated about a "critical" value of the fabric's porosity. Presumably, below this value water percolates with low velocity and above this with a high value. In contrast, the higher weft density of the twill weave results in higher porosity (cf. Table 2). The values of absorption time obtained for the twill texture are generally very low of about 2 - 5 s and are almost independent of the porosity, as illustrated in Figure 7.

The differences in the penetration behaviour of water observed on two predetermined patterns of interlacing are caused by the different topographical structure, since the chemical nature of filaments used was kept constant. It is noted, that the lateral distance between the threads is about 120 and 300 μm for the twill and plain weaves, respectively. The vertical dimension of the surface features is measured up to 20 μm for the plain topography and 40 μm for the twill topography. It is well known, that in the case of moderately hydrophobic surfaces the complex internal geometry of real porous systems could enhance liquid penetration (Bico et al., 2001). As reported earlier (Matsui, 1994), polyester is moderately hydrophobic with a water contact angle of 77° on its flat surface. The results obtained in the present study would suggest that water advanced in a stable flood (wicking regime) is observed (Kissa, 1996). The difference in the penetration behaviour (lower for the plain weave and faster for the twill weave) arises from the difference in the shape and size of the pores.

The same comments were applied in the relationship between air permeability and the absorption time in an earlier work (Calvimontes et al., 2006), where it was found that water penetration strongly depends on air permeability for the plain topography. It seems to be absolutely independent of this textile parameter for the twill structure, although no correlations between both characteristics air permeability and porosity were found.

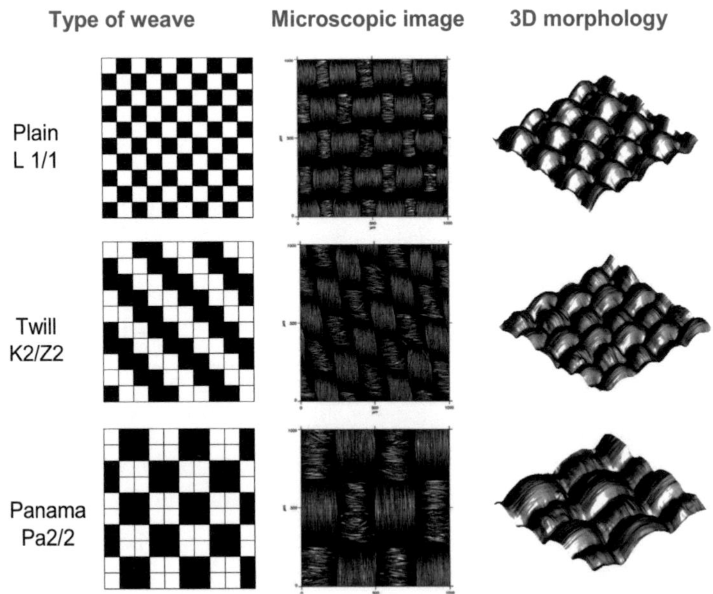

Fig. 9. Polyester fabrics used to study the influence of the type on weave on water spreading

In essence, to achieve a more hydrophobic fabric texture, the technological parameters should be changed as follows: for both plain and twill structure, the weft density and filament fineness should be increased, and the yarn fineness should be reduced. In general, the plain weave with the yarn combination A-B and density of 46 wefts/cm (desized) shows the "best" wetting properties with the longest delay of penetration.

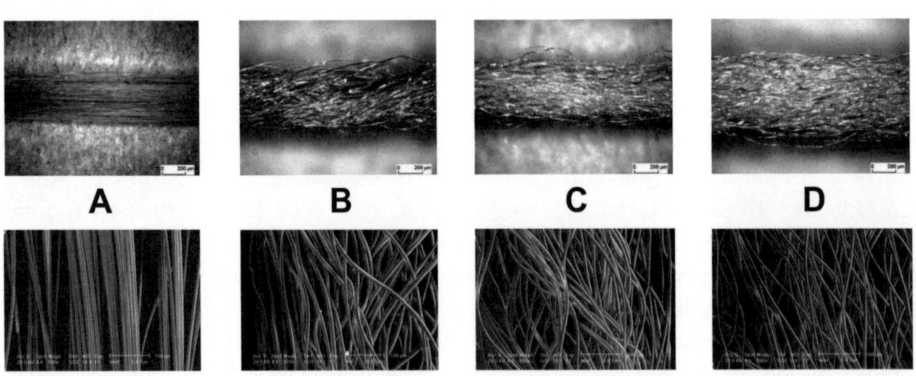

Fig. 10. Microscopic images of warp yarns (above) and filaments (below), according to parameters detailed in Table 1

Calvimontes et al. (2007) used 14 different polyester fabrics, having plain, twill and panama structures to show how the use of topographic characterisation at different scales can provide important information of the spreading behaviour.

Polyester fabrics of three different types of weave (Figure 9) were manufactured using filaments produced by spinning of the same polymer material (polyethylene therephthalate). Warp yarns were formed from flat filaments, while wefts were textured by three different processes (Table 1 and Figure 10).

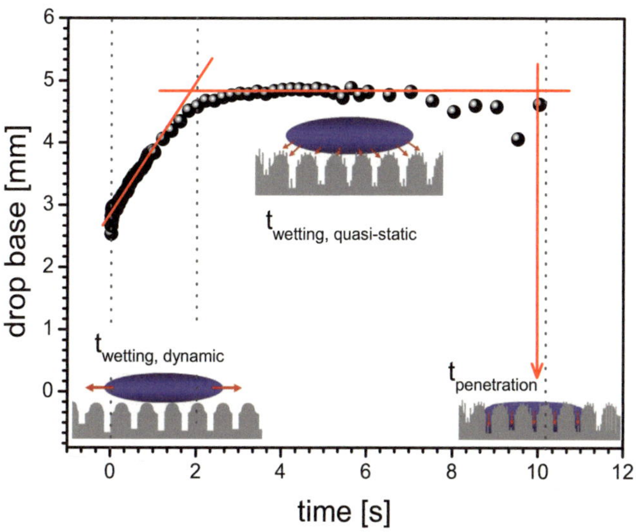

Fig. 11. Three different wetting regimes for a textile surface

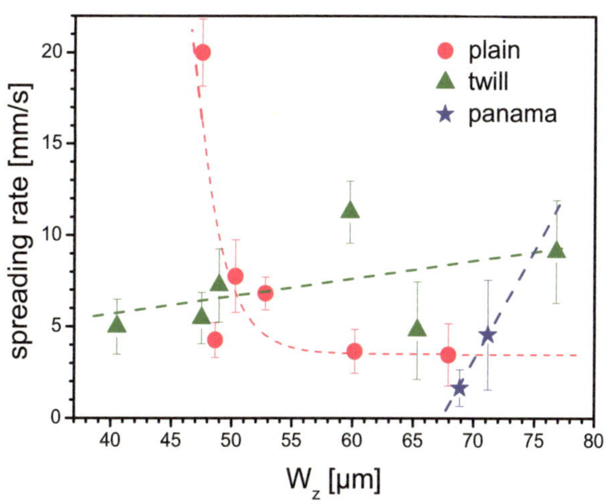

Fig. 12. Dynamic wetting: meso-morphology controls the spreading rate of a liquid drop on a textile structure

On the basis of macroscopic water drop base changes measured with a dynamic contact angle tester (Fibro DAT 1122, Fibro System, Sweden), the wetting behaviour of a water drop can be divided into three regimes (Figure 11): dynamic wetting, defined as growing of the drop diameter depending on time (also known as spreading), the quasi-static wetting, where the drop diameter remains approximately constant, and penetration, which is marked by liquid drop absorption into fabrics depending on time.
By using the waviness as a meso-topographical parameter, it is evident that the meso-topography of the fabrics controls the spreading rate of a liquid drop (Figure 12). For the plain weave, an increase of the waviness depth causes a decrease of the spreading rate; warp yarns ("hills") slow down the liquid motion (Figure 13). For twill weave, an increase of the waviness depth causes formation of deep and long domains of weft yarns ("canals") with small "islands". As a consequence, an increase of the spreading rate is observed. Finally, for the panama weave, an increase of the waviness depth causes formation of long and quasi-endless (without "islands") deep domains ("canals"). Consequently, the waviness depth and spreading rate are proportional to each other.

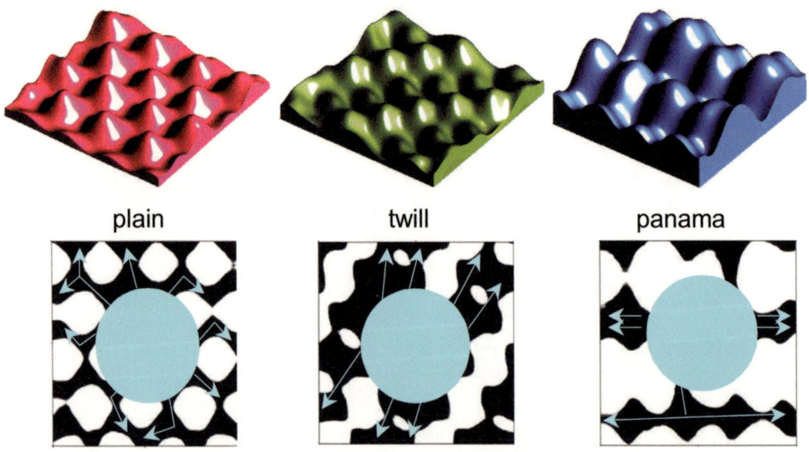

Fig. 13. Respective textile morphology at a meso-length scale controls the spreading rate. Above: morphology; below: spreading directions of a liquid drop

A thorough comparison between topographic parameters for 14 fabrics, having three different types of weaves, reveals that the respective morphology at meso length scale controls the spreading rate.

3.3 Yarns texture

By zooming of warps and wefts separately at a smaller scale, topography measurements and the characterisation concept at different length scales provide important information about changes in textile microstructures. Using this information, the behaviour of a liquid drop on fabrics, detailed in Tables 1 and 2, while wetting can be explained. On the basis of experimental results (Calvimontes, 2009), revealing differences for two basic types of woven fabrics - plain and twill - in respect to capillarity and water penetration (Figure 14), the concept of a novel wicking model was developed. This conceptual model was verified in respect to the cleanability behaviour of fabrics using paraffin oil and acetylene black soils.

Results illustrated in Figure 14 show: (i) warp yarns topography hardly affect the cleanability of fabrics; (ii) spaces between fibres make the plain weave surface oleophil (the larger they are, the more stain penetrate); (iii) spaces between fibres make the twill weave surface oleophob. The larger and deeper they are, the more stain penetrates and the worse their cleanability and (iv) the weft yarn roughness controls the hydrophobicity or hydrophilicity of fabrics and, as a consequence, their cleanability (cf. Figure 15).

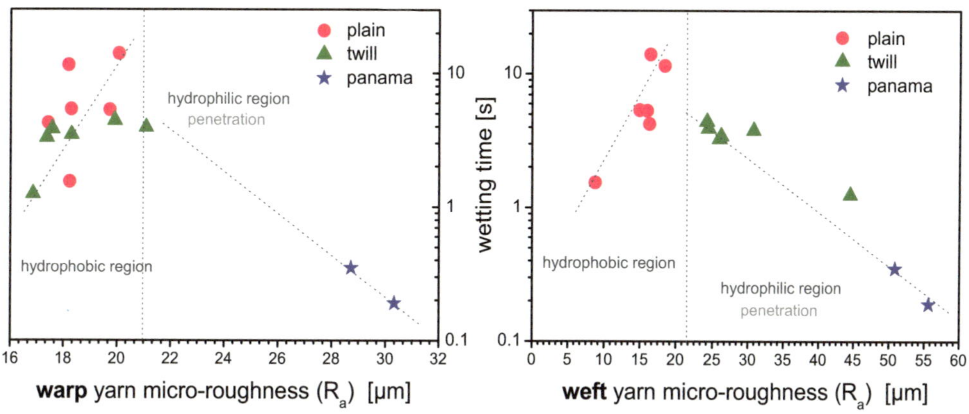

Fig. 14. Liquid flow in warp and weft directions occurs in two different regimes, depending on the micro-topography

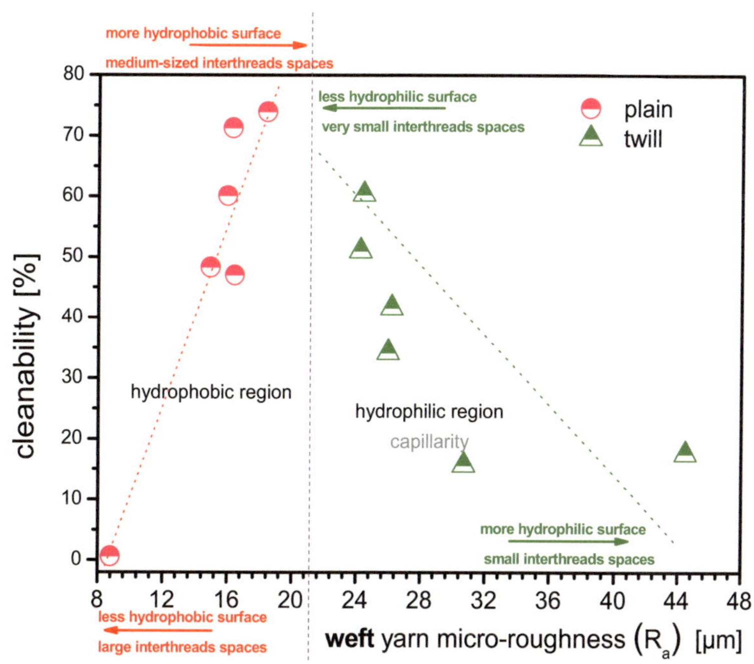

Fig. 15. The conceptual model proposed allows a better understanding of the cleanability phenomenon of polyester fabrics by using a different length scale concept for their characterisation. Soil material: paraffin oil and acetylene black in the ratio 97.98:2.02

3.4 Fibre cross sectional shape and the effect of heat-setting

Hasan et al. (2008), studied warp rib (2/2) fabrics (a derivative of the plain weave) produced using differently profiled polyester filaments - round and cruciform - in a melt-spinning process. The general manufacturing procedure is detailed by Hasan (2007).

Microscopic images of different cross sectional shapes of the fibre manufactured are shown in Figure 16. The fabrics were desized and treated by heat-setting at 190 °C for 10 s after being manufactured. Both modifications - desized with and without heat-setting - were discussed. Figure 17 illustrates different basic weaves used in this study.

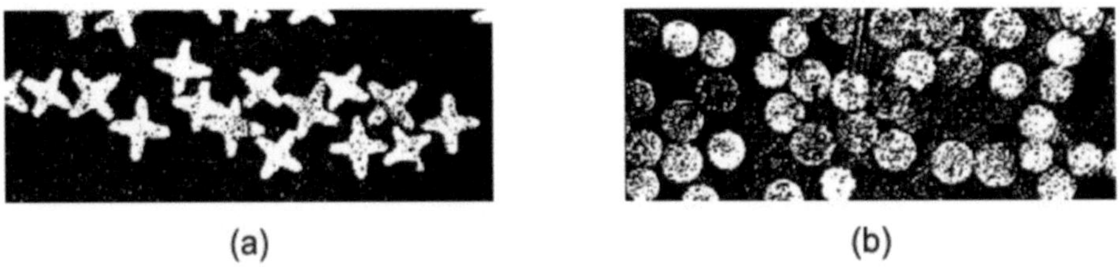

Fig. 16. Microscopic images showing different cross-sectional shapes of the polyester fiber manufactured: (a) cruciform; (b) circle-shaped

Geometric cover factor of the fabrics was calculated using their microscopic images as the ratio of the projected fabric surface area covered by yarns to the total fabric surface area given using the following equation (Sabit, A., 2001).

$$cf = c_w + c_f - c_w \cdot c_f$$

where c_w and c_f are the warp and weft cover factor, respectively. The warp cover factor is a product of the warp count and the diameter of warp yarn. Following the same logics, the weft cover factor is a product of the weft count and the diameter of weft yarn. In our calculations, the yarn diameter was replaced by the major axis length of yarn having an ellipsoid form. The relevant topographic characteristics obtained for the fabrics manufactured using differently profiled fibre as well as water contact angle for the fabrics are given in Table 4. For convenience, the fabrics analyzed are specified by identification codes, detailed in Table 5.

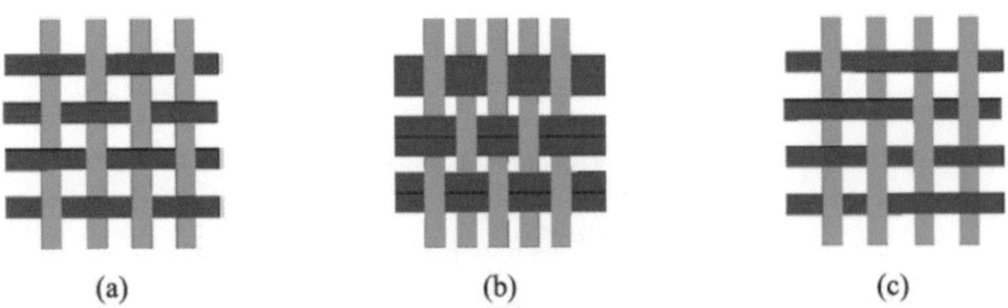

Fig. 17. Basic weaves used: (a) plain weave; (b) warp rip weave; (c) twill

Fabrics identification code	Cover factor	Water contact angle, deg
R-20-D	1.084	113 ± 5
R-30-D	1.017	106 ± 6
R-20-H	1.110	118 ± 6
R-30-H	1.029	117 ± 6
X-20-D	1.032	109 ± 5
X-30-D	0.974	114 ± 4
X-20-H	1.043	119 ± 5
X-30-H	0.998	124 ± 5

Table 4. Geometric factor and water contact angle data

Fibre cross-section	Weft density	Treatment	Fabrics identification code
Round	20/cm	desized	R-20-D
		heat-setted	R-20-H
	30/cm	desized	R-30-D
		heat-setted	R-30-H
Cruciform	20/cm	desized	X-20-D
		heat-setted	X-20-H
	30/cm	desized	X-30-D
		heat-setted	X-30-H

Table 5. Identification code of fabrics.

In contrast to wetting measurements on plain and twill fabrics, woven using commercial multi-microfilament yarn, water drops do not penetrate into the textile surface. No statistically significant changes were found in the wetting behaviour of the fabrics containing round and cruciform shaped fibres (Figure 18). The wettability of fabrics changes

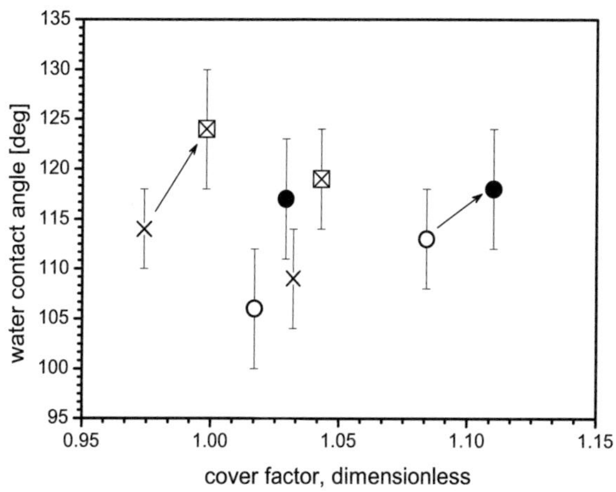

Fig. 18. Water contact angle as a function of the geometric cover factor: (O) round desized without heat-setting (●) round desized with heat-setting; (×) cruciform without heat-setting; (⊠) cruciform with heat-setting.

after their heat-setting. The largest water contact angle of 124° was observed for fabrics containing cruciform fibres with a cover factor that is smaller than that for fabrics with round fibres.
However, no relationship was found between the geometric cover factor and water contact angle. It therefore seems worth pointing out that the geometric cover factor is an idealized effective parameter, which cannot precisely describe the topography of fabrics. On the contrary, surface roughness and waviness can be used to understand the wetting behaviour of liquids on fabrics with a complex structure. A clear linear tendency was found for the water contact angle and surface roughness, as shown in Figure 19, independent of the modification applied.

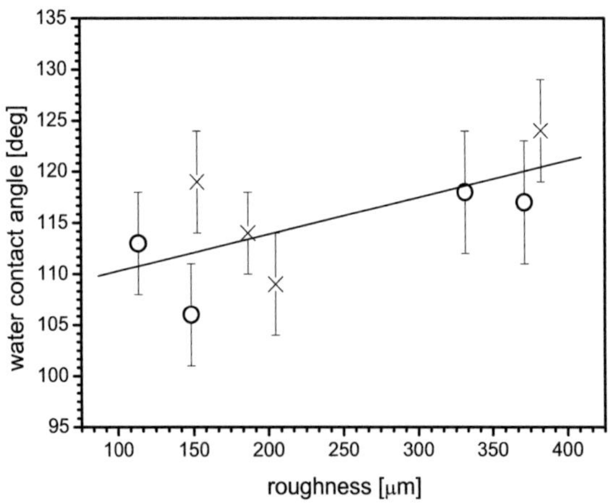

Fig. 19. Water contact angle as a function or roughness (O) round; (×) cruciform

The topography analysis allows filtering measured total profiles of a surface to split them into two analytical representations for displaying surface features: roughness and waviness. The first information represents the shorter spatial wavelengths, whereas the second one represents the longer wavelength features of the surface. Both filtered profiles as well as the schematic of a complex fabric structure are illustrated in Figure 20. Interestingly, surface waviness measured is a function of the geometric cover factor calculated, shown in Figure 21. The fabrics with cruciform fibres have smaller cover factors than those of round fibres. Generally, the maximum cover factor is 1, whereby idealised yarns touch each other. The factor can be larger than 1 if the yarns pile up on top of each other (Sabit, 2001). Real yarns, usually consisting of several single filaments, are flexible and can take different shapes from elliptic to round depending on warp and weft densities. By these means, the appearance of a minimum in the cover factor versus the porosity curve, as shown in Figure 22a, can be explained. A sketch in Figure 22b illustrates possible fabric structures depending on the warp and weft density.
Summarising, by the use of profiled fibres, e.g. the cruciform, the fabric manufacture could lead to a more hydrophobic fabric texture on the basis of different roughness length scales.

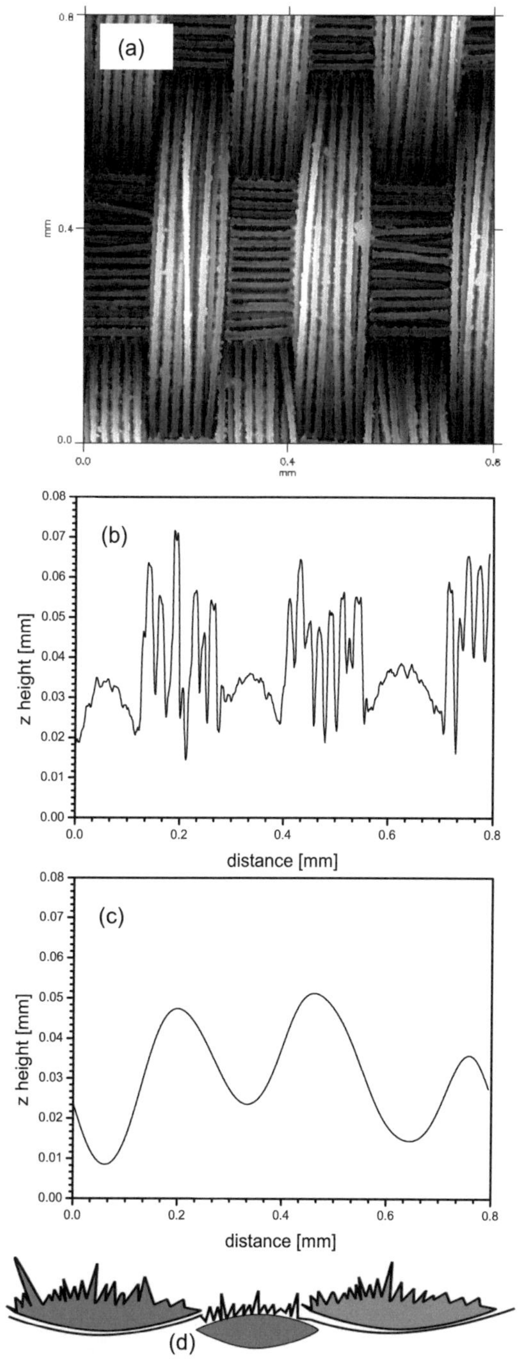

Fig. 20. (a) 2D image of a fabric scanned with chromatic confocal sensor; (b) filtered roughness profile of a plain fabric surface as an example; (c) filtered waviness profile of a plain fabric surface as an example; (d) schematic of a complex fabric structure: the total profile contains both roughness and waviness information

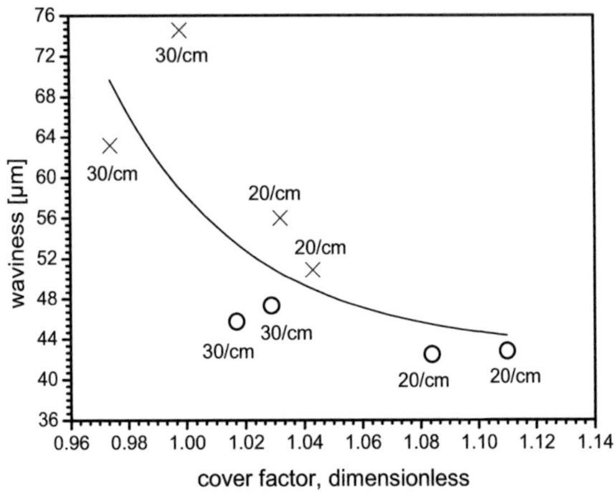

Fig. 21. Relationship between waviness measured and the cover factor calculated: (O) round; (×) cruciform

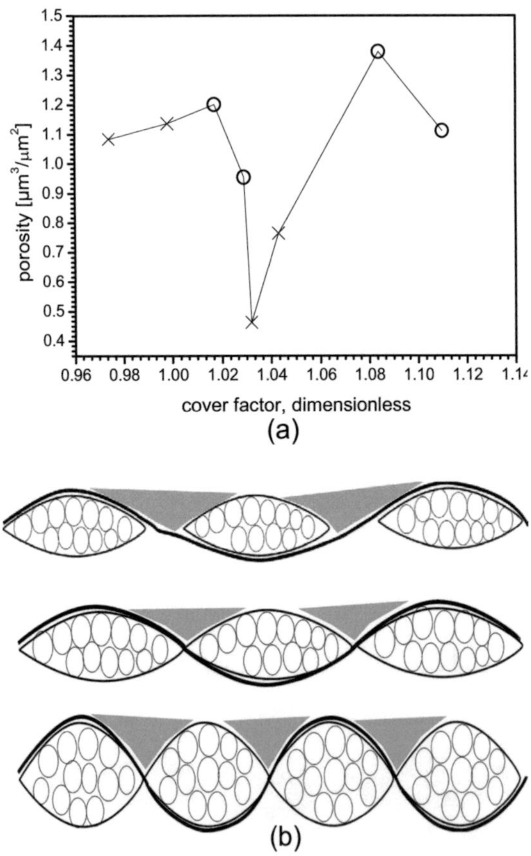

Fig. 22. (a) Fabrics surface porosity measured versus the cover factor calculated: (O) round; (×) cruciform; (b) schematic of differently shaped yarns within a fabric resulting in different surface porosity

Saha (2010) reported the effect of heat setting on the improvement of hydrophobicity of an hydrophobic polyester plain woven fabric, which was heat-setted at 170°C, 180°C, 190°C, 200°C, 210°C, 220°C and 240°C for 60 seconds. According to these results, the effect of thermosetting increased the water contact angle for fabrics thermofixed at 170°C from 113° up to 122° for the thermofixed ones at 200°C (Figure 23). Pilling factor, calculated using mean roughness variations according to Calvimontes (2009), shows a minimum value precisely at 200°C (Figure 24). At this temperature a relative maximum of micro-porosity of warps surface was also found (Figure 25a).

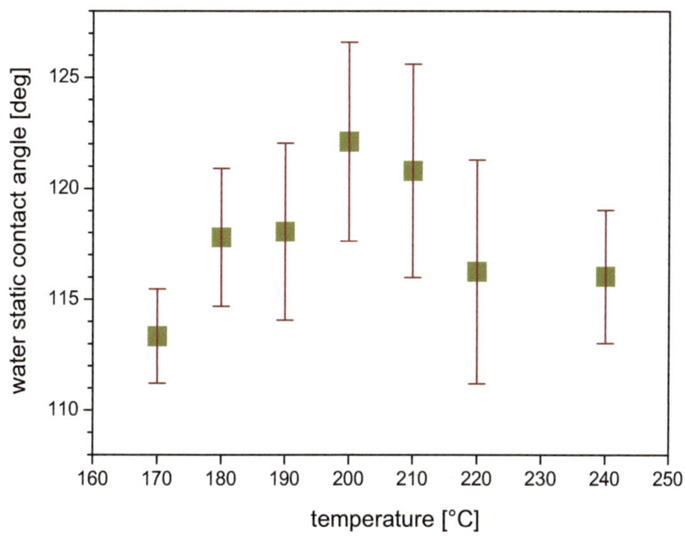

Fig. 23. Water contact angle on thermofixed polyester fabrics

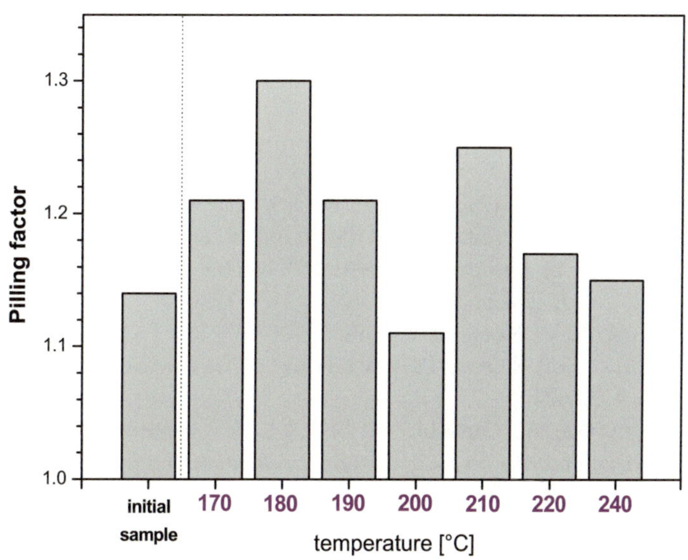

Fig. 24. Pilling factor for thermofixed polyester fabrics

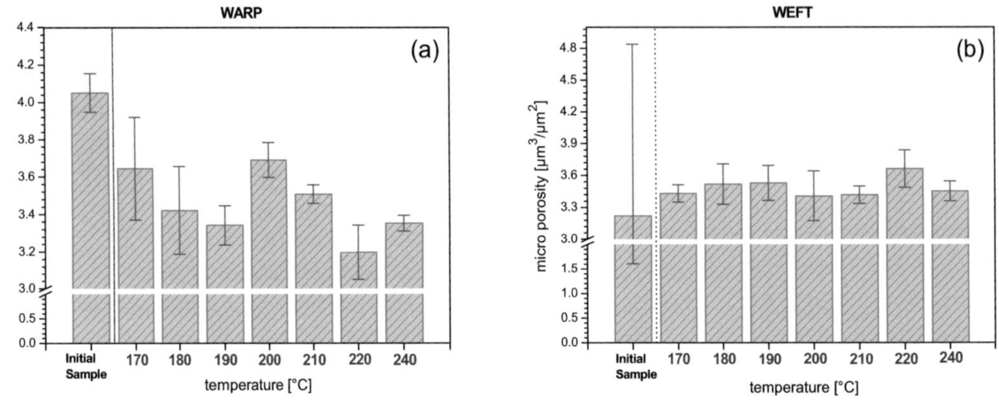

Fig. 25. Micro-porosity of thermofixed polyester fabrics: (a) warps; (b) wefts

On the contrary, the porosity of wefts surface is stabilized by thermosetting (Figure 25b). However, according to that explained previously (Section 2.3), the first contact of any liquid with the fabric surface takes place by the warps, because their higher wave amplitude.

This information allows to conclude that heat-setting smoothes the fabric surface on mesoscopic scale by decreasing pilling, but at the same time, it increases the spaces between warp filaments. The effect is a decrease of the solid fraction –contact area- between polyester fibres surface and water, which, according to Cassie & Baxter (1944), improves hydrophobicity of hydrophobic surfaces.

It was shown in this Chapter, that topographical characteristics of the fabrics strongly depend on their construction parameters such as the type and fineness of filaments, yarn fineness, yarn density, and the type of weave. This characteristics have strong influence on, and in many cases, control the wetting properties.

The topographical study of textile materials using a length scale concept allows to effectively characterize surfaces separately by considering and analyzing their specific morphologies caused by their construction parameters and to successfully find correlations between topography and wettability.

4. References

Bico, J., Tordeux, C. & Quéré, D. (2001). Rough wetting. *Europhys. Lett.*, Vol. 55, 214-220

Calvimontes, A., Dutschk, V., Breitzke, B., Offermann, P. & Voit, B. (2005). Soiling Degree and Cleanability of Differently Treated Polyester Textile Materials. *Tenside Surf. Det.*, Vol. 42, No. 1, 17-22

Calvimontes, A., Dutschk, V., Koch, H. & Voit, B. (2005). New Detergency Aspects through Visualisation of Soil Release Polymer Films on Textile Surfaces. *Tenside Surf. Det.*, Vol. 42, No. 4, 209-216

Calvimontes, A., Synytska, A., Dutschk, V., Bell, Ch. & Lehmann, B. (2006). Influence of polyester woven fabrics on their wettability. *Melliand English*, Vol. 1-2, E16-E18

Calvimontes, A., Dutschk, V., Cheriff, Ch. & Heinrich, G. (2007). Ein neues Konzept zum besseren Verständnis der Penetration von Flüssigkeiten in textilien Oberflächenstrukturen, *Proceedings of 15. Neues Dresdner Vakuumtechnisches*

Kolloquium: Beschichtung und Modifizierung von Kunstoffoberflächen, pp.48-52, ISBN 973 3 00 022604 5, Dresden, Germany

Calvimontes, A., Dutschk, V. & Stamm, M. (2009). Advances in Topographic Characterization of Textile Materials. *Textile Res. J.*, Vol. 80(11), 1004-1015, DOI: 10.1177/0040517509348331

Calvimontes A. (2009). *Topographic Characterization of Polymer Materials at Different Length Scales and the Mechanistic Understanding of Wetting Phenomena*, PhD Thesis, Technische Universität Dresden, Germany

Cassie, A.B.D. & Baxter, S. (1944). Wettability of porous surfaces. *Trans. Faraday Soc.*, Vol. 40, 546-551

Dutschk, V., Sabbatovskiy, K.G., Stolz, M. Grundke, K., Rudoy, V.M. (2003). Unusual wetting dynamics of aqueous surfactant solutions on polymer surfaces. *J. Colloid Interf. Sci.*, Vol. 267, 456-462

Dutschk, V., Myat, S., Märtin, J., Stolz, M., Breitzke, B., Cheriff, Ch. & Heinrich, G. (2007). A Comparative Analysis between Different Ether Carboxylic Acids with Respect to Wettability and Surface Topography of Abrasively Treated Polyester Fabrics. *Tenside Surf. Det.*, Vol. 44, No. 6, 348-354

Hasan, M.M.B. (2007). *Untersuchungen der Oberflächeneigenschaften von gewebten Textilmaterialien aus Polyesterfasern verschiedener Geometrie*, Master Thesis, Texhnische Universität Dresden, Germany

Hasan, M.M.B., Dutschk, V., Calvimontes, A., Hoffmann, G., Heinrich , G. & Cheriff, Ch. (2008). Influence of the Cross-sectional Geometry on Wettability and Cleanability of Polyester Woven Fabrics. *Tenside Surf. Det.*, Vol. 45, No. 5, 274-279

Hasan, M.M.B., Calvimontes, A., Dutschk, V. (2008). Effects of Topographic Structure on Wettability of Differently Woven Fabrics. *Textile Res. J.*, Vol. 78, No. 11, 996-1003

Hasan, M.M.B., Calvimontes, A. & Dutschk, V. (2009). Correlation Between Wettability and Cleanability of Polyester Fabrics Modified by a Soil Release Polymer and Their Topographic Structure. *J. Surfact. Deterg.*, Vol. 12, 285-294

Kissa, E. (1996). Wetting and wicking. *Textile Res. J.*, Vol. 66, 660-668

Kumpikaitė, E. (2007). Analysis of Dependencies of Woven Fabric's Breaking Force and Elongation at Break on its Structure Parameters. *Fibres Text. East. Eur.*, Vol. 15, No. 1, 35-38

Lukesch, M. (2009). *Vergleichende Untersuchungen mit modernen optischen Messmethoden zur Bestimmung der Mikrotopographie von Oberflächen, Praxissemesterarbeit*, Hochschule Zittau/Görlitz (FH) - University of Applied Sciences, Germany

Matsui, M. (1994). *Advanced Fiber Spinning Technology*, Nakajima, T. (ed.), Woodhead Publishing Ltd, p. 115, ISBN 1 85573 182 7, Cambridge

Milašius, V., Milašius, R., Kumpikaitė, E. & Olauškine, A. (2003). Influence of Fabric Structure on Some Technological and End-use Properties. *Fibres Text. East. Eur.*, Vol. 11, 48-51

Potluri, P. Parlak, I., Ramgulam, R. & Sagar, T.V. (2006). Analysis of tow deformations in textile performs subjected to forming forces. *Comp. Sci. Technol.*, Vol. 66, 297-305

Raja, J. & Radhakrishnan, V. (1979). Filtering of Surface Profiles using Fast Fourier Transform. *Int. J. Mach.Tool. Des. Res.*, Vol. 19, 133-141

Sabit, A. (2001). Handbook of Weaving, Technomic Publ. P. 362, ISBN 9 7815871 60134 , Lancaster

Saha, R. (2010). *Roughness-induced capillary uptake of liquids into textile structures*, Master Thesis, Institute of Textile Machinery and High Performance Material Technology, Technische Universität Dresden, Germany

Stout, K. J., Sullivan, P.J., Dong, W.P., Mainsah, E., Luo, N., Mathia, T. & Zahouani, H. (1993). *The Development of Methods for the Characterisation of Roughness in Three Dimensions*, Commission of the European Communities, ISBN 1 8571 8023 2, Brussels-Luxembourg

Trukada, T. & Sasajima, K. (1982). *An Optimum Sampling Interval of Digitizing Surface Asperity Profiles*. Wear, Vol. 83, 119-128

Walz, F. & Luibrand, J. (1947). Die Gewebedichte. *Textil Praxis*, Vol. 11-12, 330-336

5

Importance of the Cloth Fell Position and Its Specification Methods

Elham Vatankhah
Isfahan University of Technology
Iran

1. Introduction

The warp and weft densities are important parameters having significant effect on the visual, physical and mechanical properties of the fabrics. So, to have a fabric with desirable properties, these two parameters play the key role. Density has affected by many factors which appreciating them is crucial in controlling the evenness of fabric. Although, most of the warp regularity is the result of the appropriate weaving preparation, weft evenness is obtained by correct weaving process.

Therefore, pick density or pickspacing is one of the most important parameters that should be controlled continuously during weaving process to prevent any variations. pickspacing itself is influenced by different causes. Some studies have been done to investigate the effect of these factors on pickspacing.

According to previous studies, position of the cloth fell and the cloth fell distance (c.f.d.) are the effective elements on the pickspacin, the variations of which result in unevenness pickspacing.

As weaving is started, the cloth fell gradually finds its correct position by action of the take-up motion. When a low-pick density fabric is woven, the cloth fell lies on the front position of the reed; however in fabrics with more pick density, the cloth fell tends to move towards the warp direction and the distance of this position of the cloth fell from the front position of the reed is referred to as the cloth fell distance (c.f.d.). This position is maintained as long as the loom runs satisfactorily.

There are, however, many occasions when the correct cloth fell position is temporarily lost. The weft may break and the loom turn over for one or two revolutions without inserting weft; a fault may necessitate unweaving; more frequently still, the loom may be stopped for the night or for a meal break and the cloth fell may creep away from its correct position. In all these cases the weaver has to restore the cloth fell to its correct position. In addition, during weaving, the warp tension undergoes a cyclic change which is due to shedding and this, in turn, causes a cyclic variation in the position of the cloth fell ([a]Greenwood & Cowhig, 1956).

The importance of the cloth fell position is recognized and commented on by many researchers, especially from the point of view of the effecting parameters on it.

The preceding discussion in this chapter has mainly focused on the cloth fell position and the relations between different weaving parameters based on the theoretical methods and the mathematical models.

After identity the various effecting parameters on the cloth fell, designing a way to measure and control the variations of this parameter is critical.
The proposed and presented methods for measuring the cloth fell movement and their advantageous and disadvantageous are discussed later in this chapter in detail.

2. A review of the cloth fell position and the interaction between weaving factors

It is necessary that a summary of the nomenclature used in this chapter be introduced. (Table 1)

Symbol	Definition	Units
a	Difference between actual and correct pickspacing	inches/pick
D	Minimum pickspacing (theoretical)	inches/pick
E_1	Elastic modulus of warp	grammes/end
E_2	Elastic modulus of fabric	grammes/end
k	Coefficient of weaving resistance	gramme inches
K	Cloth fell coefficient	square inches
l_1	Free length of warp	inches
l_2	Free length of fabric	inches
L	Cloth fell distance (c.f.d.)	inches
2m	Sweep of reed	inches
n	Number of picks woven	
P	Rate of take-up (correct pickspacing)	inches/pick
r	Distance between reed and cloth fell	inches
R	Weaving resistance	grammes/end
S	Pickspacing	inches/pick
t	Time	seconds
T_0	Basic warp tension	grammes/end
T_1	Instantaneous warp tension	grammes/end
T_2	Instantaneous fabric tension	grammes/end
v	Angular velocity of main shaft	radians/second
X	Instantaneous distance of the reed from its front position	inches

Table 1. Definition of the nomenclatures

2.1 The effect of take-up motion on cloth fell position

In the conventional power loom the way to obtain a desired pickspacing, is to make the rate of cloth take-up per pick equal to pickspacing by suitable adjustment of the take-up motion. During normal weaving, the amount of fabric woven is increased by a length P at every beat-up, and the function of the take-up motion is to take up this length so as to ensure that the cloth fell stays in the same position ([a]Greenwood & Cowhig, 1956).

In fact, during beat-up, the pickspacing is generally controlled by the action of the take-up mechanism (Zhonghuai & Mansour,1989).

The equilibrium between the amount of fabric woven and the amount taken up will be disturbed if the instantaneous pickspacing has a value which differs from the correct pickspacing. In this case there will be a net increase in the free length of fabric of $S-P(=a)$ when the take-up has operated and the c.f.d. will have changed by an amount a. Thus

$$\frac{dL}{dn}=-a \tag{1}$$

Equation (1) will be referred to as the take-up equation and describes the function of the take-up motion in the most general way. When the pickspacing is correct, a=0 and the c.f.d. remains constant, as is the case in normal weaving. Thus, the function of the take-up motion is to maintain the c.f.d. constant when pickspacing is correct. The direct effect of the take-up motion is therefore confined to the c.f.d. and it can affect pickspacing only to the extent that the latter affects pickspacing. The relation between the rate of take-up and the c.f.d. is given by the take-up equation. The c.f.d. in turn is related to pickspacing by a cloth fell equation. This shows that the c.f.d. is a vital link between the rate of take-up and pickspacing and also explains the importance of the relation between c.f.d. and pickspacing ([a] Greenwood & Cowhig, 1956).

2.2 The effect of the cloth fell position on pickspacing

The fact that pickspacing depends on the c.f.d. is due to a simple geometrical cause and to a more complex physical cause and is peculiar to the positive reed motion.

2.2.1 The geometrical aspect of the cloth fell equation

This aspect exists because the new pick is always carried to the front position of the reed. The previous pick is, by definition, at the position of the cloth fell and therefore pickspacing, which is the distance between the new and the previous pick, will depend on the position of the latter, i.e., on the c.f.d. The geometrical aspect would be the only one if no force were required to beat up the new pick and is, in fact, important where the applied beat-up force is small, i.e., in the weaving of low-pick fabrics or with other fabrics when the cloth fell is too near the weaver so that a bad thin place is created ([a] Greenwood & Cowhig, 1956).

2.2.2 The physical aspect of the cloth fell equation

This aspect exists because in beating up the new pick a resistance is encountered, which increases as the new pick approaches the cloth fell. A higher beat-up intensity will therefore cause a closer pickspacing and vice versa. The beat-up intensity, in turn, depends on the c.f.d. ([a] Greenwood & Cowhig, 1956).

2.3 The relation between the cloth fell position and the intensity of beat-up

Three distinct theories have been put forward to account for the variation of the intensity of beat-up with the cloth fell position. The first one (velocity theory) relates to the fact that the reed velocity decreases as the reed approaches its front position. This means that the kinetic energy of slay at the impact of the reed with the cloth fell, and hence the intensity of the beat-up, depends on the c.f.d.

The second theory (contact theory) suggests that the intensity of beat-up depends on the length of the period of contact between the reed and the cloth fell. This period in turn depends on the c.f.d., and hence the intensity of beat-up depends on the c.f.d.

The third theory (excess tension theory) relates to the fact that the force of beat-up has to be balanced by an excess of warp tension over fabric tension. This excess can be created only by a displacement of the cloth fell during beat-up and this displacement in turn will depend on the c.f.d. Therefore the intensity of beat-up depends on c.f.d.

Neither the velocity nor the contact theory has ever been formulated in precise mathematical form. This makes a direct comparison on the theories somewhat difficult. It must also be noted that the three theories interpret the intensity of beat-up in terms of three different physical quantities, namely, energy, time and force, respectively.

The excess tension theory has the advantage of being capable of extension to include the other two effects.

The relation between the c.f.d. and the force of beat-up on the basis of the excess tension can be established from equations (2) and (3) for non-bumping and bumping conditions, respectively ([a]Greenwood & Cowhig, 1956).

$$R_S = (S-L)\left(E_1/l_1 + E_2/l_2\right) \tag{2}$$

$$R_S = (S-L)E_1/l_1 + T_0 \tag{3}$$

2.4 The relation between beat-up force and pickspacing

The beat-up force at any instant is equal to weaving resistance. The latter arises from many factors; the friction between warp and weft, the rigidity and tension of warp and weft, etc. The total weaving resistance, R, can be regarded as the sum of a frictional resistance and an elastic resistance, the difference between the two being as follows. The energy used in overcoming the frictional resistance is dissipated in heat and, to some extent, in the form of static electricity and, when the new pick has been forced to a point near the cloth fell, the frictional force will tend to keep it there. The energy used in overcoming the elastic resistance is stored in the form of potential energy and, when the new pick has been forced to a point near the cloth fell, the elastic resistance will tend to eject it from the fabric. In spite of these differences the elastic and frictional resistances have many common features. Both resistances increase as the new pick approaches the fell and tend to infinity for a finite value of the distance, r, of the new pick from the fell. Both increase with an increase in warp tension.

The distinction between elastic and frictional resistance is obviously of great importance in dealing fully with the phenomena which take place during beat-up, but this distinction becomes operative only after the beat-up is complete. During beat-up, the features common to both are dominant and, therefore, it was considered justifiable to simplify the treatment by dealing only with the total weaving resistance, R. The different and complex modes of action of the various components are also ignored, and it is assumed that the weaving resistance acts as a simple repulsive force between the cloth fell and the new pick.

For a mathematical treatment, it is necessary to express the weaving resistance, R, as a function of the distance, r, of the new pick from the cloth fell. Experimental technique on the running loom to determine this relationship is difficult and it would be unwise to expect too much from the results. On the other hand, it was also felt that to derive the relationship from theoretical considerations would lead to very complex mathematical expressions and be of little practical value. Therefore, to assume a simple relationship between R and r and to see how experimental results from more easily measured quantities would confirm this assumption. It is assumed that:

$$R = \frac{k}{r - D} \tag{4}$$

This equation takes account of the fact that R varies inversely with r and tends to infinity for a finite value of r. D and k are considered to be constants for a particular fabric. D can be considered the theoretical minimum pickspacing (the practical minimum will always be higher than D) and should be of the order of the yarn diameter. k is termed the coefficient of weaving resistance and is a measure of the difficulty of weft insertion into a given warp. Although as a first approximation k is considered constant, there is a probability that it will vary with warp tension and, possibly, also with loom speed, some loom settings, or atmospheric conditions. The magnitude of all these effects will have to be investigated experimentally.
Equation (4) will be called to inverse distance equation. It applies at any instant during beat-up and when the reed reaches its front position it becomes:

$$R_s = \frac{k}{S - D} \tag{5}$$

This gives the required relationship between beat-up force and pickspacing ([a] Greenwood & Cowhig, 1956).

2.5 The relation between cloth fell position and pickspacing

By substituting for R_s in equations (2) and (3), the followings relations are obtained for non-bumping and bumping conditions, respectively:

$$L = -\frac{K}{S - D} + S \tag{6}$$

$$L = -\frac{k/(S - D) - T_0}{E_1 / l_1} + S \tag{7}$$

where

$$K = \frac{k}{\left(\frac{E_1}{l_1} + \frac{E_2}{l_2}\right)} \tag{8}$$

When conditions are stable, equations (6) and (7) take on the form

$$L_p = -\frac{K}{P - D} + P \tag{9}$$

$$L_p = -\frac{k/(P - D) - T_0}{E_1 / l_1} + P \tag{10}$$

and can be used to determine the correct c.f.d. for any rate of take-up ([a] Greenwood & Cowhig, 1956).

2.6 The effect of warp tension on cloth fell position and pickspacing

2.6.1 The warp tension cycle

Fig. 1 shows a typical warp tension curve. During the whole weaving cycle, with the exception of beat-up, the warp tension is the same as the fabric tension. During beat-up, the warp rises to a peak and the fabric tension falls to a minimum. The tension value at which beat-up begins is, by definition, the basic warp tension ([a]Greenwood & Cowhig, 1956).
The fluctuation of the tension curve is the result of shedding, beat-up, take-up, and let-off motions. The effects of these factors are not simultaneous and vary in duration and magnitude. Among these, the shedding and the beat-up processes bring into being considerable tension loads, and the effect of shedding is of the longest duration compared with the others (Gu, 1984).

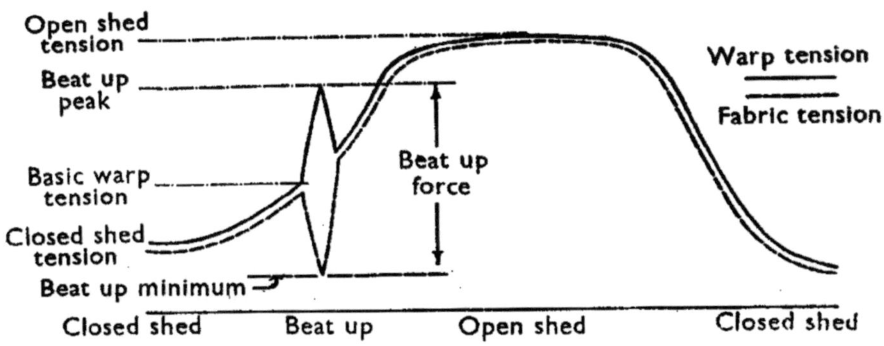

Fig. 1. Typical warp tension cycle

Outside the period of beat-up, the shape of the tension trace is determined by the shedding motion and reaches a peak at open shed and a minimum at closed shed. The whole tension cycle can thus be summed up in terms of five tension values, namely: open-shed tension, closed-shed tension, basic warp tension, beat-up peak and beat-up minimum (fabric).
The difference between the beat-up peak and the beat-up minimum represents the applied beat-up force (according to the excess tension theory). The applied beat-up force in turn determines pickspacing by the inverse distance equation ([a]Greenwood & Cowhig, 1956).

2.6.2 The effect of the warp tension cycle

The warp tension plays the contradictory effect during weaving. On the one hand, warp tension provides the only means of holding or supporting the cloth fell in position during beat-up. This is referred to as the supporting function. On the other hand, higher warp tension tends to increase the weaving resistance and can be said to oppose the entry of the new weft into the fabric. This is referred to as the opposing function. To introduce the opposing function it would be necessary to express the coefficient of weaving resistance as a function of warp tension. There is a great deal of evidence that the supporting function is by far the more important ([a]Greenwood & Cowhig, 1956).

2.6.3 The setting of the let-off motion

The relation between the let-off motion and warp tension is somewhat similar to the relation between the take-up motion and the cloth fell position. When a loom is gaited up and weaving started, a certain amount of fabric has to be woven before the warp tension attains the value for which the let-off motion is set. This value is maintained as long as the loom runs satisfactorily, but when the loom is stopped, the instantaneous warp tension may deviate greatly from the normal value because of relaxation or unsatisfactory letting back ([a] Greenwood & Cowhig, 1956).

2.6.4 The effect of the simultaneous variation in the cloth fell position and the warp tension on pickspacing

As mentioned in part 2.5, equations (6) and (7) show the instantaneous relation between c.f.d. and pickspacing under non-bumping and bumping conditions, respectively. Under normal weaving conditions, the pickspacing is always equal to the rate of take-up P, and to find the correct c.f.d. L_P for a given pickspacing, one need only to substitute P for S in equations (6) and (7) (equations (9) and (10)).

Equations (6) and (7) can be used to calculate the effect of a displacement of cloth fell from its correct position. Such a displacement is very often combined with a change in the basic warp tension from its normal value T_0 to some other value T_0'.

If the changes are small, the effect on pickspacing can be calculated by differentiating equations (6) or (7). This gives

$$dP = \frac{1}{1+\frac{K}{(P-D)^2}}dL \tag{11}$$

$$dP = \frac{dL\frac{E_1}{l_1}-dT_0}{\frac{k}{(P-D)^2}+\frac{E_1}{l_1}} \tag{12}$$

for non-bumping and bumping conditions, respectively.

The effect of large changes in the c.f.d. can be calculated as follows, for non-bumping and bumping conditions, respectively:

The correct cloth fell position L_p is given by equations (9) and (10).

$$L-L_p = -\frac{K}{(S-D)}+S+\frac{K}{(P-D)}-P \tag{13}$$

$$L-L_p = -\frac{\frac{k}{(S-D)}-T_0'}{\frac{E_1}{l_1}}+S+\frac{\frac{k}{(P-D)}-T_0}{\frac{E_1}{l_1}}-P \tag{14}$$

These equations determine the deviation from correct pickspacing in terms of the deviation from the correct c.f.d.

From these calculations, it will be seen that under non-bumping conditions, pickspacing depends on the c.f.d. alone and is independent of warp tension, whereas under bumping conditions both the c.f.d. and the warp tension are important ([b] Greenwood & Cowhig, 1956).

2.6.5 The cloth fell displacement and the warp tension variation during loom stoppage

It is a well-known fact that, during a loom stoppage, the cloth fell tends to creep away from its correct position. To examine the effect of relaxation in the warp and the woven fabric on the position of the cloth fell and hence pickspacing, it was assumed that the cloth fell is fixed in its correct position during the stoppage and that, because of the difference in the properties of warp and fabric, the tension in the two materials falls from its original value T_0 to a lower value T_1 in the warp and T_2 in the fabric, where in general $T_1 \neq T_2$. Immediately before starting the loom, the cloth fell is released to find its equilibrium position. Equilibrium exists when the warp and fabric tensions are again equal at a value T'_0 and it is necessary to determine the displacement of the cloth fell which would bring T_1 and T_2 to the same value T'_0. The displacement of the cloth fell is calculated by equation (15) and the common warp and fabric tension resulting from this displacement is given by equation (16).

$$dL = \frac{T_2 - T_1}{\frac{E_1}{l_1} + \frac{E_2}{l_2}} \tag{15}$$

$$T'_0 = \frac{T_1 \frac{E_2}{l_2} + T_2 \frac{E_1}{l_1}}{\frac{E_1}{l_1} + \frac{E_2}{l_2}} \tag{16}$$

Equations (15) and (16) apply only when a brake-type of let-off motion is used. In this case it will be seen that, if T_2 is smaller than T_1, i.e., if the fabric relaxes more than the warp, the cloth fell will move away from the weaver. This is the more usual case. Thus the direction of the movement is determined by the relative values of T_1 and T_2 alone.
The effect of the displacement of the cloth fell on pickspacing is calculated by substituting for dL from equation (15) in equation (11) and for K from equation (8). This leads to:

$$dP = \frac{T_2 - T_1}{\frac{k}{(P-D)^2} + \left(\frac{E_1}{l_1} + \frac{E_2}{l_2}\right)} \tag{17}$$

In comparing the brake-type with the dead-weight type of let-off, it is often held in favour of the latter that it keeps warp tension constant during a stoppage.
Using the same simplified assumption as before, it will be assumed that the cloth fell is fixed in its correct position during the stoppage. Owing to the action of the dead weight, the warp tension in the present case remains constant at T_0, whereas the fabric tension falls to a value T_2 as before. Immediately before starting the loom, the cloth fell is released and the dead weight brings the fabric tension back to its original value T_0 by stretching the fabric by an amount dL given by ([b] Greenwood & Cowhig, 1956):

$$dL = \frac{T_0 - T_2}{\frac{E_2}{l_2}} \quad (18)$$

2.6.6 The effect of the let-off motion type

It will be seen on comparing equation (18) with equation (15) that, when the fabric relaxes more than the warp ($T_2 < T_1$), the displacement of the cloth fell will always be greater with a dead-weight than with a break-type let-off motion.

It may be said that the constancy of the warp tension in the dead-weight let-off is obtained at the expense of a greater movement of the cloth fell. Whether this is an advantage or not depends on the type of fabric. With a lighter type of fabric, which will normally be woven under non-bumping conditions, the cloth fell position is probably more important and therefore the brake-type let-off preferable. With a heavier type of fabric, warp tension becomes more critical and the dead-weight type may give better results ([b] Greenwood & Cowhig, 1956).

2.7 The effect of free length of warp and fabric on cloth fell position and pickspacing

Equations (15) and (17) indicate that a long free length of warp and fabric will tend to increase cloth fell displacement and changes in pickspacing due to relaxation, respectively, i.e., it will tend to make setting-on places worse but equation (18) points out that just free length of fabric will affect on cloth fell displacement ([b] Greenwood & Cowhig, 1956).

In addition, during weaving increasing in length of warp and fabric will affect on K and according to equation (11), will decrease the changes in pickspacing.

2.8 The effect of loom speed on cloth fell position

Three theories explained in section 2.3 differ in their prediction of the effect of loom speed on the cloth fell position.

According to the contact theory, a reduction in loom speed would cause a movement of the cloth fell towards the weaver; according to the velocity theory, it would cause a movement of the cloth fell away from the weaver; according to the excess tension theory, it would not cause any change in the cloth fell position.

To investigate these antonym effects, some experiments were carried out and their results are shown in Table 2. The weft used was 140/26 denier bright acetate and 56 picks per inch were inserted so that P= 0.0154.

Test Number	Loom Speed (Picks per Minute)	Cloth Fell Distance L (inches)
I	128	-0.151
	158	-0.139
		Difference 0.012
II	128	-0.162
	158	-0.141
		Difference 0.021

Table 2. The results of the investigation of loom speed effect

Table 2 shows that the cloth fell moves away from the weaver when the loom is running more slowly, as would be expected from the velocity theory; whereas the following argument shows that a much larger movement would be expected from the velocity theory. The results for the cloth fell position in the two tests agree well and for condition corresponding to a speed of 158 picks per minute, the position is taken as -0.140 inch.
At the time of reed impact with the fell, the reed distance from the fell and the front position of the reed is P(=-0.154) and X_1(= -0.140- 0.154= -0.155) ([c]Greenwood & Cowhig, 1956).
it is assumed that the motion of the reed near the cloth fell is single harmonic with an amplitude, m, so that:

$$X = m(\sin \nu t - 1) \tag{19}$$

So, the speed of the reed is given by:

$$\frac{dX}{dt} = m\nu \cos \nu t \tag{20}$$

$$\cos \nu t = \sqrt{\frac{2X}{m}} \tag{21}$$

Substituting for cos vt from equation (21) in equation (20), the speed of the reed at impact moment is given by ([a, c]Greenwood & Cowhig, 1956):

$$\frac{dX}{dt} = \nu \sqrt{2Xm} \tag{22}$$

From the velocity theory, one would expect that the cloth fell position would change with any alteration in loom speed, so that the velocity of the reed at the time of the impact with the fell should remain unaltered.
Thus

$$\nu_1 \left(2X_1 m\right)^{\frac{1}{2}} = \nu_2 \left(2X_2 m\right)^{\frac{1}{2}} \tag{23}$$

so that

$$X_2 = X_1 \left(\frac{\nu_1}{\nu_2}\right)^2 \tag{24}$$

Substituting numerical values, X_2= -0.155 $(158/128)^2$= -0.237 inch. Subtracting from this the value of P (0.0154 inch), the new cloth fell position should, on this calculation, be -0.222 inch. Thus, from the velocity theory one would expect a movement of the cloth fell of -0.082 inch, which is much greater than that actually observed in the above experiments (0.012 in test I and 0.021 in test II).
It is obvious that, for this particular loom and fabric, the observed effect of changes in loom speed on the cloth fell position cannot be explained by any of the three theories alone ([c] Greenwood & Cowhig, 1956).

3. Methods of measuring the cloth fell position

In general, proposed and presented methods for measuring the cloth fell movement are divided into two main categories:

1. Methods that measure the cloth fell movement during weaving.
2. Methods that measure the cloth fell movement during loom stoppage and try to prevent start-up marks. Such movements are due to difference in visco-elastic properties of warp threads and the cloth during loom stoppage.

These mentioned methods and their advantageous and disadvantageous are discussed later in this chapter in detail.

3.1 The methods of measuring the cloth fell position during weaving

3.1.1 Microscopic method

Early measurements of the cloth fell position were carried out by means of a microscope mounted on a rigid frame on the loom ([c] Greenwood & Cowhig, 1956). This method is difficult and time consuming.

3.1.2 Optical method

A source of light was placed under the loom and focussed on the cloth fell by a 5-inch focal length lens, the illuminated area being approximately 0.5-inch in diameter. A 4-inch focal length projector lens was mounted about 5 inches above the cloth in line with the source and the first lens. The image of the fell formed by this lens was thrown horizontally on the wall by a mirror inclined at 45° mounted 20 inches above the cloth surface. A white screen fixed to the wall enabled the image to be clearly seen and its position measured. It is essential to avoid movement and vibration in the lens and mirror, and so both were mounted on a frame of 2-inch steel tubing bolted firmly to the floor. To specify the position of the cloth fell, the loom was stopped with fully open shed and cloth fell position was read off on the screen, in terms of its distance from an arbitrary zero line. A diagram of the arrangement of this method is shown in Fig. 2 ([c] Greenwood & Cowhig, 1956).

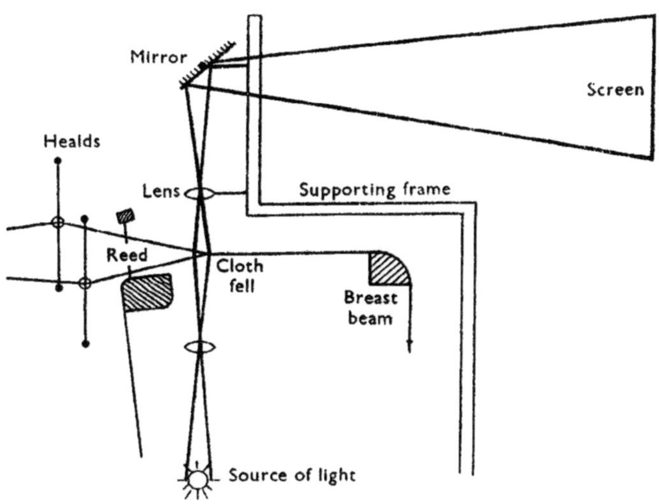

Fig. 2. Optical method for measuring the cloth fell position

The disadvantages of this method are as follows; there is no way to measure the cloth fell position dynamically during the loom running, so the loom has to be stopped to take the measurements ([a] Azzam & Büsgen, 2006). Moreover, it is impossible to measure the cloth fell position immediately before beat-up, since at this stage the race board of slay obstructs the path of the light from the source of the fell ([c] Greenwood & Cowhig, 1956; Azzam & Büsgen, 2006). In addition, the system is excessively complex, especially regarding the arrangement made to fix a reference point and adjustable pointer at the front position of the reed ([a] Azzam & Büsgen, 2006).

3.1.3 Improved optical method

From Fig. 2, it can be seen that it is impossible to measure the cloth fell position immediately before beat-up. It was, therefore, necessary to apply a correction. The loom was stopped with open shed, the slay was left stationary and the healds were brought to beat-up position by loosening and rotating the shedding tappets. The corresponding change in the cloth fell position was measured and used as shedding correction ([c] Greenwood & Cowhig, 1956).

3.1.4 Electronic needle wheel method

This method is a continuous measuring system for determining the movement of the cloth fell movement. The movement is measured in the immediate vicinity of the fell by a small wheel, placed on the fabric as close as possible to the fell, which follows the movement undergone by the cloth, as shown in Fig. 3, and is evaluated electronically (Kohlhass, 1981).
The disadvantages of this method are the difficulty in fixation on the loom, the difference between the measuring point and the cloth fell is greater than 10 mm, and a great deal of electronics were used ([a] Azzam & Büsgen, 2006).

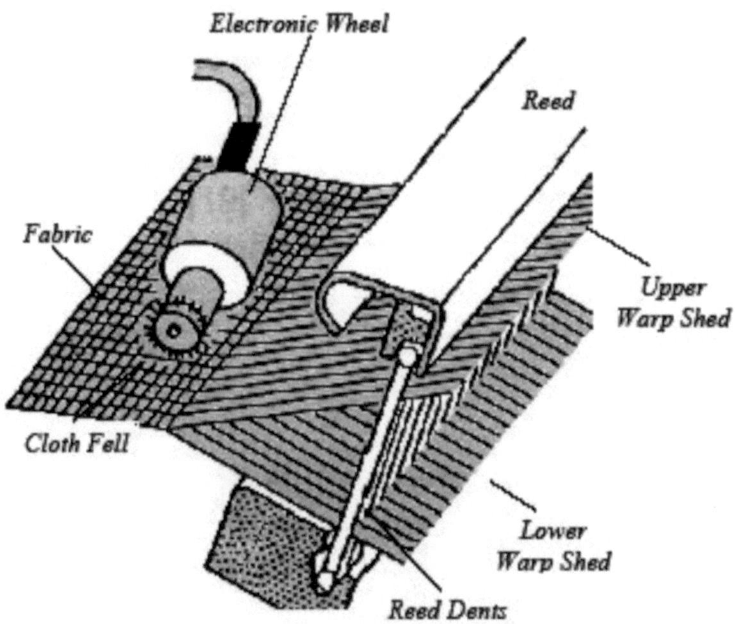

Fig. 3. Electronic needle wheel method for measuring the cloth fell movement

3.1.5 Using the high speed video system

The Kodak Ektapro High Speed Video System consists of a video camera, processor, controller, monitor, macro-objective, endoscope, personal computer, motion analysis package, and video recorder. The system provides a very sharp, clear depiction of the movement of the fell of the cloth during weaving (Weinsdörfer & Salama, 1992).

Also, a high speed camera at 1000 frames/sec has been used to record the cloth fell displacement during the beat-up. Fig. 4 shows the position of the cloth fell when the beat-up takes place. In this diagram the beat-up starts from 42° and ends around 58° (Mirjalili, 2003).

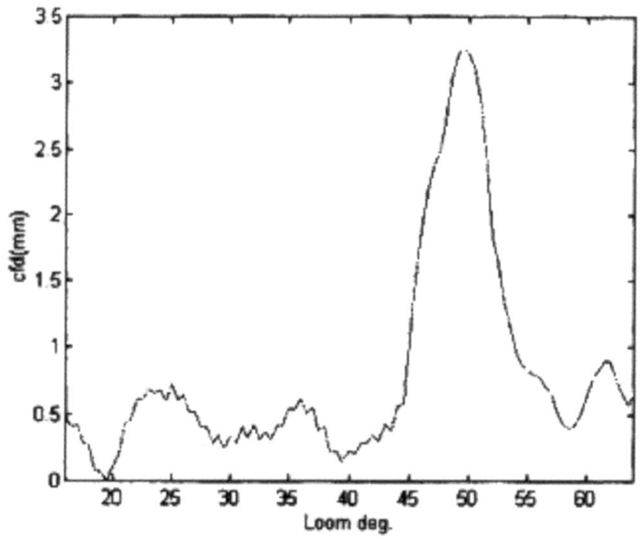

Fig. 4. Cloth fell displacement, c.f.d vs. loom degree

Uneconomical preparation and installation of this system on the loom are the great disadvantages of this method.

3.1.6 Using the lengths of cloth and warp thread

This method is based on the free lengths of warp and fabric. The desired lengths of cloth and warp thread are determined by averaging the actual values over a given number of weft insertions. The actual lengths of the cloth and warp are then continuously measured and compared with the desired values during weaving to compute warp and cloth offset values. A correction factor is determined based on the offset values and the elastic modulus of the warp thread and cloth. The cloth fell position is then corrected by adjusting the actual length of the cloth or the warp thread based on the correction factor ([a] Godert, 1996).

This method is also usable for measuring the cloth fell after loom stoppage and before loom restart and avoiding the start-up marks in fabrics.

3.1.7 Needle mechanical device

In this system, a simple and flexible needle mechanical device, fixed onto the loom frame and connected with the weave master to measure the cloth fell movement dynamically.

The measuring position for the needle was kept at 2-millimeter distance from the fabric support plate, as indicated in figure 5. This method is based on forced equilibrium. The forces affecting the cloth fell will push the needle in the direction of force equilibrium; that means, if warp tension force is greater than the fabric tension force, the needle will be pushed towards the warp direction until beat-up takes place. In addition, if the fabric tension force is greater than the warp tension , then the needle will be pushed by the fell to move towards the fabric direction until shedding takes place. By using a special force-displacement calibration the cloth fell positions during shedding and beating could be calculated. In turn, the cloth fell movement could also be calculated. In fact, the difference between the cloth fell position due to beat-up and the cloth fell position due to open shed represents the cloth fell movement ([b] Azzam & Büsgen, 2006).

However, this method is a satisfactory way to measure the cloth fell movement, it can not calculate the cloth fell distance.

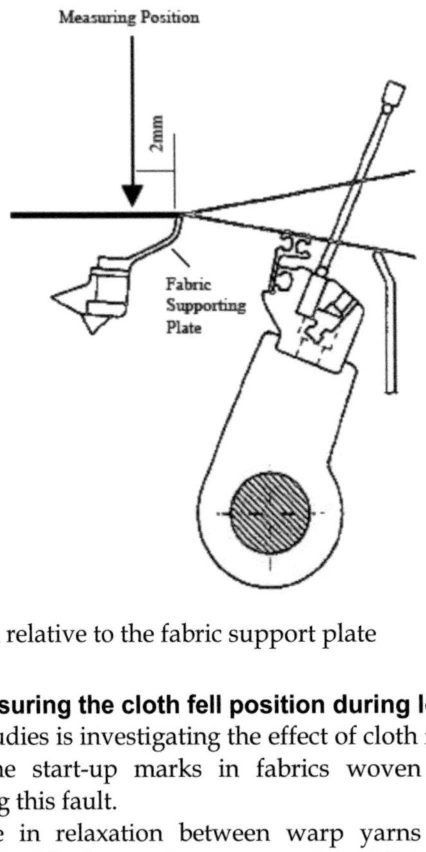

Fig. 5. Measuring position relative to the fabric support plate

3.2 The methods of measuring the cloth fell position during loom stoppage

The aim of this kind of studies is investigating the effect of cloth fell movement during loom stoppage on creating the start-up marks in fabrics woven after loom restatart and controlling and preventing this fault.

Generally, the difference in relaxation between warp yarns and fabrics will cause a displacement of the cloth fell during loom stops. In addition to relaxation, cloth fell displacement is determined by the modulus of elasticity of warp yarn and fabric and by their relative lengths within the fabric warp yarn combination that exists on the loom (Vangheluwe & Kiekens, 1995).

3.2.1 Using the sensor needle

In this method the position of the cloth fell is sensed with a sensor needle that can be moved into and out of the cloth at a location proximate the fell during interruption of the weaving operation. Upon insertion of the sensor needle into the cloth, it moves with the cloth and therewith measures cloth displacement while the loom is at rest. The sensor needle is attached to an actuation device capable of reciprocating the needle into and out of the cloth. The sensor needle is inductively coupled to a distance sensor which measures the needle displacement and therewith the displacement of the fell. In this manner, fell displacement while the loom is at rest can be measured and, before weaving is restarted, the position of the cloth, and therewith of the fell can be corrected ([b] Godert, 1996).

3.2.2 Using the photography system

This test method was presented for measuring the cloth fell displacement during relaxation of fabric yarn combination on a tensile tester. The equipment consists of a CCD-camera, a frame grabber, and appropriate software. This study has been carried out on the tensile tester rather than on the loom, where oscillations can hinder the efficient use of a camera.
Fig. 6 shows the arrangement of the equipment. The camera takes one image per second during relaxation measurements. This image is sent to the frame grabber in the PC controlling the tensile tester. In this way, the images are known as a grid (the grid points being called pixels), in which each pixel carries a light intensity value of the total image. In this digitized image, rows of pixels are in the direction of the weft yarns, and columns of pixels are in the direction of the warp yarns. At first, the sum of the pixel values is calculated over the horizontal lines (rows) of the digitized image. Next, values are filtered to eliminate noise. The results in a graph show the position of the weft yarns in the fabric clearly as peaks in the sum of intensity along the warp yarn direction (Fig. 7).

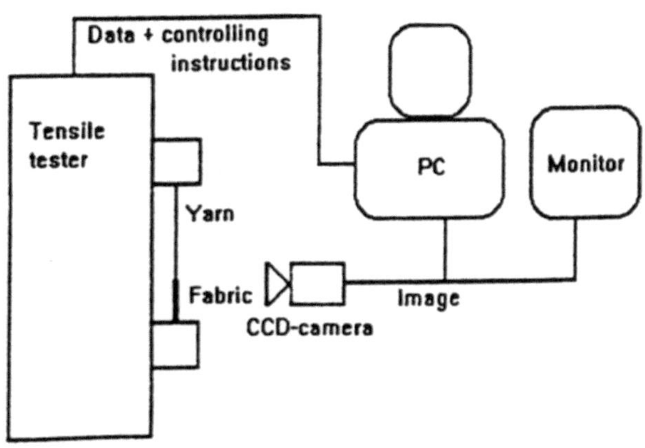

Fig. 6. Test equipment for measuring the cloth fell displacement

The position of the first weft yarn in the image determines the cloth fell. This position is compared with the cloth fell position of the first image that takes at the start of the relaxation. In fact, after calibration, cloth fell displacement is obtained in millimeters as a function of time during which relaxation is measured. This calibration is done by taking an image of a sheet of paper with lines on known distances, which is clamped in the tensile tester. A computer program converts the known distance between the lines to the number of pixels as measured between the lines. Therefore, the cloth fell displacement during relaxation will be measured (Vangheluwe & Kiekens, 1995).

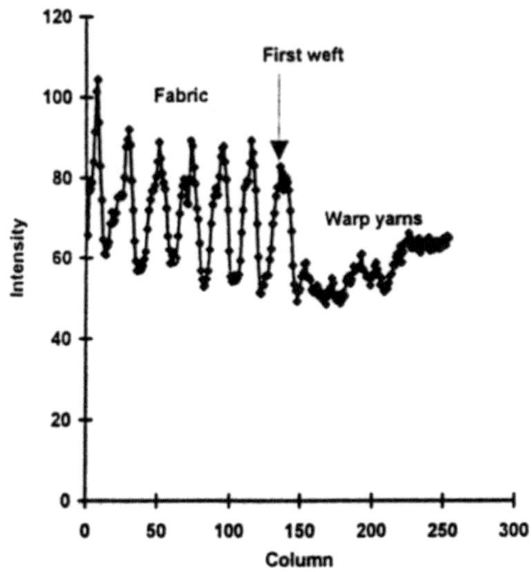

Fig. 7. Results obtained from image filtration

3.2.3 Using the laser analogue displacement sensor

A laser analouge displacement sensor was used, due to its comparatively small physical dimensions. This device, uses a fine laser beam to monitor the position of a target by detecting a spot of light falling on it. With a suitable amplifier, the device enables movements of the target position to be determined to an accuracy of 1 μm. The working distance, which is the separation between the target and the front edge of the detector is, 20±2 mm. Within this range, the output signal of the sensor varies linearly with distance. Accordingly, a small lightweight target plate was made that could be placed very close to the cloth fell in such a way that it stayed stable during the loom stoppage period. As such, the movement of the target would be very nearly the same as that of the fell. The target had to be placed with precision on the fabric as soon as the loom stopped, and be removed immediately before the loom was restarted. The output of the sensor is a current signal, which can be easily converted into a voltage and amplified with a stable DC amplifier to enable detection of any movement of the target. A safeguard was provided that prevented the loom from being restarted without the target first being removed. The position of fixing this device on the loom and the cloth fell drift signal produced by the laser sensor are shown in Fig. 8 and Fig. 9, respectively. The output signal provided by the sensor shows the drift of the cloth fell that takes place during the loom stoppage. The rise of the trace indicates a corresponding movement of the fell towards the back of the loom.

The creep of the fell is dependent on the nature and duration of the loom stoppage and other factors. Hence the amount of cloth fell correction required before start-up varies, depending on the loom stoppage duration. Since the sensor enables this amount to be determined, it was possible to move the cloth fell back to where it would have been before the loom stopped. This was done by controlling the stepper motor on the take-up roller, as the number of steps of the motor rotation required could be determined from the required fell displacement. Following this, the let-off was adjusted to restore warp tension to its normal value (Islam & Bandara, 1998).

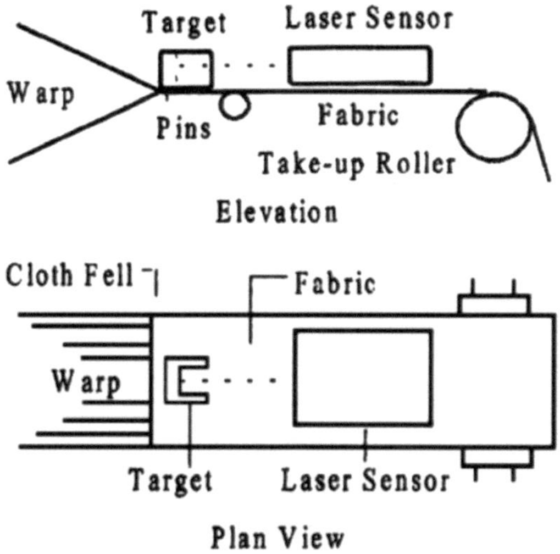

Fig. 8. Placement of the laser sensor on the loom

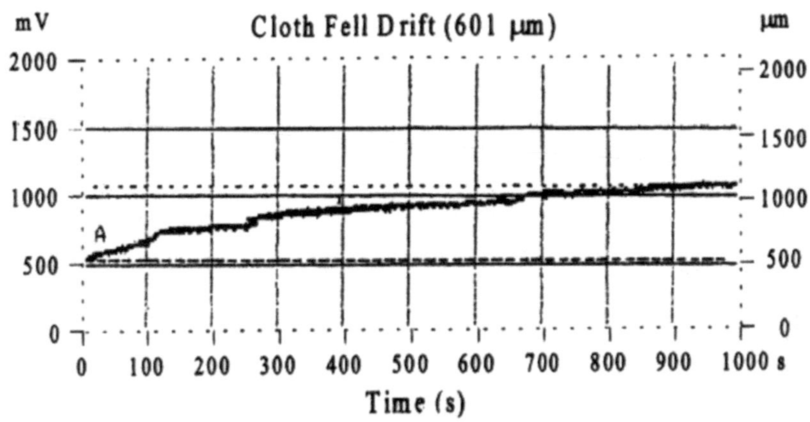

Fig. 9. Fell displacement on a narrow loom with viscose linning fabric

4. Conclusion

In this chapter the importance of the cloth fell position and the effective elements on cloth fell movement were discribed. After that some measurement methods for determining the cloth fell movement durin weaving process and during loom stoppage were presented. However, today no economic method is available to measure the cloth fell position with the required accuracy on the loom.

With regarding to developing and progressing in online electronic systems based on advanced computer programs, It is our hope that a precise method with high speed processing will be designed not only for measuring and monitoring the cloth fell displacement dynamically but also for online controlling and correcting its faults during weaving process. This chapter can be a comprehensive and complete background to help researchers to get innovative ideas.

5. References

[a] Azzam, H. A. & Büsgen, A. (2006), Dynamic Cloth Fell Movement, Part I: Critical Review, *Autex Research Journal*, Vol. 6 (No. 1): 14-22.

[b] Azzam, H. A. & Büsgen, A. (2006), Dynamic Cloth Fell Movement, Part II: New Measuring Device, *Autex Research Journal*, Vol. 6 (No. 1): 23-29.

[a] Godert, D. J. (1996), System and Method for Regulating the Cloth Fell Position in a Loom, Us 005 538 048.

[b] Godert, D. J. (1995, 1996), Method and Weaving machine for Monitoring the Fell Position Following Weaving Operation Interuption, EP 0 682 132, 1995 and US 005 520 224, 1996.

[a] Greenwood, K. & Cowhig, W. T. (1956), The Position of the Cloth Fell in Power Looms, Part I: Stable Weaving Conditions, *The Journal of the Textile Institute*, Vol. 47 (No. 5): 241-254.

[b] Greenwood, K. & Cowhig, W. T. (1956), The Position of the Cloth Fell in Power Looms, Part II: Disturbed Weaving Conditions, *The Journal of the Textile Institute*, Vol. 47 (No. 5): 255-65.

[c] Greenwood, K. & Cowhig, W. T. (1956), The Position of the Cloth Fell in Power Looms, Part III: Experimental, *The Journal of the Textile Institute*, Vol. 47 (No. 5): 274-286.

Gu, H., (1984) Reduction of Warp Tension Fluctuation and Beat-up Strip Width in Weaving, *Textile Research Journal*, Vol. 54 (No. 3):143-148.

Islam, A.T.M.S. & Bandara, M.P.U. (1999), Cloth Fell Control to Prevent Start-up Marks in Weaving, *The Journal of the Textile Institute*, Vol. 90, Part 1 (No. 3): 336-345.

Kohlhass,O. (1981), Gerät zur Ermittlung der Warenrandbewegung, *Melliand Textilberichte*, Vol. 62 (No. 6): 457-460.

Mirjalili, S. A. (2003), Computer Simulation of Warp Tension on a Weaving Machine, *Journal of Textile Engineering*, Vol. 49 (No. 1): 7-13.

Vangheluwe, L. & Kiekens, P. (1995), Test Method for Cloth Fell Displacement Caused by Relaxation, *Textile Research Journal*, Vol. 65 (No. 9): 540-544.

Weinsdörfer,H. & Salama,M. (1992), Messung der Warenrandbewegung beim Weben mit Hilfe eines High-Speed Video System, *Textile Praxis International*, Vol. 47 (No. 9): 812-814.

Zhonghuai, Z. & Mansour, H. M. (1989), Theoretical Investigations of Beat-up, *Textile Research Journal*, Vol. 59 (No. 7): 395-404.

6

Artificial Neural Networks and Their Applications in the Engineering of Fabrics

Savvas Vassiliadis[1], Maria Rangoussi[1],
Ahmet Cay[2] and Christopher Provatidis[3]
[1]*Department of Electronics, Technological Education Institute of Piraeus, Athens,*
[2]*Department of Textile Engineering, Ege University, Izmir,*
[3]*School of Mechanical Engineering, National Technical University of Athens,*
[1,3]*Greece*
[2]*Turkey*

1. Introduction

Historically the main use of the textile fabrics has been limited mainly to clothing and domestic applications. The technical uses were of minor importance. However in the last decades the use of the textile structures has started to spread over other sectors like construction, medicine, vehicles, aeronautics, etc. The increased interest in technical applications have improved the fabric design and engineeringprocedures, given that the final products must be characterized by certain mechanical, electrical etc. properties. The performance of the fabrics should be predictable right from the design phase. The design of a fabric is focusing on the materials selection as well as on the definition of its structural parameters, so that the requirements of the end use be fulfilled.

These changes in the application field of the textile structures caused a move from the esthetic design to the total technical design, where the fabric appearance and the particular properties affecting its final performance are taken in account. However, the textile structures are highly complex. A textile fabric consists of yarns; yarns in turn consist of fibres. Thus the mechanical performance of the fabrics is characterized by the structural geometrical complexity and non-linearity, as well as from the non-linearities of the materials themselves. This double non-linear bahaviour of the textile fabrics increases the difficulty in the fabric design and engineering processes. The complex structure and the difficulties introduced by the raw materials do not allow the use of precise analytical models for the technical design of the fabrics.

Fabric engineering activities are increasingly based on computational models that aim at the prediction of the properties and the performance of the fabrics under consideration. Various computational tools have been used in order to represent the fabrics in a computational environment and to predict their final properties. Among others, Finite Element Method (FEM) analysis has supported mainly the prediction of the behaviour of the complex textile structures under mechanical loads. In the case of classification problems Artificial Neural Networks (ANNs) have proved a very efficient tool for the fast and precise solution. ANNs have found an increasing application in the textile field in the classification as well as in the prediction of properties and optimization problems (Chattopadhyay & Guha 2004). In parallel or complementarily to the ANNs, fuzzy logic and genetic algorithms techniques have been used in the textile field (Guruprasad & Behera, 2010). The applications of the ANNs in the textile classification and prediction problems cover the fields of fibres, yarns and fabrics as well as color, wet processing and clothing.

2. ANN applications in the textile engineering field

It is mainly since 1990 that the applications of ANNs in the textile engineering field have become more and more popular. Gradually it was proven that they can address successfully complex engineering problems. Many researchers have turned to ANNs when they were in front of a multiparameter and non-linear problem, without an obvious or straightforward analytical solution. In the following paragraphs, a thematic overview of the uses of the ANNs in the various textile fields will be presented.

2.1 Fibres

The implementation of the ANNs presupposes an initial phase of features extraction, which will be used later to feed the ANN. It includes the processing of the given data or measurements, typically in the form of a signal, an image etc. One of the most typical problems of the animal fibres classification has been faced using Artificial Neural Networks (She et al., 2002). In the case of cotton, ANNs have been used for the grading of the color of the raw fibres (Cheng et al., 1999; Xu et al., 2000; Kang et al., 2002). An attempt for the classification of the cotton lint has been also based on the use of ANNs (Mwasiagi et al., 2009). A method for the selection of cotton bales based on certain criteria was established (Majumdar et al., 2004). In the case of synthetic fibres, the ANNs have supported the identification of the production control parameters (Allan et al., 2001) and the prediction of the properties of the melt spun fibers (Kuo, 2004). ANNs have been used in conjunction with NIR spectroscopy for the identification of the textile fibres (Jasper & Kovacs 1994). A system for the optimization of the yarn production based on the blend characteristics and the process parameters has been developed based also on the use of ANNs (Sette & van Langenhove, 2002).

2.2 Yarns

The spinning of the staple fibres for the production of the yarns is a multistage procedure including many parameters, which influence the characteristics of the end product, the spun yarn. The examination of the image of the web produced by a carding machine and the detection of faults has been made possible in an automatic sense using ANN's (Kuo et al., 1999; Shiau et al., 2000). The autolevelling and thus the linear density control in the draw frame has been achieved using ANNs (Huang & Chang, 2001; Farooq & Cherif, 2008). In the main spinning phase, the optimization of the top roller diameter as well as the study of the ring ballon has been examined via the use of ANNs (Ghane et al., 2008; Tran et al., 2010).
The overall process performance in the case of the worsted spinning was estimated (Beltran et al., 2004), while the selection of the suitable parameters was the target of other researchers (Wu et al., 1994; Yin & Yu, 2007). The spinning process and its role on the prediction of the cotton-polyester yarn properties were examined using ANNs (Lu et al., 2007; Jackowska-Strumiłło et al., 2008). The effect of the fibres properties on the yarn characteristics is a topic

of great interest for many researchers, with different points of view or dealing with specific fibres or spinning method cases (Dayik, 2009; Jayadeva et al., 2003; Majunmdar et al., 2006). A method based on a combination of Genetic Algorithms and Neural Networks has been used for the prediction and optimization of the yarns properties (Subramanian et al., 2007). In a similar sense, a combination of an adaptive neuro-fuzzy system has been developed for the prediction of fiber to yarn relation (Admuthe & Apte, 2010).

The prediction of the tensile properties of yarns is of main interests of the international research community. Many publications appeared dealing with the prediction of the tensile properties of the yarns under general (Ramesh et al., 1995; Guha et al., 2001; Majumdar & Majumdar, 2004; Nurwaha & Wang, 2008; Üreyen & Gürkan, 2008; Mwasiagi et al., 2008; Nurwaha & Wang, 2010) or under specific conditions, such as is the case of core spun yarns (Gharehaghaji et al., 2007), air-jet spun yarns (Zeng et al., 2004) or for the estimation of the torque of worsted yarns (Tran & Phillips, 2007). The ANN prediction method is compared with the Support Vector Machine (SVM) approach and conditions under which each method is better suited are investigated (Ghosh & Chatterjee, 2010). The warp breakage rate during the weaving is a complex function of the yarn properties and thus an application field of an ANN model (Yao, 2005). The prediction of the yarn evenness and hairiness is of great practical interest. ANNs have been used for the prediction of hairiness of worsted wool yarns (Khan et al., 2009) and of cotton yarns (Babay et al., 2005; Majumdar, 2010). In a similar way, ANNs have been used for the prediction of the evenness of ring spun worsted yarns (Wang & Zeng, 2008) and cotton yarns (Majumdar et al., 2008; Üreyen & Gürkan, 2008; Majumdar, 2010) or the evenness of blended rotor yarns (Demiryurek & Koc, 2009).

As it is known, when two yarn ends must be joined, instead of knotting they are subjected in the splicing process. Splicing positions are of special interest because they could affect heavily the mechanical performance of the yarn in total. Evaluation and comparison of the properties of the spliced yarns have been made based also on ANNs (Cheng & Lam, 2003). Later studies have used ANNs to predict the properties of the spliced yarns (Lewandowski & Drobina, 2008). Latest studies have contributed to the prediction of the spliced yarns tensile properties as well as to the prediction of the retained yarn diameter, thus covering the mechanical and the visible results of the presence of the splicing points in the yarn (Gürkan-Ünal et al., 2010). ANNs have also been used for the appearance analysis of false twist textured yarn packages (Chiu et al., 2001), for the prediction of yarn shrinkage (Lin, 2007) or for the modelling of the relaxation behaviour of yarns (Vangheluwe et al., 1996).

2.3 Fabrics

Textile fabrics are often the final product of the textile process. Their properties must directly meet the user requirements; obviously, the prediction of their properties and their final behaviour is very important. The fabrics are complex structures, if their micromechanical structure is considered. The structural complexity in conjunction with the materials complexity do not usually permit the development of Computer Aided Engineering tools for the support of the design phase, as it usually the case in other engineering fields such as mechanical, structural, naval, electrical, etc. Therefore, a lot of effort has been given towards the development of computational tools for the prediction of the behaviour of the fabrics (Basu et al., 2002).

The inspection of the fabrics for the detection of faults is a very important operation, traditionally carried out by skilled operators. Many attempts have been made in order to perform the inspection automatically. Consequently the task of automated defects detection is popular and many research teams have focused their interest on it, while many of them have used ANNs to support the fault detection task, (Tsai et al., 1995; Sette & Bullard, 1996; Tilocca et al., 2002, Kumar, 2003; Islam et al., 2006; Shady et al., 2006; Behera & Mani, 2007; Mursalin at al., 2008). Another similar approach is the combined use of fuzzy systems (Choi et al., 2001; Huang & Chen, 2003) or wavelet packet bases (Hu & Tsai, 2000; Jianli & Baoqi, 2007). Defects can be detected also by a dynamic gray cloth inspecting machine system (Kuo et al., 2008). The detection and recognition of the patterns on a fabric is of the same class of problems and thus a candidate for the use of ANNs (Jeon et al., 2003; Chiou et al., 2009; Liu et al., 2009). Using the same principles, stitch inspection can be achieved (Yuen et al., 2009).

Drapability is far the most complex mechanical property of the fabrics and it is essential for many applications of the textile fabrics. The prediction of the drape has been made using ANNs (Fan et al., 2006). In parallel the engineering of the drapability of the fabrics became possible though a predictive tool (Stylios & Powel, 2003). Fabric hand is a property that combines the mechanical properties of a fabric with the sensory perception of the fabric by the humans when they touch it. It is difficult to give an objective description of the fabric handle, because a subjective evaluation takes place in practice. However, there have been developed some complex systematic approaches for the definition of the fabric hand, which include the full set of the low stress mechanical properties of the fabrics. Obviously the prediction of the fabric hand is equivalent to the prediction of the low stress mechanical properties of the fabrics. The prediction of the fabric hand is a complex, highly non-linear problem and therefore an early target for the application of ANNs (Youssefi & Faez, 1999; Hui et al., 2004; Shyr et al., 2004; Matsudaira, 2006). The data from the FAST system were used to approach the hand of the fabrics (Sang-Song & Tsung-Huang, 2007), while fuzzy logic has been combined with ANN for the evaluation of the fabric hand (Park et al., 2000; Park et al., 2001). ANNs in combination with fuzzy logic have been used in the case of the prediction of the sensory properties of the fabrics, as well (Jequirim et al., 2009). Cosely related applications are the objective evaluation of the textile fabric appearance (Cherkassky & Weinberg, 2010) and the emotion based textile indexing using ANNs (Kim et al., 2007)

The prediction of the simpler mechanical properties of the textile fabrics is an essential technical requirement. ANNs have been used for the prediction of the tensile strength (Majumdar et al., 2008) and for the initial stress-strain curve of the fabrics (Hadizadeh et al., 2009). The same problem has been solved using an adaptive neuro-fuzzy system (Hadizadeh et al., 2010). The shear stiffness of the worsted fabrics (Chen et al., 2009) and their compression properties have been successfully modelled (Murthyguru, 2005; Gurumurthy, 2007). In general, the prediction of the properties of a fabric enables the support of the design phase, (Behera & Muttagi, 2004).

The prediction of bursting using ANNs for knitted fabrics (Ertugrul & Ucar, 2000) as well as for woven fabrics (Vassiliadis et al., 2010) has been achieved with satisfactory results. The permeability of the woven fabrics has been modelled using ANNs as well (Tokarska, 2004; Çay et al., 2007). Further on, the impact permeability has been studied (Tokarska & Gniotek, 2009) and the quality of the neural models has been assessed (Tokarska, 2006). The pilling propensity of the fabrics has been predicted (Beltran et al., 2005) and the pilling of the fabrics has been evaluated (Chen & Huang, 2004; Zhang et al., 2010), while the presence of fuzz fibres has been modelled (Ucar & Ertugrul, 2007). The evaluation of the wrinkle of the fabrics has been realized on an objective basis with a system based on ANNs (Su & Xu, 1999; Kim, 1999; Mori & Komiyama, 2002). Prediction of the spirality of the relaxed knitted fabrics

(Murrells et al., 2009) as well as knit global quality (Slah et al., 2006) and subjective assessment of the knit fabrics (Ju & Ryu, 2006) have been implemented.
Prediction of the thermal resistance and the thermal conductivity of the textile fabrics has been realized with the help of ANNs (Bhattacharjee & Kothari, 2007; Fayala et al., 2008). Moisture and heat transfer in knitted fabrics has been also studied similarly (Yasdi et al., 2009). Engineering of fabrics used in safety and protection applications is supported by ANNs (Keshavaraj et al., 1995; Ramaiah et al., 2010). Prediction of the fabrics end use is also possible via the same method (Chen et al., 2001). Optimization of the application of a repellent coating has also been approached by the ANN model (Allan et al., 2002).

2.4 Color

The color measurement, comparison, evaluation and prediction are major actions in the dying and finishing field of the textile process. Although color measurement is possible in the laboratory with the help of specialized equipment like the spectrophotometers, few capabilities exist for the prediction of the color changes or the final color appearance, because the problem is multivariable. A model for the prediction of color change after the spinning process has been developed (Thevenet et al., 2003). The prediction of the color and the color solidity of a jigger dyed cellulose based fabric has been achieved by using cascade ANNs. In the field of printing, the color recipee specification has been made possible using radial basis function neural networks (Rautenberg & Todesco, 1999). The pigment combinations for the textile printing can be determined (Golob et al., 2008), the color of the printed fabric images can be identified and the color separation can take place by using different ANN types (Xu & Lin, 2002; Kuo et al., 2007). The prediction of CIELAB values is possible for color changes after chemical processes (Balci & Ogulata, 2009), for nylon 6,6 and for stripped cotton fabrics (Balci et al., 2008). The optimization of the processing conditions and the prediction of the dyeing quality of nylon and lycra fabrics and the classification of dyeing defects have been carried out with the help of ANNs and fuzzy neural networks respectively (Kuo & Fang, 2006; Huang & Yu, 2001).

2.5 Making up and clothing

Clothing articles are the end product of the main stream of the textile production flow. Although precision of the prediction of properties is not that critical as in technical applications, estimation of the final properties is essential for the clothing design, the selection of raw materials and their required properties. One of the most important factors affecting the garment quality is related to the seam, the result of the sewing process. Indeed, prediction of the seam strength is very important, especially for parachutes (Oanl et al., 2009). The thread consumption is predicted via an ANN model (Jaouadi et al., 2006), while the seam puckering is evaluated and the sewing thread is optimized through ANN models, respectively (Mak & Li, 2007; Lin, 2004). The prediction of the sewing performance is also possible using ANNs (Hui & Ng, 2005; Hui et al., 2007; Hui & Ng, 2009).
The human psychological perceptions of the clothing sensory comfort and the analysis of the tactile perception of the textile materials can be carried out using ANN approaches (Wong et al., 2003; Karthikeyan & Sztandera, 2010). Prediction of the performance of the fabrics in garment manufacturing and fit garment design have been realized based on ANN systems (Hu et al., 2009; Gong & Chen, 1999). Cases of special interest, like the selection of the optimal interlinings, or of broad interest, like the simulation of a textile supply chain, have been successfully modelled by ANNs (Jeong et al., 2000; Nuttle et al., 2000).

2.6 Nonwovens

The nonwoven is a specific category of fabrics, made directly of fibres and not of yarns. The nonwoven fabrics find many technical applications and their role is essential. The nonwoven fabrics undergo a process of inspection in order to ensure quality of the delivered material. A visual inspection system has been based on wavelet texture analysis and robust bayesian ANNs (Liu et al., 2010), or similarly wavelet transforms and ANNs (Huang & Lin, 2008), while a neuro-fractal approach has been used for the recognition and classification of nonwoven web images (Payvand et al., 2010). Many quality issues are addressed via ANN methods, like the structure-properties relations of the nonwoven fabrics (Chen et al., 2007), the construction of a quality prediction system (Kuo et al., 2007), the modeling of the compression properties of needle-punched nonwoven fabrics (Debnath & Madhusoothanan, 2008), the simulation of the drawing of spunbonding nowoven process (Chen et al., 2008) and also the objective evaluation of the pilling on nonwoven fabrics (Zhang et al., 2010).

3. Artificial Neural Networks

3.1 Functionalities of interest for textile engineering

Artificial Neural Networks (ANNs) are algorithmic structures derived from a simplified concept of the human brain structure. They belong to the Soft Computing family of methods, along with fuzzy logic / fuzzy control algorithms and genetic algorithms, (Zadeh, 1994). They all share an iterative, non-linear search for optimal or suboptimal solutions to a given problem, without the presupposition of a model of any type for the underlying system or process, (Keeler, 1992). Various different ANN types have already been successfully employed in a wide variety of application fields, (Haykin, 1998). Major ANN functionalities are

i. Function approximation: this functionality is exploited in system input-output modeling and prediction, and
ii. Classification: this functionality is exploited in pattern recognition / classification problems, (Lippman, 1987).

- In their capacity as function approximators, ANNs have long been studied as to the required properties of the target function as well as to the structure of the ANN, in order to guarrantee convergence of the - typically iterative - approximation algorithm. The first brain-inspired ANN structure was proposed by McCulloch and Pitts in 1943, along with a proof that it could approximate any deterministic function, (Hertz et. al., 1991). In light of the Cybenko theorem, (Cybenko, 1989), ANNs are recognized today as 'universal approximators', i.e. they can approximate arbitrarilly closely any function on a compact subset of $\boldsymbol{R}^n$, under certain general assumptions on the function. The property was proved for a specific ANN structure (the standard multilayer feed-forward network with a single hidden layer that contains a finite number of hidden neurons, with a sigmoid activation function and a linear output layer). Similar results exist for arbitrary activation functions, (Hornik, 1989) and other ANN strucutres, as well, (e.g., Lin, 1994). A common prerequisite for the ANN to operate as approximator is the linearity of its output node(s).

  Under their function approximator form, ANNs have served as a powerful modeling tool, able to capture and represent almost any type of input-output relation, either linear or nonlinear. The shortcoming of such an ANN-based modeling solution is that the model is implicit. Indeed, rather than formulating an explicit input-output analytic expression, either linear or nonlinear, an ANN processes examples of inputs and outputs, to capture and store knowledge on the system. It can subsequently simulate the system, or predict output values; yet, it cannot offer a closed-form description of the system.

In a textiles context, the function approximation capacity of ANNs is of great practical interest because a variety of quantities that characterize yarns and/or fabrics depend on (i.e., are functions of) the yarn or fabric consistency, structure and weaving characteristics. Air permeability of a woven fabric, for example, depends on such parameters as warp and weft yarn density and mass per unit area. These dependencies do not always lend themselves to accurate description by an analytic mathematical function; yet, the ability to estimate or predict the value of such a quantity of interest given the yarn or fabirc parameters - before actually constructing the yarn or fabricating the fabric - is highly desirable in the textiles design and production phases. Research has turned to ANNs for estimation and prediction tasks in various textile applications, (Stylios & Parsons-Moore, 1993), (Stylios & Sotomi, 1996), (Ertugul & Ucar, 2000), (Majumdar, 2004), (Lin, 2007), (Bhattacharjee & Kothari, 2007), (Gurumurthy, 2007), (Guruprasad & Behera, 2010).

- In their capacity as classifiers, on the other hand, ANNs have found extensive and successful application in virtually all pattern recognition tasks, including 1-D and 2-D signal (image) processing applications, clustering, etc. In such problems, unknown input data are classified as belonging to one of a finite number of known classes or categories. ANN structures with an output layer of nodes (neurons) of the 'competitive' type are suitable for classification tasks. Individual binary outputs of the output layer nodes are vectorized in order to enumerate the class where incoming data belong, in typical classification problems. Single and multiple-layer perceptrons, self-organizing maps and other types of ANN structures serve as classifiers. Among them, of practical interest are networks that compute probabilities that a given input belongs to one of the considered classes, rather than deterministic outputs. They can thus subsititute multi-expert decision algorithms, such as majority-voting, etc. Probabilities can subsequently be handled in a variety of ways to obtain final answers in the output.

  The classification capacity of ANNs in a textiles context finds extended use in classification of yarn or fabric types or other visual properties, such as color, defects, weaving / knitting pattern, percentage consistency in various materials and the like, (Guruprasad & Behera, 2010). These tasks are affordable in time and equipment investment thanks to the recent technological advances (a) in image capturing equipment of high quality and very low cost and (b) in hardware processors of increased processing power that allow for real or quasi-real time applications.

3.2 Architecture and training algorithms

The structure or 'architecture' of an ANN contains a number of nodes, called neurons, organized in a number of layers and interconnected to form a network. Neurons are rather simple algorithmic structures that can perform parallel computation for data processing and knowledge representation, (Brasquet & LeCloirec, 2000). Weighted averaging, followed by linear or non-linear (thresholding) operations, with the possibility of feedback between layers, constitute the main processing operations in an ANN. The acquired knowledge is stored as the weight values of the nodes. The real power of this model, therefore, lies in the mesh of simple but highly interconnected nodes rather than the power or sofistication of each node. Although inspired from the way neurons are interconnected in the human brain and nervous system to transfer messages by chemo-electrical procedures, no further analogy between ANNs and the human brain operation is claimed as to the functionalities achieved.

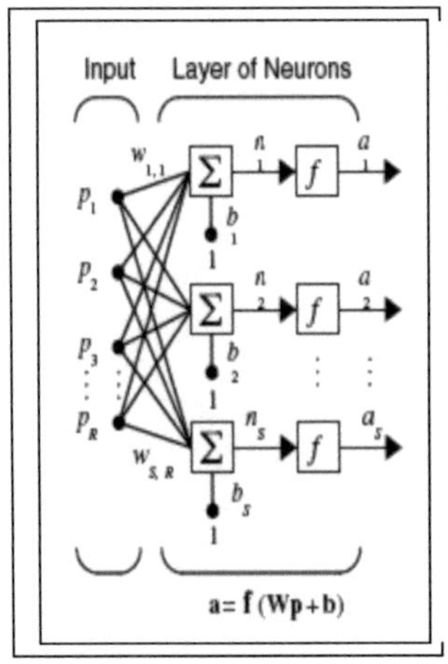

Fig. 1. A sample ANN architecture (one single layer of S nodes or neurons).

The architecture of a single layer of nodes for a sample ANN is shown in Figure 1, (Matlab, 2005). There exist S nodes in this layer. In the i-th node, a linear combination of a vector of R inputs $[p_1, \ldots .p_R]$ is computed. In it, inputs are weighted by weights $[w_{i1}, \ldots, w_{iR}]$ and a bias term b_i is added. The scalar value produced undergoes a transformation by a generally nonlinear activation function $f(\cdot)$ to yield the corresponding i-th output, for all $i = 1, 2, \ldots, S$. Sigmoid, log-sigmoid, hard-limiter or even linear types of functions are employed, resulting in accordingly varying properties of the network.

In general, the ANN architecture is characterized by:

- A large number of simple, neuron-like processing elements;
- A large number of weighted connections between elements. The weights of these connections store the 'knowledge' of the network. The adaptive adjustment of the weight values, while moving down an error surface to a minimum point, constitutes the 'learning process' of the network;
- Highly parallel and distributed control; and
- An emphasis on learning internal system representations automatically, (Rich & Knight, 1991).

Apart from the topology and node characteristics (type of activation function, etc.), an ANN model is specified by the training rules or training algorithm employed to adapt its weights. These rules define an initial set of values for the weights and indicate how weights should be iteratively adapted to improve the performance of the network by minimizing the error between actual and ideal outputs, when the network is presented with a set of known inputs. A variety of different network models have already been proposed and used in practical applications, such as the Perceptron, the Multilayer Perceptron, the Hopfield network, the Carpenter-Grossberg classifier, the Back-propagation network, the Self-organizing Map, the Radial Basis Function Network, the Probabilistic Network, etc., (Ramesh et al., 1995), (Lippman, 1987).

In terms of operation, an ANN undergoes a training phase, where a rich set of examples matched in pairs of input - output values and called the 'training set' is presented to the network input. The weights in the nodes of the network are iteratively optimized by a gradient-type algorithm, so as to produce correct outputs for all inputs in the training set. As iterations proceed, the algorithm leads to sets of weights that minimize the error between ideal and actual ANN outputs - usually, in the least squares sense. Once trained, the network enters the so-called test phase, where it is 'questioned' to produce outputs for unknown inputs presented to it. Decisions are made (i.e., outputs are produced) on the basis of the experience gained by the network over the training set. During the test phase, the network is expected to 'generalize' successfully, i.e., to exploit the experience stored in its elements during the training phase in order to make correct decisions on incoming data that are similar but not identical to the training set data. Satisfactory generalization implies

- a correct choice of the network type in relation to the problem at hand and
- a successful training phase, in the sense that:
    - the training set was rich and informative enough to represent adequately the space of the input vectors, and
    - the training algorithm was allowed an adequate number of iteration to converge.

Major reasons for non-convergence of the training algorithm are

i. the choice of an inappropriate ANN structure (e.g., one with too many / not enough nodes or too many / not enough layers of nodes), and
ii. the unavailability of a rich training set, which means that more data or measurements are necessary to solve the problem at hand.

The second problem is more crucial in practice, as it is not always straightforward how more data are to be obtained or measured in real field applications. The first problem, on the other hand, can be efficiently addressed by simulations during the design phase of the whole application.

In order to illustrate the above, an example is provided in the following section, where an ANN is employed to predict or estimate a fabric property in terms of a set of structural weaving parameters of the yarns used and of the fabric itself.

4. A sample application of ANNs in fabric air permeability prediction

4.1 Problem description

Vacuum drying is a method for removing the water content from wet fabrics by suction. In a typical industrial fabric production process, vacuum drying is a pre-drying stage, positioned before the main drying unit, which operates by heating the wet fabric (stenter). The behavior of woven fabrics during vacuum drying is primarily influenced by their air permeability. Air permeability can be calculated from the porosity of the fabric. This approach, however, is not used in practice, because of known difficulties in determining basic porosity calculation parameters, such as the shape factor. Alternatively, air permeability can be measured in the laboratory after the production of the fabric. Of practical interest, however, is the possibility to predict air permeability during the design phase, based on technical, micro-structural characteristics of the fabric. This would allow for the prediction of the behavior of a fabric during vacuum drying and would thus support a realistic and optimized planning of the production process.

Air permeability depends both on the material(s) of the yarn and on the structural parameters of the fabric, through a generally complex, nonlinear relation, (Backer, 1951). In order to 'capture' this relation and use it later as a predictor, the nonlinear approach of ANNs is adopted, (Cay et. al., 2007). Given the construction process of woven fabrics, three structural parameters are intuitively more attractive for this task, namely,

- warp yarn density,
- weft yarn density and
- mass per unit area.

They are advantageous in that their values either are predefined in the weaving process (warp and weft densities) or can be easily and accurately measured (mass per unit area), while at the same time they influence directly the properties of the woven fabric.

4.2 Network type selection

As discussed in section 3, the choice of the appropriate type of ANN depends on the peculiarities of the problem investigated. Several types of ANNs are promising candidates for the problem under consideration. In fact, any neural network that can take on the form of a universal function approximator will do. These are networks whose output layer nodes operate linearly and can therefore produce practically any real value in the output - as opposed to the limited output values possible when nonlinear / thresholding nodes are employed in the output layer.

The major network families in this class are

1. Multilayer Perceptrons,
2. Radial Basis Function Networks; more specifically, Generalized Regression Neural Networks, (Chen et al., 1991), and
3. Elman Networks, (Elman, 1990).

Among these three alternatives, Generalized Regression Neural Networks (GRNNs), are selected because

- they are known to approximate arbitrarily closely any desired function with finite discontinuities, and
- although they require an increased number of nodes in their architecture, they can be designed in far less time than necessary for the design and training of, e.g., a multiplayer perceptron.

GRNNs work by measuring how far a given sample pattern lies from patterns in the training set, in the N-dimensional space defined by the cardinality of the input set. When a new pattern is presented to the network, it is compared in the N-dimensional space to all patterns in the training set to determine how far in distance it lies from each one of them. The output that is predicted by the network is proportional to all outputs produced by the training set. The proportion value depends upon the distance between the incoming pattern and the given patterns in the training set. Euclidean or city-block distance metrics are used.

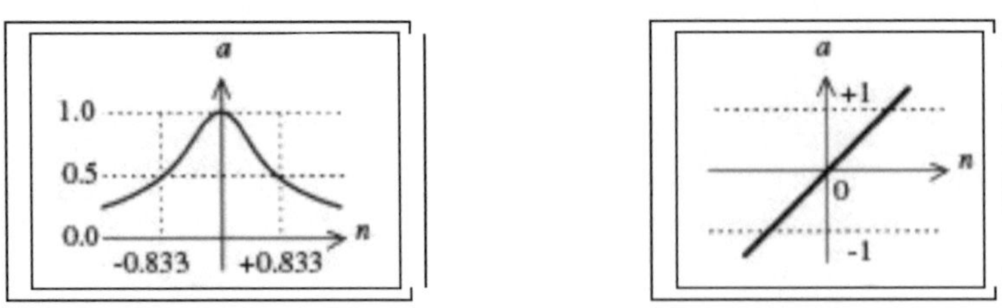

Fig. 2. A radial basis (a) and a linear (b) neuron transfer function.

In terms of structure, a GRNN contains two layers of neurons, each consisting of N neurons, where N is the cardinality of the training set (number of the input - output pairs available for training). The first, hidden layer consists of radial basis function (RBF) neurons while the second, output layer consists of linear neurons of special structure. Thanks to the later, the network can produce any real value in its output. Figure 2a shows a RBF neuron sample transfer function. An RBF neuron 'fires' or produces an output of 1 when its input lies on or fairly close to its central coordinates. Thanks to its symmetry and finite radial support controlled by a spread parameter, it responds only to local inputs, i.e. those befalling within a neighbourhood of controlled radius. Figure 2b shows a linear neuron sample transfer function. The linear layer of a GRNN is special in that its weights are set equal to the output values contained in the training set. When a value equal or close to 1 is produced by the first layer and propagated to the second layer, it produces by means of an inner product operation a significantly non-zero value at a certain output and practically zero values at the rest of the outputs.

During the design phase of the GRNN, the centres of the RBF neurons in the first layer are set equal to the input vectors in the training set. During the test phase, when an unknown input close (similar) to one of the training set inputs appears, the corresponding RBF neuron (or neighbouring set of neurons) fires a value equal or close to 1. The second layer performs an inner product between the vector of the first layer outputs and the vector of training set outputs. Training set outputs matched with values close to 1 in this inner product contribute significantly to the output, while the contribution of the rest is negligible. This is how the GRNN generalizes from the training set to unknown inputs.

The major shortcoming of a GRNN is that both its layers consist of a number of nodes equal to the number N of vectors in the training set. Given the fact that the training set should be rich enough to be representative of the input space, this number may grow large for a given application. As a counter-balance, however, it has been proved that the GRNN can be designed to produce zero error for any given training set in a very short time.

4.3 Experimental setup

A set of different samples of woven fabrics are produced and tested experimentally. All fabric types used are made of 100% cotton yarns and they are of plain weave structure. The warp and weft yarn linear densities are Ne 40 and Ne 50, respectively. The warp and weft yarn densities of the samples vary in incremental steps. Six (6) different weft densities and five (5) different warp densities are combined to produce a set of 6 x 5 = 30 different fabric types in total.

The air permeability of these fabrics is measured under constant pressure drop. Actually, rather than being directly measured, the air permeability is estimated by the measurement of air velocity. Air permeability in ($cm^3/s/cm^2$) is measured via the standard BS5636 method, using the Shirley FX 3300-5 air permeability tester. For each one of the thirty (30) fabric types constructed, the air permeability measurement is repeated across five (5) different samples of the same fabric, thus producing a total of 30 x 5 = 150 measurements. The five measurements of the same fabric are averaged to produce an average air permeability value, in order to reduce non-systematic measurement errors. The structural fabric parameters, the respective air permeability values (averages across five samples) and the standard deviation of the measurements are given in the Table I.

As it can be seen in Table 1, air permeability values decrease when warp and weft yarn densities increase - an expected behaviour because the dimensions of the pores through which air flows are getting smaller when moving from looser towards tighter fabric types, where resistance to the airflow is higher.

Warp and weft yarn densities and mass per unit area form the input vector to the ANN. Therefore, each input - output pair in the training set contains a vector of these three inputs and a single output (air permeability). As mentioned earlier, air permeability measurements are averaged across sets of five samples for each fabric type. Therefore the data set consists of thirty (30) independent combinations of values of the three independent input variables along with the respective averaged air permeability value per case.

Sample No.	Warp yarn density (ends/cm)	Weft yarn density (picks/cm)	Mass per unit area (gr/m2)	Average Air permeability (cm3/s/cm2)	STDEV of Air permeability	Sample No.	Warp yarn density (ends/cm)	Weft yarn density (picks/cm)	Mass per unit area (gr/m2)	Average Air permeability (cm3/s/cm2)	STDEV of Air permeability
1	54	20	101.0	45.74	1.80	**16**	60	35	141.4	4.56	0.41
2	54	25	110.3	27.02	2.44	**17**	60	40	151.2	2.12	0.07
3	54	30	119.5	15.68	0.68	**18**	60	45	153.7	1.70	0.05
4	54	35	127.3	8.76	0.39	**19**	63	20	109.0	27.68	0.55
5	54	40	136.7	4.37	0.15	**20**	63	25	119.1	15.24	0.64
6	54	45	144.1	2.90	0.40	**21**	63	30	129.1	7.23	0.13
7	57	20	101.2	47.94	1.83	**22**	63	35	148.5	3.22	0.42
8	57	25	111.1	27.52	0.78	**23**	63	40	150.3	1.89	0.08
9	57	30	120.5	14.84	0.97	**24**	63	42	154.4	1.61	0.05
10	57	35	130.7	8.99	0.48	**25**	66	20	111.2	27.28	1.68
11	57	40	142.0	3.98	0.24	**26**	66	25	123.0	14.42	1.39
12	57	45	146.3	2.68	0.12	**27**	66	30	134.0	6.90	0.21
13	60	20	107.9	33.98	1.16	**28**	66	35	144.4	3.19	0.17
14	60	25	118.7	17.01	0.79	**29**	66	40	155.0	1.65	0.06
15	60	30	129.5	9.75	0.41	**30**	66	42	160.2	1.38	0.05

Table 1. Structural parameters and corresponding air permeability measurements for 30 fabric types.

Twenty-four (24) out of the thirty (30) cases are used for training while the remaining six (6) cases are retained for testing. This leads to a GRNN architecture of 24 neurons in each of the two layers. Furthermore, in order to reduce the dependency of the results on a specific partition of the data into training and testing sets, a five-ways cross validation test is carried out, i.e., the data are partitioned in five (5) different ways and finally results are averaged across all five (5) partitions. Figure 4 show results of one out of the five (5) different experiments of the five-ways cross-validation. The upper plot shows results where the training set per se is used as the test set, while the lower plot shows results where the test set is disjoint to the training set. Within each plot, stars (*) and circles (o) indicate measured and estimated air permeability values, respectively, while dots (·) indicate the ANN performance error (difference between measured and estimated air permeability values).

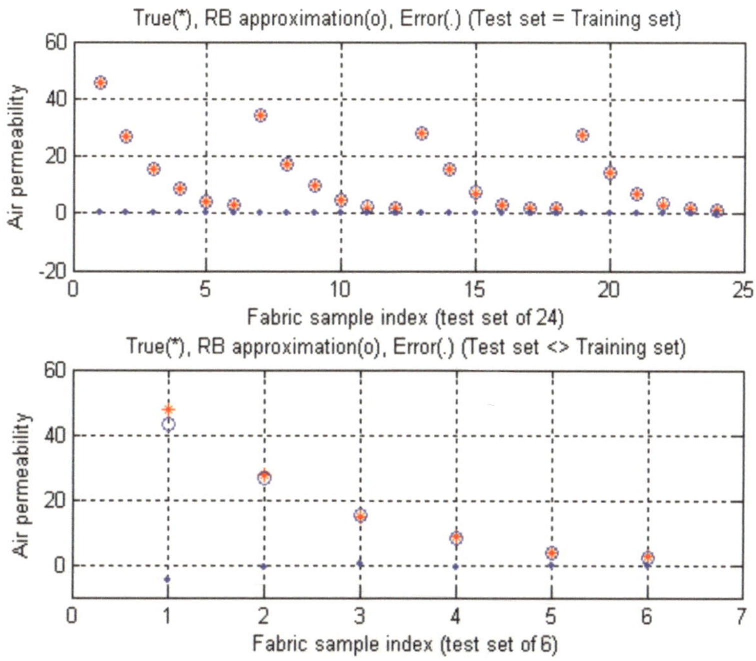

Fig. 4. Air permeability estimated (approximated) by the GRNN (partition no. 2).

As it can be seen in Figure 4, the network employed approximates air permeability values with a very low output error – a result observed regardless of the specific data partition into training and test sets. Error (sum-of-squares) values, given in Table 2, are indeed very low.

Data Partition No.	Error (sum-of-squares)
1	16.62
2	22.64
3	17.10
4	17.52
5	94.49
Average error (sum-of-squares)	33.27

Table 2. Sum-of-squares error across five partitions and average.

Although the relation under investigation is clearly non-linear, due to the complex dynamics involved in the underlying physical phenomena, a multiple linear regression is performed on the data of Table 1, in order to obtain

a. a crude linear approximation of the relation sought and
b. a clue as to the relative importance of the three intuitively chosen structural parameters within this relation.

```
The regression equation is
C4 = 105 - 0,679 C1 - 0,392 C2 - 0,355 C3

Predictor       Coef  SE Coef      T      P
Constant      105,04    16,36   6,42  0,000
C1           -0,6793   0,7280  -0,93  0,359
C2           -0,3920   0,5694  -0,69  0,497
C3           -0,3551   0,3655  -0,97  0,340

S = 5,59416   R-Sq = 84,0%   R-Sq(adj) = 82,2%

Analysis of Variance

Source          DF      SS      MS      F      P
Regression       3  4285,9  1428,6  45,65  0,000
Residual Error  26   813,7    31,3
Total           29  5099,6

Source  DF  Seq SS
C1       1  3825,8
C2       1   430,6
C3       1    29,5

Unusual Observations

Obs    C1     C4    Fit  SE Fit  Residual  St Resid
  1  20,0  45,74  34,42    2,54     11,32     2,27R
  7  20,0  47,94  33,17    2,05     14,77     2,84R

R denotes an observation with a large standardized
residual.
```

Table 3. Multiple linear regression analysis on the data of Table 1.

The results are given in Table 3. Data in columns C1, C2, C3 contain weft densities, warp densities and mass per unit area data, respectively, while column C4 contains average air permeability values.

In Table 3, the P-value in the Analysis of Variance section is less than 10^{-3}, showing that the regression per se is indeed significant, i.e. there does exist an influential relation between the dependent variable (air permeability) and the three independent variables selected here. The R2 value is 84% (adjusted R2 is 82,2%), showing that 84% of the variability in the air permeability data is 'explained' by the linear combination of the specific three independent variables. This percentage is high enough to indicate that a linear relation could certainly be used as a crude approximation of the true relation and, at the same time, low enough to justify the investigation of nonlinear alternatives τo explain the remaining variability in the data.

It is interesting to notice that the (appropriately scaled) average error produced by the ANN approach in Table 2 (33.68 * 30 / 6 = 168.4) compares favourably to the respective error produced by the multiple linear regression approach (813.7). The later is a very encouraging result, as it justifies the extra effort required by the nonlinear approach. Indeed, in terms of the multiple linear regression, it can be claimed that the ANN structure leaves unexplained only the (168.4 / 5099.6 =) 3,3% of the total variability in the air permeability data, thus offering a five times better result than the respective 16% of the linear regression approach.

An error pattern apparent in the lower plot of Figure 4 associates looser fabric types (towards the left side of the horizontal axis) with error values higher than the respective error values for dense fabrics (towards the right side of the horizontal axis). For loose fabrics the pores or openings between the yarns are bigger so the yarn mobility is higher, thus the pore dimensions become bigger because of the deformation during the airflow. On the contrary, dense fabrics have very small pores and high resistance to airflow and can preserve their compactness during airflow.
Finally, Figure 5 shows the strong correlation between the measured and the predicted by the ANN air permeability values.

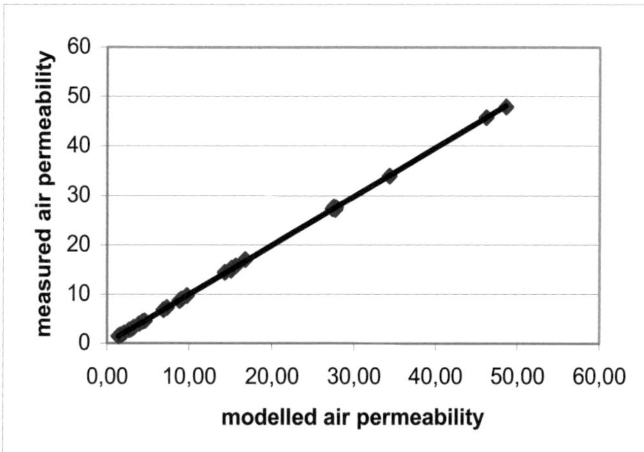

Fig. 5. Correlation between measured and ANN-predicted air permeability values.

5. Conclusions – future trends

An extended overview of the application of artificial neural networks methods for the solution of textile problems is provided in an attempt to cover this evolving field up to the current research and technology advances. It is clear to the reader that related publications on the field are produced in increasing numbers from the research community. Starting from the decade of the nineties a continuously increasing number of ANN applications has provided solutions to complex and multivariable textile problems. It seems that the textile community has been familiarized with this powerful tool and it trusts it more and more. The continuous increase of the computational power of the personal computers reduces the drawback of the computational cost that the use of ANNs requires. Therefore, it is expected that ANNs will augment their percentage of participation in solving complex textile problems and they will support the design and the computer aided engineering concepts in the textile field.

6. References

Admuthe L.S. & Apte S. (2010). Adaptive Neuro-fuzzy Inference System with Subtractive Clustering: A Model to Predict Fiber and Yarn Relationship, *Text. Res. J.*, 80(9), pp 841-846.

Allan G., Yang R., Fotheringham A. & Mather R. (2001). Neural modelling of polypropylene fibre processing: Predicting the structure and properties and identifying the control parameters for specified fibres, *J. of Materials Science,* 36 pp. 3113 - 3118.

Allan G., Fotheringham A. & Weedall P. (2002), The Use of Plasma and Neural Modelling to Optimise the Application of a Repellent Coating to Disposable Surgical Garments, *AUTEX Res. J.*, Vol. 2, No2.

Babay, A. , Cheikhrouhou, M. , Vermeulen, B. , Rabenasolo, B. & Castelain, J. M.(2005). Selecting the optimal neural network architecture for predicting cotton yarn hairiness, *J. of the Text. Inst.*, 96: 3, pp. 185 - 192.

Backer, S. (1951). The relationship between the structural geometry of a textile fabric and its physical properties, Part IV: Intercise geometry and air permeability, *Text. Res. J.*, vol. 2, pp. 703-714.

Balci O., Ogulata S.N., Sahin C. & Ogulata R.T. (2008). Prediction of CIELab Data and Wash Fastness of Nylon 6,6 Using Artificial Neural Network and Linear Regression Model, *Fibers and Polymers*, Vol.9, No.2, pp. 217-224.

Balci O., Ogulata S.N., Sahin C. & Ogulata R.Y. (2008). An Artificial Neural Network Approach to Prediction of the Colorimetric Values of the Stripped Cotton Fabrics, *Fibers and Polymers*, Vol.9, No.5, pp. 604-614.

Balci O. & Ogulata R.T. (2009). Prediction of CIELab Values and Color Changing Occurred After Chemical Finishing Applications by Artificial Neural Networks on Dyed Fabrics, Tekstil ve Konfeksiyon, 1, pp. 61-69.

Balci O. & Ogulata R.T. (2009). Prediction of the Changes on the CIELab Values of Fabric after Chemical Finishing Using Artificial Neural Network and Linear Regression Models, *Fibers and Polymers*, Vol.10, No.3, pp. 384-393.

Basu A., Chellamani, K. P. & Ramesh P.R. (2002). Fabric Engineering by Means of an Artificial Neural Network, *J. of the Text. Inst.*, 93:3, pp. 283 - 296.

Behera B.K. & Mani M.P. (2007). Characterization and Classification of Fabric Defects Using Discrete Cosine Transformation and Artificial Neural Network, *Ind. J. of Fibre & Text. Res.*, Vol. 32, pp. 421-426.

Behera B.K. & Muttagi S.B. (2004). Performance of Error Back Propagation vis-á-vis Radial Basis Function Neural Network: Part I: Prediction of Properties for Design Engineering of Woven Suiting Fabrics, *J. of the Text. Inst.*, 95:1, pp. 283 - 300.

Behera B.K. & Muttagi S.B. (2004). Performance of Error Back Propagation vis-à-vis Radial Basis Function Neural Network: Part II: Reverse Engineering of Woven Fabrics, *J. of the Text. Inst.*, 95: 1, pp. 301 - 317.

Beltran R., Wang L. & Wang X. (2004). Predicting Worsted Spinning Performance with an Artificial Neural Network Model, *Text. Res. J.*, 74(9), pp 757-763.

Beltran R., Wang L. & Wang X. (2005). Predicting the Pilling Propensity of Fabrics through Artificial Neural Network Modeling, *Text. Res. J.*, 75(7), pp 557-561.

Bhattacharjee D. & Kothari V.K. (2007). A Neural Network System for Prediction of Thermal Resistance of Textile Fabrics, *Text. Res. J.*, 77(1), pp 4-12.

Brasquet, C., LeCloirec, P. (2000). Pressure drop through textile fabrics-experimental data modeling using classical models and neural networks. *Chemical Engineering Science,* 55, pp. 2767-2778.

Çay A., Vassiliadis S., Rangoussi M. & Tarakçıoğlu I. (2007). Prediction of the Air Permeability of Woven Fabrics using Neural Networks, *Int. J. of Cloth. Sc. & Techn.* Vol. 19, No 1, pp 18-35.

Chattopadhyay R. & Guha A. (2004). Artificial Neural Networks: Applications to Textiles, *Textile Progress*, 35:1, 1-46.

Chen, S., Cowan, C.F.N., and Grant, P.M. (1991). Orthogonal least-squares learning algorithm for radial basis function networks, *IEEE Transactions on Neural Networks*, vol. 2(2), pp. 302-309.

Chen T., Li L., Koehl L., Vroman P. & Zeng X. (2007). A Soft Computing Approach to Model the Structure - Property Relations of Nonwoven Fabrics, *J. of Applied Polymer Science*, Vol. 103, pp. 442–450.

Chen T., Wang K. & Wu L., (2009). Artificial Neural Network Modeling for the Weft Shear Stiffness of Worsted Fabrics, *3rd Int. Symp. on Intell. Inf. Techn. Appl.* v. 1, pp.211-215.

Chen T., Zhang C., Chen X. & Li L.(2009). An Input Variable Selection Method for the Artificial Neural Network of Shear Stiffness of Worsted Fabrics, *Statistical Analysis and Data Mining*, Vol. 1 (5), pp 287-295.

Chen T., Zhang C., Li L. & Chen, X. (2008). "Simulating the drawing of spunbonding nonwoven process using an artificial neural network technique", *J. of the Text. Inst.*, 99: 5, pp. 479 - 488.

Chen X. & Huang X.B. (2004). Evaluating Fabric Pilling with Light-Projected Image Analysis, *Text. Res. J.*, 74(11), pp 977-981.

Chen Y., Zhao T. & Collier B. J. (2001). Prediction of Fabric End-use Using a Neural Network Technique, *J. of the Text. Inst.*, 92: 2, pp. 157 - 163.

Cheng K.P.S. & Lam H.L.I. (2003). Evaluating and Comparing the Physical Properties of Spliced Yarns by Regression and Neural Network Techniques, *Text. Res. J.*, 73(2), pp 161-164.

Cheng L., Ghorashi H., Duckett K., Zapletalova T. & Watson M. (1999). Color Grading of Cotton Part II: Color Grading with an Expert System and Neural Networks, *Text. Res. J.*, 69(12), pp 893-903.

Cherkassky A. & Weinberg A. (2010). Objective Evaluation of Textile Fabric appearance. Part 2: SET Opti-grade Tester, Grading Algorithms and Testing, *Text. Res. J.*, 80(2), pp 135-144.

Chiu S.H., Chen H.M., Chen J.Y. & Wen C.Y. (2001). Appearance Analysis of False Twist Textured Yarn Packages Using Image Processing and Neural Network Technology, *Text. Res. J.*, 71(4), pp 313-317.

Chiou Y.-C., Lin C.- S. & Chen G.-Z. (2009). Automatic texture inspection in the classification of papers and cloths with neural networks method, *Sens. Review*, 29/3, pp. 250–259.

Choi H.T., Jeong S.H., Kim S.R., Jaung J.Y. & Kim S.H. (2001). Detecting Fabric Defects with Computer Vision and Fuzzy Rule Generation. Part II: Defect Identification by a Fuzzy Expert System, *Text. Res. J.*, 71(7), pp 563-573.

Cybenko, G. (1989). Approximations by superpositions of sigmoidal functions. *Mathematics of Control, Signals, and Systems*, no. 4, pp. 303-314.

Dayik M. (2009). Prediction of Yarn Properties Using Evaluation Programing", *Text. Res. J.*, 79(11), pp. 963-972.

Debnath S. & Madhusoothanan M. (2008). Modeling of Compression Properties of Needle-Punched nonwoven Fabrics Using Artificial Neural Network, *Ind. J. of Fibre & Text. Res.*, Vol. 33, pp. 392-399.

Demiryurek O. & Koc E. (2009). Predicting the Unevenness of Polyester/Viscose Blended Open-end Rotor Spun Yarns Using Artificial Neural Network and Statistical Models, *Fibers and Polymers*, Vol.10, No.2, pp. 237-245.

Elman, J.L. (1990). Finding structure in time, *Cognitive Science,* vol. 14, pp. 179-211.

Ertugrul S. & Ucar N. (2000). Predicting Bursting Strength of Cotton Plain Knitted Fabrics Using Intelligent Techniques, *Text. Res. J.*, 70(10), pp 845-851.

Fan J., Newton E., Au R. & Chan S.C.F. (2001). Predicting Garment Drape with a Fuzzy-Neural Network, *Text. Res. J.*, 71(7), pp 605-608.

Farooq A. & Cherif C. (2008). Use of Artificial Neural Networks for Determining the Leveling Action Point at the Auto-leveling Draw Frame, *Text. Res. J.*, 78(6), pp 502-509.

Fayala F., Alibi H., Benltoufa S. & Jemni A. (2008). Neural Network for Predicting Thermal Conductivity of Knit Materials, *J. of Eng. Fibers and Fabrics*, Vol. 3, Issue 4, pp. 53-60.

Furferi R. & Carfagni M. (2010). Prediction of the Color and of the Color Solidity of a Jigger-dyed Cellulose-based Fabric: A Cascade Neural Network Approach, *Text. Res. J.*, 80, pp --.

Ghane M., Semnani D., Saghafi R. & Beigzadeh H. (2008). Optimization of Top Roller Diameter of Ring Machine to Enhance Yarn Evenness by Using Artificial Intelligence, *Ind. J. of Fibre & Text. Res.*, Vol. 33, pp. 365-370.

Gharehaghaji A.A., Shanbeh M. and Palhang M. (2007). Analysis of Two Modeling Methodologies for Predicting the Tensile Properties of Cotton-covered Nylon Core Yarns, *Text. Res. J.*, 77(8), pp 565-571.

Ghosh A. & Chatterjee P. (2010). Prediction of Cotton Yarn Properties Using Support Vector Machine, *Fibers and Polymers,* Vol.11, No.1, pp. 84-88.

Golob D., Osterman D.P. & Zupan J. (2008). Determination of Pigment Combinations for Textile Printing Using Artificial Neural Networks, *Fibr. & Text. in East. Eur.*, Vol. 16, No. 3 (68), pp 93-98.

Gong R.H. & Chen, Y. (1999). Predicting the Performance of Fabrics in Garment Manufacturing with Artificial Neural Networks, *Text. Res. J.*, 69(7), pp 477-482.

Guha A., Chattopadhyay, R. & Jayadeva (2001). Predicting Yarn Tenacity: A Comparison of Mechanistic, Statistical, and Neural Network Models, *J. of the Text. Inst.*, 92: 2, pp. 139 - 145.

Gürkan Ünal P., Özdil N. & Taskin C. (2010). The Effect of Fiber Properties on the Characteristics of Spliced Yarns Part I: Prediction of Spliced Yarns Tensile Properties, *Text. Res. J.*, 80(5), pp 429-438.

Gürkan Ünal P., Arikan C., Özdil N. & Taskin C. (2010). The Effect of Fiber Properties on the Characteristics of Spliced Yarns: Part II: Prediction of Retained Spliced Diameter, *Text. Res. J.*, 80, pp--.

Gurumurthy B.R. (2007). Prediction of Fabric Compressive Properties Using Artificial Neural Networks, *AUTEX Res. J.*, Vol. 7, No 1.

Guruprasad R. & Behera B.K. (2010). Soft Computing in Textiles, *Ind. J. of Fibre & Text. Res.*, Vol. 35, pp. 75-84.

Hadizadeh M., Jeddi A.A.A. & Tehran M.A. (2009). The Prediction of Initial Load-extension Behavior of Woven Fabrics Using Artificial Neural Network, *Text. Res. J.*, 79(17), pp 1599-1609.

Hadizadeh M., Tehran M. A. & Jeddi A.A.A. (2010). Application of an Adaptive Neuro-fuzzy System for Prediction of Initial Load/Extension Behavior of Plain-woven Fabrics, *Text. Res. J.*, 80(10), pp 981-990.

Haykin, S. (1998). *Neural Networks: A Comprehensive Foundation,* Prentice Hall, ISBN 0132733501, New York.

Hertz, J., Krogh, A. & Palmer, R.G. (1991). *Introduction to the Theory of Neural Computation,* Addison-Wesley Longman Publishing Co., Boston, MA, USA.

Hornik, K., Stinchcombe, M. & White, H. (1989). Multilayer Feedforward Networks are universal approximators. *Neural Networks,* vol. 2, pp.359-366.

Hu M.C. & I.S. Tsai I.C. (2000). Fabric Inspection Based on Best Wavelet Packet Bases, *Text. Res. J.*, 70(8), pp 662-670.

Hu Z. H., Ding Y.S, Yu X. K., Zhang W. B. & Yan Q. (2009). A Hybrid Neural Network and Immune Algorithm Approach for Fit Garment Design, *Text. Res. J.*, 79(14), pp 1319-1330.

Huang C.C. & Chen I.C. (2001). Neural-Fuzzy Classification for Fabric Defects, *Text. Res. J.*, 71(3), pp 220-224.

Huang C.C. & Chang K.T. (2001). Fuzzy Self-Organizing and Neural Network Control of Sliver Linear Density in a Drawing Frame, *Text. Res. J.*, 71(11), pp 987-992.

Huang C.C. & Lin T.F. (2008). Image Inspection of Nonwoven Defects Using Wavelet Transforms and Neural Networks, *Fibers and Polymers*, Vol.9, No.5, 633-638

Huang CC. & Yu W.H. (2001). Fuzzy Neural Network Approach to Classifying Dyeing Defects, *Text. Res. J.*, 71(2), pp 100-104.

Hui C.L., Lau T.W., Ng S.F. & Chan K.C.C. (2004). Neural Network Prediction of Human Psychological Perceptions of Fabric Hand, *Text. Res. J.*, 74(5), pp 375-383.

Hui C.-L. & Ng S.-F. (2005). A new approach for prediction of sewing performance of fabrics in apparel manufacturing using artificial neural networks, *J. of the Text. Inst.*, 96: 6, pp. 401 - 405.

Hui P.C.L., Chan K.C.C., Yeung K.W. & Ng F.S.F. (2007). Application of artificial neural networks to the prediction of sewing performance of fabrics, *Int. J. of Cloth. Sc. & Techn.*, Vol. 19 No. 5, pp. 291-318.

Hui C.L. & Ng S.F. (2009). Predicting Seam Performance of Commercial Woven Fabrics Using Multiple Logarithm Regression and Artificial Neural Networks, *Text. Res. J.*, 79(18), pp 1649-1657.

Islam M.A., Akhter S., Mursalin T.E. & Amin M.A. (2006). A Suitable Neural Network to Detect Textile Defects, King et al. (Eds.): *ICONIP 2006*, Part II, LNCS 4233, pp. 430 - 438, Springer.

Jackowska-Strumiłło L., Cyniak D., Czekalski J. & Jackowski T. (2008). Neural Model of the Spinning Process Dedicated to Predicting Properties of Cotton-Polyester Blended Yarns on the Basis of the Characteristics of Feeding Streams, *Fibr. & Text. in East. Eur.*, Vol. 16, No. 1 (66).

Jasper W.J. & Kovacs E. (1994). Using Neural Networks and NIR Spectro-photometry to Identify Fibers, Text. Res. J., August 1994; vol. 64, 8: pp. 444-448.

Jaouadi M., Msahli S., Babay A. & Zitouni B. (2006). Analysis of the modeling methodologies for predicting the sewing thread consumption, *Int. J. of Cloth. Sc. & Techn.*, Vol. 18 No. 1, pp. 7-18.

Jayadeva, Guha, A. & Chattopadhyay, R. (2003). A Study on the Capability of a Neural Network Ranking Fibre Parameters Having an Influence on Yarn Properties, *J. of the Text. Inst.*, 94: 3, pp. 186 - 193.

Jeguirim S.E.G., Dhouib A.B., Sahnoun A. Cheikhrouhou M. Schacher L. & Adolphe D. (2009). The use of fuzzy logic and neural networks models for sensory properties prediction from process and structure parameters of knitted fabrics, *J. Intell. Manuf.*, DOI 10.1007/s10845-009-0362-y.

Jeon B. S., Bae J. H. & Suh M. W. (2003). Automatic Recognition of Woven Fabric Patterns by an Artificial Neural Network, *Text. Res. J.*, 73(7), pp 645-650.

Jeong S.H., Kim J.H. & Hong C.J. (2000). Selecting Optimal Interlinings with a Neural Network, *Text. Res. J.*, 70(11), pp 1005-1010.

Jianli, Liu & Baoqi, Zuo (2007). Identification of fabric defects based on discrete wavelet transform and back-propagation neural network, *J. Text. Inst.*, 98(4), pp. 355 - 362.

Ju J. & Ryu H. (2006). A Study on Subjective Assessment of Knit Fabric by ANFIS, *Fibers and Polymers*, Vol.7, No.2, pp. 203-212.

Kang T.J. & Kim S.C. (2002). Objective Evaluation of the Trash and Color of Raw Cotton by Image Processing and Neural Network, *Text. Res. J.*, 72(9), pp. 776-782.

Karthikeyan B. & Sztandera L. M. (2010). Analysis of tactile perceptions of textile materials using artificial intelligence techniques, Part 2: reverse engineering using genetic algorithm coupled neural network, *Int. J. of Cloth. Sc. & Techn.*, Vol. 22 No. 2/3, pp. 202-210.

Keeler, J. (1992). Vision of Neural Networks and Fuzzy Logic for Prediction and Optimisation of Manufacturing Processes, In: *Applications of Artificial Neural Networks III*, vol. 1709, pp. 447-456.

Keshavaraj R., Tock R.W. & Nusholtz G.S. (1995). A Simple Neural Network Based Model Approach for Nylon 66 Fabrics Used in Safety Restraint Systems: A Comparison of Two Training Algorithms, *J. of Applied Polymer Science*, Vol. 57, pp. 1127-1144.

Khan Z., Lim A.E.K., Wang L., Wang X. & Beltran R. (2009). An Artificial Neural Network-based Hairiness Prediction Model for Worsted Wool Yarns, *Text. Res. J.*, 79(8), pp 714-720.

Kim E.H. (1999). Objective Evaluation of Wrinkle Recovery, *Text. Res. J.*, 69(11), pp. 860-865.

Kim N.Y., Shin Y. & Kim E.Y. (2007). Emotion-Based Textile Indexing Using Neural Networks, J. Jacko (Ed.): Human-Computer Interaction, Part III, *HCII 2007*, LNCS 4552, pp. 349–357, Springer.

Kumar A. (2003). Neural Network Based Detection of Local Textile Defects, *Pattern Recognition*, 36 pp.1645 - 1659.

Kuo C.F.J. & Fang C.-C. (2006). Optimization of the Processing Conditions and Prediction of the Quality for Dyeing Nylon and Lycra Blended Fabrics, *Fibers and Polymers*, Vol.7, No.4, pp. 344-351.

Kuo C.F.J., Hsiao K.I. & Wu Y.S. (2004). Using Neural Network Theory to Predict the Properties of Melt Spun Fibers, *Text. Res. J.*, 74(9), 2004, pp 840-843.

Kuo C.F. J., Su T.L., Chang C.D. & Lee C.H. (2008). Intelligence Control of On-line Dynamic Gray Cloth Inspecting Machine System Module Design. II. Defects Inspecting Module Design, *Fibers and Polymers*, Vol.9, No.6, pp. *768-775.*

Kuo C.-F. J., Su T.L., Chiu C.-H. & Tsai C.-P. (2007). Analysis and Construction of a Quality Prediction System for Needle-Punched Non-woven Fabrics, *Fibers and Polymers*, Vol.8, No.1, 66-71.

Kuo C.-F. J., Su T.L. & Huang Y.J. (2007). Computerized Color Separation System for Printed Fabrics by Using Backward-Propagation Neural Network, *Fibers and Polymers*, Vol.8, No.5, pp. 529-536.

Kuo C.F.J., Wang C.C. & Hsieh C.T. (1999). Theoretical Control and Experimental Verification of Carded Web Density Part III: Neural Network Controller Design, *Text. Res. J.*, 69(6), pp 401-406.

Lewandowski S. & Drobina R. (2008). Prediction of Properties of Unknotted Spliced Ends of Yarns Using Multiple Regression and Artificial Neural Networks. Part I: Identification of Spliced Joints of Combed Wool Yarn by Artificial Neural Networks and Multiple Regression, *Fibr. & Text. in East. Eur.*, 16, 5 (70) pp. 33-39.

Lewandowski S. & Drobina R. (2008). Prediction of Properties of Unknotted Spliced Ends of Yarns Using Multiple Regression and ArtificialNeural Networks. Part II: Verification of Regression Models, *Fibr. & Text. in East. Eur.*, 16,6 (71) pp. 20-27.

Lin, D.-T. (1994). *The Adaptive Time-Delay Neural Network: Characterization and Applications to Pattern Recognition, Prediction and Signal Processing.* PhD thesis, University of Maryland, USA.

Lin J.J. (2007). Prediction of Yarn Shrinkage using Neural Nets, *Text. Res. J.*, 77(5), pp 336-342.

Lin T.H. (2004). Construction of Predictive Model on Fabric and Sewing Thread Optimization, *J. of Textile Engineering*, Vol. 50, No. 1, pp. 6-11.

Lippman, R.P. (1987). An introduction to computing with neural nets. *IEEE ASSP Magazine,* pp. 4-22.

Liu S., Wan O. & Zhang H. (2009). Fabric Weave Identification Based on Cellular Neural Network, *The Sixth Int. Symposium on Neural Networks,* AISC 56, pp. 563–569.

Liu J., Zuo B., Vroman P., Rabenasolo B., Zeng X. & Bai L. (2010). Visual Quality Recognition of Nonwovens using Wavelet Texture Analysis and Robust Bayesian Neural Network, *Text. Res. J.*, 80, pp –.

Lu Z.-J., Yang J.-G., Xiang Q. & Wang X.-L. (2007). Support Vector Machines for Predicting Worsted Yarn Properties, *Ind. J. of Fibre & Text. Res.*, Vol. 32, pp. 173-178.

Majumdar A., Majumdar P.K. & Sarkar B. (2004). Selecting Cotton Bales by Spinning Consistency Index and Micronaire Using Artificial Neural Networks, *AUTEX Res. J.*, Vol. 4, No1, pp. 2-8.

Majumdar P.K. & Majumdar A. (2004). Predicting the Breaking Elongation of Ring Spun Cotton Yarns Using Mathematical, Statistical, and Artificial Neural Network Models, *Text. Res. J.*, 74(7), pp 652-655.

Majumdar A., Majumdar P.K. & Sarkar B. (2006). An investigation on yarn engineering using artificial neural networks, *J. of the Text. Inst.*, 97: 5, pp. 429 - 434.

Majumdar A., Ciocoiu M. & Blaga M. (2008). Modelling of Ring Yarn Unevenness by Soft Computing Approach, *Fibers and Polymers*, Vol.9, No.2, 210-216.

Majumdar A., Ghosh A., Saha S.S., Roy A., Barman S., Panigrahi D. & Biswas A. (2008). Empirical Modelling of Tensile Strength of Woven Fabrics, *Fibers and Polymers* 2008, Vol.9, No.2, 240-245.

Majumdar A. (2010). Modeling of Cotton Yarn hairiness Using Adaptive Neuro-Fuzzy Inference System", *Ind. J. of Fibre & Text. Res.*, Vol. 35, pp. 121-127.

Mak K.L. & Li W. (2007). Objective Evaluation of Seam Pucker on Textiles by Using Self-organizing Map. *IAENG International J. of Computer Science*, 35:1, IJCS_35_1.

Matlab, (2005), MATLAB 7 R14, *Neural Network Toolbox User's Guide,* The MathWorks Inc., Natick, MA, USA.

Matsudaira M. (2006). Fabric Handle and Its Basic Mechanical Properties, *J. of Textile Engineering*, Vol 52, No 1, pp. 1-8.

Mori T. & Komiyama J. (2002). Evaluating Wrinkled Fabrics with Image Analysis and Neural Networks, *Text. Res. J.*, 72(5), pp. 417-422.

Murrells C.M., Tao X.M., Xu B.G. & Cheng K.P.S. (2009). An Artificial Neural Network Model for the Prediction of Spirality of Fully Relaxed Single Jersey Fabrics, *Text. Res. J.*, 79(3), pp 227-234.

Mursalin T.E., Eishita F.Z. & Islam A. R. (2008). Fabric Defect Inspection System Using Neural Network and Microcontroller, *J. of Theoretical and Applied Information Technology*, Vol4, No7.

Murthyguru, (2005). Novel Approach To Study Compression Properties in Textiles, *AUTEX Res. J.*, Vol. 5, No 4.

Mwasiagi J.I., Huang X. & Wang X. (2008). Performance of Neural Network Algorithms during the Prediction of Yarn Breaking Elongation, *Fibers and Polymers*, Vol.9, No.1, pp. 80-86.

Mwasiagi J. I. , Wang X. H. & Huang X. B. (2009). The Use of K-means and Artificial Neural Network to Classify Cotton Lint, *Fibers and Polymers*, Vol.10, No.3, 379-383.

Nurwaha D. & Wang X.H. (2008). Comparison of the New Methodologies for Predicting the CSP Strength of Rotor Yarn, *Fibers and Polymers*, Vol.9, No.6, pp. 782-784.

Nurwaha D. & Wang X.H. (2010). Prediction of Rotor Spun Yarn Strength from Cotton Fiber Properties Using Adaptive Neuro-Fuzzy Inference System Method, *Fibers and Polymers*, Vol.11, No.1, pp. 97-100.

Nuttle, H.L.W., King R.E., Hunter N.A., Wilson J.R. & Fang S.C. (2000). Simulation Modeling of the Textile Supply Chain. Part I: The Textile-plant Models, *J. of the Text. Inst.*, 91: 1, pp. 35 - 50.

Onal L., Zeydan M., Korkmaz M. & Meeran S. (2009). Predicting the Seam Strength of Notched Webbings for Parachute Assemblies Using the Taguchi's Design of Experiment and Artificial Neural Networks, *Text. Res. J.*, 79(5), pp 468-478.

Park S.W., Hwang Y.G., Kang B.C. & Yeo S.W. (2000). Applying Fuzzy Logic and Neural Networks to Total Hand Evaluation of Knitted Fabrics, *Text. Res. J.*, 70(8), pp 675-681.

Park S.W., Hwang Y. G., Kang B.C. & Yeo S.W. (2001). Total handle evaluation from selected mechanical properties of knitted fabrics using neural network, *Int. J. of Cloth. Sc. & Techn.*, Vol. 13 No. 2, pp. 106-114.

Payvandy P., Yousefzadeh-Chimeh M. & Latifi M. (2010). A note on neurofractal-based defect recognition and classification in nonwoven web images", *J. of the Text. Inst.*, 101: 1, pp. 46 – 51.

Ramaiah G.B., Chennaiah R.Y. & Satyanarayanarao G.K. (2010). Investigation and modeling on protective textiles using artificial neural networks for defence applications, *Materials Science and Engineering* B 168 pp.100–105.

Ramesh M.C., Rajamanickam R. & Jayaraman S. (1995). The Prediction of Yarn Tensile Properties by Using Artificial Neural Networks, *J. Text. Inst.*, 86: 3, pp. 459 – 469.

Rautenberg S. & Todesco J.L. (1999). Color Recipe Specification in the Textile Print Shop Using Radial Basis Function Networks, Engineering Applications of Bio Inspired ANN, pp 884-892, Springer.

Rich, E. & Knight, K. (1991). *Artificial Intelligence*, McGraw-Hill, New York, USA, pp. 487-524.

Sang-Song L. & Tsung-Huang L. (2007). FAST System Approach to Discriminate the Characterized Generic Hand of Fabrics, *Ind. J. Fibr. & Text. Res.*, Vol. 32, pp. 344-350.

Sette S. & Boullart L. (1996). Fault detection and quality assessment in textiles by means of neural nets, *Int. J. of Cloth. Sc. & Techn.*, Vol. 8 No. 1/2, pp. 73-83.

Sette S. & van Langenhove L. (2002). Optimising the Fibre-to-Yarn Production Process: Finding a Blend of Fibre Qualities to Create an Optimal Price/Quality Yarn, *AUTEX Res. J.*, Vol. 2, No2, pp 57-63.

Shady E., Gowayed Y., Abouiiana M., Youssef S. & Pastore C. (2006). Detection and Classification of Defects in Knitted Fabric Structures, *Text. Res. J.*, 76(4), 2006, pp 295-300.

She F.H., Kong L.X., Nahavandi S. & Kouzani A.S. (2002). Intelligent Animal Fiber Classification with Artificial Neural Networks, *Text. Res. J.*, 72(7), pp 594-600.

Shiau Y.-R., Tsai I-S. & Lin C.-S. (2000). Classifying Web Defects with a Back-Propagation Neural Network by Color Image Processing, *Text. Res. J.*, 70(7), 2000, pp. 633-640.

Shyr T.W., Lin J.Y. & Lai S.S. (2004). Approaches to Discriminate the Characteristic Generic Hand of Fabrics, *Text. Res. J.*, 74(4), pp 354-358.

Slah M., Amine H. T. & Faouzi, S. (2006). A new approach for predicting the knit global quality by using the desirability function and neural networks, *J. of the Text. Inst.*, 97:1, pp. 17 - 23.

Stylios, G. & Parsons-Moore, R. (1993). Seam pucker prediction using neural computing. *Intl. J. of Clothing Science and Technology,* vol. 5, no.5, pp. 24-27.

Stylios G.K. & Powell N.J. (2003). Engineering the drapability of textile fabrics, *Int. J. of Cloth. Sc. & Techn.*, Vol. 15 No. 3/4, pp. 211-217.

Stylios, G. & Sotomi, J.O., (1996). Thinking sewing machines for intelligent garment manufacture. *Intl. J. of Clothing Science and Technology,* vol. 8 (1/2), pp. 44-55.

Su J. & Xu B. (1999). Fabric wrinkle evaluation using laser triangulation and neural network classifier, *Opt. Eng.* 38, pp. 1688.

Subramanian S.N., Venkatachalam A. & Subramaniam V. (2007). Prediction and Optimization of Yarn Properties Using Genetic Algorithm/Artificial Neural Network, *Ind. J. of Fibre & Text. Res.*, Vol. 32, pp. 409-413.

Thevenet L., Dupont D. & Jolly-Desodt A.M. (2003). Modeling Color Change after Spinning Process Using Feedforward Neural Networks, *Color Research and Application,* Vol. 28, No 1, pp. 50-58.

Tilocca A., Borzone P., Carosio S. & Durante A. (2002). Detecting Fabric Defects with a Neural Network Using Two Kinds of Optical Patterns, *Text. Res. J.*, 72(6), pp 545-550.

Tokarska M. (2004). Neural Model of the Permeability Features of Woven Fabrics, *Text. Res. J.*, 74(12), pp 1045-1048.

Tokarska M. (2006). Assessing the Quality of Neural Models Using a Model of Flow Characteristics of Fabrics as an Example, *AUTEX Res. J.*, Vol. 6, No 3.

Tokarska M. & Gniotek K. (2009). Determination of Woven Fabric Impact Permeability Index, *Ind. J. of Fibre & Text. Res.*, Vol. 34, pp. 239-244.

Tran C.-D. & Phillips D. G. (2007). Predicting torque of worsted singles yarn using an efficient radial basis function network-based method, *J. of the Text. Inst.*, 98: 5, pp. 387 - 396.

Tran C.-D. , Phillips D. G. & Fraser W. B. (2010). Stationary solution of the ring-spinning balloon in zero air drag using a RBFN based mesh-free method, *J. of the Text. Inst.*, 101: 2, pp. 101—110.

Tsai I-S., Lin C.H. & Lin J.J. (1995). Applying an Artificial Neural Network to Pattern Recognition in Fabric Defects, *Text. Res. J., vol. 65, 3: pp. 123-130.*

Vangheluwe L., Sette S. and Kiekens, P.(1996). Modelling Relaxation Behaviour of Yarns Part II: Back Propagation Neural Network Model, *J. of the Text. Inst.*, 87: 2, pp. 305 - 310.

Vassiliadis S., Rangoussi M., Kremenakova D., Belkhir M., Boukhris M.A. & Louda O. (2010). Prediction of Bursting Strength of Fabrics using Neural Networks, *International Conference of Applied Research in Textile*, CIRAT-4, Monastir, Tunisia.

Ucar N. & Ertugrul S. (2007). Prediction of Fuzz Fibers on Fabric Surface by Using Neural Network and Regression Analysis, *Fibr. & Text. in East. Eur.*, 15, 2 (61) pp. 58-61.

Üreyen M.E. & Gürkan P. (2008). Comparison of Artificial Neural Network and Linear Regression Models for Prediction of Ring Spun Yarn Properties. I. Prediction of Yarn Tensile Properties, *Fibers and Polymers*, Vol.9, No.1, pp. 87-91.

Üreyen M.E. & Gürkan P. (2008). Comparison of Artificial Neural Network and Linear Regression Models for Prediction of Ring Spun Yarn Properties. II. Prediction of Yarn Hairiness and Unevenness, *Fibers and Polymers*, Vol.9, No.1, pp. 92-96.

Xu B. & Lin S. (2002). Automatic Color Identification in Printed Fabric Images by a Fuzzy-Neural Network, AATCC Review, 2(9), 42-45, 2002.

Xu B., Su J., Dale D.S. & Watson M.D. (2000). Cotton Color Grading with a Neural Network, *Text. Res. J.*, 70(5), pp 430-436.

Yao G., Guo J. & Zhou Y. (2005). Predicting the Warp Breakage Rate in Weaving by Neural Network Techniques, *Text. Res. J.*, 75(3), pp. 274-278.

Yazdi M.M, Semnani D. & Sheikhzadeh M. (2009). Moisture and Heat Transfer in Hybrid Weft Knitted Fabric with Artificial Intelligence, *J. of Applied Polymer Science*,Vol. 114, pp. 1731–1737.

Yin X.G. & Yu W. (2007). Selection and Evaluation of Input Parameters of Neural Networks Using Grey Superior Analysis, *Text. Res. J.*, 77(6), pp 377-386.

Youssefi M., & Faez K. (1999), "Fabric Handle Prediction Using Neural Networks", *Proceedings of the IEEE-EURASIP Workshop on Nonlinear Signal and Image Processing (NSIP'99)*, Antalya, Turkey,, Bogaziçi University Printhouse, ISBN 975-518-133-4.

Yuen C.W.M., Wong W.K., Qian S.Q., Fan D.D., Chan L.K. & Fung E.H.K. (2009). Fabric Stitching Inspection Using Segmented Window Technique and BP Neural Network, *Text. Res. J.*, 79(1), pp 24-35.

Wang K.F. & Zeng Y. (2008). A Wavelet Neural Network Applied to Textile Spinning, Aspects of Mathematical Modelling, Mathematics and Biosciences in Interaction, pp. 363-369.

Wong A.S.W., Li Y., Yeung P.K.W. & Lee P.W.H. (2003). Neural Network Predictions of Human Psychological Perceptions of Clothing Sensory Comfort, *Text. Res. J.*, 73(1), pp 31-37.

Wu, P., Fang, S.-C., Nuttle, H.L.W., King, R.E. & Wilson, J.R. (1994). Decision surface modeling of textile spinning operations using neural network technology, Textile, Fiber and Film Industry Techn. Conf., IEEE 1994 Annual,pp.0-19, 4-5 May 1994.

Zadeh, L. (1994). *Soft Computing and Fuzzy Logic,* IEEE Software, vol. 11, no.1-6, pp. 48-56.

Zeng Y.C., Wang K.F. & Yu C.W. (2004). Predicting the Tensile Properties of Air-Jet Spun Yarns, *Text. Res. J.*, 74(8), pp 689-694.

Zhang J., Wang X. & Palmer S. (2010). Performance of an Objective Fabric Pilling Evaluation Method, *Text. Res. J.*, 77(6), pp –.

Zhang J., Wang X. & Palmer S. (2010). Objective Pilling Evaluation of Nonwoven Fabrics, *Fibers and Polymers*, Vol.11, No.1, pp. 115-120.

b. Porous Properties Engineering

7

Prediction of Elastic Properties of Plain Weave Fabric Using Geometrical Modeling

Jeng-Jong Lin
Department of Information management, Vanung University
Taiwan, R.O.C.

1. Introduction

Fabrics are typical porous material and can be treated as mixtures of fibers and air. There is no clearly defined boundary and is different from a classical continuum for fabrics. It is complex to proceed with the theoretical analysis of fabric behavior. There are two main reasons (Hearle et al., 1969) for developing the geometrical structures of fabrics. One is to be able to calculate the resistance of the cloth to mechanical deformation such as initial extension, bending, or shear in terms of the resistance to deformation of individual fibers. The other is that the geometrical relationships can provide direct information on the relative resistance of cloths to the passage of air or light and similarly it can provide a guide to the maximum density of packing that can be achieved in a cloth. The most elaborate and detailed account of earlier work is contained in a classical paper by Peirce (Peirce, 1937). A purely geometrical model, which involves no consideration of internal forces, is set up by Peirce for the determination of the various parameters that were required. Beyond that, the geometrical structures of knits are another hot research issue, for instances, for plain-knitted fabric structure, Peirce (Peirce, 1947), Leaf and Glaskin (Leaf & Glaskin, 1955), Munden (Munden, 1961), Postle (Postle, 1971), DemirÖz and Dias (Demiröz & Dias, 2000), Kurbak (Kurbak, 1998), Semnani (Semnani et al., 2003), and Chamberlain (Chamberlain, 1949) et al. Lately, Kurbak & Alpyildiz propose a geometrical model for full (Kurbak & Alpyildiz, 2009) and half (Kurbak & Alpyildiz, 2009) cardigan structure. Both the knitted and woven fabrics are considered to be useful as a reinforcing material within composites. The geometrical structure of the plain woven fabric (WF) is considered in this study.

Woven fabric is a two-dimension (2-D) plane formation and represents the basic structural element of every item of clothing. Fabrics are involved to various levels of load in transforming them from 2-D form into 3-D one for an item of clothing. It is important to know the physical characteristics and mechanical properties of woven fabrics to predict possible behavior and eventual problems in clothing production processes. Therefore, the prediction of the elastic properties has received considerable attention. Fabric mechanics is described in mathematical form based on geometry. This philosophy was the main objective of Peirce's research on tensile deformation of weave fabrics. The load-extension behavior of woven fabrics has received attention from many researchers. The methods used to develop the models by the researches are quite varied. Some of the developed models are theory-based on strain-energy relationship e.g., the mode by Hearle and Shanahan (Hearle &

Shanahan, 1978), Grosberg and Kedia (Grosberg & Kedia, 1966), Huang (Huang, 1978), de Jong and Postle (Jong & Postle, 1977), Leaf and Kandil (Leaf & Kandil, 1980), and Womersley (Womersley, 1937). Some of them are based on AI-related technologies that have a rigorous, mathematical foundation, e.g., the model by Hadizadeh, Jeddi, and Tehran (Hadizadeh et al., 2009). Artificial neural network (ANN) is applied to learn some feature parameters of instance samples in training process. After the training process, the ANN model can proceed with the prediction of the load-extension behavior of woven fabrics. The others are based on digital image processing technology, e.g., the model by Hursa, Rotich and Ražić (Hursa et al., 2009). A digital image processing model is developed to discriminate the differences between the image of origin fabric and that of the deformed one after applying loading so as to determine pseudo Poisson's ratio of the woven fabric.

However, the above-mentioned methods have their limitations and shortcomings. The methods based on extension-energy relationship and system equilibrium need to use a computer to solve the basic equations in order to obtain numerical results that can be compared with experimental data. The methods based on AI-related technologies (i.e., ANN model) need to prepare a lot of feature data of samples for the model training before it can work on the prediction. Thus, the developed prediction models need quite a lot of tedious preparing works and large computation.

In this study, a unit cell model based on slice array model (SAM) (Naik & Ganesh, 1992) for plain weave is developed to predict the elastic behavior of a piece of woven fabric during extension. Because the thickness of a fabric is small, a piece of woven fabric can be regarded as a thin lamina. The plain weave fabric lamina model presented in this study is 2-D in the sense that considers the undulation and continuity of the strand in both the warp and weft directions. The model also accounts for the presence of the gap between adjacent yarns and different material and geometrical properties of the warp and weft yarns. This slice array model (i.e., SAM), the unit cell is divided into slices either along or across the loading direction, is applied to predict the mechanical properties of the fabric. Through the help of the prediction model, the mechanic properties (e.g., initial Young's modulus, surface shear modulus and Poisson's ratio) of the woven fabric can be obtained in advance without experimental testing. Before the developed model can be applied to prediction, there are parameters, e.g., the sizes of cross-section of the yarns, the undulation angles of the interlaced yarns, the Young's modulus and the bending rigidity of the yarns, and the unit repeat length of the fabric etc., needed to be obtained. In order to efficiently acquire these essential parameters, an innovative methodology proposed in this study to help eliminate the tedious measuring process for the parameters. Thus, the determination of the elastic properties for the woven fabric can be more efficient and effective through the help of the developed prediction model.

2. Innovative evaluation methodology for cross-sectional size of yarn

2.1 Definitions and notation for fundamental magnitudes of fabric surface

A full discussion of the geometrical model and its application to practical problems of woven fabric design has been given by Peirce (Peirce, 1937). The warp and weft yarns, which are perpendicular straight lines in the ideal form of the cloth, become curved under stress, and form a natural system of curvilinear co-ordinates for the description of its deformed state. The geometrical model of fabric is illustrated in Fig. 1. The basic parameters consist of two values of yarn lengths l, two crimp heights, h, two yarn spacings, P, and the sum of the diameters of the two yarns, D, give any four of these, the other three can be calculated from the model. There are three basic relationships as shown in equations 1~3 among theses parameters. The definitions of the parameters set in the structural model are denoted as follows.

$$h = (l - D\theta)\sin\theta + D(1 - \cos\theta) \tag{1}$$

$$p = (l - D\theta)\cos\theta + D\sin\theta \tag{2}$$

$$h_w + h_f = D \tag{3}$$

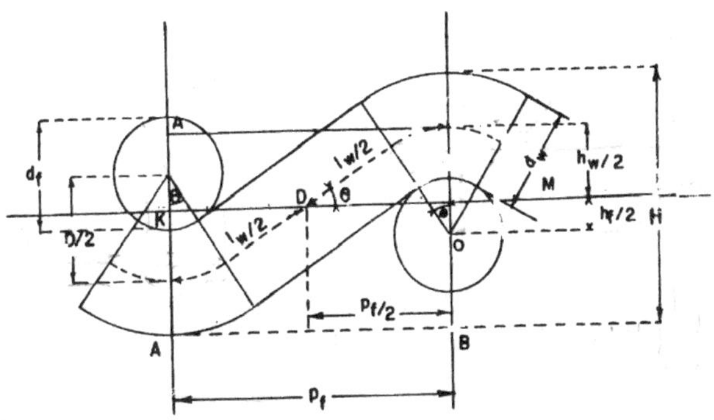

Fig. 1. Geometrical model (Hearle et al., 1969)

- Diameter of warp d_w, diameter of weft yarn d_f, and d_w+d_f=D.
- Distance between central plane of adjacent warp yarns P_w
- Distance between central plane of adjacent weft yarns P_f
- Distance of centers of warp yarns from center-line of fabric, $h_w/2$
- Distance of centers of weft yarns from center-line of fabric, $h_f/2$
- Inclination of warp yarns to center-line of fabbric, θ_w
- Inclination of weft yarns to center-line of fabbric, θ_f
- Length of warp between two adjacent weft yarns l_w
- Length of weft between two adjacent warp yarns l_f
- Warp crimp $C_w = l_w / P_f - 1$
- Weft crimp $C_f = l_f / P_w - 1$

The woven fabric, which consists of warp and weft yarns interlaced one another, is an anisotropic material (Sun et al., 2005). In order to construct an evaluation model to help determine the size of the deformed shape (i.e., eye shape) of cross section, Peirce's plain weave geometrical structure model is applied in this study. Because both the warp and weft yarns of the woven fabric are subject to the stresses during weaving process by the shedding, picking and beating motions, the shapes of cross section for the yarns are not actually the idealized circular ones (Hearle et al., 1969). The geometrical relations, illustrated in equations 1 and 2, can be obtained by projection in and perpendicular to the plane of the fabric. From these fundamental relations between the constants of the fabrics, the shape and the size of the cross section of the yarns can be acquired. Through the assistance of the proposed evaluation model, the efficiency and effectiveness in acquiring the size of section for warp (weft) yarn can be improved.

2.2 Yarn crimp

The crimp (Lin, 2007)(i.e., C_w) of warp yarn and that (i.e., C_f) of weft yarn can be obtained by using equation 4. The measuring of yarn crimp is performed according to Chinese National Standard (C.N.S.). During measuring the length of the yarn unravelled from sample fabric (i.e., with a size of 20 cm × 20 cm), each yarn was hung with a loading of 346/N (g), where N is the yarn count (840 yds/1lb) of the yarn for testing.

$$C=(L-L')/L' \tag{4}$$

Where L denotes the measured length of the warp (weft) yarn, L′ denotes the length of the fabric in the warp (weft) direction.

2.3 Cross sectional shape and size

Both the warp and weft yarns of the woven fabric are subject to the stresses from weaving process during the shedding, picking and beating motions. Due to subjecting to stresses, the shapes of cross section for the yarns are not actually the idealized circular ones. Fig. 2 shows the deformed eye shape of the yarn with a long diameter "a" and a short diameter "h". The sizes of warp and weft yarn are of denoted as a_w, h_w and a_f, h_f, respectively.

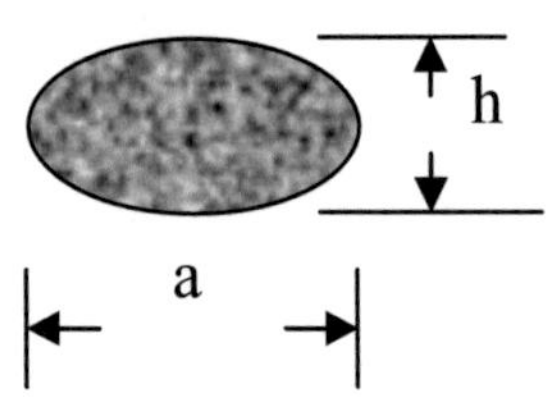

Fig. 2. Deformed shape of yarn

The Length of warp l_1 (weft l_2) between two adjacent weft (warp) yarns can be acquired using equation 5. The inclination of warp θ_1 (weft θ_2) yarns to center-line of fabric, can be obtained from equation 6, which is proposed by Grosberg (Hearle et al., 1969) and verified to be very close to the accurate inclination degree.

$$l = \frac{1+1/2nC}{N} \tag{5}$$

$$\theta = 106\sqrt{C} \tag{6}$$

where

C: Crimp
n: number of the warp and weft yarns in one weave repeat
N: Weaving density (ends/in; picks/in)

By Putting the measured values of l and θ into equations 1 and 2, the summation of the sizes of the short diameter for the warp and weft yarns (i.e., $D=h_w + h_f$ for the warp and weft in the thickness direction) and that of the sizes of the long diameter for the warp and weft yarns (i.e., $D_1=a_w^1 + a_f^1$ ($D_2=a_w^2 + a_f^2$) calculated from the known distance between central plane of adjacent warp P_w (weft P_f) yarns). Because the obtained summation values calculated from the known distances between central plane of adjacent warp yarns P_w and weft yarns P_f are different, the average value $\overline{D}$ of them is calculated. The obtained $\overline{D}$ represents the sum of the long diameters of the warp and weft yarns. The larger the value of $\overline{D}$ is, the more flattened shape the warp and weft yarns are.

Although the summation for the diameter sizes of the warp and weft yarn in the length (thickness) direction of the woven fabric is obtained, the individual one for warp (weft) yarn is still uncertain. In order to estimate the individual diameters of warp and weft yarn, the theoretical diameter (Lai, 1985) is evaluated using equation 7 in the study. The diameter of the individual yarn can be estimated by the weigh ratios shown in equations 8~11.

$$d\ (\mu m) = 11.89\sqrt{\frac{Denier}{\rho}} \tag{7}$$

where

Denier: denier of yarn
ρ: specific gravity of yarn

$$a_w = \overline{D} \times \frac{d_w}{d_w + d_f} \tag{8}$$

$$a_f = \overline{D} \times \frac{d_f}{d_w + d_f} \tag{9}$$

$$h_w = D \times \frac{d_w}{d_w + d_f} \tag{10}$$

$$h_f = D \times \frac{d_f}{d_w + d_f} \tag{11}$$

where

a_w: Long diameter of eye-shaped warp yarn
a_f: Long diameter of eye-shaped weft yarn
h_w: Short diameter of eye-shaped warp yarn
h_f: Short diameter of eye-shaped weft yarn
$D = h_w + h_f$
$\overline{D} = (D_1 + D_2)/2$
d_w: Theoretical diameter of circular warp yarn
d_f: Theoretical diameter of circular weft yarn

3. Geometrical model and properties of spun yarn

The idealized staple fiber yarn is assumed to consist of a very large number of fibers of limited length, uniformly packed in a uniform circular yarn. The fibers are arranged in a helical assembly, following an idealized migration pattern. Each fiber follows a helical path, with a constant number of turns per unit length along the yarn, in which the radial distance from the yarn axis increases and decreases slowly and regularly between zero and the yarn radius. A fiber bundle illustrated in Fig. 3a, which is twisted along a helical path as shown in Fig. 3b, is manufactured into a twisted spun yarn.

In order to describe the distributed stresses on the body of yarn, a hypothetical rectangular element from is proposed and illustrated in Fig. 4. The stresses acting on the elemental volume dV are shown in Fig. 4. When the volume dV shrinks to a point, the stress tensor is represented by placing its components in a 3×3 symmetric matrix. However, a six-independent-component is applied as follows.

$$\sigma = \left[\sigma_x, \sigma_y, \sigma_z, \tau_{yz}, \tau_{zx}, \tau_{xy} \right]^T \tag{12}$$

Where $\sigma_x, \sigma_y, \sigma_z$ are normal stresses and $\tau_{yz}, \tau_{zx}, \tau_{xy}$ are shear stresses.

The strains corresponded to the acting stresses can be represented as follows.

$$\varepsilon = \left[\varepsilon_x, \varepsilon_y, \varepsilon_z, \gamma_{yz}, \gamma_{zx}, \gamma_{xy} \right]^T \tag{13}$$

Where $\varepsilon_x, \varepsilon_y, \varepsilon_z$ are normal strains and $\gamma_{yz}, \gamma_{zx}, \gamma_{xy}$ are engineering shear strains.

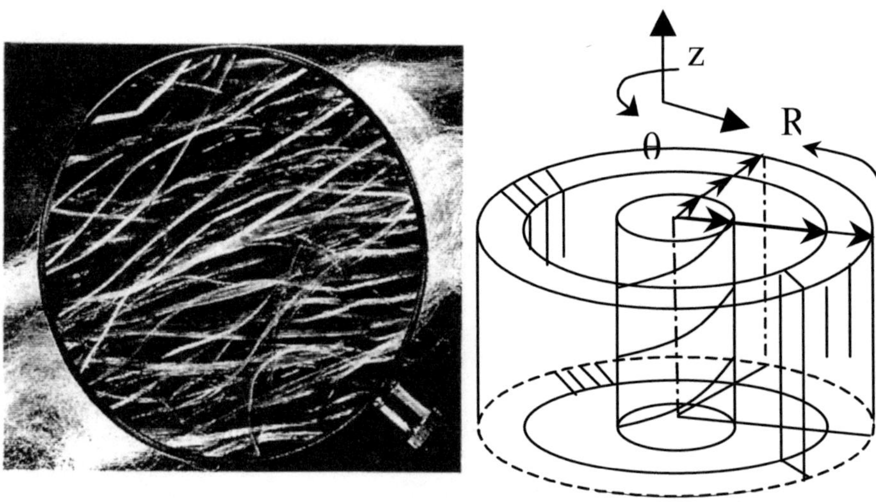

(a) A fiber bundle as seen under a magnifying (Curiskis & Carnaby, 1985) b) fiber bundle twisted along a helical path

Fig. 3. A fiber bundle

In the continuum mechanics of solids, constitutive relations are used to establish mathematical expressions among the variables that describe the mechanical behavior of a material when subjected to applied load. Thus, these equations define an ideal material response and can be extended for thermal, moisture, and other effects. In the case of a linear elastic material, the constitutive relations may be written in the form of a generalized Hooke's law:

$$\sigma = [S] \varepsilon \tag{14}$$

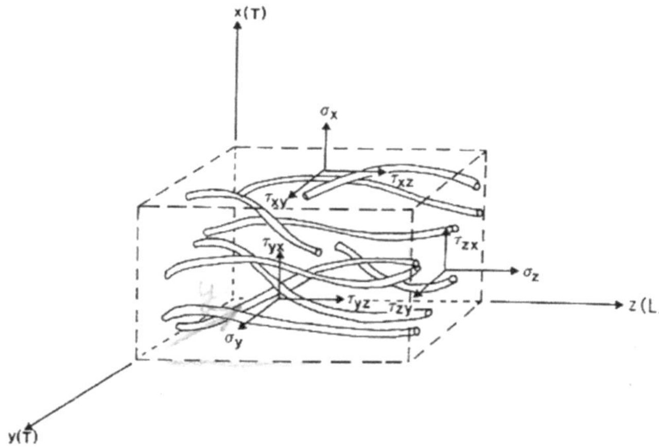

Fig. 4. A rectangular element of a fiber bundle (Curiskis & Carnaby, 1985)

That is

$$\begin{Bmatrix} \sigma_x \\ \sigma_y \\ \sigma_z \\ \tau_{xy} \\ \tau_{yz} \\ \tau_{zx} \end{Bmatrix} = \begin{bmatrix} S_{11} & S_{12} & S_{13} & S_{14} & S_{15} & S_{16} \\ S_{21} & S_{22} & S_{23} & S_{24} & S_{25} & S_{26} \\ S_{31} & S_{32} & S_{33} & S_{34} & S_{35} & S_{36} \\ S_{41} & S_{42} & S_{43} & S_{44} & S_{45} & S_{46} \\ S_{51} & S_{52} & S_{53} & S_{54} & S_{55} & S_{56} \\ S_{61} & S_{62} & S_{63} & S_{64} & S_{65} & S_{66} \end{bmatrix} \begin{Bmatrix} \varepsilon_x \\ \varepsilon_y \\ \varepsilon_z \\ \gamma_{xy} \\ \gamma_{yz} \\ \gamma_{zx} \end{Bmatrix} \tag{15}$$

$$\varepsilon = [C]\sigma \tag{16}$$

That is

$$\begin{Bmatrix} \varepsilon_x \\ \varepsilon_y \\ \varepsilon_z \\ \gamma_{xy} \\ \gamma_{yz} \\ \gamma_{zx} \end{Bmatrix} = \begin{bmatrix} C_{11} & C_{12} & C_{13} & C_{14} & C_{15} & C_{16} \\ C_{21} & C_{22} & C_{23} & C_{24} & C_{25} & C_{26} \\ C_{31} & C_{32} & C_{33} & C_{34} & C_{35} & C_{36} \\ C_{41} & C_{42} & C_{43} & C_{44} & C_{45} & C_{46} \\ C_{51} & C_{52} & C_{53} & C_{54} & C_{55} & C_{56} \\ C_{61} & C_{62} & C_{63} & C_{64} & C_{65} & C_{66} \end{bmatrix} \begin{Bmatrix} \sigma_x \\ \sigma_y \\ \sigma_z \\ \gamma_{xy} \\ \gamma_{yz} \\ \gamma_{zx} \end{Bmatrix} \tag{17}$$

Where σ and ε are suitably defined stress and strain vectors (Carnaby 1976) (Lekkhnitskii , 1963), respectively, and [S] and [C] are stiffness and compliance matrices, respectively, reflecting the elastic mechanical properties of the material (i.e., moduli, Poission's ratios, etc.) There are four possible models (Curiskis & Carnaby, 1985) (Carnaby & Luijk, 1982) for the continuous fiber bundle, i.e., the general fiber bundle, Orthotropic material, square-symmetric material, and transversely isotropic material. The orthotropic material model is adopted in this study.

Thwaites (Thwaites, 1980) applied his equations subject to the further constrain of incompressibility of the continuum, that is,

$$\varepsilon_x + \varepsilon_y + \varepsilon_z = 0 \tag{18}$$

In which case the two Poisson's ratio terms are no longer independent：

$$G_{TT} = \frac{E_T}{2(1 + \upsilon_{TT})} \tag{19}$$

$$\upsilon_{TT} = 1 - \upsilon_{TL} \tag{20}$$

And

$$\upsilon_{TL} = \upsilon_{LT} E_T / E_L \tag{21}$$

Thus, for the incompressible material of a spun yarn, whose elastic properties can be described using the seven elastic constants, i.e., G_{TT}, G_{LT}, E_T, E_L, v_{LT}, v_{TL}, and v_{TT}, an orthotropic material model is adopted to depict it in this study. The orthotropic material model as shown in Fig. 4, the fiber packing in the xy plane and along the z axis is such that the xz and yz planes are also planes of elastic symmetry. Furthermore, the continuum idealization then allows application of the various mathematical techniques of continuum mechanics to simplify the setting-up of physical problems in order to obtain useful results for various practical situations. For the study, the yarn (fiber bundle) is mechanically characterized as a degenerate square-symmetric homogeneous continuum. The elastic compliance relationship (Carnaby, 1980) can be described using the moduli and Poisson's ratio parameters illustrated as follows.

$$\begin{Bmatrix} \varepsilon_x \\ \varepsilon_y \\ \varepsilon_z \\ \gamma_{xy} \\ \gamma_{yz} \\ \gamma_{zx} \end{Bmatrix} = \begin{bmatrix} \frac{1}{E_T} & -\frac{V_{TT}}{E_T} & -\frac{V_{LT}}{E_L} & 0 & 0 & 0 \\ -\frac{V_{TT}}{E_T} & \frac{1}{E_T} & -\frac{V_{LT}}{E_L} & 0 & 0 & 0 \\ -\frac{V_{TL}}{E_T} & -\frac{V_{TL}}{E_T} & \frac{1}{E_L} & 0 & 0 & 0 \\ 0 & 0 & 0 & \frac{1}{G_{TT}} & 0 & 0 \\ 0 & 0 & 0 & 0 & \frac{1}{G_{LT}} & 0 \\ 0 & 0 & 0 & 0 & 0 & \frac{1}{G_{LT}} \end{bmatrix} \begin{Bmatrix} \sigma_x \\ \sigma_y \\ \sigma_z \\ \gamma_{xy} \\ \gamma_{yz} \\ \gamma_{zx} \end{Bmatrix} \tag{22}$$

Where E_L is the longitudinal modulus governing uniaxial loading in the longitudinal (z) direction. v_{LT} is the associated Poisson ratio goving induced transverse strains, E_T is the transverse modulus goving uniaxial loading in the transverse (x or y) direction. v_{TT} is the associated Poisson ratio governing resultant strains in the remaining orthogonal transverse (y or x) direction. v_{TT} is the associated Poisson ratio governing the induced strain in the longitudinal direction, G_{LT} is the longitudinal shear modulus goving shear in the longitudinal direction, and G_{TT} is the transverse shear modulus governing shear in the transverse plane.

The theoretical equation for Young's modulus of the spun yarn developed by Hearle (Hearle et al., 1969) is adopted in the study. It is illustrated in equation 23. The fibers are assumed to have identical dimensions and properties, to be perfectly elastic, to have an axis of symmetry, and to follow Hooke's and Amonton's laws. The strains involved are assumed to be small. The transverse stresses between the fibers at any point are assumed to be the same in all directions perpendicular to the fiber axis. Beyond these, there are other assumptions for the developed equation. Thus, it can not expected to be numerically precise because of the severe approximations, can be expected to indicate the general form of the factors affecting staple fiber yarn modulus. However, despite the differences between the idealized model and actual yarns, it is useful to have a knowledge of how an idealized assembly would behave.

$$y_M = f_M \times \frac{1-\dfrac{2}{3L_f}\left\{\dfrac{a\gamma W_y^{1/2}(1+4\pi v_f\varphi^{-1}\tau^2 10^{-5})^{1/2}}{4\tau\mu[1-(1+4\pi v_f\varphi^{-1}\tau^2 10^{-5})^{-1/2}}\right\}^{1/2}}{(1+4\pi v_f\varphi^{-1}\tau^2 10^{-5})} \tag{23}$$

Where

f_M: modulus of fiber

L_f: fiber length

a: fiber radius

γ : migration ratio (γ=4 for spun yarn)

W_y: yarn count (tex)

v_f: specific volume of fiber

ϕ: packing fraction

τ: twist factor (tex$^{1/2}$ turn/cm)

μ: coefficient of friction of fiber

The flexural rigidity of a filament yarns is the sum of the fiber flexural rigidities under the circumstance that the bending length of the yarn is equal to that of a single fiber. It has been confirmed experimentally by Carlen (Hearle et al., 1969) (Cooper, 1960). The spun yarn is regarded as a continuum fiber bundle in the study, so the flexural rigidity of it is approximately using the same prediction equation illustrated in equation 24.

$$G_y=N_fG_f \tag{24}$$

Where

N_f: cross-sectional fiber number

G_f: flexure rigidity of fiber

The change of yarn diameter and volume with extension has been investigated by Hearle etc. (Hearle et al., 1969) Through the experimental results for the percentage reductions in yarn diameter with yarn extension by Hearle, the Poisson's ratio v_{LT} in the extension direction can be estimated to be at the range of 0.6 ~ 1.1. The Poisson's ratio v_{LT} is set to be 0.7 for the spun yarn in the study.

Young's modulus E_L of the yarn in the (length) extension direction can be estimated using equation 23. Equation 24 can be applied to estimate the flexure rigidity G_{TL} of the yarn. Through putting the obtained E_L, G_{TL}, and the set value of 0.7 for the Poisson's ratio v_{LT} of the yarn into equations 19~21, the other four elastic properties (i.e., G_{TT}, E_T, v_{TL}, v_{TT}) can be acquired, respectively.

Now that the elastic properties of a spun yarn can be represented using the above-mentioned matrix. The simplification for the setting-up of physical problems using various mathematical techniques of continuum mechanics can thus be achieved. For the study, the yarn (fiber bundle) is mechanically characterized as a degenerate square-symmetric homogeneous continuum. The complex mechanic properties of the combination of the warp and weft yarns interlaced in woven fabric can be possible to be constructed as follows.

4. Construction of unit cell model

4.1 Mechanical properties of unit cell of fabric

Fig.5a illustrates a unit cell (Naik & Ganesh, 1992) of woven fabric lamina. There is only one quarter of the interlacing regin analysed due to the symmetry of the interlacing regin in plain weave fabric.

The analysis of the unit cell, i.e., slice array model (SAM), is performed by dividing the unit cell into a number of slices as illustrated in Fig. 5b. The sliced picess are idealized in the form of a four-layered laminate, i.e., an asymmetric crossply sandwiched between two pure matrix (if any) layers as shown in Fig. 5c. The effective properties of the individual layer considering the presence of undulation are used to evaluate the elastical constants of the idealized laminate. Because there is no matrix applied, the top and the bottom layer of the unit cell are not included in this study.

There are two shape functions proposed in the study, one as shown in Fig. 6a for the cross-section in the warp direction and the other as illustrated in Fig. 6b for the one in the weft direction.

Along the warp direction, i.e., in the Y-Z plane (Fig. 5(a))

$$z_{y_1}(y) = -\frac{h_f}{2}\cos\frac{\pi y}{a_{yt}} \tag{25}$$

$$z_{y2}(y) = \frac{h_f}{2}\cos\frac{\pi y}{a_f + g_f} \tag{26}$$

Where

$$a_{yt} = \frac{\pi a_f}{2\left[\pi - \cos^{-1}\left(\frac{2z_{yt}}{h_f}\right)\right]}, \quad z_{yt} = \frac{h_f}{2}\cos\frac{\pi a_f}{2(a_f + g_f)}$$

and

$$\left.\begin{aligned} hy_1(y) &= \frac{h_f + h_m}{2} - zy_2(y) \\ hy_2(y) &= h_w \\ hy_3(y) &= zy_2(y) - zy_1(y) \text{ (when } y = 0 \to a_f/2) \\ &= 0 \text{ (when } y = a_f/2 \to (a_f + g_f)/2) \\ hy_4(y) &= \frac{h_f + h_m}{2} - zy_1(y) \end{aligned}\right\} \tag{27}$$

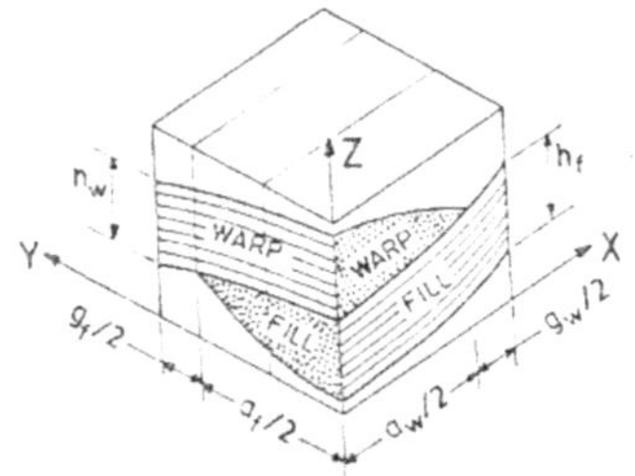

(a) Unit Cell

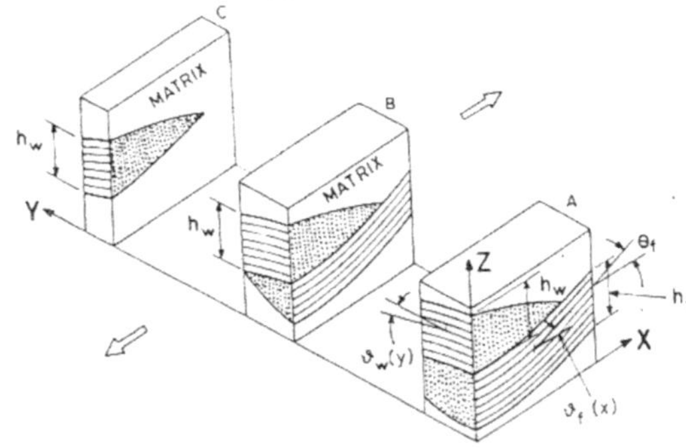

(b) Actual Slice

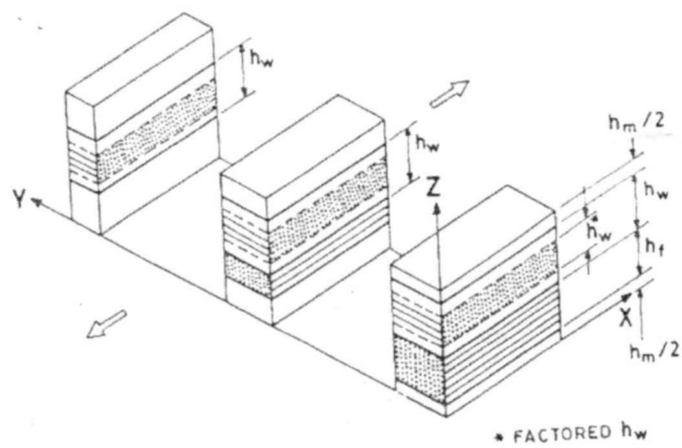

(C) Idealized Slice

Fig. 5. Illustration for the slicing of unit cell and the idealized slice (Naik & Ganesh, 1992)

Along the weft (fill) direction, i.e., in the X-Z plane (Fig. 5(a))

$$zx_{1_1}(x,y) = \frac{h_w}{2}\cos\frac{\pi x}{a_{xt}} - hy_1(y) + \frac{h_m}{2} \tag{28}$$

$$zx_2(x,y) = -\frac{h_w}{2}\cos\frac{\pi x}{a_w + g_w} - hy_1(y) + \frac{h_m}{2} \tag{29}$$

where

$$a_{xt} = \frac{\pi a_w}{2\cos^{-1}(\frac{2z_{xt}}{h_w})}, \quad z_{xt} = -\frac{h_w}{2}\cos\frac{\pi a_w}{2(a_w + g_w)};$$

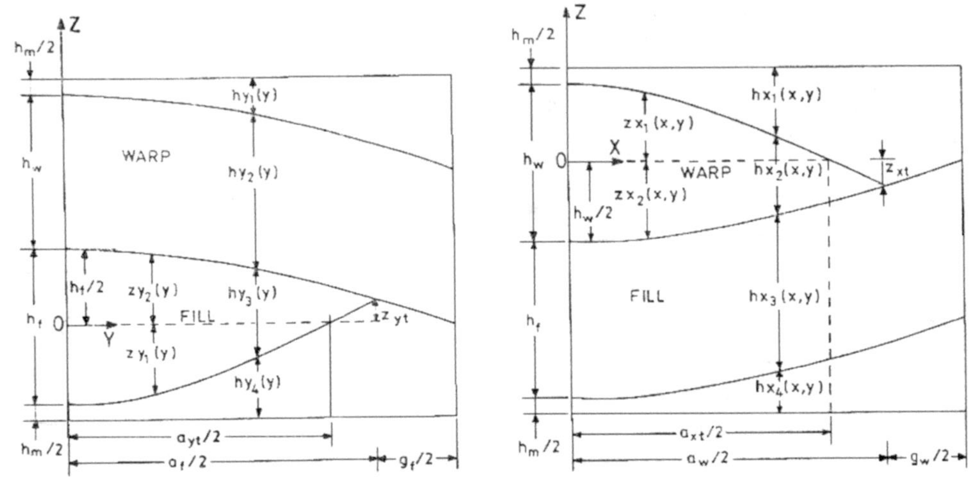

(a) Plain weave cross-section: warp (b) Plain weave cross-section: weft (fill)

Fig. 6. Illustration for the shape functions (Naik & Ganesh, 1992)

And

$$\left.\begin{aligned} hx_1(x,y) &= \frac{h_w + h_m}{2} - zx_1(x,y) \\ hx_2(x,y) &= zx_1(x,y) - zx_2(x,y) \text{ (when } x = 0 \rightarrow a_w / 2) \\ &= 0 \text{ (when } x = a_w / 2 \rightarrow (a_w + g_w) / 2) \\ hx_3(x,y) &= hy_3(y) \\ hx_4(x,y) &= zx_2(x,y) - hx_3(x,y) + (h_w + h_m) / 2 + h_f \end{aligned}\right\} \quad (30)$$

The local off-axis angles in the weft (i.e.,fill) and warp direction can be calculated using equations 31 and 32, respectively.

$$\vartheta_f(x) = \tan^{-1} \frac{d}{dx}[zx_2(x,y)] = \tan^{-1}\left(\frac{\pi h_w}{2(a_w + g_w)} \sin \frac{\pi x}{(a_w + g_w)}\right) \quad (31)$$

$$\vartheta_w(y) = \tan^{-1} \frac{d}{dy}[zy_2(y)] = \tan^{-1}\left(\frac{\pi h_f}{2(a_f + g_f)} \sin \frac{\pi y}{(a_f + g_f)}\right) \quad (32)$$

Because the woven fabric is manufactured by the interlacing of warp and weft yarn, there exists a certain amount of gap between two adjacent yarns. It is obvious that the presence of a gap between two the adjacent yarns would affect the stiffness of the WF lamina. Furthermore, the warp and weft yarns interlaced in fabric are undulated. It can be expected that the elastic properties of the yarn under the straight form and the undulated one are definitely different.

4.2 Mechanical properties of the undulated spun yarn

The respective off-axis angles reduce the effective elastic constants in the global X and Y directions. The increased compliance can be evatuated as follows. (Lekhnitskii, 1963).

$$C_{11}(\vartheta)=\frac{1}{E_L(\vartheta)}=\frac{m^4}{E_L}+(\frac{1}{G_{LT}}+\frac{2\upsilon_{LT}}{E_L})m^2n^2+\frac{n^4}{E_T} \tag{33}$$

$$C_{22}(\vartheta)=\frac{1}{E_T(\vartheta)}=\frac{1}{E_T} \tag{34}$$

$$C_{12}(\vartheta)=\frac{\upsilon_{TL}(\vartheta)}{E_T(\vartheta)}=\frac{\upsilon_{TL}m^2}{E_T}+\frac{\upsilon_{TT}n^2}{E_T} \tag{35}$$

$$C_{66}(\vartheta)=\frac{1}{G_{LT}(\vartheta)}=\frac{m^2}{G_{LT}}+\frac{n^2}{G_{TT}} \tag{36}$$

Where $m=\cos\vartheta$, $n=\sin\vartheta$; E_L and E_T are Young's moduli of yarns in the length direction and the cross-sectional direction, respectively; G_{LT} and G_{TT} are the flexure rigidity and torsion one, respectively. The value of E_L is calculated by the theoretical equation 23 developed by Hearle (Hearle et al., 1969). Through the experimental results for the percentage reductions in yarn diameter with yarn extension by Hearle, the values of E_T , G_{LT} , and G_{TT} for the yarns are determined based on the orthotropic material model proposed by Curiskis and Carnaby (Curiskis & Carnaby, 1985).
The compliance of yarn is related to the angle of undulation of the yarn crimped in the fabric. The off-axis angle for each specific location at the warp and weft yarn can be acquired from equation 31 and 32. In order to precisely evaluate the changed compliances for the warp and weft yarn, the mean value of the compliance is applied and illustrated in equation 37.

$$\bar{C}_{ij}=\frac{1}{\theta}\int_0^{\theta}C_{ij}(\vartheta)d\vartheta \tag{37}$$

where θ is the angle of undulation for the yarn at $x=a_w/2+g_w/2$.

4.3 Evaluation of mechanical properties of slices and unit cell

After evaluating the changed elastic constants of the warp and weft yarn using equation 37, the extensional stiffness of the slice can be obtained from equation 38. The integration used in the equation is fulfilled by neumatic method in the study.

$$A_{ij}^{sl}(y)=\frac{1}{H}\sum_{k=1}^{2}hx_k(x,y)(\bar{S}_{ij})_k \tag{38}$$

Where, $hx_k(x,y)$ and $(\bar{S}_{ij})_k$ are the thickness and mean transformed stiffness of the kth layer in the nth slice.
The sliced pieces are idealized in the form of a two-layered lamina, i.e., warp and weft asymmetric crossply sandwiched between two pure matrix (if any) layers as shown in Fig. 5c. If there is no matrix applied on the fabric, i.e., the 1st and the 4th layers are vacant; the extensional stiffness of the slice consisting of a warp and a weft yarn can still be estimated from equation 38. The effective properties of the individual layer considering the presence of undulation are used to evaluate the elastical constants of the idealized woven fabric lamina.

Based on Fig. 5 and Fig. 6, $hx_k(x,y)$ is evaluated at constant x, for different values of y. The thickness of the warp yarn is maximum at x=0 and zero from $x=a_w/2$ to $x=(a_w+g_w)/2$. In order to acquire the mean thickness of each layer of different material, the coordinate of x is set to be at the middle (i.e., $x=(a_w/2+g_w/2)/2$) of the unit cell in the study. The extensional stiffness of the unit cell is evaluated from those of the slices by assembling the slices together under the isostrain condition in all the slices. In other words, the in-plane extensional stiffness of the unit cell is evaluated and can be expressed as follows.

$$A_{ij} = \frac{2}{(a_f + g_f)} \int_0^{(a_f+g_f)/2} A_{ij}^{sl}(y)dy \tag{39}$$

According to Fig. 5a, the unit cell is obviously not symmetric about its midplane, so there exist the coupling stiffness terms. However, the coupling terms in two adjacent unit cells of the woven fabric lamina would be opposite signs due to the nature of interlacing of yarns in the plain weave fabric. Thus, the elastic constants of the unit can be obtained and expressed as follows.

$$\left.\begin{aligned} E_x &= A_{11}\left(1 - \frac{A_{12}^2}{A_{11}A_{22}}\right) \\ G_{xy} &= A_{66} \\ \upsilon_{yx} &= \frac{A_{12}}{A_{22}} \end{aligned}\right\} \tag{40}$$

Where, E_x is the Young's modulus, G_{xy} is the flexure rigidity, and v_{yx} is the Poisson's ratio for the fabric, respectively.
Accordingly, the Young's modulus in the warp direction can be calculated using the above-mentioned steps as well.

5. Experiments

5.1 Characteristics of sample fabrics

The measured characteristics of the sample fabric are shown in table 1. The theoretically generalized elastic properties of cotton fiber are given in Table 2. Base on the data of the raw material cotton fiber, Young's modulus of the cotton spun yarn (i.e., y_M) is predicted to be 6694 (N/mm^2) using equation 23 developed by Hearle (Hearle et al., 1969). The flexure rigidity (i.e., G_{LT}) of the spun yarn can be acquired as 0.0031 (N/mm^2) using equation 24 as well.

Weave	Yarn count (warp×weft)	Yarn specific volume, cm^3/g	Density, yarns/inch (warp×weft)	Material (warp×weft)
plain	20'S × 20'S	1.22	60 × 60	C × C

Yarn count 'S=840 yd/ 1lb, Material: C=cotton

Table 1. Characteristics of woven fabric sample

5.2 Preprocessing and procedures

The sample fabrics are scoured at 30°C for one hour in sodium carbonate. Then they are washed and dried at room temperature. Static tensile test specimens are prepared according to Chinese National Standard (C.N.S.). The testing size is 25mm×100mm. The specimens are tested at room temperature (25°C) at a crosshead speed of 10 mm/min. A total of ten specimens are tested, five of which are the samples made for testing in warp direction and the other five are for testing in weft direction direction. An experimental program is designed by C language to calculate the elastic constants of the woven fabric lamina along the warp and weft directions in the study. The experiment is performed on cotton woven fabric lamina according to the essential requirements proposed by Bassett et al. (Bassett et al., 1999).

f_M (N/mm^2)	L_f (mm)	a (mm)	γ	W_y (tex)	v_f	ϕ	τ ($tex^{1/2}$ turns/cm)	μ	E_L (N/mm^2)	G_{LT} (Nmm^2)
8000	40	0.0130	4	31.9200	0.6500	0.5300	33.36	0.22	6694	0.0031

f_M: Young's modulus of fiber, L_f: fiber length, a: fiber radius, γ:migration ratio, W_y: yarn count, v_f: specific volume of fiber, ϕ:packing fraction, τ:twist factor($tex^{1/2}$turn/cm), μ:coefficient of friction

Table 2. Characteristics of the cotton fiber

6. Results and discussion

6.1 Cross-sectional size of yarn

Woven fabric, which consists of warp and weft yarns interlaced one another, is an anisotropic material. Peirce's plain weave geometrical structure model is used to set up a prediction model for the shapes and sizes of warp and weft yarn. Both the warp and weft yarns of woven fabric are subject to the stresses from weaving process during the shedding, picking, beating motions. Due to the occurred stresses, the shapes of section for yarns are not actually the idealized circular ones. It shows that the theoretically calculated results are pretty consistent to the experimental. Through the evaluation methodology for cross-sectional size of yarn based on Peirce's structure model, the efficiency and effectiveness in acquiring the sectional size for warp (weft) yarn can be improved.

The geometrical scales of the fabric are determined by means of an optical microscope at a magnification of 20. The obtained results are used to compare with the calculated ones for validation of the innovative evaluation methodology proposed in this study. The measured and calculated results are illustrated in Table 3 and Table 4, respectively. Table 5 shows there are errors less than 5% for each between the calculated and the tested results. It reveals that the proposed method is of good accuracy and can more efficiently acquire the geometrical sizes, i.e., the long and short diameters of the warp and weft yarns in the fabrics.

Crimp		Undulation angle (degree)		Length of repeat unit (mm)		Crimped length (mm)	
C_w	C_f	θ_w	θ_f	P_w	P_f	l_w	l_f
0.06	0.06	25.9646	25.9646	0.4305	0.4305	0.4487	0.4487

Table 3. Measured and induced results of the basic sizes for the fabric

Long diameter $p=(l-D\theta)\cos\theta+D\sin\theta$		Short diameter $h=(l-D\theta)\sin\theta+D(1-\cos\theta)$	Diameter (circular shape)		Diameter (actual eye shape)			
			warp	weft	warp		weft	
D_1(mm)	D_2(mm)	D (mm)	d_w (mm)	d_f (mm)	a_w (mm)	h_w (mm)	a_f (mm)	h_f (mm)
0.8856	0.8856							
$\overline{D}$ (mm)		0.3288	0.2140	0.2140	0.4428	0.1644	0.4428	0.1644
0.8856								

Table 4. Calculated results by evaluation methodology based on Peirce's model

Predicted				Measured			
warp		weft		warp		weft	
a_w (mm)	h_w (mm)	a_f (mm)	h_f (mm)	a_w(mm)	h_w (mm)	a_f(mm)	h_f (mm)
0.4428	0.1644	0.4428	0.1644	0.4348	0.1625	0.4348	0.1625

Table 5. Comparison between the predicted and the measured sizes

6.2 Extensional behavior of fabric

The generalized load-extension curve as illustrated in Fig. 7 (Hearle et al., 1969) shows three actions, as in the initial decrimping region the load-extension curve possesses a point of inflexion. The initial high modulus of the fabric is probably due to frictional resistance to bending of the thread. Once the frictional restraint is overcome, a relatively low modulus is obtained which is mainly governed by the force needed to unbend the threads in the direction in which the force is being applied, and at the same time, the need to increase the curvature in the threads at the right angles to the direction of application of the force. As the crimp is decreased the size of this force rises very steeply and, as a result, the fibers themselves begin to be extended and in the final region, the load-extension properties of the cloth are almost entirely governed by the load-extension properties of the yarns themselves. According to the description of the load-extension process, it can be concluded that the initial modulus of the fabric is determined by the first part. In other words, the resistance to bending of the thread (including frictional forces affected by the surface features of the warp (weft) yarns and the bending rigidity of the warp (weft) yarns) governs the initial modulus.

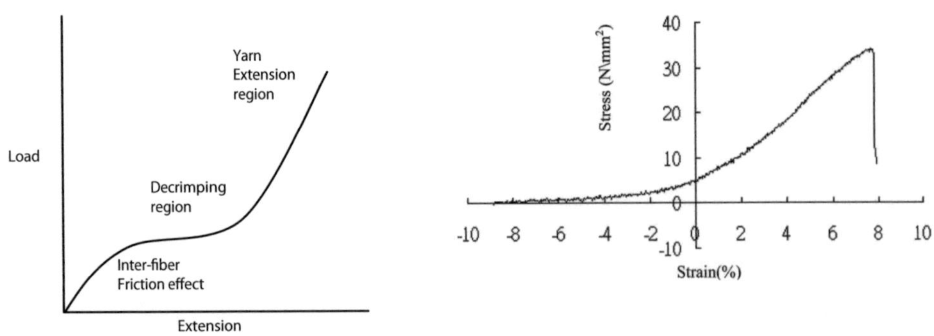

(a) Generalized load-extension curve (b) Sample fabric stress-strain curve

Fig. 7. Comparison between generalized and tested curve

This above-mentioned viewpoints on extension-load curve by Grosberg is quite in accordance with the equation deduced for Young's modulus by Leaf & Kandil (Leaf & Kandil, 1980). According to the deduced equation, the initial modulus is related to the bending rigidities of warp and weft yarns. However, the tested results for the sample fabric shown in Fig. 7b, in which there is a lack of level out appearance in the load-extension curve for the sample fabric, are different from the generalized load-extension curve. This is mainly because the sample fabric used in the study is of a small crimped angle of 26°. It brings about that once the resistances of the friction force occurred from the rough contact surfaces of the adjacent yarns and the bending rigidities of the warp and weft yarns are conquered, the second stage (i.e., lever out region) is skipped and directly move to the third region (i.e., the applied force used to extend the fibers themselves in the yarn).

6.3 Validation of slice array model

The slice array model (i.e., SAM) (Naik & Ganesh, 1992), which considers the actual yarn cross-sectional geometry and the presence of a gap between the adjant yarns, is presented for the elastic analysis of 2-D orthogonal plain weave fabric lamina. The shape functions agree well with the actual geometry of the woven fabric lamina. The assumption that the locally bending deformations are constrained is realistic considering the nature of interlacing of the plain weave fabrics.

In order to examine the micromechanical approaches for the prediction of the elastic constants of a woven fabric lamina, a plain woven fabric with warp and weft spun yarn of cotton fibers is selected. The elastic properties of the cotton fiber are given in Table 1. Based on the mechanical properties of the raw material cotton fiber, Young's modulus of the cotton spun yarn (i.e., y_M) is predicted to be 6694 (N/mm^2) using equation 23 developed by Hearle (Hearle et al., 1969). However, it is much higher than the actual measured value of 581 (N/mm^2). As Hearle (Hearle et al, 1969) said the prediction equation can not be expected to be numerically precise because of severe approximations.

E_T (N/mm^2)	G_{TT} (N/mm^2)	G_{LT} (N/mm^2)	v_{LT}	v_{TT}	v_{TL}
5592^a	1969^a	0.0031^a	0.70^a	0.42^a	0.58^a
484^b	171^b	0.0031^b	0.70^b	0.42^b	0.58^b

a: Calculated based on E_L=6694 (N/mm^2); b: Calculated based on E_L=581 (N/mm^2)

Table 6. Calculated elastic properties of the straight yarn based on Predicted and Measured E_L

E_x (N/mm^2)	G_{xy} (N/mm^2)	v_{yx}
3622^a	669^a	1.95×10^{-5} ^{a}
316^b	58^b	2.11×10^{-4} ^{b}
363^c	---	---

a: Calculated based on E_L=6694 (N/mm^2); b: Calculated based on E_L=581 (N/mm^2); c: measured

Table 7. Elastic properties of plain weave fabric lamina: Comparison of predicted and experimental results

Once the value of E_L is determined and the value of v_{LT} is set at 0.7, those of E_T, G_{LT} and G_{TT} for the yarns can be calculated based on the orthotropic material model (Curiskis &

Carnaby, 1985) proposed by Curiskis and Carnaby. The obtained results of the mechanical properties of the yarn in straight form are illustrated in Table 6. Thus, the elastic properties (i.e. compliance coefficients) for the undulated warp and weft yarn in the fabric can be estimated by equations 33~36. The compliance matrix for each slice of the unit cell shown in Fig. 5b can thus be obtained and the one of the unit cell can be acquired from the summation of each of the slices as shown in Fig. 5c by equation 39. Furthermore, the stiffness matrix of the unit cell can be calculated from the inverse matrix of the obtained compliance matrix. Young's modulus of the woven fabric in the extension direction can be evaluated using equation 40. The predicted results are illustrated in Table 7, in which it reveals that the predicted Young's modulus E_x in the weft extension direction based on the actual measured E_L (= 581 N/mm^2) of yarn is much closer to the measured one than based on the predicted E_L (=6694 N/mm^2) of yarn.

7. Conclusions

In this study, a unit cell model for plain weave is developed to predict the elastic behavior of woven fabric during extension. A piece of woven fabric is regarded as a thin lamina because of its thickness is small. The plain weave fabric lamina model presented in this study is 2-D in the sense that considers the undulation and continuity of the strand in both the warp and weft directions. The model also accounts for the presence of the gap between adjacent yarns and different material and geometrical properties of the warp and weft yarns. This slice array model (i.e., SAM), which is used to predict the elastic properties of WF composites by Naik and Ganesh, is applied to evaluate the mechanical properties of woven fabric in this study. The applicability of SAM to prediction of the elastic properties of fabrics is as good as to that of the composites. However, it is necessary to have accurate elastic constants, i.e., Young's moduli of the warp and weft yarn, for the model in order to obtain a promising predicted result. It is found that the accuracy of a predicted Young's moduli of warp and weft yarn obtained from a deduced equation by Hearle is not as good as expected. In order to help eliminate the tedious measuring process but to obtain the exact sizes of the yarns in the fabric, an innovative methodology based on Peirce's geometrical model is developed in the study. It reveals that the proposed method is of good accuracy and can more efficiently acquire the geometrical sizes, i.e., the long and short diameters of the warp and weft yarns in the fabrics. Through the help of the modified SAM prediction model, the mechanic properties (e.g., initial Young's modulus, surface shear modulus and Poisson's ratio) of the woven fabric can be obtained in advance without being through experimental testing. Thus, the determination of the woven fabric can be more efficient and effective through the help of the modified SAM model. Another weave structure of woven fabrics, e.g., twill and satin, is to be selected to construct the geometrical model and a close examination into the stress-strain relations of the unit cell by using finite element analysis is interesting to be followed in our further study.

8. References

Bassett, R.J., Postle, R. & Pan, N. (1999). Experimental Methods for Measuring Fabric Mechanical Properties: A Review and Analysis, Textile Research Journal, Vol.69, No.11, pp. 866-875

Curiskis, J.I., Carnaby, G. A. (1985). Continuum Mechanics of the Fiber Bundle, Textile Research Journal, Vol.55, 334-344

Carlene, P.W. (1950). The Relation between Fiber and Yarn Flexural Rigidity in Continuous Filament Viscose Yarn, The Journal of The Textile Institute, Vol.41, T159-172.

Cooper, D.N.E. (1960). The Stiffness of Woven Textiles, The Journal of The Textile Institute, Vol.51, T317-335.

Carnaby, G. A. (1980). The Compression of Fibrous Assemblies, with Applications to Yarn Mechanics, in "Mechanics of Flexible Fiber Assemblies," J.W.S. Hearle, J.J. Thwaites, and J. Amirbayat, Eds., Sijthoff and Noordhoff, Alphen aan den Rijn, The Netherlands, , pp. 99-112.

Carnaby, G.A. & Luijk, C.J. Van (1982). The Mechanical Properties of Wool Yarns, in "Objective Specification of Fabric Quality, Mechanical Properties and Performance, S.Kawabata, R. Postle, & M. Niwa, Eds., Textile Machinery Society of Japan, Osaka, , 239-250.

Chamberlain, J. (1949). "Hosiery Yarns and Fabrics", Vol. 2., Leicester College of Technology and Commerence, Leicester, , pp. 107.

Curiskis, J.I., G. A. Carnaby (1985). Continuum Mechanics of the Fiber Bundle, Textile Research Journal, Vol.55, 334-344

Carnaby, G.A. (1976). The Structure & Mechanics of Wool Carpet Yarns, Doctoral thesis, University of Leeds,

De Jong, S. & Posite, R. (1977). Energy Analysis of Woven-Fabric Mechanics By Means of Optimal-Control Theory, Part I: Tensile Properties,Journal of Textile Institute, , Vol. 68, p350-362.

Demiröz, A., & Dias, T. (2000). A Study of the Graphical Representation of Plain-Knitted Structures, Part I: Stitch Model for the Graphical Representation of Plain-Knitted Structures, The Journal of The Textile Institute, Vol. 91 (4), No. 1, pp. 463-480.

Demiröz, A., & Dias, T. (2000). A Study of the Graphical Representation of Plain-Knitted Structures, Part II: Experimental Studies and Computer Generation of Plain-Knitted Structures, The Journal of The Textile Institute, Vol. 91 (4), No. 1, pp. 481-492.

Grosberg, P. and Kedia, S. (1966). The Mechanical Properties of Woven Fabrics, Part I: The Initial Load Extension Modulus of Woven Fabrics, Textile Research Journal, Vol.36 (1), pp71-79.

Hadizadeh, M., Jeddi, A.A.A. & Tehran, M. A. (2009). The Prediction of Initial Load-extension Behavior of woven Fabrics Using Artificial Neural Network, Textile Research Journal, vol.79, No.17, pp. 1599-1609.

Hearle, J.W.S., Grosberg, P., & S. Baker (1969). "Structural Mechanics of Fibers, Yarns, & Fabrics, " Vol. 1, John Wiley& Sons, New York, USA.

Hearle, J.W.S. & Shanahan, W.J. (1978). An Energy Method for Calculations in Fabric Mechanics, Part I: Principles of the Method, Journal of Textile Institute, , 69 (4), pp81-91.

Hearle, J.W.S. & Shanahan, W.J. (1978). An Energy Method for Calculations in Fabric Mechanics, Part II: Examples of Application of the Method to Woven Fabrics, Journal of Textile Institute, 69 (4), 92-100.

Huang, N.C. (1978). Technical Report SM7801, Solid Mechanics Group, Department of Aerospace and Mechanical Engineering, University of Notre Dame, U.S.A., Sep.,

Hursa, A., Rotich, T. & Ražić, S. E. (2009). Determining Pseudo Poisson's Ratio of Woven Fabric with a Digital Image Correlation Method, Textile Research Journal, vol.79, No.17, pp. 1588-1598.

Kurbak, A. (1998). Plain-Knitted Fabric Dimentions, Part II, Textile Asia, April, pp. 36-40, pp. 45-46.

Kurbak, A. & Alpyildiz, T. (2009). Geometrical Models for Cardigan Structures, Part I: Full Cardigan, Textile Research Journal, Vol. 79, No. 14, pp. 1281-1300.

Kurbak, A. & Alpyildiz, T. (2009). Geometrical Models for Cardigan Structures, Part II: Half Cardigan, Textile Research Journal, Vol. 79, No. 18, pp. 1635-1648.

Lekkhnitskii, S.G., Theory of Elasticity of an Anisotropic Elastic Body, Holden-Day, San Francisco,

Leaf, G.A.V., & Glaskin, A. (1955). Geometry of Plain-Knitted Loop, The Journal of The Textile Institute, Vol. 46, pp. 587-605.

Leaf, G.A.V. & Kandil, K.H. (1980). The Initial Load-extension Behaviour of Plain-woven Fabrics, Journal of Textile Institute, , Vol. 71, No. 1, pp. 1-7.

Lai, T.-P. (1985). "Practical Fiber Physical Chemistry", Tai-Long Publishing Co. Taipei, ROC,

Lekhnitskii, S.G. Theory of Elasticity of an Isotropic Body, Holden-Day, San Francisco, USA.

Lin, J. J. (2007), Prediction of Yarn Shrinkage Using Neural Nets, Textile Research Journal, Vol.77 (5), pp336-342,

Munden, D. L. (1961). The geometry of a Knitted Fabric in its Relaxed Condition, Hosiery Times, April, Vol. 43.

Naik, N.K. & Ganesh, V.K. (1992). Prediction of on-axes elastic properties of plain weave fabric composites, Composite Science and Technology, Vol. 45, 135-152.

Peirce, F.T. (1937), The Geometry of Cloth Structure, Journal of Textile Institute, Vol 28, T45-96.

Peirce, F.T. (1947). Geometrical Principles Applicable to the Design of Functional Fabrics, Textile Research Journal, Vol. 17, pp. 123-147.

Sun, H., Pan, N. & Postle, R. (2005). On the Poisson's ratios of a Woven fabric, Composite Structures Vol.68, pp. 505-510.

Postle, R. (1971). Structure Shape and Dimensions of Wool Knitted Fabrics, Applied Polymer Symposium, No. 18, John Wiley, Chichester, pp. 149.

Semnani, D., Latifi, M., Hamzeh, S., & Jeddi, A. A. A. (2003). A New Aspect of Geometrical and Physical Principles Applicable to the Estimation of Textile Structures: Ideal Model for Plain knitted Loop, The Journal of The Textile Institute, Vol. 99, No. 1, pp. 204-213.

Thwaites, J.J. (1980). A Continuum Model for Yarn Mechanics, in "Mechanics of Flexible Fiber Assemblies," J.W.S. Hearle, J.J. Thwaites, & J. Amirbayat, Eds., Sijthoff and Noordhoff, Alphen aan den Rijn, The Netherlands,, pp. 87-97.

Womersley, J.R. & D.I.C., B.Sc. (1937). The Application of Differential Geometry to the Study of the Deformation of Cloth under Stress, Journal of Textile Institute, , Vol. 28, T99-113.

8

Prediction of Fabric Tensile Strength by Modelling the Woven Fabric

Mithat Zeydan
Erciyes University, Department of Industrial Engineering
Turkey

1. Introduction

The variety of fabric structures is divided into four parts as wovens, knitts, braids and nonwovens. Comparing with other fabrics, woven fabrics display both good dimensional stability in the warp and weft directions and highest cover yarn packing density. One of the most important features for the characterization of woven fabric quality and fabric performance is tensile properties of fabric strength. Even though the end products of spinning and weaving are woven fabrics, they are raw materials for clothing and other industries such as composites and medical textiles. Every piece of woven fabric is an integration of warp and weft yarns through intersection. The extent of this intersection is largely dependent on the friction between fibres and yarns.

The study of woven fabric mechanics is came across in a work reported by Haas in the German aerodynamic literature in 1912 during development of airships. In English literature, the paper by Peirce (1937) who is the pioneer in the investigation of tensile deformation of woven fabrics presented a geometrical and a mathematical force model of the plain-weave structure which is highly theoretical, both of which have been used extensively and modified by subsequent workers in the field. Many researchers modified his geometrical model to analyze tensile behaviour. Considerable progress has been made over the last century in the development of the theory of geometrical structure and mechanical properties of fabrics.

The mathematical modelling of fabric stress- strain relationships is a very tough topic. During the last 60 years, many outstanding textile scientists have dedicated to their talents to this field. The development of mathematical models for woven fabrics is an extremely complicated and difficult task due to the large numbers of factors on which the behaviour of the fabric depends. Usually, a mathematical model requires a large number of assumptions, covering missing knowledge or inability to express some of the relevant factors.

Mathematical models based on the fundamental mechanics of woven fabrics often fail to yield satisfactory results, as it is hardly possible to incorporate all the complexities in the model. Moreover, the application range of mathematical models is also very specific. Therefore, it is necessary to introduce a different approach for the mathematical modelling of fabric constitutive equations. With fabric, fundamental distinctions may be made between three kinds of modelling, namely: predictive, descriptive and numerical models. The predictive models which form most of the existing research into fabric mechanics are based on the consideration of at least the most important of the relevant factors, while the effect of the remaining ones is covered by suitable assumptions, defining the limits of validity and the accuracy of the resulting theories. Numerical models may ignore the exact mechanism taking place within the structure but emphasize the numerical relations of two variables such as stress- strain relations.

This method is based on statistical considerations; it needs fewer assumptions and provides, perhaps, an approach more relevant to real situations. There exist various methods for fitting a curve in many industrial or science fields. The descriptive models are largely empirical and reflect the need for simple mathematical relations, expressing the phenomenological behaviour of a fabric from the point of view of a particular property (Hu, 2004).

In addition to aforementioned models, in woven fabrics, parallel yarns are only in contact with each other over a fraction of their lengths, and crossover contact may act over relatively complex curved surfaces. Hence to produce an analytical model, a number of simplifications are required. Woven fabrics are well known to have non-linear mechanical properties. The tensile behaviour of woven fabrics is non-linear at low tensions, even if the yarn is linear in tension.

2. Importance of the study

Prediction of fabric mechanical properties such as strength, elongation, bending and shear is an intricate task, as it requires complete understanding of fabric structural mechanics and the interaction between warp and weft threads. Therefore, the solution of the fabric strength prediction problem could be performed by employing the empirical and computational models such as artificial neural network (ANN) or classical regression analysis (Majumdar et al, 2008).

In this study, the data obtained from Zeydan's paper (Zeydan, 2008) will be used for finding both the effect of some fibre, yarn and fabric parameters on the strength of jacquard woven mattress fabric and level configuration of parameters providing maximum fabric tensile strength. A new modelling methodology in the prediction of woven fabric strength will be introduced in this chapter and compared by using TDOE (Taguchi Design of Experiment), ANN, GA-ANN (Genetic Algorithm based Artificial Neural Network) Hybrid structure and multiple regression methodology. Initially, parameters affecting the fabric strength are chosen from experimental design perspective and then fabric strength is modelled based on the given parameters with TDOE, ANN, GA-ANN and multiple regression modelling approach. Besides, a hybrid model structure depending on GA-ANN is used to verify optimum woven fabrics manufacturing parameter configuration of TDOE. ANN and GA are two of the most important computational techniques of Artificial Intelligence. While ANN is a very powerful modelling method used in complex non-linear systems, GA can be suitable for parameter optimization. The performance criteria assessing appropriate model for the four approaches are root mean square error (RMSE) and MAE (Mean Absolute Error).

The following parameters collected from a Textile Factory producing jacquard woven bedding fabric have been identified as potentially important parameters affecting the strength of woven fabric as shown in the following Table 1.

While determining parameters of this study, parameters related to weaving process have been considered rather than yarn-based parameters. The firm that the research was carried out purchases the yarn from its suppliers. But, it performs all weaving and treatment processes in its plant. Because of this reason, production process parameters of the firm were adjusted according to desired conditions during the sample production. Thus, parameters significantly affecting fabric strength such as yarn strength and twist could not be taken into consideration. Consequently, it was not possible to manufacture additional (extra) samples that are suitable for the aforementioned parameters. Fabric strength was tested at Titan fabric strength tester machine according to the ASTM D5035 testing method.

	Levels	
Factors	1	2
A: Number of warp yarns at fabric width	7040	8658
B: Weft density (weft/cm)	8	16
C: Weft yarn count(denier)	300	600
D: Fibre Type of weft yarn	PF	CF
E: Warp density(warp/cm)	33	38
F: Warp yarn count (denier)	150	354
G: Fibre type of warp yarn	PF	CF

Table 1. Factors and Levels

Strip method (ASTM D5035) was adopted for the evaluation of breaking force of narrow fabrics in Titan fabric strength tester. Any fabric slippage from the tester jaws was recorded. Breaking force was recorded for evaluation. Five samples were tested at each group, which were 20 cm in length and 25mm in ravelled width. Gage length was set to 75mm and none the samples were failed at or close to the grip region. Only warp-wise testing was performed. Testing machine was set for a loading rate of 300 mm/min. All tests were performed under standard atmospheric conditions (per cent 65 +-2 relative humidity and 20+- 28C temperature) and the samples were conditioned hours under such conditions for 24 before testing (Zeydan, 2008). Jacquard woven fabrics are widely used in various sections of upholstery industry, where mattress cover is one of them. Although Jacquard fabrics are most often used for upholstery, they are becoming more and more popular in the apparel trades. Strength of jacquard woven mattress fabric depends on several factors. The objective of this study is to model the relationship between fibre, yarn and fabric parameters on the strength of fabric using artificial neural network (ANN), Taguchi design of experiment (TDOE), Multiple Regression and ANN-GA modelling methodologies. There have been some studies in the literature of Textile about the usage of Fabrics and woven Fabrics modelling with ANN. Keshavaraj et al. (1996) modelled air permeability of woven fabrics for airbags. Ogulata et al. (2006) used regression and ANN models to predict elongation and recovery test results of woven stretch fabric for warp and weft direction using different test points. Behera and Muttagi (2004) reported the possibility of woven fabric engineering. Majumdar et al. (2008) employed ANN to forecast the tensile strength of a woven fabric. Hadizadeh et al. (2009) predicted initial load-extension behaviour of plain weave and plain weave derivative fabrics. Tilocca et al. (2009) detected fabric defects using two kinds of optical patterns. Gong and Chen (1999) predicted the performance of fabrics in garment manufacturing. Behera and Karthikeyan (2006) made a design of canopy fabrics. However, any research about comparing ANN, TDOE, multiple regression and ANN-GA in the literature hasn't been conducted on the strength prediction of woven fabric from fibre, yarn and fabric parameters using woven fabric modelling approaches with all together so far. Modelling the woven fabrics comes into existence in many forms.

In this study, traditional and computational modelling techniques are compared between each other. Compared the other classical modelling techniques, computational modelling methodology seems to have been more robust and appropriate. This study has many advantages to reduce the waste and scrap ratio before and during manufacturing. Therefore production planning will become more efficient in a textile plant. This study presents a new approach given in the following figure 1 about which woven fabric modelling is more efficient than the others.

2.1 Production cycle of weaving department at a textile factory

This study was made in a textile Factory in Turkey. The textile factory was founded in 2002. With over 1,000 employees, state of the art plants are made up of 120.000 m^2 open area of which 83.000 m^2 is closed area. It is a fully integrated facility consisting of weaving and knitting plants including narrow weaving crochet knitting, fabric finishing, chenille yarn and Bulk continuous filament polypropylene yarn manufacturing sections. The company is the world's largest upholstery fabric manufacturer. The Factory ships (delivers) the great bulk of its output (approximately 80 %) to world markets. Its plants are regarded as the Europe's most modern Jacquard weaving facility comprising 150 full automatic Jacquard weaving looms and composes of several departments related to woven and knitted upholstery and bedding fabrics. Annually weaving production capacity are 30.000.000 meters. Average scrap of production is 5/1000. The reason of returning the production back is generally originated from vision-based.

Textile Factory composes of several departments related to woven and knitted upholstery and bedding fabrics. Production cycle of woven bedding fabrics starts with warping process. Warp and weft yarns are sourced from external companies and stored at yarn warehouse at the factory. Warp yarns needs to be wound into beams in order to weave fabrics in order. Conic warping machine is used to arrange warp yarns in order at the beam. After that, warp yarns are drawn-in and the weaving machine is ready for weaving. Weft yarns are fed into weaving process in the cone form. Design office in the company is responsible for preparing the jacquard design of fabric according to the customer preferences. The fabric design is transferred into jacquard system of weaving machine via floppy disk. Woven fabrics are wound into larger beams at the greige fabric control section of the factory. This initial quality control enables to evaluate the quality of weaving process and checking for the weaving faults. Depending on the demands of the customers, chemicals and finishing additives are applied to the greige fabric at the finishing department. Some of the widely preferred finishing processes are bulkiness finish, fire retardancy, soil and oil repellency, and water resistance. Stenter is the next step before packaging and final quality control. Dimensional stability and skewness of fabric are adjusted at the stenter. Chemicals and additives are also applied to fabrics at stenter. Fabrics are dried and further stability is generated at the drying section of the stenter called calendaring. Finished fabric quality control and packaging is performed based on the customer specifications and instructions. Finally fabrics are shipped to the customer. This production lay-out is schematized at Figure 2.

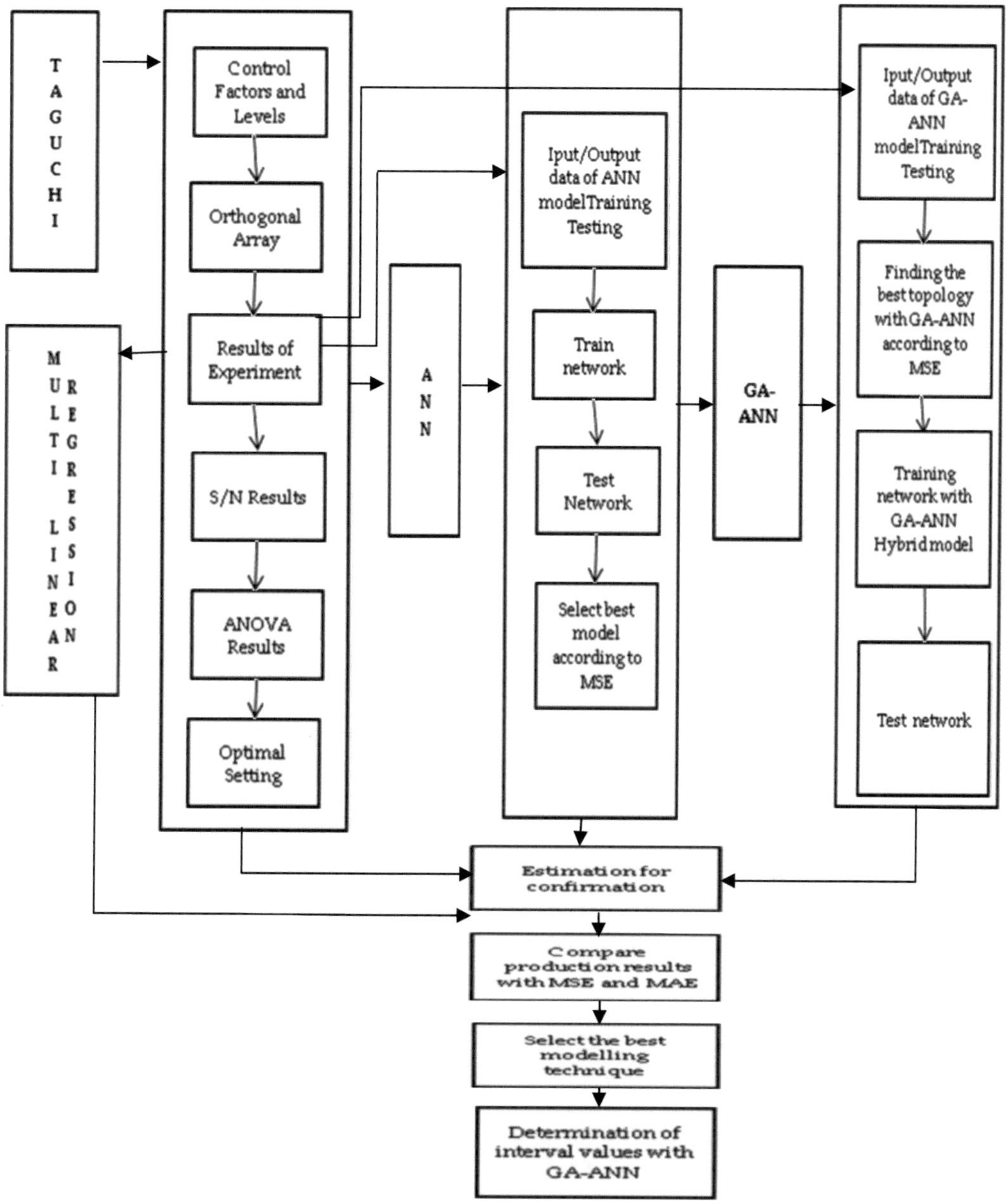

Fig. 1. Methodology of application

3. Modelling techniques

Here, Taguchi Design of Experiment Methodology will not be explained in detailed since this analyze was performed in Zeydan's paper (2008). Orthogonal matrix is given in Table 2. According to the result of this analyze, the most efficient parameter affecting the woven fabric strength was warp density in terms of S (signal) to N (Noise) Ratio. Optimal parameter setting is $A_2B_2C_1D_2E_2F_2G_1$ which means A (8658), B (16), C (300/DN), D (PF),

E(38), F (30/2 DN), G (PF) the following stages will be considered according to the results obtained from TDOE.

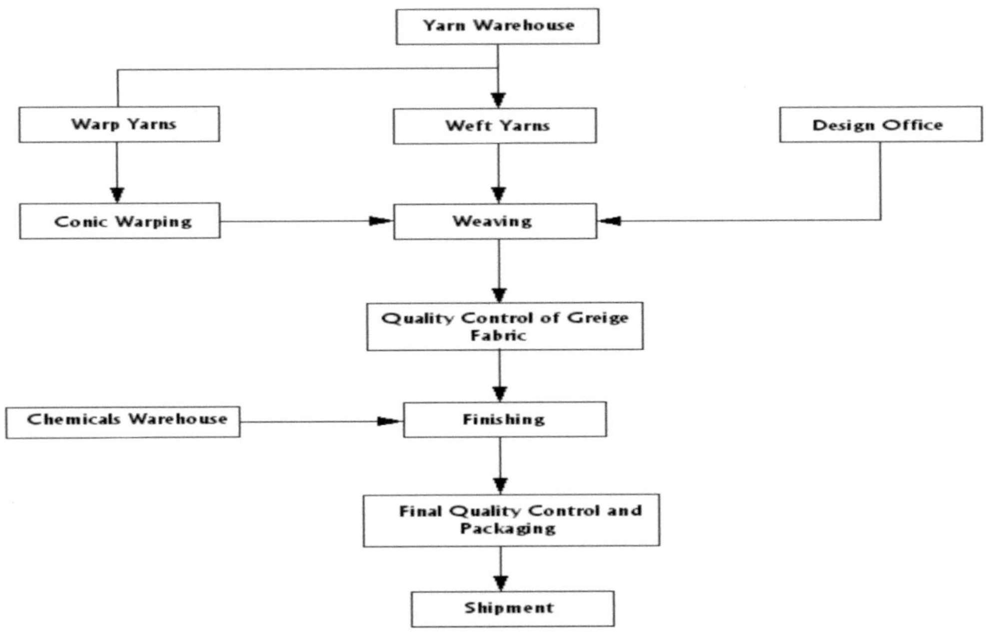

Fig. 2. Production work flow chart for jacquard woven bedding fabric

	Parameters							
Order	A	B	C	D	E	F	G	Average Fabric Strength (N/m)
1	7040	8	300	PF	33	150	PF	1026 (21.6)
2	7040	8	300	CF	38	354	CF	1313 (32.7)
3	7040	16	600	PF	33	354	CF	1057 (26.9)
4	7040	16	600	CF	38	150	PF	1350 (32.7)
5	8658	8	600	PF	38	150	CF	1148 (34.2)
6	8658	8	600	CF	33	354	PF	1161 (38.0)
7	8658	16	300	PF	38	354	PF	1669 (36.3)
8	8658	16	300	CF	33	150	CF	1117 (34.9)

Table 2. Orthogonal matrix

L8 experimental design orthogonal matrix formed related with fabric strength is given in table 2. Total amount of data about fabric strength collected from the factory is 120.

3.1 Multiple Linear Regression

Multiple Linear Regression (MLR) is a well known statistical procedure trying to find a linear relationship between two or more explanatory variables and a dependent variable by observing data. It can be used for forecasting output values. Dependent variable (y) can be explained by the equation below:

$$y = a + \sum_{i=1}^{n} b_i x_i + \varepsilon \quad (1)$$

Before using ANN, multiple linear regression model is constructed. MLR is used as a verification and comparison model of ANN in the literature (Noori et al., 2010). It is claimed that ANN generally gives better results than MLR (Valvarde Ramirez et al., 2005). Fabric strength is defined as dependent variable and explanatory variables are; number of warp yarns at fabric width, weft density, weft yarn count, fiber type of weft, warp density, warp yarn count and fiber type of warp yarn. Multiple Linear Regression equation defined as below by considering the mean fabric strength data:

$$y = 618 + 87.2A + 136B - 102C + 10.2D + 280E + 140F - 143G \quad (2)$$

3.2 Artificial Neural Network (ANN)

ANN is a nonlinear mapping system based on principles observed in nervous systems (Co, 2008). Complex nonlinear systems can be modelled efficiently by using ANN. In this study, multilayer perception is used for solving difficult predictive modelling problems. Multilayer perceptions networks consist of typically an input layer, single or more hidden layers, and one output layer. Hidden layers have one or more hidden neurons which performs nonlinear mapping between inputs and outputs (Lin and Choou, 2008). Activation functions are used to activate nodes. Here, sigmoid activation function is used, because generally MLP neural networks uses the logistic sigmoid function (Co, 2008) and using the sigmoid function in ANN topology provides a good nonlinear input-output mapping capability (Lo and Tsao, 2002). Choosing the proper learning algorithm is also very important while training the networks. For solving nonlinear optimization problems Levenberg-Marquardt algorithm is employed because of its efficient method (Sanjari et al., 2009). Besides, minimizing the MSE is the best known advantage of Levenberg-Marquardt (Purwanto et al., 2008). There is not a common sense about the number of the network layers, hidden nodes and generally a trial-and-error process approach is used for predetermining the optimal number of nodes in the hidden layer (Tu, 1996). Networks which contain different number of hidden layers and hidden neurons are being compared to find the best one (Wu et al., 2009). Mean square error (MSE) is generally used to judge the capability of networks while selecting the best one. Beside that, the simplest architecture is better than others (Didier, 2009). Therefore, the most important issue is to find the proper number of hidden neurons. Too much hidden neurons causes too much flexibility and this leads over fitting. But, too few hidden neurons prevent the learning capacity and decrease approximation performance of network (Haykin, 1994; and Xu and Chen, 2008). A single hidden layer can be chosen and it is sufficient for any continuous nonlinear mapping. But also, there have been many applications using two hidden layers (Behera and Karthikeyan, 2006). Recently, Artificial Neural Network became a popular modelling tool used in textile engineering area. It is commonly used in the applications of various types of fabrics. Defect detection and strength determination are some of the most known usage areas of ANN. In this study, ANN is also used to determine fabric strength. In Table 3, various options by comparing between 1 hidden layer and 2 hidden layers from one neuron to 10 neurons in terms of RMSE and MAE performance values, were tried to obtain the best ANN topology using trial and error method by considering literature claims. Results show that two hidden layers and three neurons are better than the other conditions. RMSE is used to compare topologies. Minimum RMSE is found for 7 (input)-2 (first hidden layer) - 3 (Second Hidden Layer) -1 (output) topology.

Performance						
	1 Hidden Layer			2 Hidden Layer		
Neuron	MSE	MAE	R	MSE	MAE	R
1	55939.1	1783289	0.558749	43301.66	1676.322	0.552495
2	15146.21	8729215	0.88216	12243.38	7986.922	0.871783
3	5953217,0	460774	0.956546	**4485.351**	5224.486	0.947723
4	8482484,0	6068052	0.94131	6852.638	6452.752	0.949635
5	12120.4	6657714	0.88593	12905.09	7121.598	0.887804
6	10212.87	5989459	0.900652	13256.95	7198.211	0.884492
7	11772.05	6277868	0.887777	24378.05	1187.689	0.808354
8	13014.74	7426107	0.871812	28423.24	1272.778	0.732697
9	11243.09	6145195	0.894136	16736.9	9074.597	0.852666
10	19056.53	9939427	0.856076	11566.42	653.403	0.896996

Table 3. Performance Values of 1-2 Hidden Layers

3.3 Genetic algorithm based artificial neural network

Genetic algorithm is a combinatorial optimization technique which models natural biologic evaluation process. GA searches for finding global optimal but cannot guarantee to find best solution (Núñez-Letamendia, 2007). It has a special terminology. Population includes alternative solutions set, genes mean variables which build up a solution, and chromosome is the name of all individual in population, so it represents an alternative solution. Generation means iteration. Also, fitness function represents an objective function.

To create a population, genetic algorithm chooses randomly a chromosome from search space. Populations are evolved until the best fitness value can be found. These evolutions are done by selection, crossover and mutation operators in which chromosomes will be used in reproduction process is decided in selection process. In this step, best chromosomes are tried to be chosen depending on the fitness function that is regarded as objective function. By this way, a decision can be made whether a solution is bad or not (Shopova and Vaklieva-Bancheva, 2006).

After selection process, recombination operator is applied to alter solutions. It combines two selected parent chromosomes' features with a probability to form new children. When recombination process finishes, mutation starts. The mutation refers creating of a new chromosome from an individual by changing some genes of chromosome with a predefined probability. After selection, crossover and mutation operators are applied, the newly created offspring is inserted into the population. Parent chromosomes in which they were derived from are replaced and so a new generation is created. Until the optimization criterion is met, this cycle is performed. Stopping criteria can be a generation number.

GA is used to determine the best network architecture and training parameters of ANN in this study. Trial and error method takes long time and sometimes cannot determine best topology. But genetic algorithm has become a popular tool in neural networks, recently. It is generally used in three ways in the literature of ANN. The first one is to optimize hidden neurons, learning rate and momentum rate (Mohebbi at al., 2008 ; Torres, 2005). By this way, time and effort required to find optimal architecture is minimized [Taheri, 2008; Kim et al., 2004; Saemi, 2007). The second usage of GA in ANN is to train parameters of neural networks (Liu et al., 2004; Heckerling, 2004; Versace, 200). And the final way is to set the weights in fixed architecture (Whitley, 1995). NeuroSolutions 5 is used to apply GA in ANN. It is an efficient software ANN optimization (Kim et al., 2004; Cheung et al., 2006).

In the first stage, GA is used to find best topology of ANN. Multilayer perception algorithm and 1 hidden layer is used. Sigmoid as an activation function and Levenberg Marquardt learning algorithm is chosen again, because of the mentioned advantages before. Under this circumstance, optimal processing element (PE) is determined by using GA which optimizes the architecture until the lowest error is found.

Results show that 1 hidden layer and four PEs are the best topology for our data. ANN-GA topology is 7-1-4-1. It was shown in Figure 3. After determining the optimal topology by using GA, next step is to perform genetic Training. Genetic training firstly creates an initial population of networks, randomly. All networks have different parameters. These are trained by considering minimum square error (MSE) that is defined as a threshold value. Threshold value is taken 0.01 for MSE. Properties of good networks are crossover and mutated to create better networks which include less MSE value than others. While selecting good chromosomes, Roulette wheel is used as selection operator. Selection depends on best fitness value.

The characteristics of the good networks are then combined and mutated to create a new population of networks. Again, the networks in this population are evaluated and the characteristics of the best networks are passed along to the next generations of networks. This process is stopped until the maximum generation is reached. Maximum generation number is 60 and maximum epoch is 1000 for this study. Population number is 50. Crossing over and mutation are done between 2 point with a probability of 0,9 and in uniform form with a probability of 0,01, respectively. Results show that GA-ANN gives better results than ANN.

Especially, testing MSE and correlation coefficiency (R) of GA-ANN has a significant difference rather than ANN. Especially, testing MSE and r of GA-ANN has a significant difference rather than ANN. According to the table 3, since all performance measurement values for GA-ANN topology are better than ANN topology, we will use the GA-ANN topology for modelling studies.

Performance	ANN topology	GA-ANN topology
MSE	5953.217	1906.985069
MAE	46.0774	29.96099284
R	0.956546	0.981429488

Table 4. Comparison of ANN and GA-ANN topologies

GA-ANN procedures generally gives better results rather than ANN. 10 experimental parameter sets collected from the production process are tested to compare the modelling performance of all methods that are used. Table 4 shows the production (forecasting) values of methods obtained from modelling. Experimental results for fabric strength are the real values. According to the experimental values, all forecasting values obtained from modelling is compared with each other. At the end of this process, Optimal modelling technique for the smallest RMSE and MAE value is determined as GA-ANN.

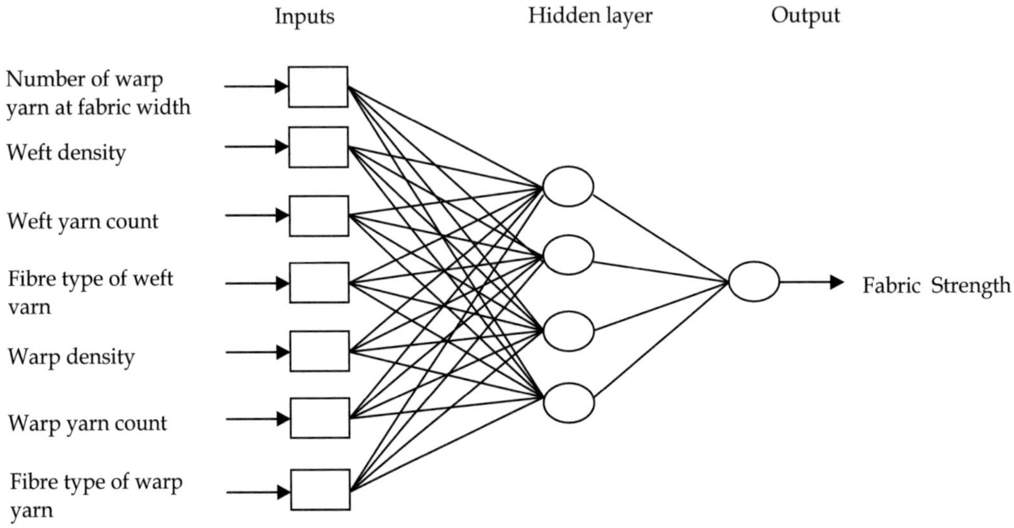

Fig. 3. ANN-GA Topology

Exp.	A	B	C	D	E	F	G	Experimental Results	Predicted Linear Regression	Predicted TDOE	Predicted ANN	Predicted GA-ANN
1	1	1	2	2	2	1	1	1250	1214.6	1213.75	1295879	1250
2	2	2	1	1	1	1	2	1098	1106.6	1106.75	1029.38	1098
3	1	1	2	1	1	2	2	1005	921.4	920.75	1035336	1005
4	2	2	1	2	2	2	2	1428	1536.8	1536.5	1518434	1428
5	1	1	2	2	2	1	2	1302	1071.6	1071	1252.7	1302
6	2	1	2	1	1	2	2	1052	1008.6	1008	1034977	1052
7	1	2	1	1	2	2	1	1637	1582.4	1581.75	1641925	1516912
8	2	1	1	1	2	1	2	1301	1250.6	1250.25	1294558	1224259
9	1	2	1	1	2	2	2	1544	1439.4	1439	1403969	1402.06
10	2	2	1	2	1	2	1	1498	1399.8	1399.5	1274407	1513988

Table 5. Comparison of experimental results with Linear Regression, TDOE, ANN and GA-ANN

RMSE and MAE values of linear regression, TDOE, ANN and GA-ANN are shown in Table 5. Also, Graphical representation of all modelling methods used is given by comparing with each other in figure 4 The best results are obtained with GA-ANN model.

	LINEAR REGRESSION	TDOE	ANN	GA-ANN
RMSE	100.6116	100.9521	93.97	63.80663
MAE	81.8	82.225	67.6584	35.47569

Table 6. RMSE and MAE values of modelling methods

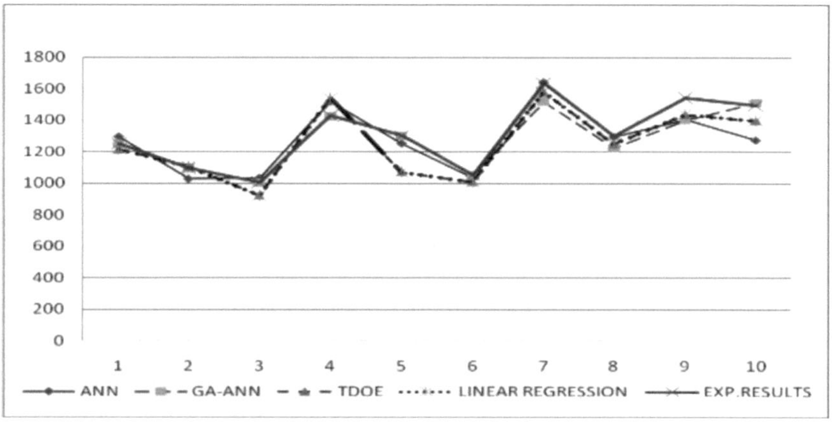

Fig. 4. Comparison of modelling methods

The final stage of the new modelling methodology is to make a verification of finding the optimum fabric strength with GA-ANN hybrid modelling technique as the best methodology. Warp density as the most important factor affecting the fabric strength is found with the Taguchi Design of Experiment Methodology and whether there have been in interval values of optimum parameter setting is tested by increasing from 33 to 38. The verification of TDOE results with GA-ANN hybrid modelling technique for interval values of warp density from 33 (warp/cm) to 38 (warp/cm) is shown in figure 5.

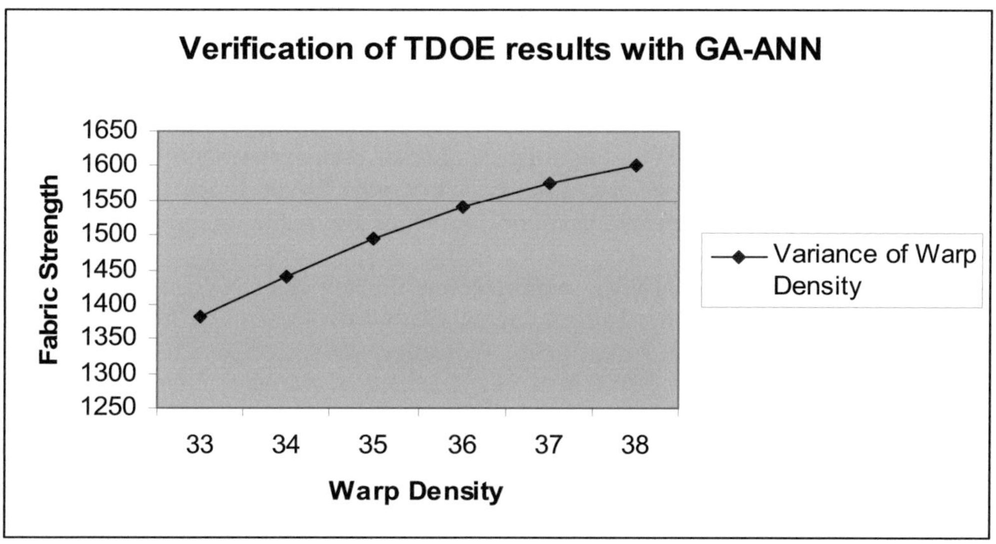

Fig. 5. Interval values of warp density

4. Conclusion

In this study, traditional and computational modelling techniques are compared between each other to predict woven fabric strength that is one of the main features for the characterization of woven fabric quality and fabric performance. Compared the other classical modelling techniques, computational modelling methodology seems to have been more robust and appropriate. This study made in a textile Factory producing jacquard woven bedding fabric in Turkey has many benefits for textile manufacturers to reduce waste and scrap ratio before and during manufacturing. Firstly, production planning function in the plant will be able to predict the woven fabric strength easily to be known optimal parameter setting before manufacturing. Secondly, The significant parameter in the manufacturing was found as Warp Density. Thirdly, after finding the optimum parameter setting with TDOE, interval values of the sensitive parameters in the production was found with ANN approach. According to the data collected from manufacturing Process of factory in Zeydan's paper (2008), Taguchi Design of Experiment methodology was applied to find the most significant parameters. Seven significant parameters affecting the Woven Fabric tensile strength was considered. Warp density was found the most important factor affecting the Fabric strength by using S/N Ratio. The main purpose of this study is modelling the woven fabric strength by comparing different modelling techniques. However, any research about comparing ANN, TDOE, multiple regression and ANN-GA in the literature hasn't been conducted on the strength prediction of woven fabric from fibre, yarn and fabric parameters using woven fabric modelling approaches with all together so far. ANN, GA-ANN hybrid approach, Multiple-Linear regression, TDOE based on RMSE and MAE modelling performance criteria which is frequently used, are compared with each other. Finally, GA-ANN hybrid methodology was found as a suitable modelling technique. At the last stage of modelling methodology, verification of TDOE results with GA-ANN hybrid modelling technique for interval values of warp density was performed by increasing from 33 (warp/cm) to 38 (warp/cm). Parameter value giving optimum fabric strength for Warp Density was determined as 38 (warp/cm).

5. References

Behera, B.K. & Muttagi, S.B.(2004). Performance of Error Back Propagation vis-á-vis Radial Basis Function Neural Network: Part I: Prediction of Properties for Design Engineering of Woven Suiting Fabric. Journal of the Textile Institute, Vol. 95, No 1, 283-300

Behera, B. K.& Karthikeyan, B.(2006). Artificial Neural Network-embedded Expert System for the Design of Canopy Fabrics. *Journal of Industrial Textiles*, Vol. 36, No. 2, 111-123

Cheung, S.O.; Wong, P.S.P.; Fung, A.S.Y. & Coffey, W.V. (2006). Predicting project performance through neural networks. *International Journal of Project Management*, Vol. 24, No. 3, 207–215

Co, H. C. (2008). Confirmation testing of the Taguchi methods by artificial neural-networks simulation, *International Journal of Production Research*, Vol. 46, No. 17, 4671 – 4685

Didier, C.; Forno, G.; Etcheverrigaray, M.; Kratje, R. & Goicoechea H. (2009). Novel chemometric strategy based on the application of artificial neural networks to crossed mixture design for the improvement of recombinant protein production in continuous culture. *Analytica Chimica Acta* , Vol. 650, 167–174

Geman, S.; Bienenstock, E. & Doursat, R. (1992). Neural networks and the bias/variance dilemma. *Neural Computation*, Vol.4, 1–58,

Gong, R.H. & Chen, Y. (1999).Predicting the Performance of Fabrics in Garment Manufacturing with Artificial Neural Networks. *Textile Research Journal*, Vol. 69, No.7, 477-482

Hadizadeh, M.; Jeddi, Ali A.A. & Tehran, M. A. (2009). The Prediction of Initial Load-extension Behavior of Woven Fabrics Using Artificial Neural Network. *Textile Research Journal*, Vol. 79, No. 17, 1599-1609

Haykin, S. (1994). *Neural networks-a comprehensive foundation*. Macmillan College Publishing, ISBN, New York

Heckerling, P. S.; Gerber, B. S.; Tape, T.G. & Wigton, R. S. (2004). Use of genetic algorithms for neural networks to predict community-acquired pneumonia. *Artificial Intelligence in Medicine*, Vol. 30, No. 1, 71-84

Hu, J. (2004). *Structure and Mechanics of Woven Fabrics*, Woodhead Publishing Limited, 1 85573 904 6 Cambridge

Keshavaraj, R. ; Tock, R.W. & Haycook, D. (1996). Airbag fabric material modeling of nylon and polyester fabrics using a very simple neural network architecture. *Journal of Applied Polymer Science*, Vol. 60, No 13, 2329-38

Kim, G.-H.; Yoon, J.-E.; An, S.-H.; Cho, H.-H. & Kang, K.-I. (2004). Neural network model incorporating a genetic algorithm in estimating construction costs, *Building and Environment*, Vol.39, 1333–1340

Majumdar, A.; Ghosh, A.; Saha, S.S.; Roy, A.; Barman, S.; Panigrahi, D.& Biswas, A. (2008). Empirical Modelling of Tensile Strength of Woven Fabrics. *Fibers and Polymers*, Vol.9, No.2, 240-245

Lin, H.L.; Chou, C.P.(2008). Modeling and optimization of Nd:YAG laser micro-weld process using Taguchi Method and a neural network. *International Journal of Advanced Manufacturing Technology*, Vol. 37, 513–522

Liu, Z.; Liu, A.; Wang, C.& Niu, Z.(2004) Evolving neural network using real coded genetic algorithm (GA) for multispectral image classification. *Future Generation Computer Systems*, Vol. 20, No. 7, 1119–1129

Lo, Y.L. & Tsao C.C. (2002). Integrated Taguchi Method and Neural Network Analysis of Physical Profiling in the Wirebonding Process, *IEEE Transactions On Components And Packaging* Technologies, Vol.25, No.2, 270-277

Mohebbi, A.; Taheri, M. & Soltani, A. (2008). A neural network for predicting saturated liquid density using genetic algorithm for pure and mixed refrigerants, *International Journal of Refrigeration*, Vol. 31, No. 8, 1317-1327

Noori, R. ; Khakpour A. ; Omidvar, B. ; Farokhnia, A. (2010). Comparison of ANN and principal component analysis-multivariate linear regression models for predicting the river flow based on developed discrepancy ratio statistic, *Expert Systems with Applications*, (article in press : doi:10.1016/j.eswa.2010.02.020)

Nunez-Letamendia, L. (2007). Fitting the control parameters of a genetic algorithm: An application to technical trading systems design. *European Journal of Operational Research*, Vol. 179, No. 3, 847-868

Ogulata, S.N. ; Sahin, C. ; Ogulata, T.O.& Balci, O. (2006), The prediction of elongation and recovery of woven bi-stretch fabric using artificial neural network and linear regression models, *Fibres & Textiles in Eastern Europe*, Vol. 14, No. 2(56), 46-9

Park, S.W.; Hwang, Y.G.; Kang, B.C. & Yeo S.W. (2000). Applying Fuzzy Logic and Neural Networks to Total Hand Evaluation of Knitted Fabrics. *Textile Research Journal*, Vol. 70, No. 8, 675-681

Purwanto, D. ; Agustiawan, H. & Romlie, M.F.; (2008). The Taguchi-neural networks approach to forecast electricity consumption. *In: IEEE Canadian Conference on Electrical and Computer Engineering*, pp.1941-1944, 4-7 May 200, Niagara Falls

Saemi, M.; Ahmadi, M. & Varjani, A.Y. (2007). Design of neural networks using genetic algorithm for the permeability estimation of the reservoir. *Journal of Petroleum Science and Engineering*, Vol. 59, 97- 105

Sanjari, M. & Taheri, A. K. & Movahedi, M. R. (2009). An optimization method for radial forging process using ANN and Taguchi method. *The International Journal of Advanced Manufacturing Technology*, Vol.40, No. 7-8, 776–784

She, F.H.; Kong, L.X.; Nahavandi, S. & Kouzani, A.Z.(2002). Intelligent Animal Fiber Classification with Artificial Neural Networks. *Textile Research Journal*, Vol. 72, No. 7, 594-600

Shopova, E.G.& Vaklieva-Bancheva, N.G.(2006). A genetic algorithm for engineering problems solution, Comput. *Chem. Eng. Vol.* 30, No.8, 1293–1309

Taheri, M. & Mohebbi, A. (2008). Design of artificial neural networks using a genetic algorithm to predict collection efficiency in venturi scrubbers, *Journal of Hazardous Materials*, Vol. 157, 122–129

Tilocca, A.; Borzone, P.; Carosio, S. & Durante, A. (2002). Detecting Fabric Defects with a Neural Network Using Two Kinds of Optical Patterns. *Textile Research Journal*, Vol. 72, No.6, 545–551.

Torres, M.; Hervás, C. & Amador, F. (2005). Approximating the sheep milk production curve through the use of artificial neural networks and genetic algorithms, *Computers & Operations Research*, Vol. 32, 2653–2670

Tu, J.V. (1996). Advantages and disadvantages of using artificial neural networks versus logistic regression for predicting medical outcomes. *Journal of Clinical Epidemiology*, Vol. 49, No. 11, 1225–1231

Valverde Ramirez, M.C. & De Campos Velho, H.F. ; Ferreira, N.J. (2005). Artificial neural network technique for rainfall forecasting applied to the São Paulo region, *Journal of Hydrology*, Vol. 301 (1-4), 146-162

Versace, M.; Bhat, R.; Hinds, O. & Shiffer M. (2010). Predicting the exchange traded fund DIA with a combination of genetic algorithms and neural networks. *Expert Systems with Applications* (article in press)

Whitley, D.(1995); Genetic algorithms and neural networks, In: J. Perioux, G. Winter, M. Galan and P. Cuesta (eds), *Genetic Algorithms in Engineering and Computer Science*, John Wiley & Sons Ltd.

Wu, S.J.; Shiah, S.W. & Yu W.L. (2009). Parametric analysis of proton exchange membrane fuel cell performance by using the Taguchi method and a neural network. *Renewable Energy*, Vol. 34, 135–144

Xu, S. & Chen, L.(2008) A Novel Approach for Determining the Optimal Number of Hidden Layer Neurons for FNN's and Its Application in Data Mining. *Proceedings of 5th International Conference on Information Technology and Applications (ICITA 2008)*, Pp. 683-686, ISBN 978-0-9803267-2-, Cairns, Queensland, 23-26 June 2008, AUSTRALIA

Zeydan, M. (2008), Modelling the woven fabric strength using artificial neural network and Taguchi methodologies, *International Journal of Clothing, Science and Technology*, Vol. 20, No. 2, 104-118

9

Data Base System on the Fabric Structural Design and Mechanical Property of Woven Fabric

Seung Jin Kim and Hyun Ah Kim
Yeungnam University and Seoul National University
Korea

1. Introduction

The structure of fabrics is very important, because fabric geometry gives considerable effects on their physical properties. Therefore, the studies for fabric structure have been carried out with following areas:

1. prediction of fabric physical and mechanical properties
2. education and understanding related to the fabric structural design
3. the area related to the fabric and garment CAD systems

Among them, the researches for the prediction of fabric physical and mechanical properties with fabric structure have been performed by many textile scientists. But the education and understanding related to the fabric structural design have been emphasized on the theoretical aspects. But the optimum fabric design plan is recently needed with the relevant fabric shrinkage in dyeing and finishing processes for making the various emotional fabrics for garment. For responding this need, the difference of fabric design plan such as fabric density, yarn count and finishing shrinkage has to be surveyed with weaving looms such as water jet, air-jet and rapier looms, and also has to be analyzed with weave patterns such as plain, twill and satin. On the other hand, recently, there are many commercial CAD systems such as fabric design CAD for fabric designers and pattern design CAD including visual wearing system for garment designers. But there is no fabric structural design system for weaving factories, so the data base system related to the fabric structural design for weaving factories is needed. Many fabric weaving manufacturers have some issue points about fabric structural design. The 1st issue point is that there is no tool about how to make fabric design according to various textile materials such as new synthetic fibers, composite yarns, and crossed woven fabrics made by these new fibres and yarns. As the 2nd issue point, they also don't have the data about what is the difference of fabric structural design such as fabric densities on warp and weft directions according to the weaving looms such as WJL, RPL and AJL. And 3rd issue point is that there is no data about how the difference of fabric structural design is among weaving factories even though they have same looms and they use same materials. Therefore, in this topic, a data base system which can easily decide warp and weft fabric densities according to the various yarn counts, weave construction and materials is surveyed by the analysis of design plan for synthetic fabrics such as nylon and PET and worsted and cotton fabrics. Furthermore, the analyses for easy deciding of fabric design from new materials and for making data base related to this fabric structural design are carried out as the objectives of this topic.

2. Background of fabric structural design

The first study for the fabric structural design was started in 1937 by Peirce paper(Peirce, 1937), which is the Peirce's model of plain-weave fabrics with circular yarn cross section. And he also proposed fabric model with an elliptic yarn cross section. In 1958, Kemp proposed a racetrack model(Kemp, 1958). Hearle and Shanahan proposed lenticular geometry (Hearle & Shanahan, 1978) for calculation in fabric mechanics by energy method in 1978. And many researches related to the fabric mechanical properties under the base of fabric structural model were carried out by Grosberg (Grosberg & Kedia, 1966), Backer (Backer, 1952) Postle(Postle et al., 1988). Lindberg(Lindberg et at., 1961) extensively studied fabric mechanical behavior related to the tailorability. Then the sophisticated measurement system of fabric mechanical properties was developed by Kawabata and Niwa(Kawabata et al., 1982) which is called KES-FB system. Another fabric mechanical measurement system called the FAST was developed by CSIRO in Australia(Ly et al., 1991). Recently new objective measurement systems(Hu, 2004) such as Virtual Image Display System(VIDS) and Fabric Surface Analysis System(FabricEye®) have been developed for the analysis of fabric geometrical properties. On the other hand, nowadays there are many CAD systems(i-Designer, Texpro) related to the fabrics design such as weave construction, color and pattern. And also there is pattern design CAD(Texpro, Harada & Saito, 1986) including visual wearing system(VWS) for garment designer. But there is no fabric structural design system related to the decision of the fabric density according to the fibre materials, yarn linear density, and weave pattern. Therefore, a data base system which can easily decide warp and weft densities according to the various yarn counts, weave constructions and materials is required through the analysis of design plan for worsted, cotton, nylon and polyester fabrics as shown in Figure 1(Kim, 2002).

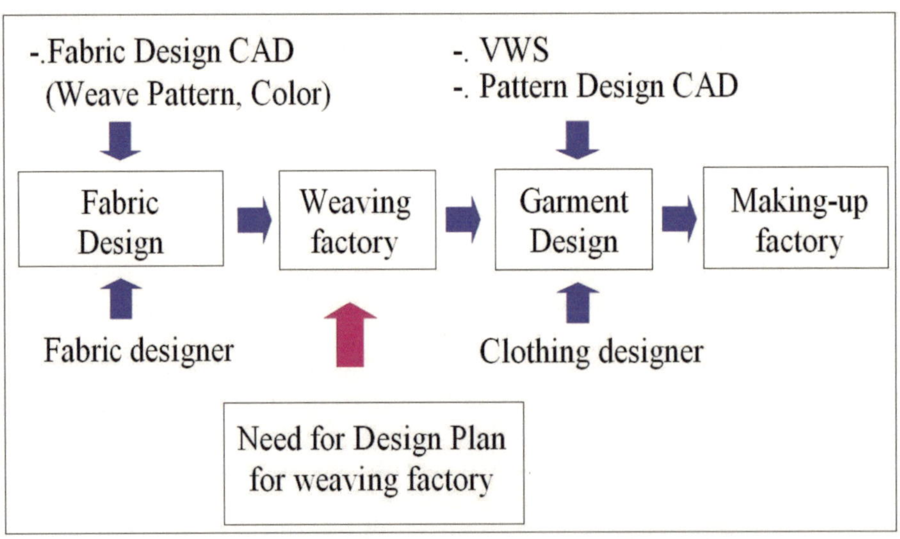

Fig. 1. Diagram for need of fabric structural design system for weaving factory

Figure 2 shows milestone of detail analysis steps related to the data-base system of the fabric structural design in relation with existing fabric design and wearing systems of garment(Kim, 2005). The final goal of this analysis is aiming to link with virtual wearing system, pattern design CAD and drape analyzer. As shown in Figure 2, in the 1st step, the data base of weave pattern and fabric factors has to be made using yarn count, fabric density and weave pattern from which weave density coefficient (WC) and warp and weft density distributions are calculated. And weave density coefficient can be analyzed according to weaving factories and loom types. Furthermore, weave density coefficient and yarn density coefficient (K) can be analyzed with cover factor of fabrics. In the 2nd step, the data base of various physical properties of fabrics is made with dyeing and finishing process factors, which affects fabric hand and garment properties measured by KES-FB and FAST systems. In the 3rd step, these data bases have to be linked with visual wearing system (VWS), pattern design CAD and drape analyzer. In this topic, the case study of data-base system of the fabric structural design in the 1st step shown in Figure 2 is introduced and analyzed with various kinds of fabric materials and structural factors.

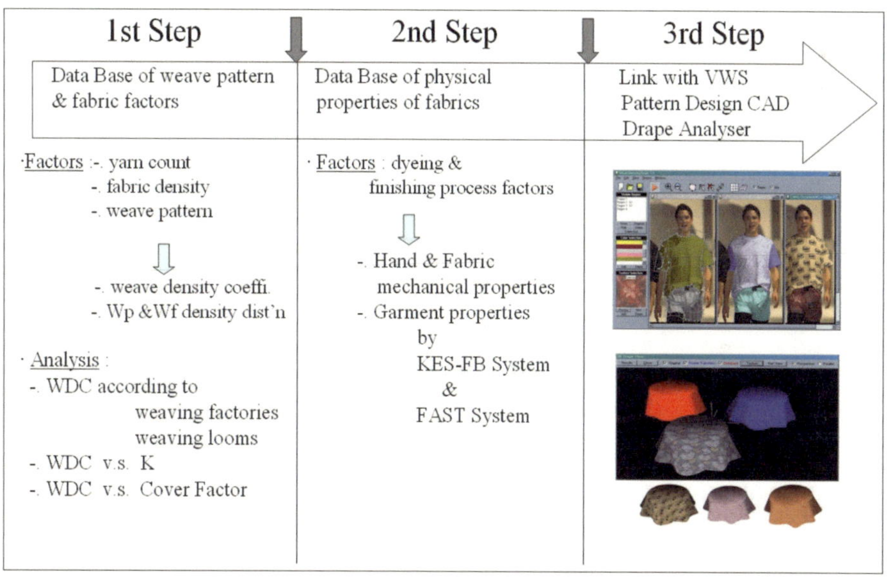

Fig. 2. Detail milestone of analysis steps in relation with existing fabric design and wearing systems of garment

3. Major issues of the mechanical property of the woven fabric related to the fabric structural design

Many researches about mechanical property of the woven fabric according to the yarn and fabric parameters were carried out using KE-FB and FAST systems (Oh & Kim,1993, 1994). Among them, the PET synthetic fabric mechanical properties according to weft filament yarn twists, yarn denier and fabric density were analysed and discussed with these yarn and fabric structural parameters. On the other hand, the worsted fabric mechanical properties according to the looms such as rapier and air jet were also analysed and discussed with weaving machine characteristics (Kim & Kang, 2004, Kim & Jung, 2005). Similar studies were also performed using the PET and PET/Tencel woven fabrics (Kim et al., 2004). The researches related to the fabric mechanical property according to the dyeing and finishing processes were also carried out (Kim et al., 1995, Oh et al., 1993). These are the discrete research results such as 1st and 2nd step shown in Figure 2. There are no informations about how these mechanical properties affect to the garment properties shown on step 3 in Figure 2. This is major issue point of the mechanical property of the woven fabric related to the fabric structural design. Fortunately, in i-designer CAD system, visual weaving performance is available by input the fabric mechanical properties measured by KES-FB system. So, the data base in 1st and 2nd step shown in Figure 2 is needed and these data bases have to be linked with existing visual wearing system, pattern design CAD and drape analyzer shown on 3rd step in Figure 2.

4. Current trends of the data base system of the fabric structural design

4.1 Procedure of data base system of the fabric structural design

Figure 3 shows the procedure of data base system of the fabric structural design. In Figure 3, yarn diameter is calculated using yarn count and weave factor is also calculated by weave structure using number of interlacing point and number of yarn in one repeat weave pattern. Then the weave density coefficient is decided using yarn diameter, weave factor and warp and weft densities. And conversely the warp and weft density distribution is made by yarn diameter, weave factor and weave density coefficient. Peirce(Peirce, 1937) proposed equation 1 as a fabric cover factor which is recommended to weaving factories by Picanol weaving machinery company(Picanol, 2005). In equation 1, yarn and fabric correction factors are shown in Table 1 and 2, respectively.

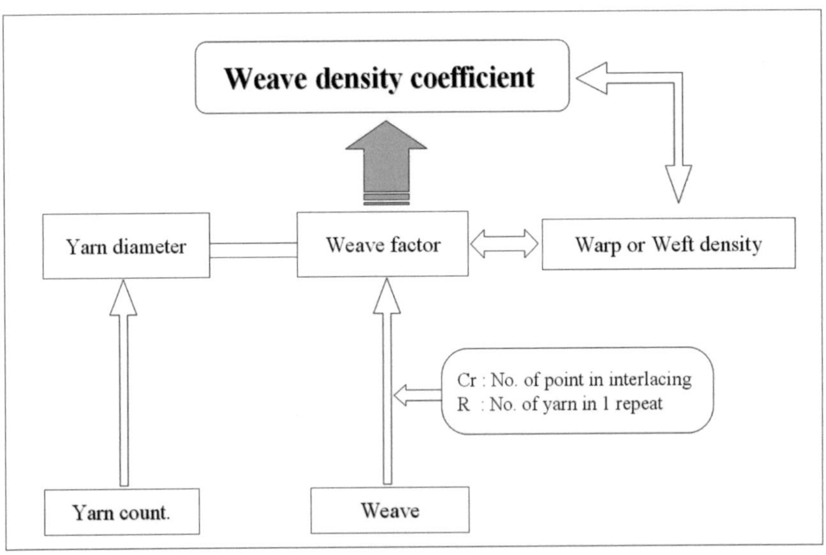

Fig. 3. Procedure diagram of data base system of the fabric structural design

$$\left(\frac{\text{ends/in}}{\sqrt{\text{Ne}}}+\frac{\text{picks/in}}{\sqrt{\text{Ne}}}\right)\times \text{yarn correction factor} \times \text{fabric correction factor} \tag{1}$$

Type of yarn	Correction factor
metallic	0.3
glass	0.6
carbon	0.9
cotton, flax, jute, viscose, polyester	1.0
acetate, wool	1.1
polyamide	1.2
polypropylene	1.4

Table 1. Yarn correction factor

Drill/twill weave		Satin weave	
Pattern	Peirce	Pattern	Peirce
2/1	0.819	1/4	0.709
3/1	0.769	1/5	0.662
2/2	0.746	1/6	0.629
4/1	0.763	1/7	0.599
5/1	0.714	1/8	0.578
6/1	0.694		
7/1	0.689		
4/4	0.671		

Table 2. Fabric correction factor

On the other hand, Prof. M. Walz(Park et al., 2000) proposed equation 2 as a little different equation form, but which is applicable to the various fabrics made by all kinds of textile materials. In equation 2, yarn and fabric correction factors are also shown in Table 3 and 4, respectively.

$$C(\%) = (\mathrm{dw} + \mathrm{df})^2 \times \mathrm{Dw} \times \mathrm{Df} \times b \tag{2}$$

$$\textit{where,} \quad d_{w,f} = \frac{a}{\sqrt{Nm}} = \frac{a\sqrt{dtex}}{100}\text{: yarn diameter(warp, weft)}$$

where

C(%): cover factor
Dw: warp density (ends/inch)
Df: weft density (picks/inch)
a: yarn correction factor (Table 3)
b: fabric correction factor (Table 4)

Basilio Bona (Park et al., 2000) in Italy proposed empirical equation 3 for deciding fabric density on the worsted fabrics.

$$D = K \times \sqrt{Nm} \times C_f \tag{3}$$

where, D: fabric density (ends/m)
K: density coefficient
Nm: metric yarn count
C_f: weave coefficient

Type of yarn	Correction factor
metallic	0.39
glass	0.71
carbon	0.86
cotton, flax, jute, viscose	0.95
polyester	0.92
acetate, wool	0.98
polyamide	1.05
polypropylene	1.17

Table 3. Yarn correction factor

Drill/twill weave		Satin weave	
Pattern	Walz	Pattern	Walz
2/1	0.69	1/4	0.50
3/1	0.58	1/5	0.45
2/2	0.56	1/6	0.42
4/1	0.49	1/7	0.39
5/1	0.43	1/8	0.38
6/1	0.41		
7/1	0.40		
4/4	0.39		

Table 4. Fabric correction factor

$$\left(C_f = \frac{R}{R + C_r} \times f_c \times f_f \times f_j \right)$$

f_c: cover factor
f_f: floating factor
f_j: jumping factor

Equation 3 is modified as equation 4 for the cotton fabrics.

$$D = K_c \times 0.0254 \times \sqrt{Ne \times 1.694} \times C_f \tag{4}$$

where, Ne: English cotton count
Kc: Yarn density coefficient (cotton)
where:
- Comber yarns : 425~350 (12 ~17 MICRONAIRE)
- Sea & Island cotton : 425, American cotton : 375
- Card yarns : 350~290 (14 ~22 MICRONAIRE)

But, in synthetic filament yarn fabrics such as nylon and polyester, more effective parameter is needed. So, weave density coefficient, WC is made by equation 5.

$$WC = \left[\frac{d_w + d_f}{25.4} \right]^2 \times D_w \times D_f \times WF \tag{5}$$

where, $d_{w,f}$: yarn diameter (warp, weft)

WF : weave factor

$D_{w,f}$: warp, weft density

In equation 5, assuming that $D_w \times D_f$ is constant, it becomes as equation 6.

$$D_w \times D_f = \frac{WC}{WF} \times \left[\frac{25.4}{d_w \times d_f} \right]^2 = const. \tag{6}$$

WC in equation 5 can be converted to K and Kc in equation 3 and 4, conversely K is converted to WC and also WC in equation 5 can be compared with cover factor, C given in equation 1 and 2, which is shown in next case study.

4.2 Calculation of fabric structural parameters

In equation 6, d_w and d_f are calculated by yarn linear density, equation 7 as shown in Figure 4. WF is calculated by equation 8 as shown in Figure 5. In Figure 4, calculated yarn diameter by equation 7 is shown in polyester, nylon and rayon yarns, respectively. As shown in Figure 5, calculated weave factors by equation 8 are shown according to the various weave patterns. For plain weave, weave factor (WF) is calculated as 1 using R=2 and Cr = 2. In a little complicated weave pattern as a derivative weave, weave factor (WF) is calculated as 0.76 using R=4 and Cr=3 as an average value by two types of repeat pattern in the weft direction. And in a very complicated weave pattern, Moss crepe, weave factor is calculated as 0.538 using R=120 and Cr=56.06.

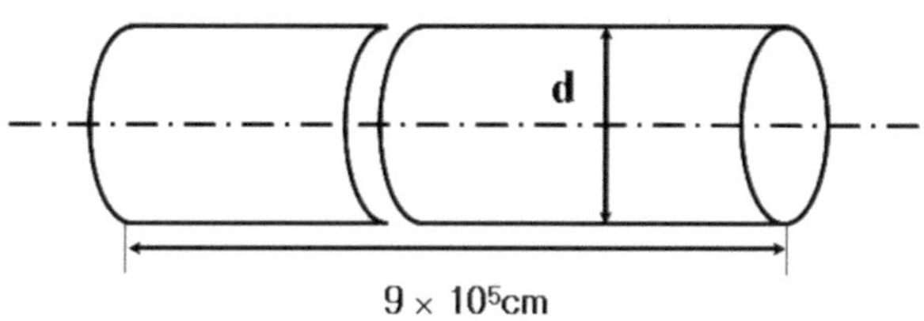

Yarn	Density	Yarn diameter (mm)
PET	1.38	$0.01246\sqrt{Den.}$
Nylon	1.14	$0.01371\sqrt{Den.}$
Rayon	1.52	$0.01189\sqrt{Den.}$

$$Den.(g) = \rho_f(g/cm^3) \times V(cm^3) = \rho_f \times \frac{\pi d^2}{4} \times 9 \times 10^5 \tag{7}$$

where, d: yarn diameter

ρ_f: fibre density

Den: denier

V: volume

Fig. 4. Diagram between yarn count and diameter

$$WF = \left[\frac{R + C_r}{2R} \right]^2 \tag{8}$$

where, WF: weave factor

R: No. of yarn in 1 repeat

Cr: No. of point in interlacing

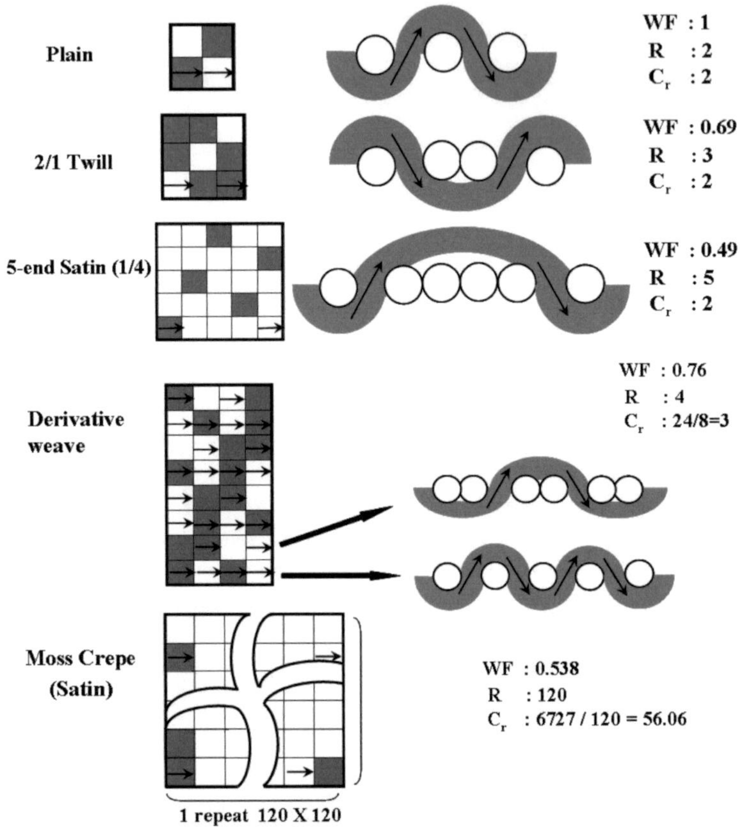

Fig. 5. Diagram of various weave constructions.

4.3 Case study of synthetic fabrics

Design plan sheets of polyester and nylon fabrics woven by various looms were selected as a specimens from various weaving manufacturers such as A, B, C, D, E and F as shown in Table 5, respectively, Table 5 shows the distribution of these specimens.

	PET fabrics						Nylon fabrics
	A company	B company	C company	D company	E company	Sub -total	F company
Loom	WJL	RPL	AJL+RPL	WJL+RPL	WJL+RPL		
Plain	26	4	14	46	5	95	516
Satin	10	41	20	4	8	83	24
Twill	60	28	33	4	9	134	113
Other	-	25	51	-	32	108	185
Sub-total	96	98	118	54	54	420	838

Table 5. Distribution of specimens

For calculation weave density coefficient as shown in equation 5, yarn diameter is first calculated using equation 7.

$$Den = \rho_f \times \frac{\pi d^2}{4} \times 9 \times 10^5 \tag{9}$$

where, ρ_f: fibre density (g/cm^3)
d: yarn diameter (mm)
Den: yarn count (denier)

For polyester filament, yarn diameter, d is $0.01246\sqrt{Den}$ and for nylon filament, that is $0.01371\sqrt{Den}$. On the other hand, weave factor, WF is also calculated using equation 8 and R, Cr in the one repeat weave pattern of fabrics. Through this procedure, yarn diameter, d and weave factor, WF are calculated for all the specimens of nylon and polyester fabrics. Finally weave density coefficient, WC is calculated using d, WF and warp and weft fabric densities, Dw and Df of the all the nylon and polyester fabrics. And WC is plotted against various yarn counts using equation 5 and conversely warp and weft density distribution is presented with various weave density coefficients and weave patterns using equation 6.

1. The distribution of weave density coefficient according to the looms

For four hundreds twenty polyester fabrics, the diameters of warp and weft yarns were calculated using deniers by equation 7, and weave factor was calculated by one repeat weave construction. The weave density coefficient was calculated using equation 5. Figure 6 shows the diagram between weave density coefficient and yarn count for the polyester fabrics woven by water jet loom. And Figure 7 shows that for rapier loom. As shown in Figure 6, the weave density coefficients of PET fabrics woven by WJL were widely ranged from 0.2 to 1.8, on the other hand, for rapier loom, was ranged from 0.4 to 1.4 as shown in Figure 7. And in Figure 6, the values for satin fabrics were ranged from 0.6 to 1.0, which were lower than those of the plain and twill fabrics. Around the yarn count 150d, 300d and

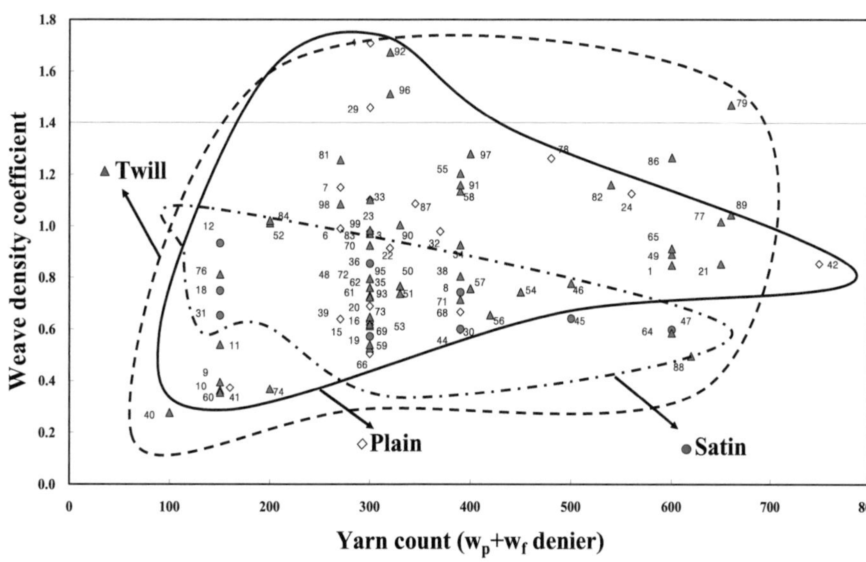

Fig. 6. The diagram between weave density coefficient and yarn count for PET fabrics (WJL). (——— : Plain, ... : Twill, ‒ · ‒ · : Satin)

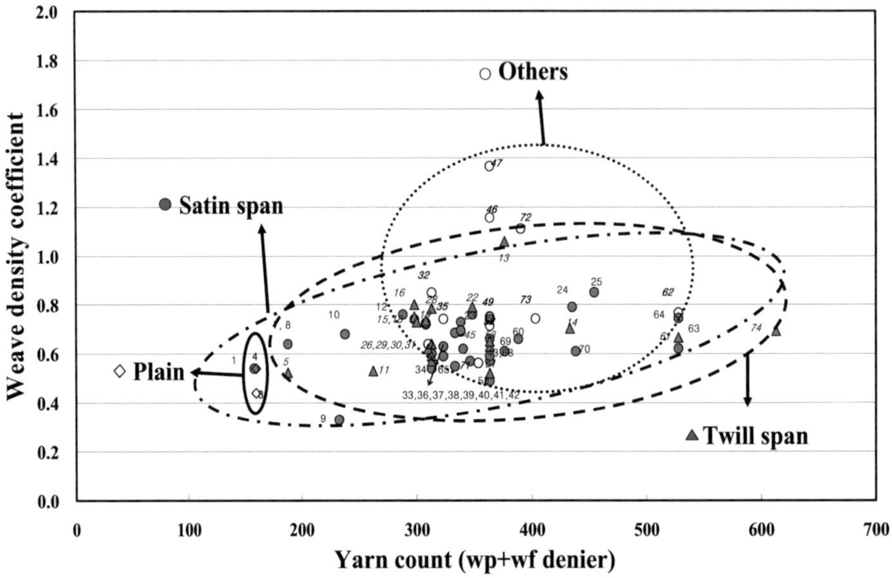

Fig. 7. The diagram between weave density coefficient and yarn count for PET fabrics (RPL). (———— : Plain, ... : Twill, ▬ . ▬ . : Satin, ... : Others)

400d for the twill fabrics, it is shown that the weave density coefficients are ranged from 0.4 to 1.0 for 150d, ranged from 0.5 to 1.7 for 300d and also from 0.6 to 1.3 for 400d. This demonstrates that the weave density coefficients of fabrics woven by water jet loom were widely distributed according to the end use of fabrics for garment.

2. The comparison of the weave density coefficient between polyester and nylon fabrics

Figure 8 shows the diagram between weave density coefficient and yarn count for polyester and nylon fabrics woven by water jet loom for the specimens of higher weft yarn count than warp. As shown in Figure 8, the weave density coefficient of nylon fabrics are widely ranged from 0.5 to 3.0, and comparing to polyester fabrics, the weave density coefficients of nylon fabrics are higher than those of PET fabrics. Especially, in polyester fabrics, plain, twill and satin weave patterns were widely divided to each other on weave density coefficient and yarn count, on the other hand, in nylon fabrics, it was shown that plain was most popular and many specimens were concentrated around yarn count 200d region. Figure 9 shows the weave density coefficients of polyester and nylon fabrics according to the weaving looms. As shown in Figure 9 (a), (b) and (c), the weave density coefficients of polyester fabrics woven by water jet loom were ranged from 0.4 to 1.5, those woven by air jet loom are ranged from 0.7 to 2.0 and woven by rapier loom was ranged from 0.5 to 2.8. And yarn count also showed wide distribution in water jet and rapier looms, but air jet loom showed a little narrow distribution. This phenomena demonstrate that the versatility of rapier loom was the highest comparing to the other weaving looms. On the other hand, comparing Figure 9 (a) with Figure 9 (d), the weave density coefficients of nylon fabrics were ranged from 0.5 to 3.0, while in polyester fabrics they were ranged from 0.4 to 1.5. Nylon fabric showed much wider distribution and much larger values of the weave density coefficient.

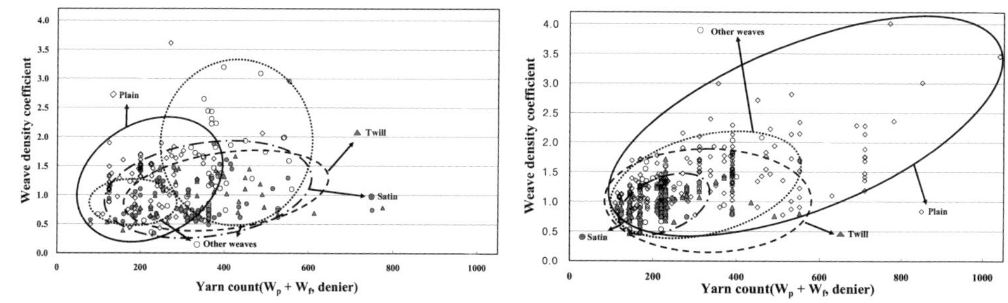

Fig. 8. Comparison of weave density coefficient between PET and Nylon fabrics (Wp<Wf).
(———— : Plain, ... : Twill, ▬ . ▬ . : Satin, ... : Others)

PET

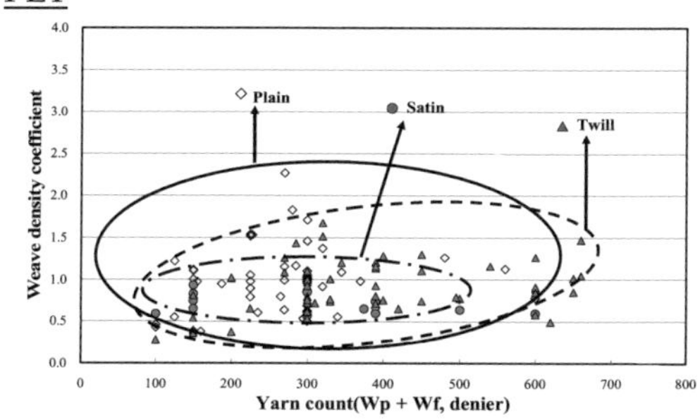

(a) WJL

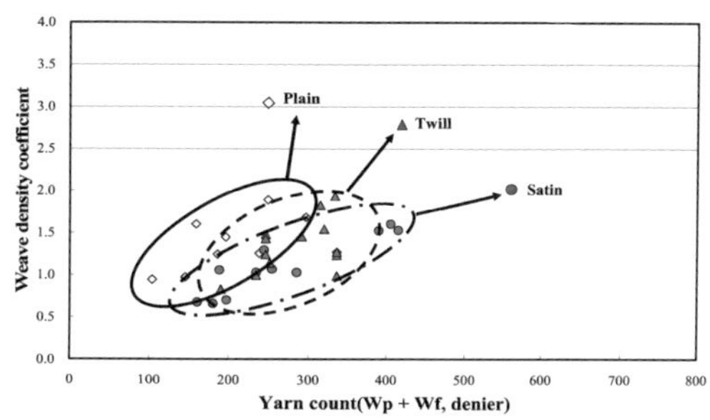

(b) AJL

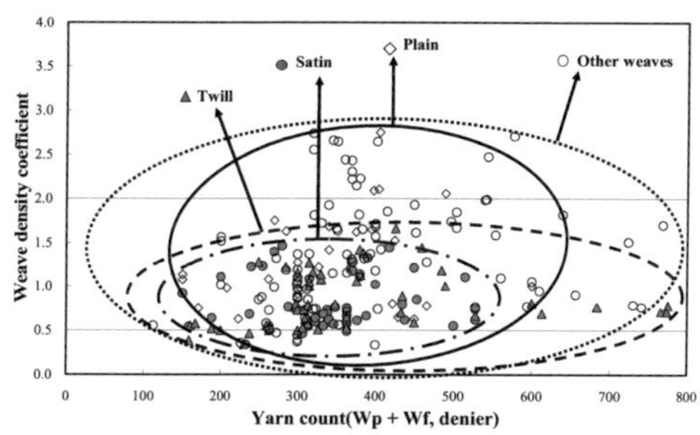

(c) RPL

Nylon

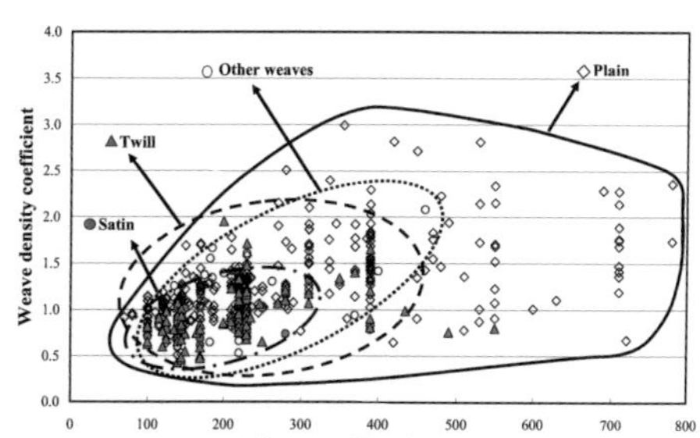

(d) WJL

Fig. 9. The weave density coefficients of polyester and nylon fabrics according to the weaving looms. (———— : Plain, ... : Twill, ▬ . ▬ . : Satin, ... : Others)

3. The density distribution

Figure 10 shows fabric density distribution calculated and simulated by equation 6 for polyester and nylon fabrics with 2 kinds of yarn counts. Figure 10 (a) shows warp and weft density distributions of polyester fabrics with various weave density coefficients and various weave patterns with warp and weft yarn counts 150 deniers. As shown in this Figure 10 (a), specimen no. 21 and 29, satin crepe fabrics, have almost same weft density of fabrics, but warp density of fabrics were different according to the end use of fabric for garment. And as shown in Figure 10 (b), many specimens of plain fabrics have same weave density coefficient, but it was shown that warp and weft densities were different one another according to the end use of fabric for garment. Then it was shown that it was very convenient to decide warp and weft fabric densities for good hand of fabrics.

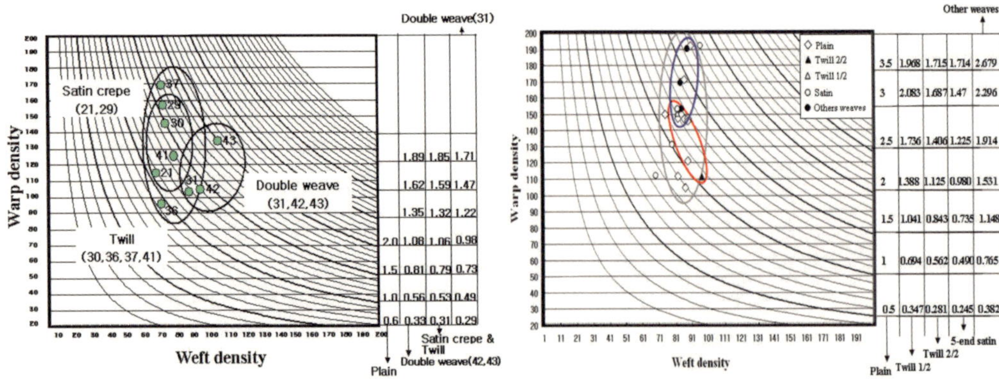

Fig. 10. The diagram between fabric density of PET and Nylon fabrics

4. Comparison between weave density coefficient and cover factor

Figure 11 shows the diagram of weave density coefficient (WC), cover factors by Picanol and Prof. Walz which are calculated by equation 5, equation 1 and equation 2 using the specimens shown in Table 5, respectively. As shown in Figure 11(a), weave density coefficients of PET plain fabrics are widely ranged from 0.5 to 3.0. On the other hand, stain fabrics are distributed from 0.5 to 1.5, and for twill fabrics, ranged from 0.3 to 2.0. This phenomena demonstrate that plain fabrics show broad and wide distribution of weave pattern, and satin shows narrow distribution, which means the versatility of plain weave pattern. And also it is shown that 90% of all specimens' weave density coefficient is ranged from 0.5 to 1.5, which shows similar distribution to cover factor shown in Figure 11(b), proposed by Prof. Walz as equation 2. On the other hand, cover factors proposed by Picanol, which are calculated by equation 1, are distributed from 25% to 90% as shown in Figure 11(c). It is shown that Picanol's cover factor is much lower than those of WC and Prof. Walz equations, And comparing between WC and Prof. Walz equation, WC is about 30% higher than that of Prof. Walz equation. The reason seems to be due to the yarn correction factor 'a' and fabric weave correction factor 'b' in equation 2.

Figure 12 shows the same diagram for nylon fabrics. As shown in Figure 12(a), the weave density coefficients of all Nylon fabric specimens are distributed from 0.5 to 4.0 which are much wider than those of PET fabrics comparing with Figure 11(a). It is shown that weave

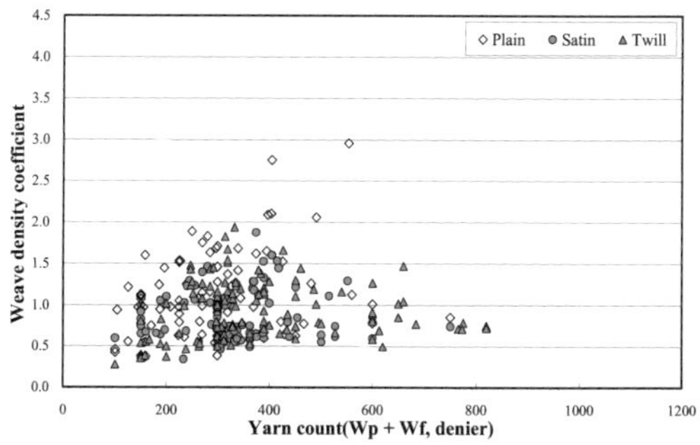

(a) Weave density coefficient

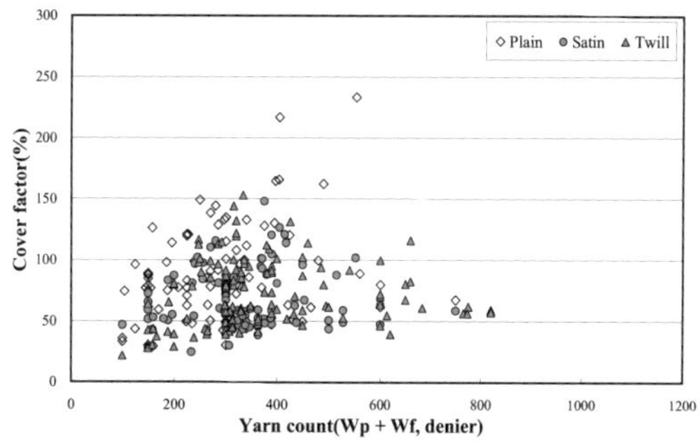

(b) Cover factor by Prof. Walz

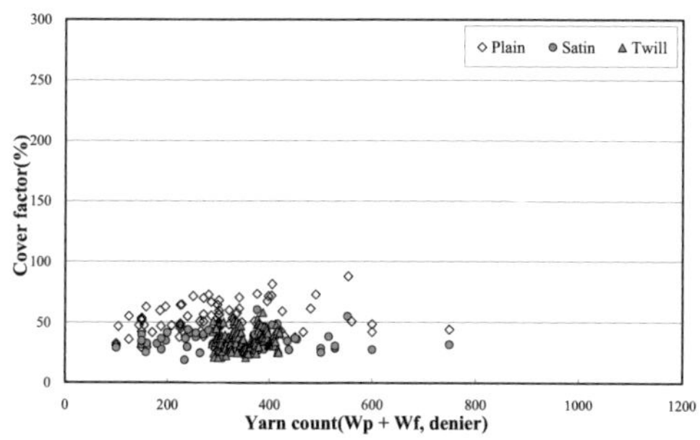

(c) Cover factor by Picanol

Fig. 11. Comparison among WC, cover factors of Picanol and Prof. Walz for PET fabrics

density coefficients of plain fabrics are widely distributed from 0.5 to 4.0. On the other hand, the weave density coefficients of twill and satin fabrics are ranged from 0.5 to 1.5, which is much lower and narrower than that of plain. As shown in Figure 12(b), cover factors by Prof. Walz are distributed from 50% to 200% which shows lower distribution than that of weave density coefficient as shown in Figure 12(a). It is shown that cover factor values by Picanol equation shown in equation 1 are distributed from 30% to 100% which is much lower than those of WC and Prof. Walz equations. And comparing between PET and nylon fabrics as shown in Figure 11(b) and Figure 12(b), in nylon fabrics, cover factors of satin and twill are distributed from 50% to 100%, but plain is widely distributed from 30% to 200% as shown in Figure 12(b). In PET fabrics shown in Figure 11(b), cover factors of all weave patterns such as plain, twill and satin are widely distributed from 30% to 150%. This phenomena demonstrate that plain weave patterns of nylon have higher density than those of satin and twill weave patterns, in one hand, the density of all weave patterns such as plain, twill and satin in the PET fabrics has almost same level. The cover factors of the nylon fabrics proposed by Picanol which are shown in Figure 12(c) ranged from 30% to 100% are much higher than those of PET fabrics which are shown in Figure 11(c).

Figure 13 shows density coefficient, K of the polyester and nylon fabrics calculated by equation 3. As shown in Figure 13, the density coefficient, K is distributed between 400 and 1600 both polyester and nylon fabrics. Mario Bona (Park et al., 2000) in Italy is recommending this value as 800 for synthetic fabrics. Comparing to this recommended value, both polyester and nylon fabrics show much higher values than recommended value,

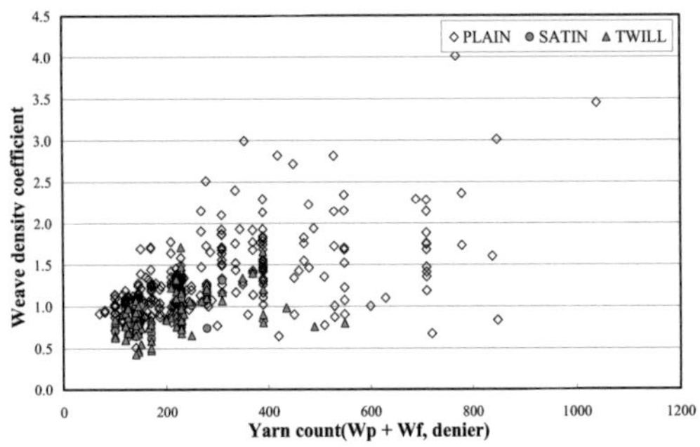

(a) Weave density coefficient

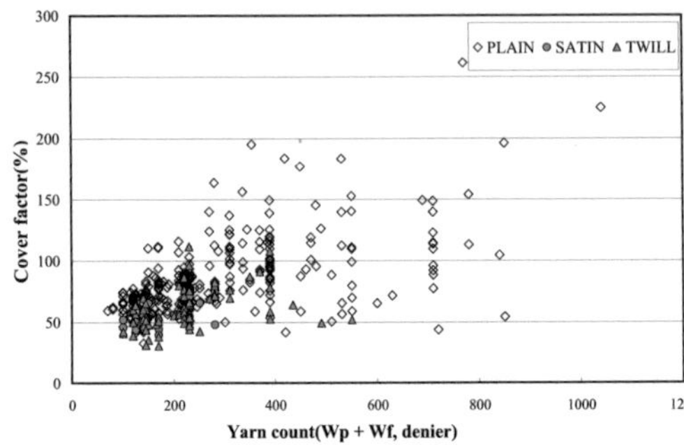

(b) Cover factor by Prof. Walz

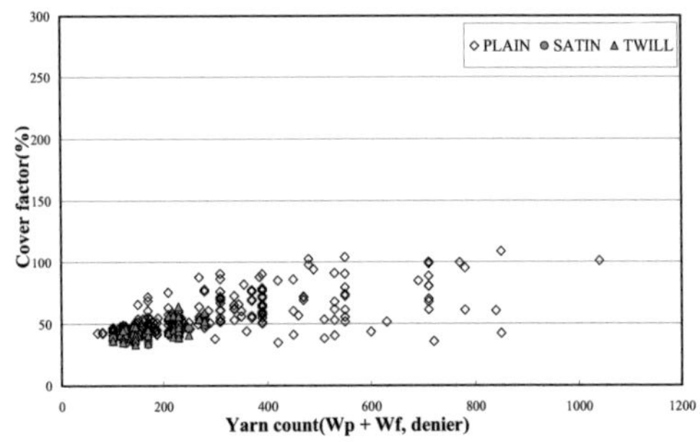

(c) Cover factor by Picanol

Fig. 12. Comparison among WC, cover factors of Picanol and Prof. Walz for nylon fabrics

800. As well known to us, the equation 3 proposed By M. Bona is based on density calculation of the worsted fabrics. Applying to synthetic fabrics as shown in Figure 13, the density coefficient distribution of the PET fabrics is mainly ranged between 600 and 1000 and for nylon fabrics, which is much more concentrated at this region. This results demonstrate the validity of the recommended value, 800 by M. Bona.

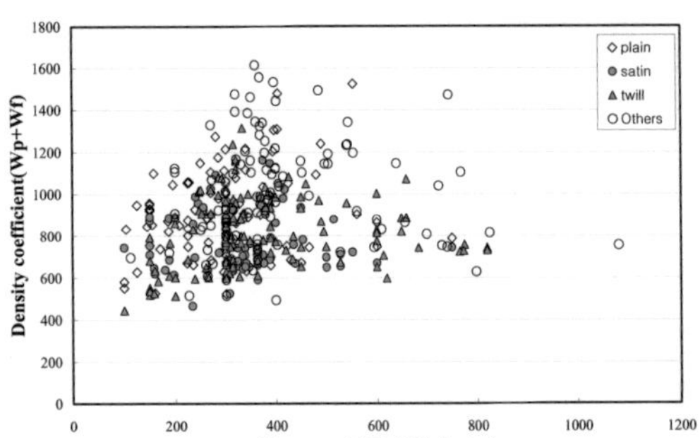

(a) PET

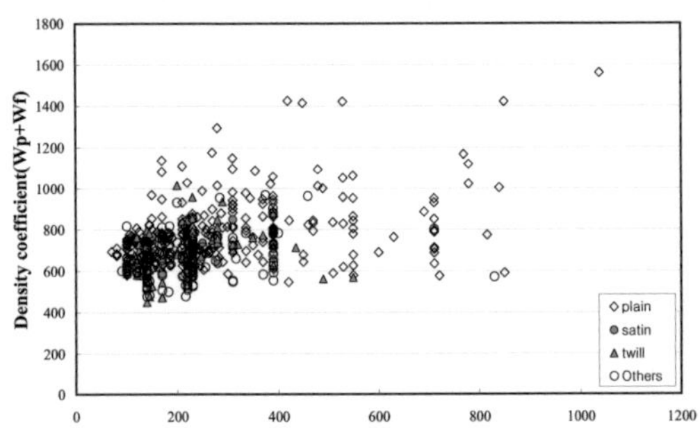

(b) Nylon

Fig. 13. Diagram of K against yarn count of polyester and nylon fabrics

4.4 Case study of worsted and cotton fabrics

Various fabrics woven by worsted and cotton staple yarns were selected as specimens, respectively. Table 6 shows these specimens. For the worsted fabrics of one hundred thirteen, density coefficient, K was calculated using equation 3. For the cotton fabrics of four hundreds seventy nine, density coefficient Kc was calculated using equation 4.

Weave pattern \ Materials & Loom	Worsted fabrics	Cotton fabric
	Sulzer	Air-jet
Plain	35	243
Twill	48	156
Others	30	80
Total	113	479

Table 6. Specimens of worsted and cotton fabrics

Figure 14 shows the diagram between density coefficient and yarn count for worsted and cotton fabrics. It is shown that the density coefficient, K of worsted fabrics is ranged from 600 to 1000, for cotton fabrics, almost same distribution is shown. Comparing to synthetic fabrics such as polyester and nylon in which were ranged from 400 to 1600, as shown in Figure 14, natural fabrics such as worsted and cotton show lower values. Figure 15 shows weave density coefficients, WC calculated by equation 5, of worsted and cotton fabrics. As shown in Figure 15(a), the weave density coefficients of worsted fabrics are ranged from 0.4 to 0.8, for cotton fabrics, they are ranged from 0.2 to 1.0. Comparing to synthetic fabrics, which were shown in Figure 11(a) and 12(a) and ranged from 0.5 to 3.0, WC of the worsted and cotton fabrics show much lower values as below 1.0. Figure 16 shows weave density coefficient WC calculated by equation 5 and cover factors, calculated by equation 1 and 2 for worsted fabrics. As shown in Figure 16(a), weave density coefficients of worsted fabrics show the values below 1.0, and cover factors also show below 100%, especially the cover factor by Picanol shows lower values than Prof. Walz as below 50%. These values are much lower than those of synthetic fabrics shown in Figure 11(a) and 12(a). Figure 17 shows the diagram for cotton fabrics. The same phenomena are shown as worsted fabrics.

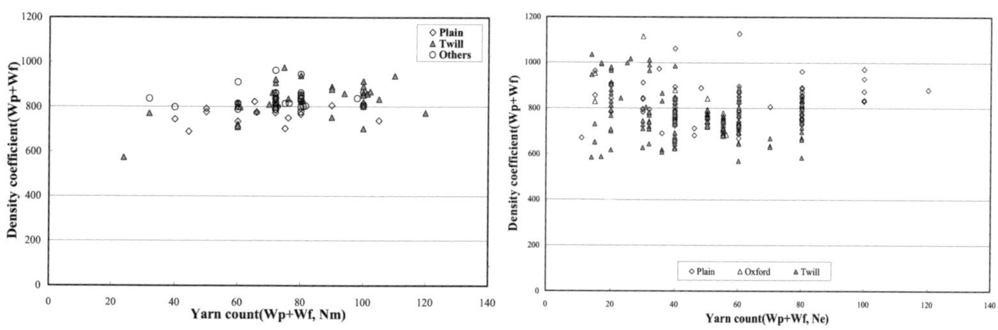

(a) Worsted (b) Cotton

Fig. 14. Diagram between density coefficients and yarn counts for worsted and cotton fabrics

4.5 Relationship between weave density coefficient and shrinkage of dyeing and finishing processes

Figure 18 shows relationship between weave density coefficient and finishing shrinkage in dyeing and finishing processes of PET fabrics woven in the weaving company as shown in Table 5. The finishing shrinkages are distributed from 2% to 40% as shown in Figure 18. It is

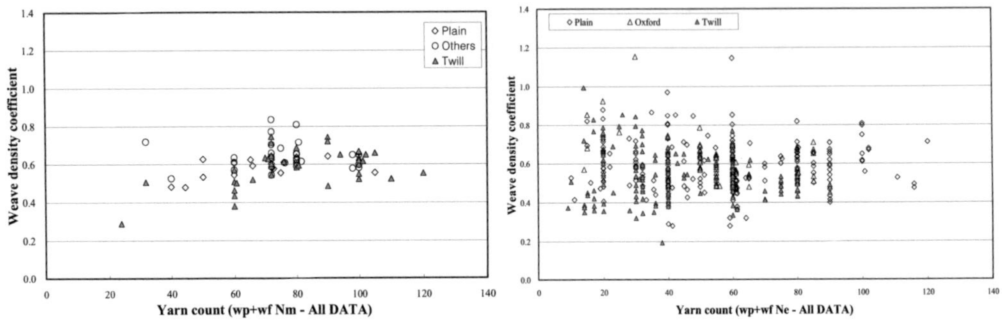

(a) Worsted (b) Cotton

Fig. 15. Diagram of weave density coefficients of worsted and cotton fabrics

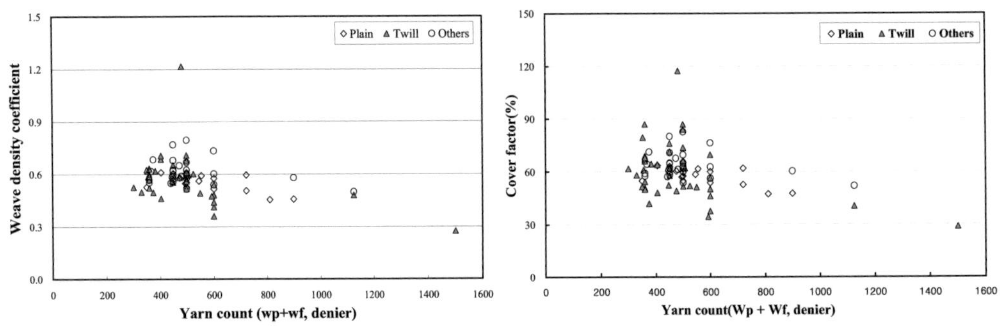

(a) Weave density coefficient (b) Cover factor by Prof. Walz

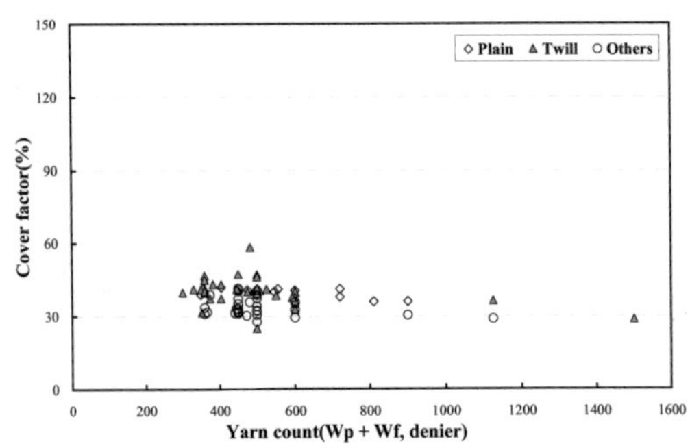

(c) Cover factor by Picanol

Fig. 16. Diagram of weave density coefficients and cover factors for the worsted fabrics.

shown that finishing shrinkage varies according to the weave pattern such as plain, twill and satin. The shrinkages of plain fabric are ranged from 5% to 20%, for twill fabrics, three types of shrinkages levels are divided, one group is below 8%, 2nd group is ranged from 12% to 20%, 3rd group is ranged from 25% to 40%. The finishing shrinkages of the satin weaves are ranged from 12% to 23%(Kim et al., 2005). Figure 19 shows finishing shrinkages distributions from data-base of polyester plain fabrics manufactured by each company fabrics manufactured in A company is ranged from 5% to 20% and for C company, it is shown in the Table 5. As shown in Figure 19, the distribution of finishing shrinkage of PET

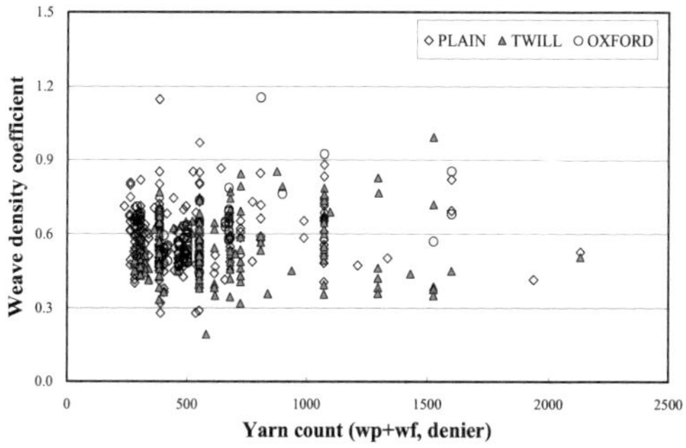

(a) Weave density coefficient

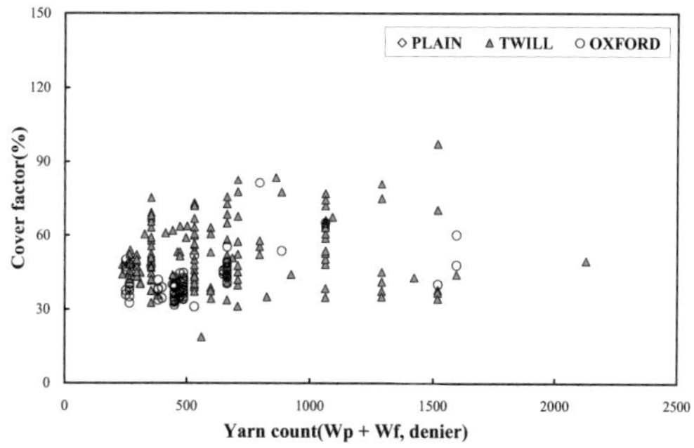

(b) Cover factor by Prof. Walz

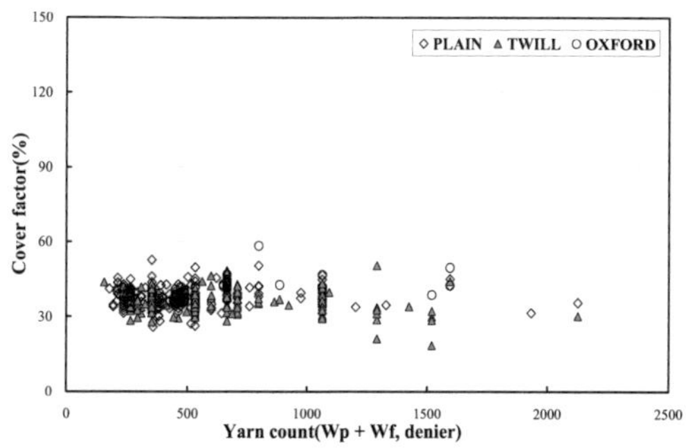

(c) Cover factor by Picanol

Fig. 17. Comparison among WC cover factors by Picanol and Prof. Walz for cotton fabrics

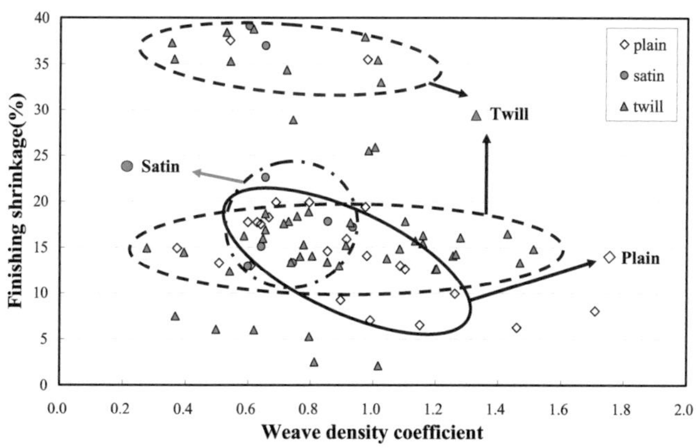

Fig. 18. Diagram between weave density coefficient and fabric shrinkage of PET fabrics woven in A company

ranged from 10% to 25%. This result gives us important information for fabric quality by getting finishing shrinkage according to the fabric manufacturers and weave density coefficients. Figure 20 shows weave shrinkages distributions of nylon fabrics manufactured by F company shown in Table 5. As shown in Figure 20, the weave shrinkages of nylon fabrics vary with weave patterns such as plain, satin and twill, which weave shrinkage values are shown as 7%, 8% and 10%. Figure 21 shows weave and finishing shrinkages of worsted fabrics shown in Table 6. As shown in Figure 21, the weave and finishing shrinkages of worsted fabrics are also distributed with weave patterns such as plain and twill, which are ranged from 2% to 10%.

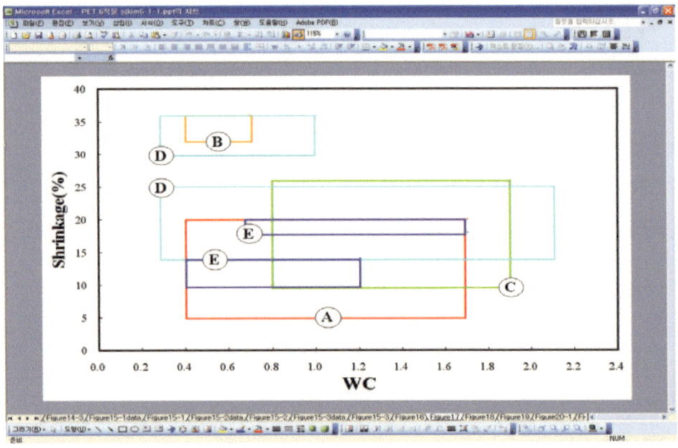

Fig. 19. Diagram between weave density coefficient and finishing shrinkage of PET fabrics woven by each company.

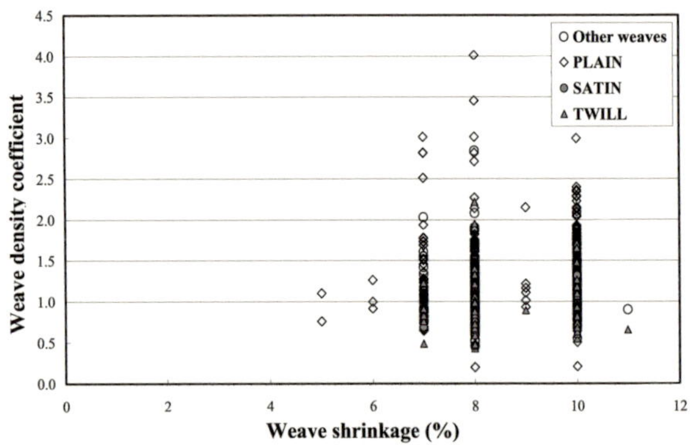

Fig. 20. Relationship between weave shrinkage and WC.

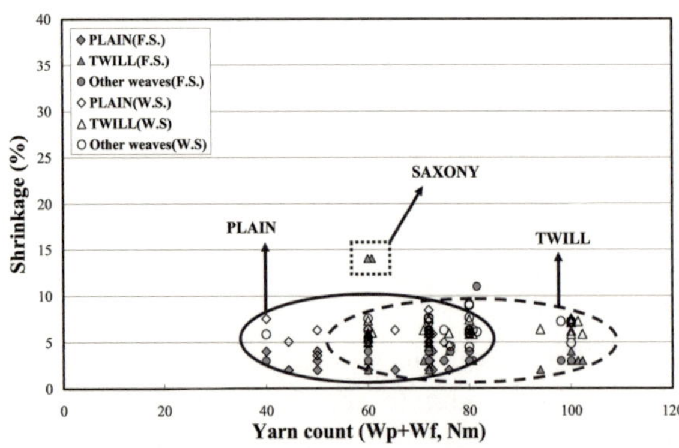

Fig. 21. Weave and finishing shrinkages according to the yarn count (F.S. : finishing shrinkage, W.S. : weave shrinkage)

Figure 22 shows finishing and weave shrinkages of cotton fabrics shown in Table 6. As shown in Figure 22, finishing shrinkages of cotton fabrics are distributed from 2% to 20%, on the one hand, weave shrinkages are ranged from 1% to 10%. It is shown that these shrinkages vary with weave patterns.

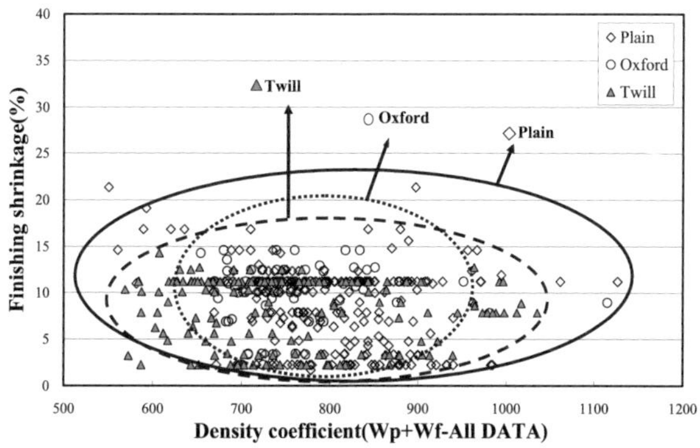

(a) Finishing shrinkage

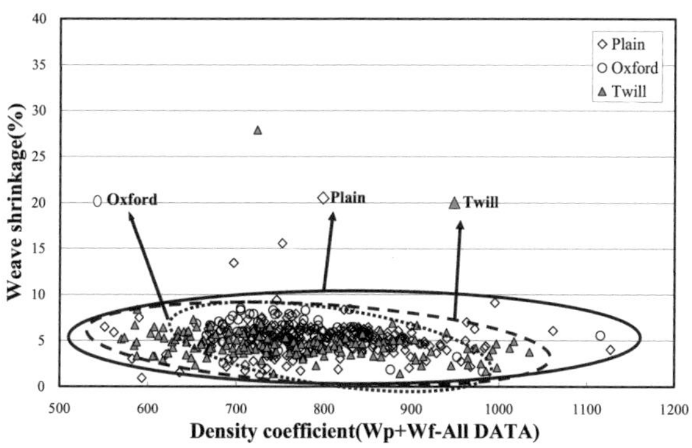

(b) Weave shrinkage

Fig. 22. Diagram between density coefficient and shrinkage of cotton fabrics

5. Future challenges of the data base system for the fabric structural design

Even though a lot of commercial CAD systems(i-Designer, Texpro) for both fabric and pattern have been introduced, any system for weaving factories has not been developed. Therefore, a data base system related to the fabric structural design for weaving factory is needed to be explored. The yarn count, weave pattern and fabric density of 420 polyester fabrics and 838 nylon fabrics shown in Table 1 were used for making data base system, which were divided by weave patterns, weaving looms and weaving manufacturers. The reason why makes data base system according to the weaving manufacturers is explained as for examining the difference of fabric design according to each weaving factory. Figure 23 shows the diagram from data base between weave density coefficient and yarn count according to the weaving manufacturers. As shown in Figure 23, weave density coefficient is easily found according to the weaving manufacturers. It is shown that the distribution of weave density coefficients of PET fabrics manufactured in A company by water jet loom (WJL) is ranged from 0.2 to 1.8 according to the yarn linear density distributed between 100 and 800 denier. For the PET fabrics manufactured in C company by air-jet loom (AJL) and rapier loom (RPL), it is distributed between 0.6 and 2.4 according to the yarn linear density distributed between 100 and 850 denier. On the other hand, the weave density coefficients for the B, D and E fabric manufacturers are differently distributed with narrow distribution of the yarn linear density. This result from data base related to the fabric structural design gives us important information for the weave density coefficients according to the yarn denier and fabric manufacturers. Figure 24 shows the diagram from data base between weave density coefficient and yarn count according to the looms. It is shown that the distribution of weave density coefficients and yarn denier of PET fabrics woven by rapier

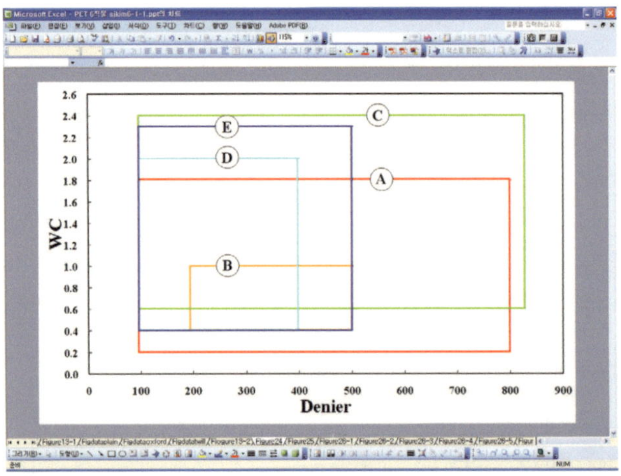

Fig. 23. Data base diagram between weave density coefficient and yarn count according to the weaving company. (PET)

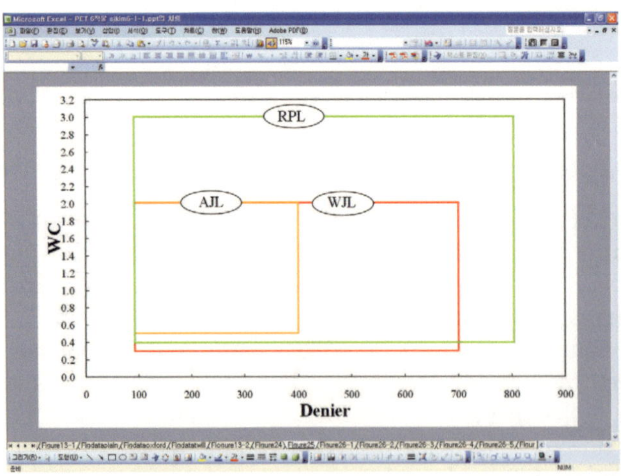

Fig. 24. Data base diagram between weave density coefficient and yarn count according to the looms. (PET)

loom is the widest and air-jet loom is the narrowest. Figure 25 shows the diagram from data base between weave density coefficient and yarn count according to the weave pattern of each weaving manufacturers. It is shown that the distribution of weave density coefficient of twill fabrics of the A company is ranged from 0.3 to 1.6, and for plain weave pattern, it is ranged from 0.6 to 1.6, and the distribution of the satin is very narrow. These phenomena as shown in B, C, D and E company are differently distributed according to the weave pattern. Figure 26 shows the diagram of shrinkage of polyester fabrics according to the weaving companies (A, B, C, D and E) and weave patterns (plain, twill and satin) from data base. This result from data base related to the weave density coefficient gives us important information for the finishing shrinkage according to the fabric manufacturers and weave pattern.

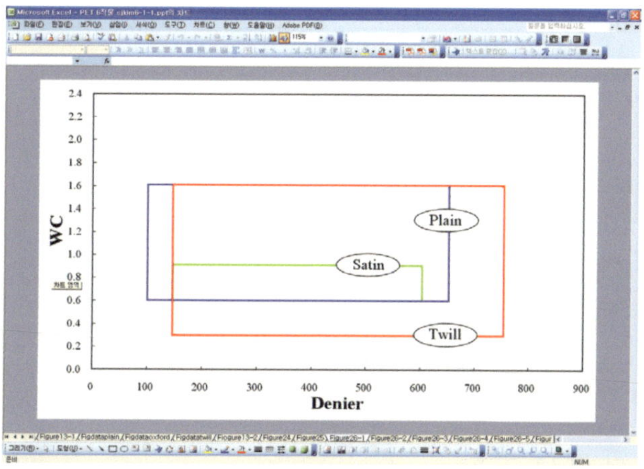

(a) A company

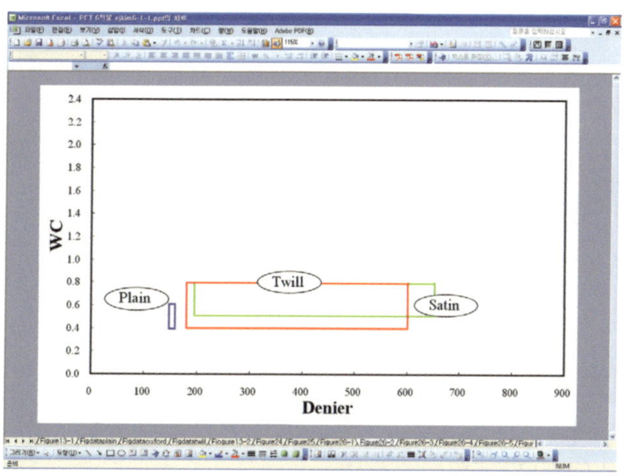

(b) B company

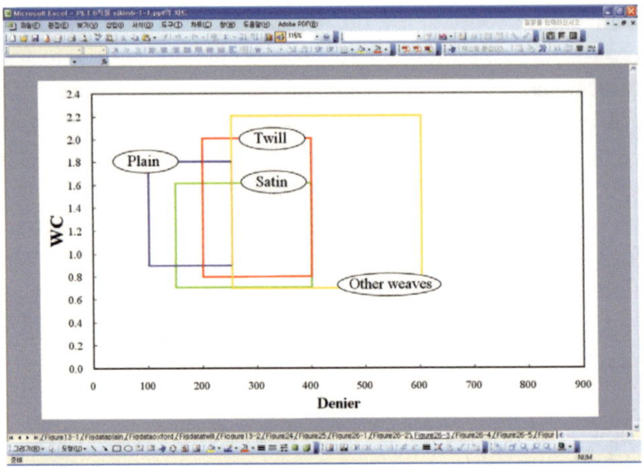

(c) C company

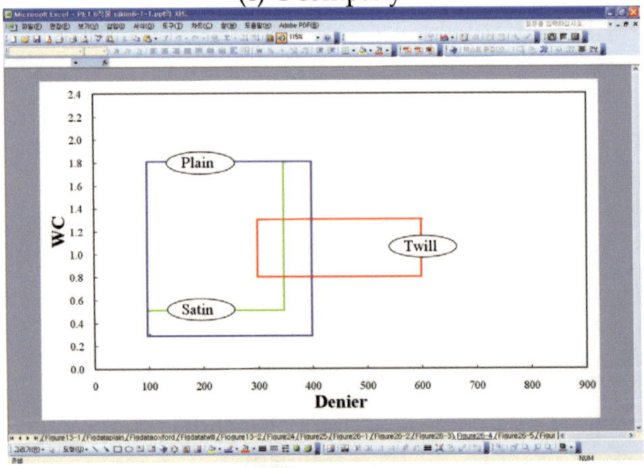

(d) D company

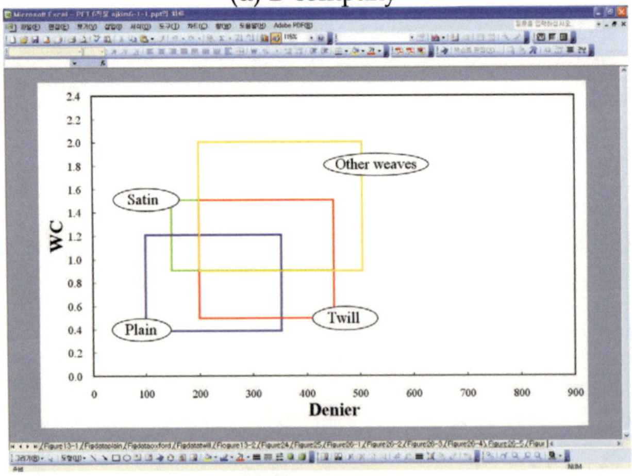

(e) E company

Fig. 25. Data base diagram between weave density coefficient and yarn count according to the weave patterns of each weaving manufacturers. (PET)

Figure 27 shows the application fields of fabric structural design data base system. As shown in Figure 27, final objectives of this topic is aiming to make a data base system with connection of the existing systems such as virtual wearing, pattern design CAD and drape analyzer, i.e. for getting some virtual wearing effect and some drape properties, this data base system is to give the answer about what is the best decision for woven fabric structural design component such as weave density coefficient, weave factor and yarn count. This topic is the first step for wide spreading this application fields to the existing woven fabric and clothing CAD systems.

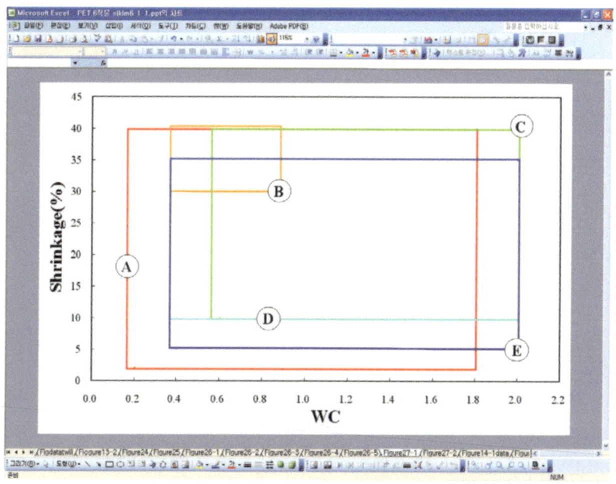

(a) according to company

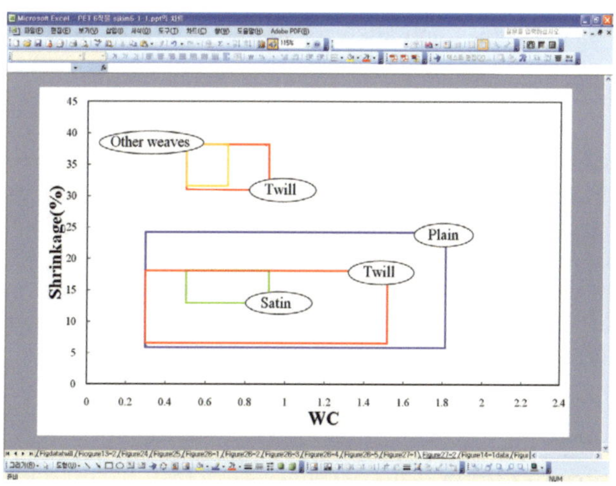

(b) according to weave pattern

Fig. 26. Data base diagram of shrinkage of polyester fabrics.

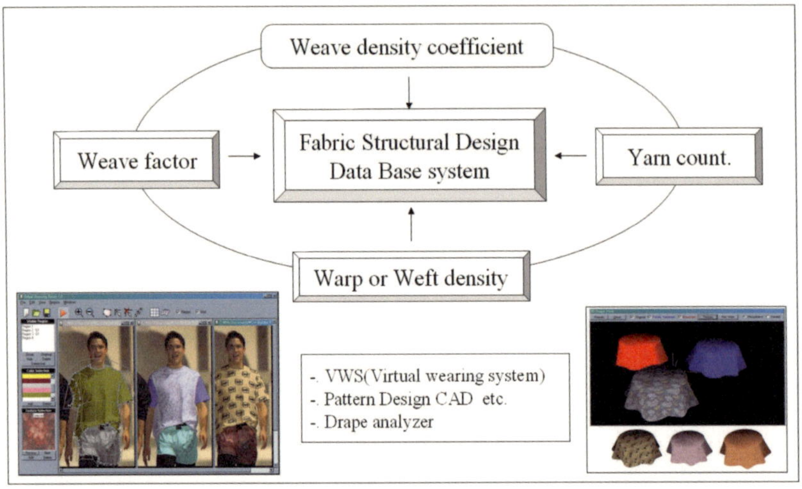

Fig. 27. The application fields of fabric structural design data base system

6. Summary

There were many tries for linking with visual wearing system of garment using fabric mechanical properties, and also there are many CAD systems such as fabric design CAD and pattern design CAD using many data bases on the computer. But there is no fabric structural design data base system linked with fabric physical properties and process conditions. The reason is due to too many factors considered for making such kind of data base system. As a 1st step for making data base system related to the fabric structural design, the data base between fabric structural parameters such as yarn count, fabric density, weave pattern and cover factor and process parameters such as weave and finishing shrinkages has to be constructed and analyzed using various kinds of fabric materials such as worsted, cotton, nylon and polyester fabrics. Through this procedure the estimation of the fabric density with given warp and weft yarn counts and weave construction seems to be possible. It makes easy application for new fabric design and also makes it possible to estimate the weavable fabric density according to the various types of looms for loom machinery maker. For getting the final goal of this topic, further study as follows is needed. 1st is to make data accumulation such as fabric structural design parameters and dyeing and finishing process parameters according to the various weaving companies and looms they are using. 2nd is to make data base for measurement of the physical properties of fabrics such as drape coefficient and mechanical properties. Finally, these have to be applied to the existing virtual wearing system and pattern design CAD.

7. References

Peirce, F. T. (1937). *The geometry of cloth structure*, J. Text. Inst., 28, pp. 45-96

Kemp, A. (1958). *An extension of Peirce's cloth geometry to the treatment of nonlinear threads*, J. Text. Inst., 49, pp. 44-48

Hearle, J. W. S. & Shanahan, J. W. (1978). *An energy method for calculations in fabric mechanics, part* I: *principles of the method*, J. Text. Inst., 69, pp. 81-89

Grosberg, P. & Kedia, S. (1966). *The mechanical properties of woven fabrics, part I: the initial load-extension modulus of woven fabrics*, Text. Res. J., 36, pp. 71-79

Backer, S. (1952). *The mechanics of bent yarns*, Text. Res. J., 22, pp. 668-681

Postle, R.; Carnady, G. A. & Jong, De. S. (1988). *The Mechanics of Wool Structures(eds)*, Chichester Ellis Horwood, U.K

Lindberg, J.; Behre, B. & Dahiberg, B. (1961). *The Mechanical properties of textile fabrics part III: shearing and buckling of various commercial fabrics*, Text. Res. J., 31(2), pp. 99-122

Kawabata, S.; Postle, R. & Niwa, M. (1982). *Objective Specification of Fabric Quality*, Textile Machinery Society of Japan

Ly, N. G.; Tester, D. H. Buckenhan, P. Roczniok, A. F. Adriaansen, A. L. Scazsbrook, F. & De, J. (1991). *Simple instruments for quality control by finishers and tailors*, Text. Res. J., 61(7), pp. 402-406

Hu, J. (2004). *Structure and Mechanics of Fabrics*, The Textile. Institute., CRC press http://www.i-designer.co.kr, i-Designer. http://www.texclub.com, Texpro.

Harada T, Saito M. (1986). *Japanese Journal of Textile System, 45, pp. 305*

Kim, S. J. (2002). Data Base System and Its Application of PET Woven Fabric, *Proceedings of 2nd International Fibre Symposium*, pp. 10-17, Fukui University, September and 2002, Fukui University, Fukui

Kim, S. J. (2005). The Preliminary Study for Data-Base System of The Various Fabrics, *Proceedings of 3rd International Conference on advanced Fiber/Textile Materials 2005 in Ueda*, pp. 225-226, Shinshu University, August and 2005, Research Center for Advanced Science and Technology Faculty of Textile, Ueda

Oh, A. G. & Kim, S. J. (1993). Study on the Mechanical Properties of Polyester Woven Fabric (1) - Tensile Behavior under Low Sterss. *Journal of The Korean Fiber Society, 30.*, 9., 641-651, 1225-1089

Oh, A. G. & Kim, S. J. (1993). Study on the Mechanical Properties of Polyester Woven Fabric (2) - Nonlinearity of Shear Properties. *Journal of The Korean Fiber Society, 30.*, 10., 719-730, 1225-1089

Oh, A. G. & Kim, S. J. (1993). Study on the Mechanical Properties of Polyester Woven Fabric (3) - Nonlinearity of Bending Properties. *Journal of The Korean Fiber Society, 30.*, 12., 919-927, 1225-1089

Oh, A. G. & Kim, S. J. (1994). Study on the Mechanical Properties of Polyester Woven Fabric (4) - Compressional Properties. *Journal of The Korean Fiber Society, 31.*, 5., 361-368, 1225-1089

Oh, A. G. & Kim, S. J. (1994). Study on the Mechanical Properties of Polyester Woven Fabric (5) - Surface Properties. *Journal of The Korean Fiber Society, 31.*, 6., 425-433, 1225-1089

Kim, S. J. & Kang, J. M. (2004). Effects of Rapier Weaving Machine Characteristics on the Physical Properties of Worsted Fabrics for Garment (1). *Journal of The Korean Society for Clothing Industry, 6.*, 6., 765-771

Kim, S. J. & Kang, J. M. (2004). Effects of Rapier Weaving Machine Characteristics on the Physical Properties of Worsted Fabrics for Garment (2). *Journal of The Korean Society for Clothing Industry, 6.*, 6., 772-777

Kim, S. J. & Tung, K. J. (2005). Effects of the Projectile and the Air-jet Weaving Machine Characteristics on the Physical Properties of Worsted Fabrics for Garment (1). *Journal of The Korean Society for Clothing Industry, 7.*, 1., 101-105

Kim, S. J. & Tung, K. J. (2005). Effects of the Projectile and the Air-jet Weaving Machine Characteristics on the Physical Properties of Worsted Fabrics for Garment (2). *Journal of The Korean Society for Clothing Industry, 7.*, 1., 106-110

Kim, S. J.; Sohn, J. H. Kang, J. M. & Park, M. H. (2004). Effects of Weaving Machine Characteristics on the Physical Properties of PET Fabrics (I). *Journal of The Korean Society of Dyers and Finishers, 16.*, 4., 206-215

Kim, S. J.; Sohn, J. H. Kang, J. M. & Park, M. H. (2004). Effects of Weaving Machine Characteristics on the Physical Properties of PET Fabrics (II). *Journal of The Korean Society of Dyers and Finishers, 16.*, 4., 216-224

Kim, S. J.; Sohn, J. H. Kang, J. M. & Park, M. H. (2004). Effects of Weaving Machine Characteristics on the Physical Properties of PET Fabrics (III). *Journal of The Korean Society of Dyers and Finishers, 16.*, 5., 278-283

Kim, S. J.; Sohn, J. H. Kang, J. M. & Park, M. H. (2004). Effects of Weaving Machine Characteristics on the Physical Properties of PET Fabrics (IV). *Journal of The Korean Society of Dyers and Finishers, 16.*, 5., 284-291

Kim, S. J. & Kang, J. M. (2004). Effects of Weaving Tension Characteristics on the Surface Properties of PET Fabrics for the Sensitive Garment (I). *Journal of The Korean Society for Emotion & Sensibility, 7.*, 4., 25-33

Kim, S. J. & Park, K. S. (2005). Effects of the Rapier Weaving Tension Characteristics on the Surface Properties of PET Fabrics. *Journal of The Korean Society for Clothing Industry, 7.*, 6., 673-679

Kim, S. J. (2008). Effects of the Air-Jet Loom Characteristics on the Hand Properties of the Sensitive Mixture Fabrics. *Journal of The Korean Society of Dyers and Finishers, 20.*, 6., 63-68

Kim, S. J.; Oh, A. G. Cho, D. H. Chang, D. H. & Song, J. S. (1995). Study on Correlation between Fabric Structural Parameter and Processing Shrinkage of Polyester Woven Fabric. *Journal of The Korean Fiber Society, 32.*, 5., 480-487, 1225-1089

Kim, S. J.; Oh, A. G. Cho, D. H. Chang, D. H. & Song, J. S. (1995). Study on Correlation between Mechanical Properties and Warp Density of Polyester Woven Fabric. *Journal of The Korean Fiber Society, 32.*, 5., 488-493, 1225-1089

Oh, A. G.; Kim, S. J. Cho, D. H. Chang, D. H. Kim, S. K. Kim, T. H. & Seo, M. H. (1993). Study on Correlation between Mechanical Properties and Processing Shrinkage of Polyester Woven Fabric. *Journal of The Korean Fiber Society, 30.*, 11., 803-816, 1225-1089

Instruction of Manual (1937). Picanol

Park, S. H.; Kim, S. J. Shin, B. J. & Lee, M. H. (2000). Theory and Application of Woven Fabric Design for Garment, *ic Associates Co. LTD.*, Seoul

Kim, S. J. & Hong, S. G. (2005). The Comparison Between Nylon and PET Fabrics, *Proceedings of the 34th Textile Research Symposium*, pp. 33-34, Fuji Institute of Education and Training, August and 2005, Textile Science Research Group in the Textile Machinery Society of Japan, Fuji

Kim, S. J. & Park, K. S. (2005). The Fabric Hand Analysis between Domestic and Foreign Worsted Fabrics, *Proceedings of the 34th Textile Research Symposium*, pp. 85-86, Fuji Institute of Education and Training, August and 2005, Textile Science Research Group in the Textile Machinery Society of Japan, Fuji

Kim, S. J.; Park, K. S. & Hong, S. K. (2005). A Study on the Relationship Between Fabric Design Condition for Garment and Shrinkage on the Dyeing and Finishing Process. *Journal of The Korean Society of Dyers and Finishers, 17.*, 5., 267-274

Kim, S. J. (2006). A Study on the Synthetic Fabric Design System. *Journal of The Korean Society for Emotion & Sensibility, 9.*, 3., 243-249

Hong, S. G. & Kim, S. J. (2004). A Study on the Data-Base of Fabric Design System on the Nylon Woven Fabrics, *Proceedings of The Korean Society for Clothing Industry Conference*, pp. 253-256, EXCO, October and 2004, The Korean Society for Clothing Industry, Daegu

c. Surface Properties Engineering

10

Surface Unevenness of Fabrics

Eva Moučková, Petra Jirásková and Petr Ursíny
Technical University of Liberec
The Czech Republic

1. Introduction

Unevenness of plain textile is counted among qualitative parameters of fabric still more often. It shows itself, for example, in the appearance of plain textile (fluttering, cloudy appearance with thick and thin places) as well as in a mass variation of fabric samples. The appearance of plain textiles is influenced by irregularity of yarns that plain textiles are made from and by manufacturing process of plain textile, i.e. by weaving or knitting.

The yarn mass irregularity displays itself in the plain textile by specific known ways (stripiness and a moiré effect). These faults are caused by a periodical irregularity of yarns. A non-periodical yarn irregularity gives cloudiness in the woven or knitted fabric.

Parameters and characteristic functions of mass irregularity (a spectrogram, a variance length curve) are usually used for the evaluation of unevenness of longitudinal textiles (yarns) (Slater, 1986). The parameters indicate a value of irregularity. The characteristic functions describe a structure of mass irregularity and enable to find the causes of irregularity. We can predicate unevenness of plain textile (surface unevenness) on the base of course of the spectrogram as well as the variance length curve. Knowledge of these problems are already known and verified (Zellweger Uster, 1971); (Zellweger Uster, 1988). Currently, there are other possibilities for the prediction of surface unevenness. One of them is the application of so called a *DR* function (Deviation Rate). It is determined, for example, by means of the Uster Tester IV-SX. Today, studies of relation between the magnitude *DR* and surface unevenness are in progress.

Instrumentation used for mass irregularity measurement (for example, the system Oasys from Zweigle, the apparatus Uster-Tester IV-SX from Zellweger Uster) makes, among others, simulation of surface appearance of plain textile (knitted and woven fabric of selected weave) possible. This image is simulated on the basis of signal of measured yarn mass irregularity. This way, the surface appearance of plain textile can be visually evaluated without plain textile manufacturing. But the image evaluation is only subjective in practice because it is realized as a visually judgment of the plain textile appearance.

In the literature (Militký, 2005); (Wegener & Hoth, 1958); (Ursíny et al., 2008); (Suh, 2005), the surface unevenness of plain textile is described by means of the variation coefficient (*CV*) of various properties of plain textile or by means of derived statistical functions. A sample of plain textile is, in these cases, divided into square fields, where individual properties, e.g. mass, are measured. On the basis of results, so-called an area-variation curve is constructed as a parallel to the variance length curve. The area variation curve is constructed also in the works (Suh, 2005); (Moučková & Jirásková, 2006); (Moučková & Jirásková, 2007).

Other statistical functions, by means of them the surface variability is possible to be described, use the fact, that magnitude z(x,y) is a random function of two variables (random field). For example, the co-variation function or so-called directional semivariograms belong to these functions (Militký & Klička, 2005); (Militký et al., 2000); (Militký & Bajzik, 2000).
This chapter summarises obtained experimental knowledge from the problem area of surface unevenness prediction and evaluation. The behaviour of the parameter *DR* in dependence on other parameters and characteristic functions of mass irregularity is studied here. The possibility of utilization of the parameter *DR* for prediction of surface unevenness is analysed. The simulated image of plain textile as well as the image of real woven fabric is used for the surface unevenness evaluation. The simulated appearance of plain textile, obtained from the measuring instrument, is in the greyscale with various intensity of greyness according to yarn irregularity. The image of real woven fabric is obtained by scanning the fabric sample and then is converted into the greyscale. Thus, unevenness of plain textiles (simulated or real) can be converted into unevenness of coloration, which is interpreted by various intensity of grey. A fluctuation of greyness degree in the image is evaluated by means of area variation curves and semivariograms, constructed by means of a special programme created by Militký, J. (Technical University of Liberec) in the programming environment Matlab. Courses of semivariograms are studied in dependence on the woven fabric parameters (the fabric sett, the fabric weave) as well as woven fabric "quality".

2. Structure of yarn mass irregularity and surface unevenness

We find the term "structure of mass irregularity" as components of periodical irregularity expressed by the spectrogram and as non-periodical irregularity in a certain range of yarn length-sections, which expressed external mass irregularity (the variance length function). Newly, the structure of mass irregularity is possible to be described by the *DR* function (Deviation Rate Function) too. The characteristic functions can be used for prediction of some typical forms of surface unevenness (the moiré effect, stripiness, cloudiness).
In following part, we focus on the utilization of *DR* function, eventually its individual values, with the aim of clearing up the relation between this function and other characteristic functions, especially the variance length curve. Thus, we will also be able to illuminate its connection with surface unevenness. The application of *DR* function in mentioned area and also the possibility of surface unevenness quantification is an important assumption for extension of possibilities of surface unevenness prediction based on characteristic functions representing structure of yarn mass irregularity.

2.1 Definition of DR function

The magnitude *DR* and the *DR* function are one of the outputs of the apparatus Uster-Tester IV-SX. The value of *DR* determinates what percentage of the total yarn length exceeds or falls below a pre-set limit of yarn mass deviation (Zellweger Uster, 2001). It is calculated for a certain yarn cut-length. The definition of deviation rate (Zellweger Uster, 2001):

$$DR(x,y)[\%] = \frac{\sum_{i=1}^{k} l_i}{L_{TOT}} \cdot 100 \quad (1)$$

Where: *DR(x,y)* is the deviation rate, sum of parts length l_i [m] of all mass deviation, which are same or higher than ± *x* [%], relative to total length L_{TOT}[m]; *x* is the set limit of mass deviation [%]; *y* is the length of section of fibrous product (yarn), which is used - so-called "cut length" [m]; l_i is the length of "i -th part" of fibrous product (yarn), which surpass the limits ± *x* [%]; L_T is the total length of fibrous product (yarn), *k* is number of parts (i = 1, 2, ..., *k*).

A definite relation between the *DR*-value and the variance-length function (*CV(L)*) results from the definition of *DR* function (Ursíny et al., 2008); (Pinčáková, 2006). It is possible to observe the deviation rate and amount of mass variability in various length sections (cut lengths).

2.2 Definition of area variation curve

The area variation curve describes the variability of greyness degrees (i.e. unevenness of plain textile image) in dependence on square field area. It can be expressed as an external or an internal curve. The curve is a certain analogy of the variance-length curve, because it has similar character of behaviour. The internal area variation curve is expressed by the variation coefficient of greyness degree inside square area in dependence on the area of observed square field. This curve increases with growing area of square field. The external variation curve shows the variability of greyness degree between square field areas of image. The curve slopes down with growing area of square field (see Fig. 1.).

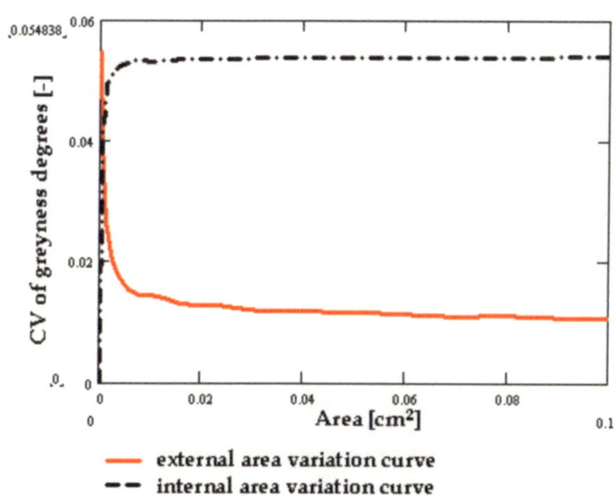

Fig. 1. Area variation curves - example

In this work, the external area variation curve is calculated by the formula:

$$CV(A) = \frac{S(A)}{\overline{\overline{X}}(A)} \tag{2}$$

Where: *CV(A)* is the external variation coefficient of average greyness degrees between square fields of the area *A* in the fabric image; *S(A)* is the standard deviation of mean values of greyness degrees in square fields of the area *A* included in a fabric image; $\overline{\overline{X}}(A)$ is the mean value from all mean values of greyness degrees in square fields of the area *A*.

2.3 Experimental results

Within the experiment, a combed yarn (100 % CO, count of T = 16.5 tex) and a carded yarn (100 % CO, count of T = 25 tex) have been used for the evaluation of unevenness in plane (surface unevenness). The possibility of utilization of parameter *DR* for prediction of surface unevenness is analysed too. The yarns have been measured on the apparatus Uster Tester IV-SX, where parameters $CV_m(1m)$ [%] and *DR(5%;1.5m)* [%], the spectrogram and the variance-length curve have been observed. It has been done 20 measurements for each type of yarn (Ursíny et al., 2008), (Pinčáková, 2006).

The dependence of the *DR (5%; 1.5 m)* [%] values on values of $CV_m(1\ m)$ [%] has been studied. Selected results are mentioned in the Fig. 2. The linear dependence is evident between observed magnitudes. The correlation coefficient ***r*** is equal to 0.9725 in the case of tested combed yarns. In the case of carded yarns the correlation coefficient is 0.6929.

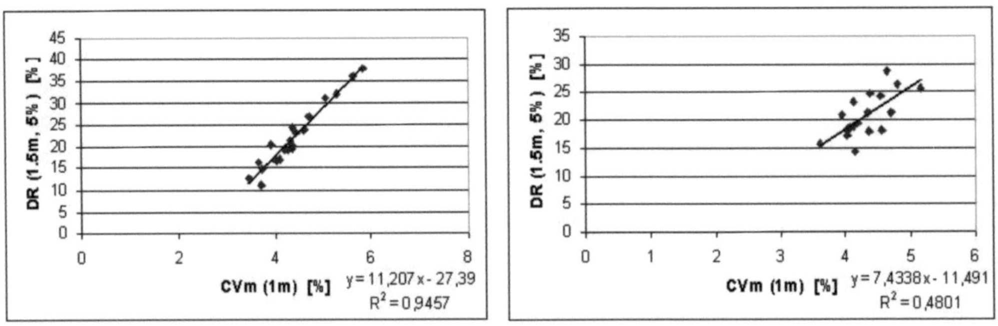

(a) Combed yarn (b) Carded yarn

Fig. 2. Relation between *DR (5%; 1.5 m)* [%] and CV_m *(1m)* [%] values

The relation between *DR*-value and the spectrograms and the variance-length function of combed yarns has been observed too. The results have been confronted with simulated appearances of woven fabrics generated by the Uster-Tester IV-SX. The courses of characteristic functions for selected combed yarns (see the Table 1) are mentioned in the Fig. 3. The examples of simulated fabric appearances are shown in the Fig. 4.

Measurement No.	CV_m (1m) [%]	DR (5%; 1,5 m) [%]
2071	5.85	38
2070	4.33	21.2
2069	4.37	19.4
2068	4.23	19
2067	4.38	20.2

Table 1. Selected parameters of mass irregularity - selected measurements - combed yarn

From the courses of the variance-length curves for the selected set of 5 tested combed yarns (Fig. 3a), it is evident, that the yarn measurement No. 2071 shows an accrual of irregularity (the cut length of 1 m - 10 m). The yarn shows worse irregularity also in the spectrogram (Fig. 3b), where the periodical irregularity is recorded on the wave- lengths of 3 m and 7 m.

The simulated images created from combed yarns, which have higher mass irregularity (*CV*), worse spectrogram as well as the variance length curve, shows worse appearance. It is more unsettled (level of greyness degree fluctuates). In the case of weaves denim and satin there were visible differences in the appearance of individual images.

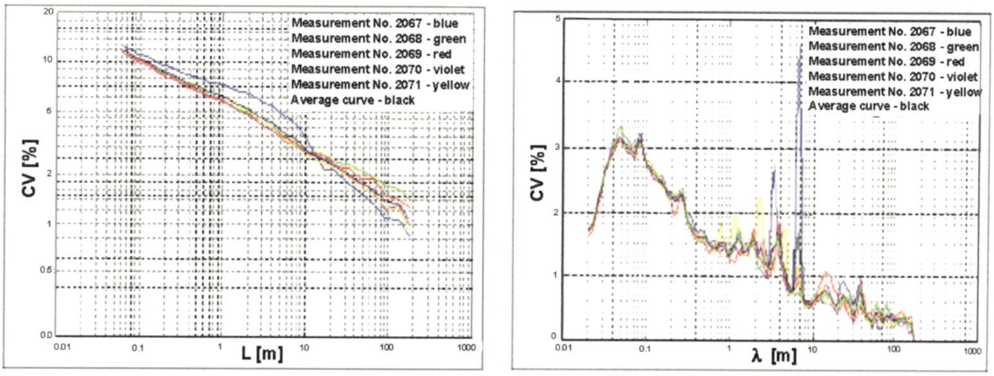

(a) Variance- length curves (b) Spectrograms

Fig. 3. Variance-length curves and spectrograms of combed yarn (100%CO, yarn count of 16.5 tex)

(a) Simulated fabric appearance - denim weave (b) Simulated fabric appearance - plain weave

Fig. 4. Simulated appearances of woven fabrics. Combed yarn. Measurement No. 2071. Real size of image - 15.54 x 9.29 cm. Resolution 300 dpi.

The visual assessment of yarn taper board simulation, generated by the apparatus Uster Tester IV-SX (for example see the Fig. 5.), has been used as an auxiliary evaluation.

Fig. 5. Simulated yarn board from the Uster-Tester IV - SX. Combed yarn. Measurement No. 2071

In the case of the yarn No. 2071 (see the Table 1), a moiré effect tendency has been registered there (Fig. 5). The appearance of this yarn seems to be the worst. The moiré effect has not been observed on the other yarn boards, total yarn appearance seems to be better (less unsettled). Higher number of neps was evident from appearances of all yarns.
Obtained images of fabrics appearances have been evaluated not only visually (the subjective method) but by means of the area variation curve too. The curve is one of results of the mentioned special script made by Militký. The program constructs this curve according to the formula (2). An influence of yarn mass irregularity on the appearance of simulated woven fabric image has been observed. Selected area variation curves of greyness degrees of simulated woven fabric appearance are mentioned in the Fig. 6 and Fig. 7.

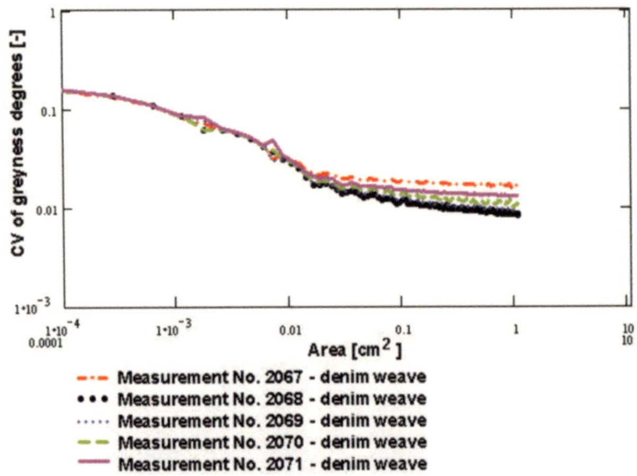

Fig. 6. Area variation curves of greyness degrees of fabric appearances simulated from irregularity measurements - combed yarn. Curves of fabrics with denim weave

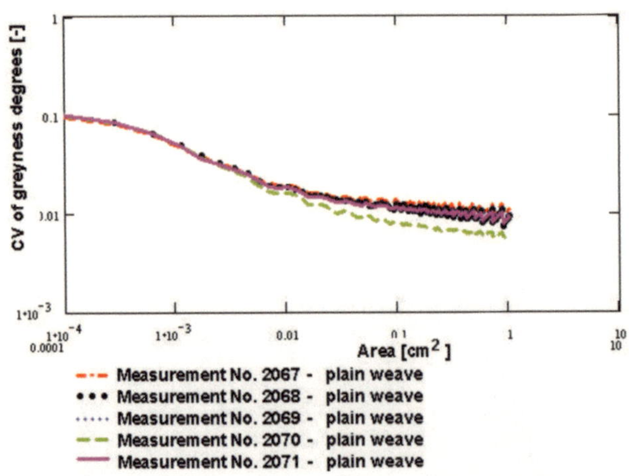

Fig. 7. Area variation curves of greyness degrees of fabric appearances simulated from irregularity measurements - combed yarn. Curves of fabrics with plain weave.

Differences between courses of area variation curves were insignificant in the case of the plain weave. High density of this weave is probably a reason of difficult surface unevenness identification, because the plain weave does not have so called a float thread and so mass irregularity of yarn hides up. In the case of weave, that are not so dense (the denim weave, the satin weave), differences in the appearance of flat textile are visible and identifiable. The appearance of flat textile corresponds with measured values of yarn irregularity and yarn appearance more. The yarn, that showed higher *CV* value, worse spectrogram as well as the course of the variance-length curve, had worse appearance of simulated fabric too - see the measurement No. 2071 where the curve is deflected up. In the case of these weaves, yarn irregularity does not hide and it is identifiable on the float thread. If courses of both variance-length curve and spectrogram are faultless, behaviours of area variation curves are nearly congruent.

Total observed area of simulated fabrics image has been divided into square fields during construction of the area-variation curves. The area of square field gradually increased (from several pixels to several thousands of pixels). The area of evaluated square has an influence on the value of variability of greyness degree. This value decreases with increasing area of square field, but simultaneously number of square fields, i.e. number of measurements, grades down. Stability of ascertained results corresponds with this fact. It shows itself by "a saw-toothed" course of area variation curve. For results reliability, a certain minimal number of square fields is necessary; therefore the evaluated area of one square was at the most of $1cm^2$.

3. Utilization of semivariograms for surface unevenness evaluation

3.1 Definition of semivariograms

The semivariogram expresses spatial dissimilarity between values at point x_i and x_j. Generally, it is defined as one-half variance of differences ($z(x_i)$ - $z(x_i$+lag)) (Cressie, 1993); (Militký et al, 2000); (Březina & Militký, 2002); (Militký & Klička, 2005):

$$\Gamma(lag) = 0,5\,.D(z(x_i) - z(x_i + lag)) \tag{3}$$

The magnitude ***lag*** is a directional vector (0°; 90°, 45°) representing separation between two spatial locations. For uniformly distributed points, x values of vector ***lag*** express the multiples of distance between squares in direction of columns (0°), rows (90°) and diagonals (45°) (Militký & Klička, 2005). Thus, 3 types of semivariograms are obtained (in direction of columns, rows and diagonals). Omni-directional semivariogram is calculated by averaging of all 3 types of semivariograms. For stationary random field the mean value is constant in individual locations. Then this formula holds (Cressie, 1993); (Militký et al, 2000):

$$\Gamma(lag) = 0,5\,.E(z(x_i) - z(x_i + lag))^2 \tag{4}$$

If $\Gamma(lag)$ = const., the magnitude $z(.)$ is not correlated in the given direction. When a random field is non-stationary (average value in each field is not constant) it is possible to construct so called a centred sample semivariogram (Militký et al., 2000), which has been used in this work:

$$G(lag) = \frac{1}{2N(lag)} \sum_{i=1}^{N(lag)} (z_c(x_i) - z_c(x_i + lag))^2 \tag{5}$$

Where: $z_c(x_i)$ is the centred average greyness degree defined as:

$$z_c(x_i) = z(x_i) - \frac{\sum_{i=1}^{n(x_i)} z(x_i)}{n(x_i)} \qquad (6)$$

N*(lag)* is number of pairs of observations separated by distance *lag; $z(x_i)$* is greyness degree in the location x_i. The woven fabric image is divided into square fields like a net. The centres of fields are the locations x. The average value of greyness degree in the given square field is assigned to the location x ($z(x_i)$).

3.1.1 Exemplary courses of semivariograms

For the prefaced of semivariograms problems, semivariograms from greyness degrees of exemplary images, made by authors, have been constructed. These images are mentioned in the Fig. 8. Size of each image is 200 x 200 pixels. The resolution is 200 dpi. The fabric images without frame have been processed by means of the mentioned special script made by Militký. The programme converts the fabric image to the greyness degrees and, in the case of the semivariogram, divides it in to square fields of selected size *step x step* pixels. The average greyness degree ($z(x_i)$) is calculated in each field. From obtained values the centred semivariogram in given direction is calculated according to the formula (5), see Fig. 8.
From semivariograms, it is possible to identify stripiness of the image pursuant to courses of the semivariogram in rows direction together with the semivariogram in direction of columns. So, it was decided to use semivariograms for analysis of surface unevenness of woven fabric.

3.2 Experiment and results

For experiment there were used:

- Woven fabric images simulated by means of the Uster-Tester IV-SX apparatus on the basis of measurement results of yarn mass irregularity. Yarns with various level of irregularity have been used.
- Real fabric samples with various weft sett, weave and quality.

The images of real fabrics have been obtained by scanning of fabric samples. The samples have been covered with the black as well as the white underlay during scanning for better identification of surface unevenness. All obtained fabric images have been processed by means of the mentioned special script. An influence of the fabric sett, the fabric weave as well as fabric quality on the behaviour of semivariograms has been observed.

3.2.1 Semivariograms of fabric images simulated on the Uster-Tester apparatus

The instrumentation Uster Tester IV-SX enables to simulate woven and knitted fabric appearances as well as a yarn board on the base of yarn mass irregularity measurement. Obtained appearances are in the grey scale, which has various intensity of greyness degree according to structure of yarn mass irregularity.
For experiment 100%CO rotor yarns have been used. Count of these yarns was 55 tex, machine twist was 625 tpm. Three yarns had been manufactured. Two of them had been produced purposely with faults. For the first case, a bad sliver had been used (the measurement No. 3398) and for the second case, an impurity has been inserted into the rotor groove of machine to produce yarn with moiré effect (the measurement No. 4192). Yarn mass irregularity has been measured on the apparatus Uster Tester IV-SX. Selected parameters of yarn mass irregularity are mentioned in the Table 2.

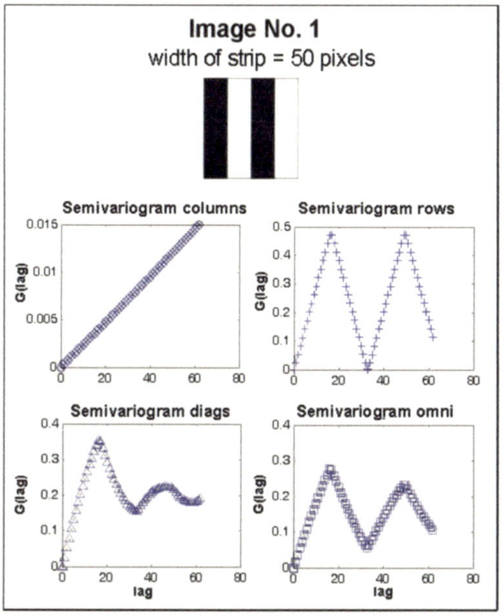

(a) Exemplary image and corresponding semivariograms

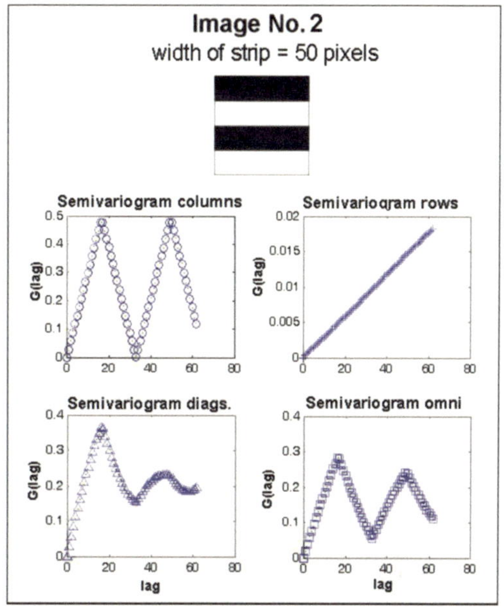

(b) Exemplary image and corresponding semivariograms

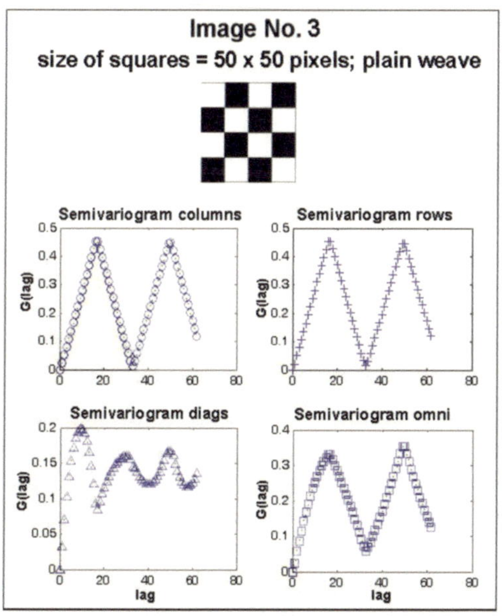

(c) Exemplary image and corresponding semivariograms

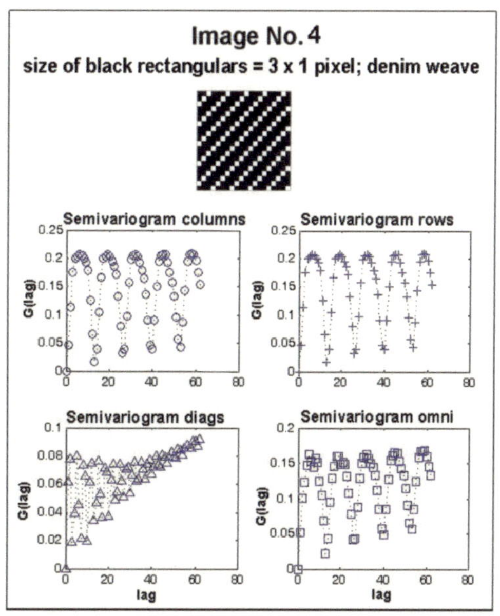

(d) Exemplary image and corresponding semivariograms

Fig. 8. Semivariograms from greyness degrees of images – used whole images without frame, set step = 3 pixels

Measurement No.	U [%]	CV [%]	CV(1m) [%]	CV(3m) [%]	CV(10m) [%]	Thin places -50% [1/km]	Thick places +50% [1/km]	Neps +280% [1/km]
3396	10,86	13,71	4,29	3,93	3,70	2,5	57,5	42,5
3398	11,13	14,17	7,98	6,79	5,18	2,5	77,5	57,5
4192	25,30	38,02	3,43	2,75	2,48	2373	6368	5738

Table 2. Selected parameters of yarn mass irregularity

For spectrograms of these yarns see the Fig. 9a-c, for variance length curves see the Fig. 10a-c.

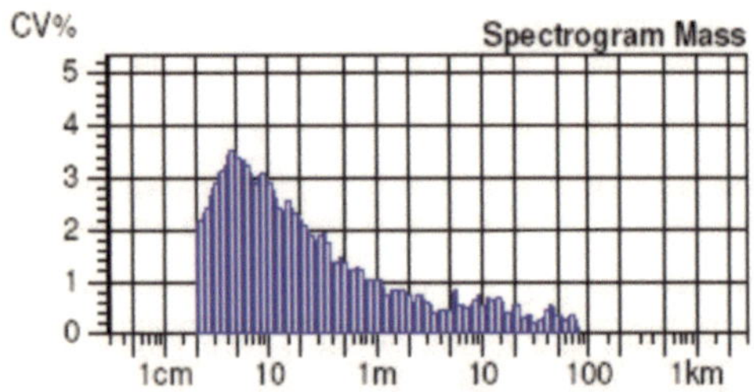

(a) Measurement No. 3396

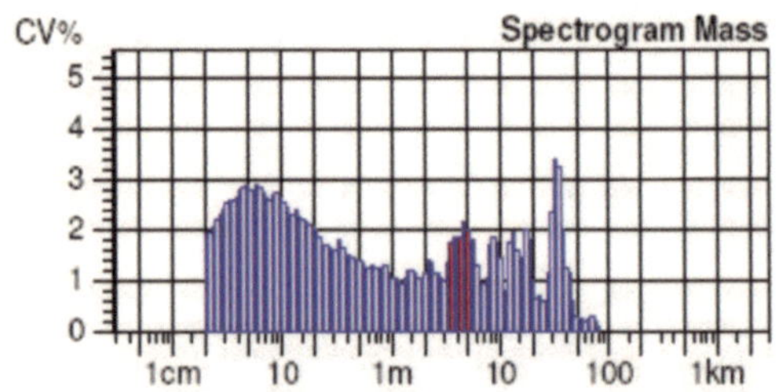

(b) Measurement No. 3398

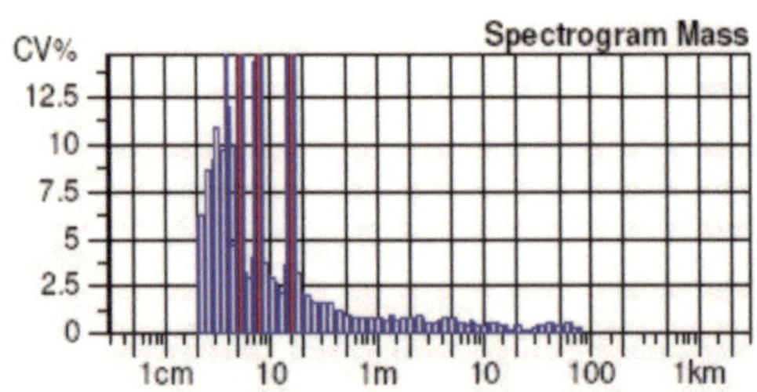

(c) Measurement No. 4192

Fig. 9. Spectrograms of yarns

It is evident (Fig. 9a), the yarn No. 3396 has short-term irregularity on wavelengths $\lambda = (4 - 6)$ cm, the shape of spectrogram embodies no other faults. The spectrogram of yarn No. 3398 (Fig. 9b) has increased amplitude on wavelength $\lambda = 35$ m and draft waves on wavelengths $\lambda = (4; 9; 15)$ m. Because of drafting waves in the spectrogram, yarn wound on the board as well as the image of flat textile should show disturbed appearance, so called cloudiness. The stripiness should be shown in the flat textile due to higher periodic irregularity on the wavelength $\lambda = 35$ m.

From the spectrogram of yarn No. 4192 (Fig. 9c) it is evident the moiré effect - higher amplitudes on wavelengths $\lambda = (16; 8; 5)$ cm. Increased amplitude on the basic wavelength (16 cm) corresponds to the rotor circumference (rotor diameter $d = 53$ mm), wavelengths of other higher amplitudes correspond to wavelengths $\lambda/2$ and $\lambda/3$. It means the yarn wound on the black board will have the moiré effect caused by impurities in the rotor groove.

The variance-length curve of yarn No. 3396 (Fig. 10a) shows gradual decrease of CV values with increasing cut length. This decrease is rapider on the cut lengths $L = (2 - 20)$ cm. The curve of yarn No. 3398 (Fig. 10b) falls more slowly up to the cut length L = 10 m, then rapid decrease of the curve follows. Increased values of CV on higher cut lengths (L = 1 - 10 m) indicate cloudiness of future flat textile. The variance-length curve of yarn No. 4192 (Fig. 10c) has markedly higher values of CV on cut lengths L = 2 cm - 1 m, its decrease is the rapidest up to the length of 1m compare to the previous curves. Higher values of CV up to the cut length L = 1 m predicate short disturbing faults in the flat textile.

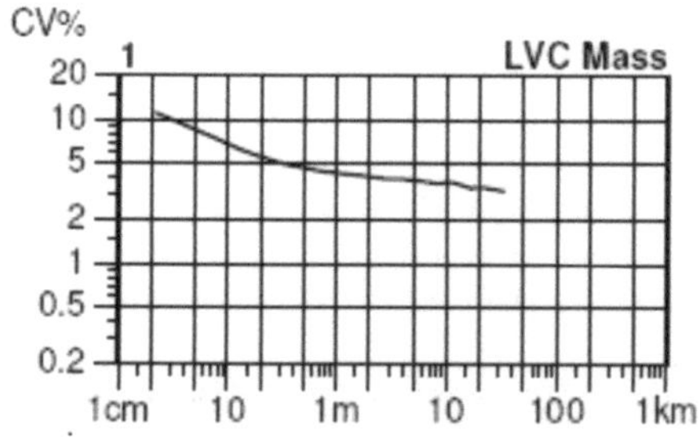

(a) Measurement No. 3396

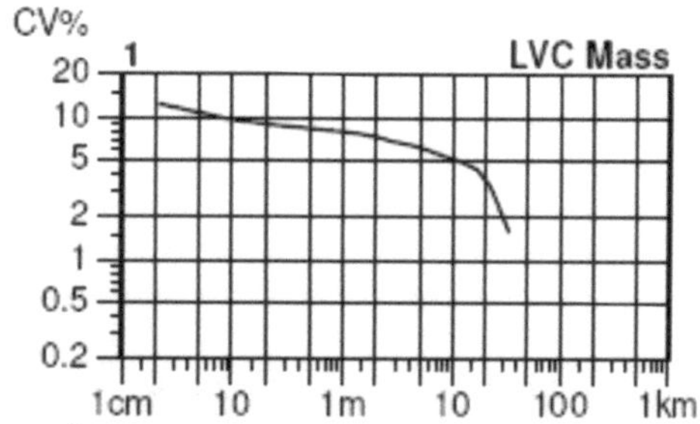

(b) Measurement No. 3398

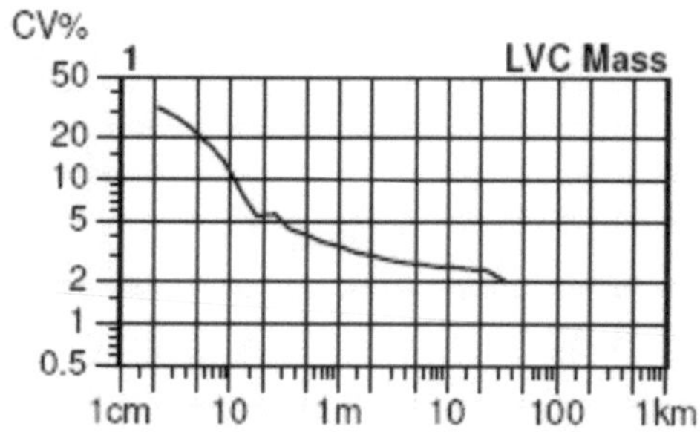

(c) Measurement No. 4192

Fig. 10. Variance-length curves of yarns

The appearance of the flat textile - the woven fabric (plain, satin and denim weave) and the yarn board has been simulated on the basis of measured data of yarn mass irregularity by the apparatus Uster-Tester IV-SX. There are yarn boards and appearances of woven fabrics in the denim weave for selected measurement in the Fig. 11 - Fig. 13.

(a) Yarn board

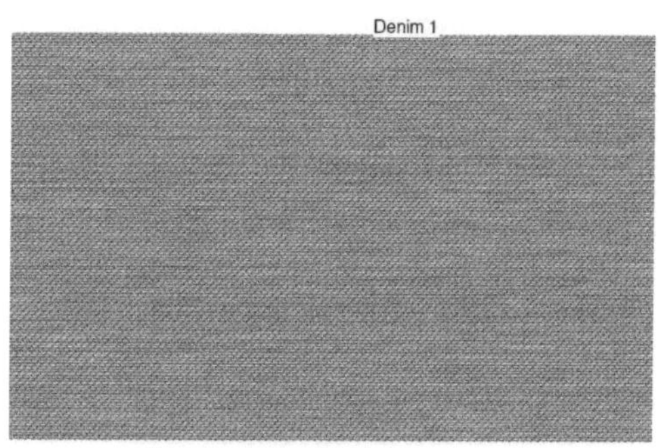

(b) Woven fabric appearance

Fig. 11. Simulated images of yarn board and woven fabric appearance with denim weave - (Real size of image - 15.54 x 9.29 cm. Resolution 300dpi) - Measurement No. 3396

Visually, the appearance of woven fabric with denim weave from the measurement No. 3396 seems to be similar to the appearance of woven fabric from the measurement No. 3398 at first sight. But seen in close-up, the woven fabric from the yarn No. 3398 has slightly worse appearance. This yarn shows slightly higher *CV* values, worse shape of the spectrogram and the variance-length curve in comparison to the yarn No. 3396. Woven fabric from the yarn No. 4192 has the worst appearance clearly caused by higher yarn mass irregularity (*CV*) and by the worst spectrogram as well as the variance length curve. The moiré effect is obvious on the yarn board (see Fig. 13a), but it is disturbed by weave in the fabric. The fabric appearance is unsettled (Fig. 13b).

(a) Yarn board

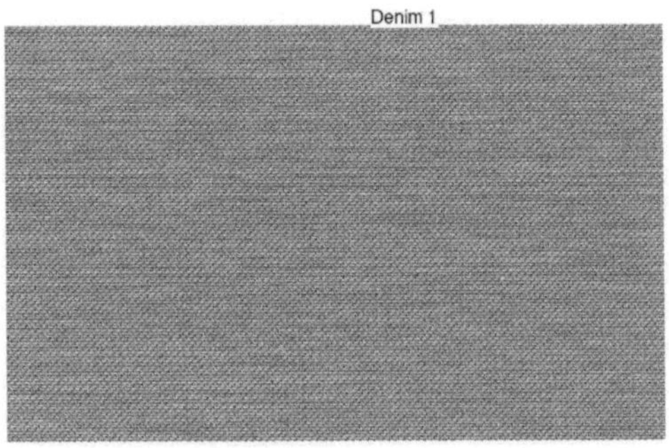

(b) Woven fabric appearance

Fig. 12. Simulated images of yarn board and woven fabric appearance with denim weave - (Real size of image – 15.54 x 9.29 cm. Resolution 300 dpi) - Measurement No. 3398

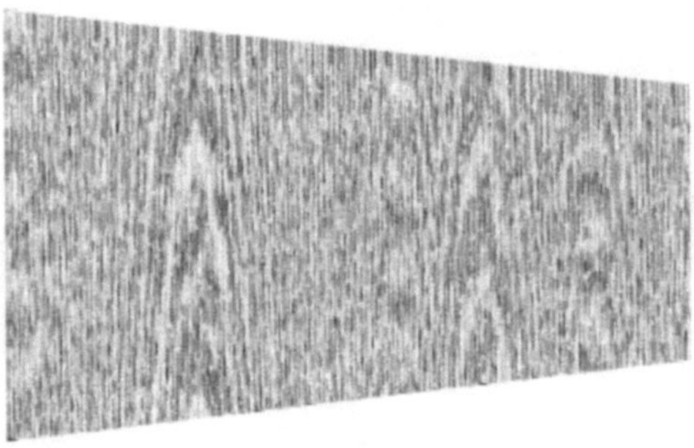

(a) Yarn board

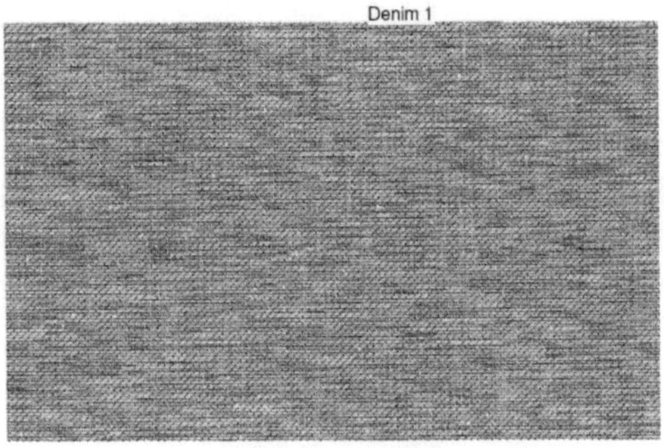

(b) Woven fabric appearance

Fig. 13. Simulated images of yarn board and woven fabric appearance with denim weave - (Real size of image – 15.54 x 9.29 cm; resolution 300 dpi) – Measurement No. 4192

These images have not been evaluated only visually, but also by means of the above-mentioned script. The size of observed image was 1000 x 1000 pixels (resolution 300 dpi, i.e. c. 8.5 x 8.5 cm).

Two types of semivariograms in the given direction have been constructed. In the first case, section of each image with size of 1000 x 1000 pixels has been observed. The step of 60 pixels has been chosen. It corresponds to real size of c. 0.5 cm. See the Fig. 14, where semivariograms of fabric image with the denim weave are mentioned. From the semivariograms it is evident, that the curve of image from the yarn No. 3396 has the best course. This yarn has got the best values in term of parameters of yarn mass irregularity. The semivariograms of this image are nearly constant from *lag* = 3 in all directions. It means, observed square fields of the image are similar to each other in term of average centred greyness degree. So, neither cloudiness nor stripiness was not record. It corresponds to visual evaluation of the image. The semivariograms of image from the yarn No. 3398 shows higher values compared to the curve of yarn No. 3396. Its course is similar to the course of the curve of yarn No. 3396 in direction of columns. They are nearly identical in direction of rows. By visual evaluation of woven fabric appearances, any marked differences between images from yarns No. 3396 and No. 3398 have not been found. But semivariograms were probably able to record colour differences in the images caused by slowly increased mass irregularity and drafting waves of the yarn. The curves of all types of semivariograms of the yarn No. 4192 very fluctuate, also show markedly higher values compare to curves of the image from other two yarns. It is possible to say, the character of curve corresponds to visual evaluation of the image - strongly unsettled appearance of the woven fabric.

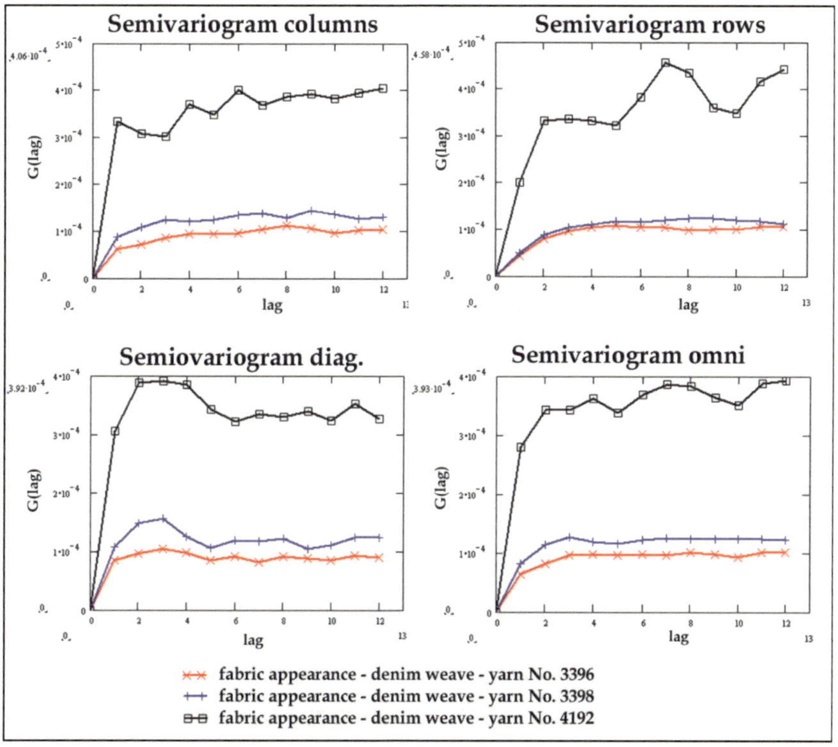

Fig. 14. Semivariograms from greyness degrees - simulated image of woven fabric - denim weave 3/1 - observed size: 1000 x 1000 pixels; step: 60 pixels

In the second case, sections of each image with size of 118 x 118 pixels from the centre of image have been observed. The step of 2 pixels has been chosen. An influence of fabric weave on the course of semivariogram has been observed (see the Fig. 15).

Semivariograms mentioned in the Fig. 15 does not record whole image, but they analyse only area of 118 x 118 pixels, i.e. 1 x 1 cm of the image. By observing of small section of the fabric image, it is possible to identify the fabric weave from courses of semivariogram in direction of rows and columns - in this case the denim (twill 3/1). It has been verified. You can compare semivariograms in the direction of row and columns in the Fig. 15 and the course of semivariograms in the Fig. 8, where exemplary semivariograms for the denim weave are shown. The denim is a weave characteristic by line spacing. Maximums and minimums in the semivariograms (Fig. 15) correspond to the spacing. The position of semivariogram curve is influenced by yarn irregularity. The yarn No. 4192 has the highest *CV* on short wavelengths. It expresses itself as quickly changing of dark and white sections in the image. That way the semivariogram curves of these images have the highest values. Courses of semivariograms of all yarns in all directions are similar.

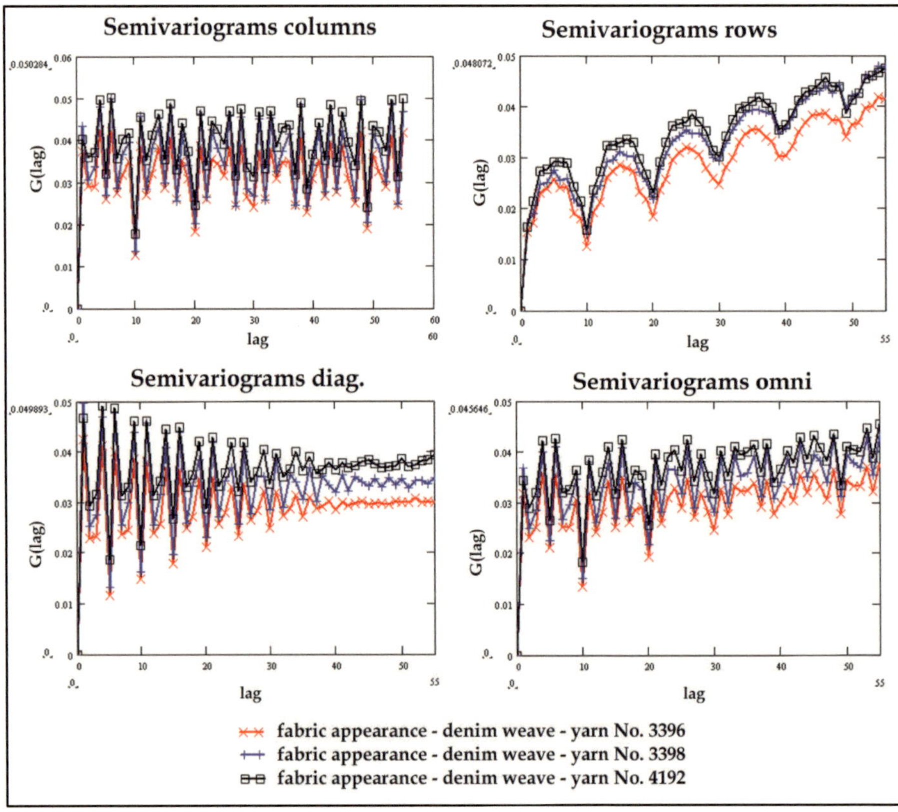

Fig. 15. Semivariograms from greyness degrees – simulated image of woven fabric – denim weave 3/1 – observed size: 118 x 118 pixels; step: 2 pixels

3.2.2 Semivariograms of real tested fabric samples

In this part, the courses of semivariograms of greyness degrees of real fabric samples with various weft sett, weave and quality are presented. For the first experiment with real fabric samples, white colour woven fabrics (100 % CO) of the plain weave have been used. Warp and weft yarn count was 33 tex. The fabrics with three level of weft density have been used. The density was: 16 threads/1cm (the fabric marked B16), 20 threads/1cm (the fabric marked B20) and 24 threads/1cm (the fabric marked B24). The yarn used for these fabrics show neither bad course of spectrogram or variance-length curve. Therefore we can expect settled fabric appearance with low greyness degree fluctuation. The images of woven fabric necessary for semivariogram construction have been obtained by scanning of 12 samples from each fabric. The image resolution was 200 dpi. The samples have been scanned with a black as well as a white underlay. By putting of a black paper onto the scanned sample the black underlay has been created, whereas for the white underlay the sample has been covered by sheet of white paper. The real size of observed image was 15 x 21 cm (i.e. 1181 x 1653 pixels), for illustration see the Fig. 16.

Fig. 16. Image of the real woven fabric B16; sample 1.1; the black underlay. Size of whole image: 1181 x 1653 pixels, resolution: 200 dpi

The real fabric images have been processed by mentioned special script to the „centred" semivariograms of greyness degree be obtained. The area of 1170 x 1170 pixels has been observed in the sample. The step of 20 pixels (corresponding to 0.25 cm) had been selected. The average semivariograms have been constructed in given direction (columns, rows, diagonals, omni) for each type of woven fabric. They are shown in the Fig. 17 and Fig. 18. An influence of the fabric sett on the course of the semivariogram has been observed in this experiment. These semivariograms show, that the methodology of scanning influences their courses. The white underlay of scanning fabrics does not seem to be suitable due to low contrast of image. The fabric trans-illumination does not evince itself on the background. The black underlay is better, thus the image with this underlay have been evaluated. From the semivariograms in the Fig. 18 it is evident the average greyness degrees in the squares of area 20 x 20 pixels (size of the step) are not correlated from c. *lag* = 10 in any direction. In the case of smaller distance of squares (*lag* < 10), semivariograms are convex ascending. These semivariograms do not show any stripiness. By visual evaluation of fabric samples any stripiness has not been evident too.

It was found out the fabric sett influences the level of semivariograms values. The values corresponding to the fabric of higher weft sett (the fabric B24) were lower in comparison with semivariograms of fabric with lower weft sett (the fabric B16). It is due to the black underlay, which was put on the fabric before scanning. In the case of the fabric of lower weft sett, this underlay strikes more through the fabric during scanning than in the case of the fabric of higher weft sett. This problem is described in the authors` work (Moučková & Jirásková, 2008).

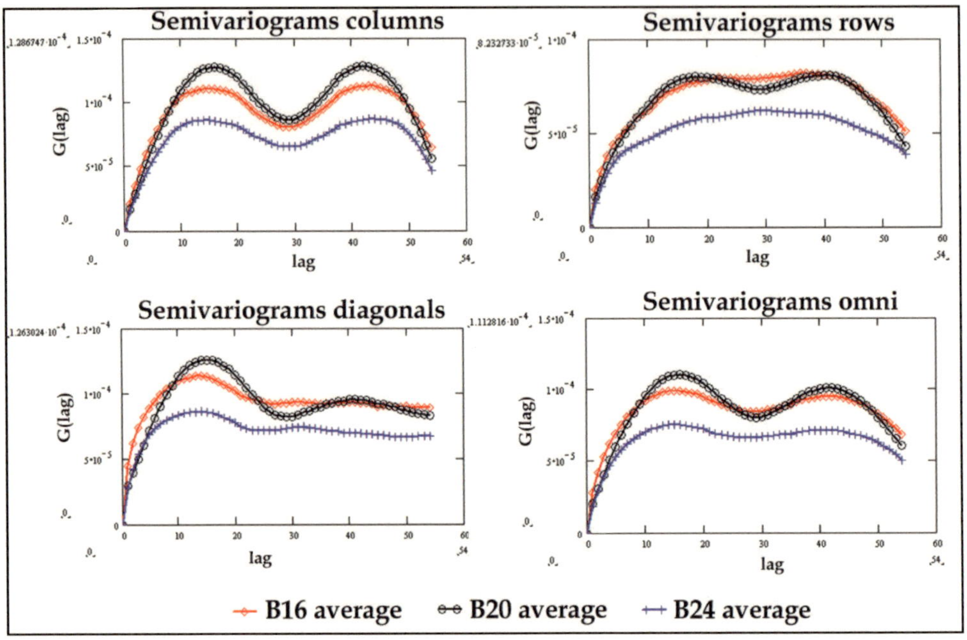

Fig. 17. Average semivariograms – real fabric images - plain weave – observed area: 1170 x 1170 pixels, step: 20 pixels, the white underlay

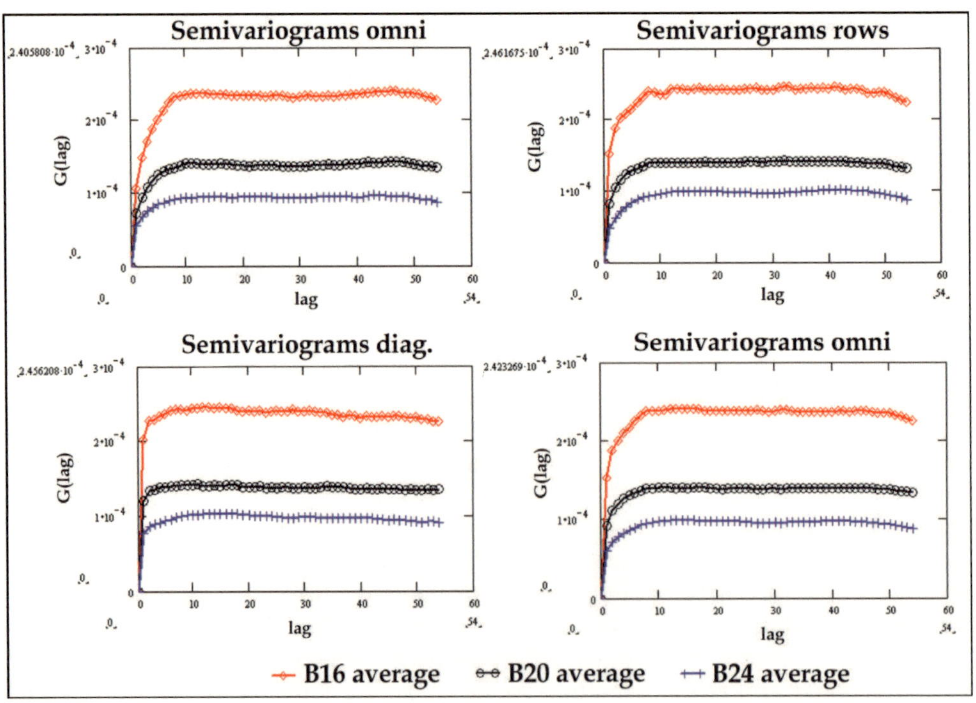

Fig. 18. Average semivariograms – real fabric images - the plain weave – observed area: 1170 x 1170 pixels, step: 20 pixels, the black underlay

Woven fabrics of various weaves have been used for the second experiment. Selected parameters of these fabrics are mentioned in the Table 3.

Fabric weave	Weft sett (threads/10cm)	Warp sett (threads/10cm)	Yarn fineness in the warp and weft [tex]	Raw material
Satin 1/7 (5)	350	388	14,5	100%CO
Twill ½ (Z)				
Twill 5/5 (Z)				
Plain				
Hopsack 2/2				

Table 3. Selected parameters of used woven fabrics

The fabrics had been manufactured both in a standard way and with a fault (stripiness in the direction of warp) on the same loom. The stripes had been obtained during warping. The yarn bobbins, produced in different spinning lot, had been set on the half of the creel. The parameters of these yarns had been the same, but the colour shade of cotton has differed. Six samples have been taken from each fabric for experiment

The fabric samples of size 15 x 21 cm have been scanned with resolution 300 dpi from the face of fabric with black underlay to obtain fabric image – for example see the Fig. 19 and the Fig. 20.

Fig. 19. Image of normal fabric - the weave satin 1/7. Real size of image: 15 x 21 cm; resolution: 300 dpi

The semivariograms in the given direction have been constructed. The section of each image with size of 1700 x 1700 pixels has been observed. The step of 60 pixels has been chosen. The courses of semivariograms according to fabric quality have been observed - see Fig. 21. From this figure it is evident that semivariograms in the direction of rows record stripes in the fabrics. The fabric samples with stripes had c. 6.5 stripes in direction of warp in evaluated area. The width of two stripes next to each other was 46 mm. It corresponds to

544 pixels in the fabric image. At the step of 60, the distance between outside squares of repetitious stripes of the same colour shade was equal to 9 (544/6) - which corresponds to period Δlag = 9 in the curve.

Fig. 20. Image of fabric with stripes – the weave satin 1/7. Real size of image: 15 x 21 cm; resolution: 300 dpi

A specimen example has been constructed for verification of this assertion. The square of size 1700 x 1700 pixels with resolution of 300 dpi containing vertical stripes of width 23 mm (272 pixels) has been drown. This image has been treated by mentioned programme at the same setting as in the case of fabric. See results on the Fig. 22.
The semivariogram in the direction of rows has a periodical course. At the step of 60 pixels, the period is 9. This period corresponds to 544 pixels in the image (i.e. 46 mm) and so to distance between edges of stripes of the same colour shade. The semivariogram in the direction of columns has linearly growing course.
It was verified that combination of semivariograms in the direction of rows and in the direction of columns seems to be suitable tool for recording of periodical unevenness – stripiness of woven fabric.

4. Conclusion

The structure of mass irregularity described by characteristic functions (the spectrogram, the variance-length curve, the *DR* function) influences surface unevenness by specific way. Periodical irregularity, presented by the spectrogram, causes surface unevenness, which is distinguished by a certain geometrical regularity (moiré effect, stripiness). Non-periodical irregularity expressed by the variance-length curve and by the *DR* function leads to a surface unevenness, which, on the contrary, can be characterized by geometrical non-uniformity (cloudiness). The experimental measurements have showed that the linear dependence exists between measured values of *DR (5%, 1.5m)* and $CV_m(1m)$. Ascertained high value of correlation coefficient confirms it. This value highly exceeds a critical value

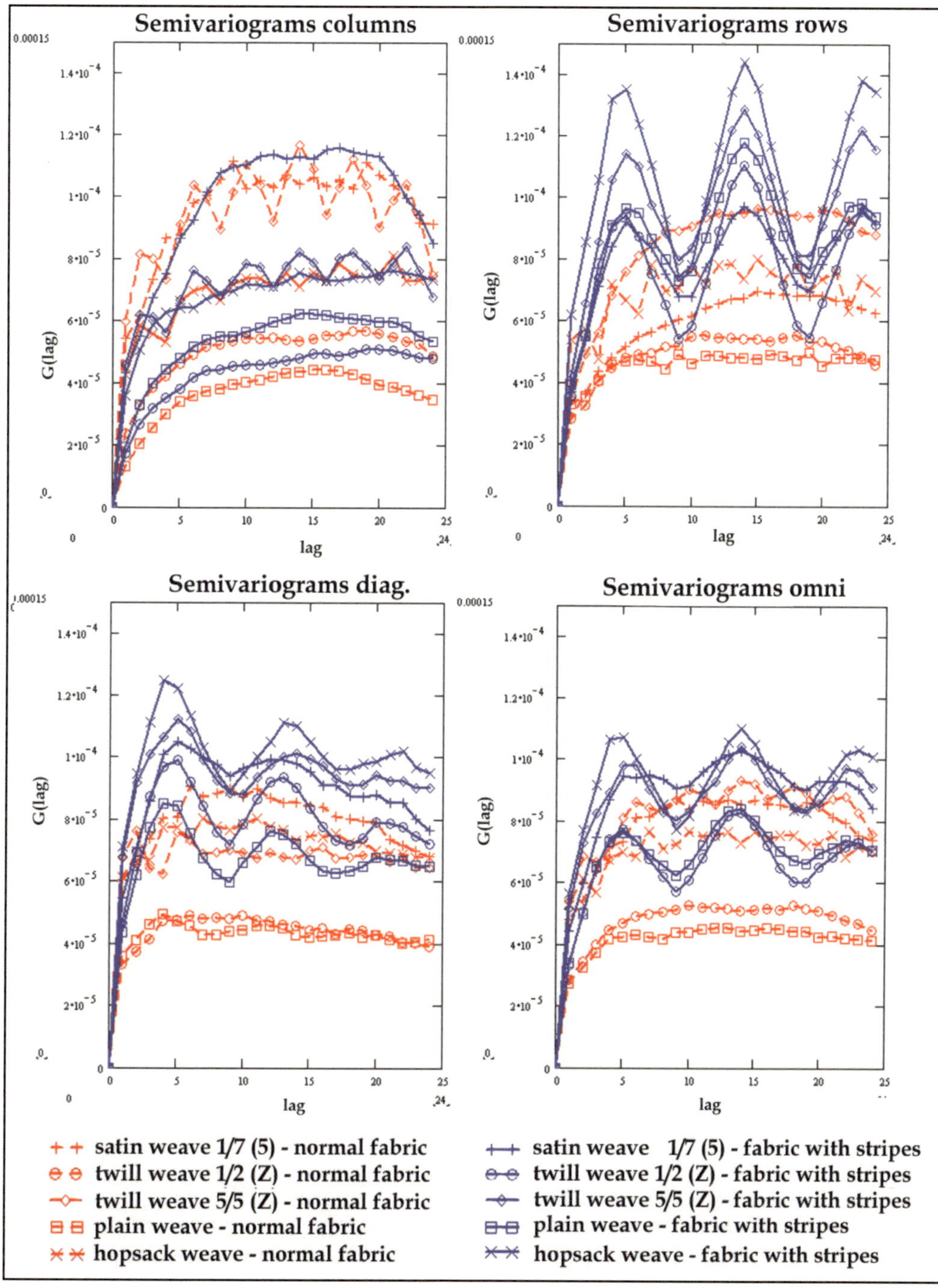

Fig. 21. Semivariograms of normal and stripped fabrics – observed sample area: 1700 x 1700 pixels; step: 60 pixels

also in the case of the confidence limit of 99%. Hitherto knowledge about relation between the variance-length curve and surface unevenness (cloudiness) will be possible to amplify on relation between the *DR* function and surface unevenness. On the basis of course of individual curves, the level and the character of surface unevenness can be judged.

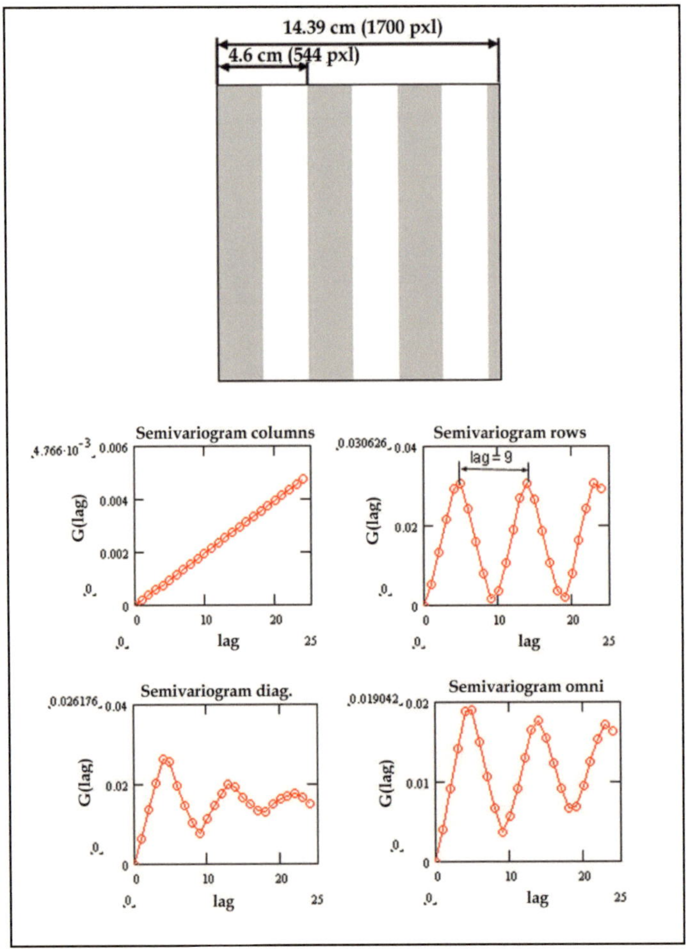

Fig. 22. Semivariograms from greyness degree of image - used whole image (1700 x 1700 pixels) without frame, step: 60 pixels

The surface unevenness of woven fabric can be evaluated by means of area variation curves and semivariograms of greyness degrees of fabric image. The area variation curve has been mentioned theoretically also in previous works (Wegener & Hoth, 1958); (Suh, 2005). From our experiment it has been found out that both the spectrogram and the variance-length curve of yarn used in fabric are basic indicators determining conceivable negative fabric appearance and this corresponding adverse course of area variation curve. Next indicator, which can be used in given context, is the *DR* function. The application of semivariograms has not been used by the other authors for the evaluation of surface unevenness of woven fabric yet. Semivariograms in the direction of both rows and columns seems to be a suitable tool for fabric stripiness evaluation.

Most of experiments have been done with fabrics (real or simulated images of fabric appearance) till this time. Thus, it would be interesting to verify obtained piece of knowledge on knitted fabrics. Next, it would be possible to find other functions suitable for the expression of unevenness of woven or knitted fabrics.

5. Acknowledgement

This work was supported by the project Textile Research Centre II No. 1M0553.

6. References

Březina, M.; Militký, J. (2002). Complex characterization of textile surface, *Robust'2002 - proceeding of twelfth winter school JČMF,* pp.50 –58, Hejnice, January 2002, Jednota českých matematiků a fyziků, Prague.

Cressie, N.A.C. (1993) *Statistics for spatial data,* J. Wiley, ISBN 0-473-00255-0, New York.

Militký, J.; Bajzik.V. (2000) : Description of thickness variation by fractal dimension, *Proceeding of 7th national conference Strutex,* pp. 165-170, ISBN 80-7083-668-7, Liberec, December 2000, Technical university of Liberec, Liberec.

Militký, J.; Rubnerová, J. & Klička, V. (2000). Spatial statistics and unevenness of surface mass of non-woven textiles, *Proceeding of 7th national conference Strutex,* pp. 199-203, ISBN 80-7083-668-7, Liberec, December 2000, Technical university of Liberec, Liberec.

Militký, J.; Klička, V. (2005). Characterization of textile mass variation in plane, *Proceeding of 5th world textile conference Autex 2005,* pp.750–755, ISBN 86-435-0709-1, Portorož, June 2005, University of Maribor, Faculty of Mechanical Engineering, Department of Textile, Maribor.

Moučková, E.; Jirásková, P. (2006): Area-variation curve of real woven fabric, *Proceedings of 13th international conference STRUTEX,* pp. 87-92, ISBN 80-7372-135-X, Liberec, November 2006, Technical University of Liberec, Liberec.

Moučková, E.; Jirásková, P. (2007): Influence of weft sett on course of area variation curve on woven fabric, *Proceedings of 6th international conference TEXSCI 2007, CD-rom edition,* ISBN 978-80-7372-207-4, Liberec, June 2007, Technical University of Liberec, Liberec.

Moučková, E.; Jirásková, P. (2008). Utilization of semivariogram for evaluation of surface unevenness, *Book of proceedings of 4th International Textile, Clothing & Design Conference - Magic World of Textiles,* pp. 848 - 853, 978-953-7105-26-6, Dubrovnik, October 2008, University of Zagreb, Faculty of Textile Technology, Zagreb.

Pinčáková, Z. (2006): Classification selected parameters and characteristic functions of mass irregularity, *Diploma Work* (Supervisor: Ursíny, P.), Technical University of Liberec, Liberec.

Slater, K. (1986). *Yarn evenness,* The Textile Institute, ISBN 0 900739 85 1, Manchester.

Suh, M., W. (2005). An electronic Imagining of Fabric Qualities by on-line yarn data, Available from www.ntcresearch.org/pdf-rtps/AnRp01/I01-A1.pdf Accessed: 2005-02-01.

Ursíny, P.; Moučková, E. & Jirásková, P. (2008). New knowledge about relation between yarn mass irregularity and surface unevenness. *Book of Proceedings of the 4th International Textile, Clothing & Design Conference - Magic World of Textiles*, pp. 915-920, ISBN 953-7105-92-, Dubrovnik, October 2008, University of Zagreb, Faculty of Textile Technology, Zagreb.

Wegener, W. & Hoth, E. G. (1958). Die CD(F)-Flächenvariation, *Textil-Praxis*, Vol. 1958, No. 13, (1958), 485- 488.

Zellweger Uster (1971). Zusammenhänge zwischen den Ergebnissen der Gleichmässigkeitsprüfung und den Aussehen der fertigen Gewebe und Gewirke. *Uster News Bulletin*, No. 15. (January 1971), 3 - 36.

Zellweger Uster (1988). Neue Möglichkeiten der Analyse von Masseschwankungen an Garnen, Vorgarnen und Bändern, *Uster News Bulletin*, No. 35. (August 1988), 6 - 18.

Zellweger Uster (2001). *Uster Tester IV Application Handbook*, V1.0/400 106-04010, Uster.

11

Detection of Defects in Fabric by Morphological Image Processing

Asit K. Datta[1] and Jayanta K. Chandra[2]
[1]*Department of Applied Optics and Photonics, University of Calcutta, Kolkata 700009*
[2]*Future Institute of Engineering and Management, Kolkata 700150*
India

1. Introduction

Defects are generated in woven fabric due to improper treatments in weaving machines, spinning errors and inadequate preparations of fiber at the spinning stage. The economic viability of a weaving plant is significantly influenced by the extent of its success in eliminating defects in fabric. Detection of defects is generally carried out by time consuming and tedious human inspection. Such manual inspection procedures are commonly agreed upon to be inefficient with detection efficiency suffering from deterioration due to boredom and lack of vigilance. The problem is accentuated by the presence of several types of defects those may occur in woven fabric at random.

In textile industry, imaging and image processing techniques are investigated for off-line and on-line visual inspection of fabric for the detection of defects (Zhang & Bresse, 1995; Drobino & Mechnio, 2006). The basic philosophy of detection of defects by such techniques is guided by the analysis of the image of fabric for distinguishing properties, those can be used to discriminate between defective and first quality fabric. In most cases, measurements are made on the first quality fabric and are then compared with the measurements made on the test fabric. Severe deviations in the measured parameters are used to indicate the presence of defects. Defects are then categorized into several types. However, the recognition of a particular type of defect amongst various classified types always remains a problem even in the context of presently available advanced image processing technology. Moreover, massive irregularities in periodic structures of woven fabric (particularly for fabrics manufactured from natural fibers) introduce very high degree of noise, which make identification and classification of defects difficult. The problem is accentuated very much due to the hairiness of natural fibers.

Elaborate image processing algorithms are usually adopted for detection and recognition of defects (Sakaguchi et al, 2001). Recent reviews are available on various techniques, those can be applied for such tasks (Xie, 2008). In this chapter we are interested to explore one of such techniques which can be termed as morphological image processing, for the detection of defects in woven fabric.

The techniques of morphological image processing are widely used for image analysis and have been a valuable tool in many computer vision applications, especially in the area of automated inspection (Haralick et al, 1987). Many successful machine vision algorithms used in character recognition, chromosome analysis and finger print classification are based on morphological image processing techniques (Maragos, 1987). It has relevance to conditioning, labeling, grouping, extracting and matching operations on image (Dougherty, 1993). Thus, from low level to high-level computer vision, morphological techniques are important in operation such as noise removal and median filtering (Heijmans, 1994). Indeed many successful machine vision algorithms employed in industry, processing thousands of images are based on morphological techniques.

For the quality control in the textile industry, the technique of morphological image analysis has been used for assessing fiber composite material (Serra & Verchery, 1973) and carpet appearance (Xu et al, 1991) and texture analysis (Wermen & Peleg, 1985). Morphological image processing based technique has also been investigated for detection and classification of defects in fabric (Chetverikov & Hanbury, 2002). The morphological operations for defect detection in fabric are inherently sensitive to the size and shape of the defect. Therefore, while applying morphological image processing technique on the fabric image for the detection of defects, the software-based morphological operations may give poor result when the defects are relatively small in comparison to the fabric structure. The testing involves two stages of operations. First a structuring element is selected from the heuristic knowledge of the likely defects. Secondly, the test image is thresholded and then morphological operations are applied on the thresholded image of the test fabric for the identification of defects.

1.1 Types of commonly occurring defects in woven fabric

The defects in fabric are generally classified into three subdivisions according to the occurrence in the fabric. They are, (i) weft-way defects (ii) warp-way defects and (iii) defects with no directional dependence. There are about twenty-two types of defects usually associated with woven fabric due to various processing irregularities. Out of these twenty-two, only few are severe defects and need elimination by rejection at the production stages.

These severe defects are:

a. **Broken pick, short pick, double pick and pick' inhomogeneity**: These defects are sometimes referred to as mispick (float) or missing pick and result in a void in the fabric due to missing yarns. Sometimes, two adjacent threads with same interlacing are beaten up together as a result of broken picks. Double pick defect is mainly due to extraneous thread in the fabric.

b. **Slub (fly):** These defects usually appear in the fabric as abnormally thick places of varying length.

c. **Knot:** These defects usually appear in fabric as hardly discernible change in yarn thickness. Knots in fabric usually arise due to fastening of yarns.

d. **Snarl and snag**: A defect, where warp or weft yarns are twisted on itself and the loop is termed as snarl. A snag is produced when one or more pick is snagged on a knitted yarn.

e. **Reed mark (crack):** It is due to improper irregular interspacing of yarns and appears as a transparent strip.

f. **Thin place**: This defect arises out of insufficient density of yarns.

Apart from the above-mentioned major defects, mechanical defects such as hole piling, oil marks and other anomalies manifest themselves as defects in woven fabric.

In general, all defects alter the normal regular structure of fabric pattern and also modify the statistical and physical properties of the first quality fabric. The effects of defects are also dependent on the textural types of woven fabric.

2. Morphological image processing operations

Mathematical morphology is a tool for extracting image components that are useful in the representation and description of region shape (Serra, 1988). As a discipline within imaging, mathematical morphology concerns with the applications of its basic operators in all aspects of image processing. Conceptually, morphological operations are rooted in planar geometric structure which is altered by probing with a structuring element. Each operation uses the structuring element to determine the geometrical filtering process. satisfying four properties: translation invariance, antiextensivity, increasing monotonically and idempotence. The structuring element of a morphological operator is therefore a function defined in the domain of the spatial pattern. The value of each pixel of the domain is the weight or coefficient employed by the morphological operator at the pixel position.

The concepts of image processing using mathematical morphology were developed mainly as an application of set theory (Matheron, 1975; Serra 1988; Heijmans & Ronse 1990). The evolution of the theory closely follows the evolution and applications of pipeline and cellular computing techniques (Danielsson & Levialdi, 1981). By nineties many image-processing algorithms for morphological operations are evolved using electronic and optical techniques (Liu, 1989; Casasent, 1990; Botha et al, 1989; Mallick-Goswami & Datta, 2000). Optical morphological image processing was also applied for feature extraction and shape description (Gracia et.al, 1993).

The morphological image processing technique has its roots in texture analysis and in the problems of determining properties of the texture. Its underlying idea is to use a so-called structuring element to define neighbourhood of points in an image. Operations with the structuring element may smooth the contours of the objects or may decompose an image into its fundamental geometrical shapes. Morphological operations can be applied not only on binary image but also on gray level images (Vincent, 1993).

Fundamental morphological operations are erosion and dilation. These operations remove or add pixels from a binary image according to rules that depend on the pattern of the neighboring pixels. The erosion operation reduces the size of an image, while the dilation operation enlarges geometrical size the image. Two other derived morphological operations are opening and closing. An opening is an erosion operation followed by dilation, and a closing operation is a dilation followed by an erosion operation. Each of these operations uses the structuring element to determine the geometrical filtering process.

In morphological operations, the choice of structuring element is very critical and is generally guided by apriori knowledge of processing tasks. However, there are several studies, where the selection of structuring element is guided by a training phase for a particular situation (Camps el al, 1996).

2.1 Binary morphological image processing operations

We restrict ourselves to the specific case of a gray fabric image $F:Z \rightarrow \bar{R}$ defined on a sub-section of the two-dimensional Euclidean space R^2. The notation $\bar{R}$ indicates the set R∪{-

$\infty,+\infty\}$. In other words, the function assigns a grey-level to each point $x \in Z$ of the fabric image. In practice, however, the Euclidean space is replaced by a discrete space Z^2. The gray values taken by F are often limited to integers and the pixel value at position x is denoted by F(x). For 8 bit gray scale images, $F(x):Z \rightarrow \{0.1.2.\ldots\ldots\ldots\ldots\ldots,255\}$.
The morphological operations are defined as set operations and can be applied for processing of fabric images by defining another selected image S called structuring element. Therefore S is a subset in the two dimension Euclidean space. S can be a binary image or a gray image. The classical Minkowsky set addition and subtraction of these two images F and S are defined in set notation as,

$$\text{Addition:} F \oplus S = \{x+a : x \in F, a \in S\} = \cup F_a \big|_{a \in s} = \cup S_a \big|_{F \in S} \tag{1}$$

$$\text{Subtraction}: F \Theta S = \bigcap F_a \big|_{a \in s} \tag{2}$$

where F_{-a} is the translate of the image set along a and is defined as,

$$F_a = \{ F + a : x \in F\} \tag{3}$$

The transformation $F \rightarrow F \oplus S$ and $F \rightarrow F \Theta S$ are called a dilation and erosion operation by structuring element S. According to standard convention the morphological dilation operation on the image F by the structuring element image S is given by,

$$\text{Dilation:} F \oplus \bar{S} = \left\{ h \in E : (\bar{S}) h \cap \frac{F}{\varphi} \right\} \tag{4}$$

$$\text{Erosion:} F \Theta \bar{S} = \{ h \in E : S_h \subseteq F \} \tag{5}$$

where h is an element in Euclidean space and, φ denotes the empty set.
The reflected or the symmetric set $\bar{S}$ is related to S by the following equation,

$$S = \{ -a : a \in S\} \tag{6}$$

Self complementation of the set yields a duality relation and the dilation operation is also expressed as:

$$F \oplus \bar{S} = \left(F^C \Theta \bar{S} \right)^C \tag{7}$$

where, F^C is the complement of the set constituting the image F.
By definition, morphological opening is erosion followed by dilation and the closing operation is viewed as the complimentary process of opening. Therefore, the opening and closing operation are denoted in set notation as,

$$\text{Open}: (F,S) = \left(F \Theta \bar{S} \right) \oplus \bar{S} \tag{8}$$

$$\text{Close}: (F,S) = \left(F \oplus \bar{S} \right) \Theta \bar{S} \tag{9}$$

The opening is the union of all translates of the structuring element that is included in the set F. It may be noted that both opening and closing operations are translation invariant. Opening and closing are related to Boolean duality by,

$$\text{Close: } (F, \bar{S}) = \text{open} : (F^c, S)^c \tag{10}$$

2.2 Gray morphological image processing operations

Binary morphological operations are performed for the sets whose elements are vectors corresponding to pixel positions and therefore are *set-set operations*. A grayscale image can be considered as a three-dimensional set where the first two elements are the x and y coordinates of a pixel and the third element is the gray-scale value. The key issue is to use the infima/suprema (minima and maxima in discrete cases) to define gray-scale morphological operators (Chu-Song et al. 1999; Heijmans 1999). The structuring elements of the gray-scale morphological operations could have the same domains as those in binary morphology. However, a gray-scale structuring element is also possible having certain values instead of having only value 1 or 0. Therefore, morphological operation of gray scale images with binary structuring element is a function-set operations and the operation for gray images and gray structuring element are function-function operations (Sternberg, 1986). The pixel value of a gray fabric image F at position x is expressed by a function f(x). The gray scale erosion and dilation with as binary structuring element *S* can then be defined as,

$$\text{Function set erosion: } \left[f \Theta \bar{S} \right](x) = \inf_{x \hat{I} S_X} f(x) \tag{11a}$$

$$\text{Function set dilation: } \left[f \Theta \bar{S} \right](x) = \sup_{x \hat{I} S_X} f(x) \tag{11b}$$

However, in case of operations with gray scale structuring element g^S, the gray morphological operations are defined as,

$$\text{Function} - \text{function} - \text{erosion: } [f \oplus g^S](x) = \max_{y \in G} \{f(x+y) + g(y) \tag{12a}$$

$$\text{Function} - \text{function} - \text{dilation} : [f \Theta g^S](x) = \min_{y \in G} \{f(x+y) g(y)\} \tag{12b}$$

Gray-scale opening and closing are defined in a similar manner as the binary case. The only difference is, when the operations are carried out, these opening and closing operations use gray-scale dilation and erosion. As binary morphological operations do, gray-scale opening is anti-extensive and gray-scale closing is extensive. Both operations make an original image smooth along to the nature of minimum and maximum functions.

3. Morphological operations on image of fabric

It is evident that F is the image of test fabric and *S* is the image of structuring element, which in most of the cases is a binary image. Selection of structuring element image is important in morphological image processing for defect detection in fabric. A particular defect in the fabric image can be detected by eroding the image with a structuring element that is slightly smaller than the shape of the defect. For example, if the dimension of the structuring element is made slightly smaller than the average dimension of a knot, then the erosion operation of the fabric image with the structuring element will result in the complete elimination of the weft and warp structure. However, the selection of structuring element is not easy for the detection of thick yarn as the yarn may run through out the entire length of the fabric.

While applying the binary morphological image processing operations as defined, it is necessary to obtain binary image of the fabric. Segmenting the gray level pixel values of fabric image into two gray level values are all that is required to produce a binary image of fabric. Let, I(i, j) is the gray level pixel value at point (i,j) of the input fabric image and F(i,j) is the gray level of point (i,j) of the output fabric image. The binary image is obtained from the gray level image by converting the pixel value to 1 (white pixel) if the value is greater than the preselected threshold value: otherwise the pixel value is returned to 0 (black pixel). It is assumed that the threshold value does not depend on the spatial coordinates (i, j) and also threshold value is independent of local properties of the point.

3.1 Extracting defects from image of fabric

Detection of defects by morphological operators are carried out for few test samples of woven fabric. For example, three types of typical defects such as a knot (fig. 1a), thick weft (fig.1b) and missing weft (fig. 1c) are subjected to morphological operations with a 3×3 structuring element. Since the fabric images are gray images, gray to binary image conversion by thresholding is necessary for defect detection. The thresholded binary images are shown in figures 1d, 1e and 1f. The morphological erosion and opening operations on the thresholded images are shown in set of figures 1g, 1h and 1i and figures 1j, 1k and 1l. It has been observed that a rectangular structuring element of 3×3 pixel size is sufficient for extracting the defects such as knot from the test fabric either by erosion of by opening operations. However, in case of detection of thick weft by opening operation is not very efficient though the detection is better than erosion operation. The detection of missing weft by erosion and opening operations needs higher size structuring element. A 5×5 rectangular structuring element gaves better results than using a 3×3 structuring element. This is because of the continuity of the defective yarn throughout the length / breadth of the test fabric.

From the results, it is shown that the presence of knot and thick yarns are detected more clearly by the erosion operation than the by opening operation. However for opening operation a structuring element of 5×5 is used. The detection of missing weft by erosion and by opening operations, however, is hopelessly useless even by using structuring elements of higher areas.

3.2 Morphological operations by removal of interlaced grating structure of fabric

While detecting defects of various sizes in test fabrics, the results of morphological operations are much better, if the basic grating structure of the fabric due to weft and warp interlacing is removed by spatial filtering. To accentuate the gray values of the pixel positions of defects and to properly distinguish them from the cross-points, it is necessary to process the fabric image with a view to smooth out or remove the cross points. The processing is done by optical or digital spatial filtering of the de-noised image F_{dn} to yield the spatial filtered image F_{sf}. A binary fabric image F_b can then be obtained from the spatial

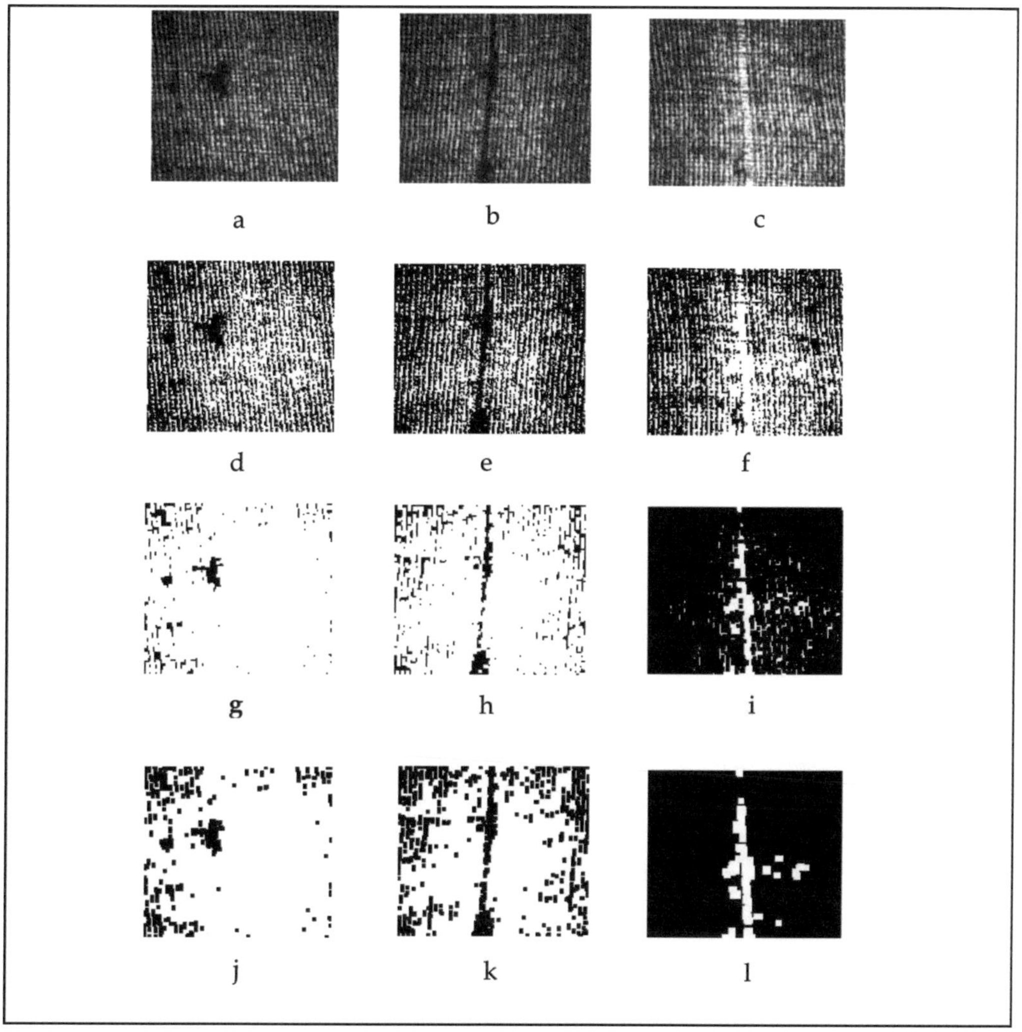

Fig. 1. Test fabrics with defects (row 1), thresholded images (row 2) and results of defect detection by morphological erosion (row 3) and opening operations (row 4).

filtered image, when the pixels are converted to either 1 or 0, if it is above or below a pre-selected threshold value P_{th}. Therefore, F_b=1, if F_{fs}>P_{th} and F_b=0 if F_{sf}<P_{th}. The threshold value P_{th} is however, depended on the type of fabric.

The test results of morphological operations after the removal of grating structures of fine, medium and coarse fabric are shown in Fig. 2. Some general conclusions can be drawn from the results. It is not possible to remove the grating structure of warp and weft yarns by digital spatial filterig, particularly for coarse fabric when the size of defects are comparable or slightly bigger than the yarn diameter (row 2 of fig. 2). In fine and medium fabrics, it is possible to remove the interlaces grating completely by suitably selecting the diameter of spatial filter and the threshold value. For defect detection opening, dilation, erosion and closing operations are performed. The type of morphological operations to be performed depends on the types of defects present in the test fabric. The result follows the expected pattern for long defects, since the connectivity of the yarn in the whole image set is higher than that of the defects. In fig. 2, for the detection of defect present in the first type of test fabric opening operations are carried out. For defects in the second, third and fourth types, dialtion, erosion and opening operations are carried out respectively.

4. Morphological filtering operations for noise removal from fabric image

Before applying the morphological operations for the detection of defects in woven fabric, the gray fabric image F is *de-noised* by using morphological filtering technique. This preprocessing is necessary particularly for cotton fabric, where the presence of hairiness is more likely to occur. Since noise in an image should be removed wherever it is located, the translation invariance property of filter is required for noise removal. Translation invariant operations are justifiable also for a fabric image where noise in the image is dimensionally

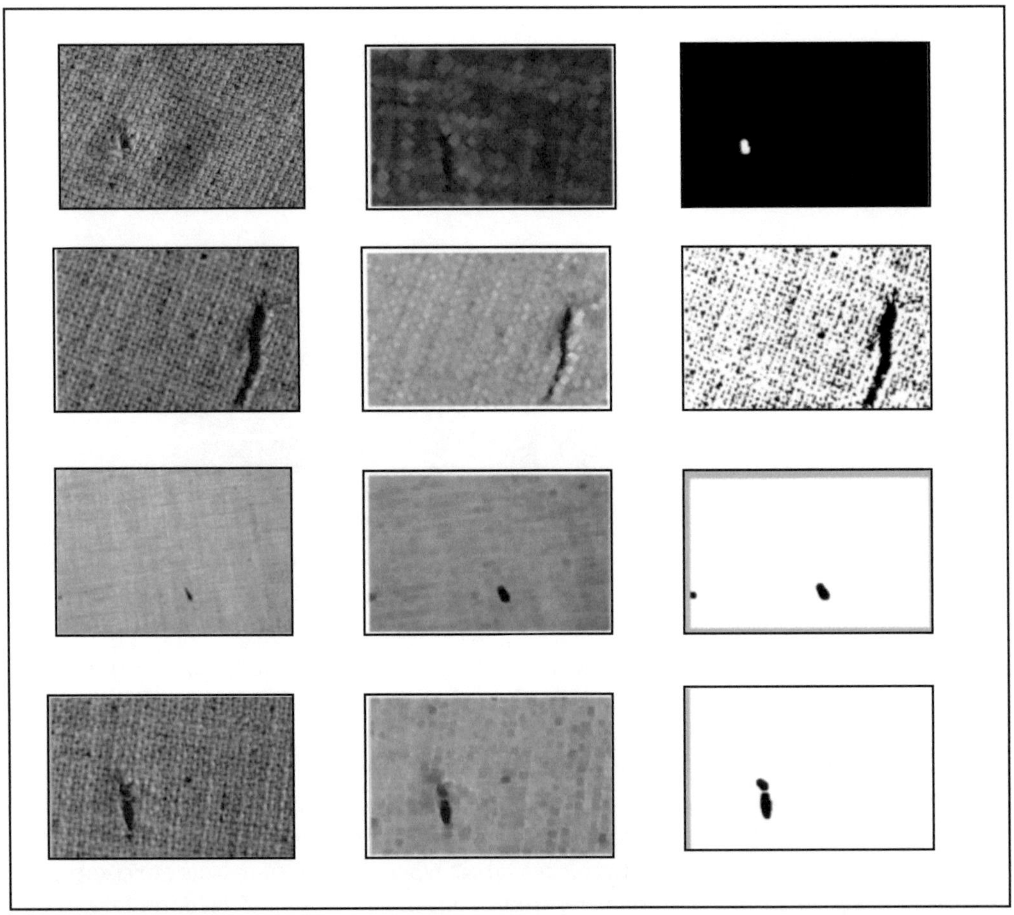

Fig. 2. Test results of morphological operations after removal of interlaced grating structure. Column 1 shows the images of test fabrics, column 2 shows the spatially filtered images of column 1. Column 3 shows the detection of defects by morphological operations on images of column 2.

smaller than the defects. The morphological filter, which can be constructed on the basis of the underlying morphological operations, are more suitable for noise removal than the standard linear filters since the latter sometimes distort the underlying geometric form of the image (Maragos & Schafer, 1987). Therefore, these filters are more suitable for noise removal (Podaru & Stanomir, 2003).

The morphological filtering is restricted to all image-image transformations that are translation invariant and increasing idempotent operations characterized by $\varphi[\varphi(F)]=\varphi(F)$, where φ is an operation. The filter $\varphi(F)$ is decomposed into OR operation of erosions by a structuring element S and is expressed as,

$$\varphi(F)= \bigcup_{S\in \ker[\varphi]} F\Theta\bar{S} \quad (13)$$

where F is a fabric image and $\ker[\varphi]$ is the kernel of φ given by, $\ker[\varphi]=\{F \mid 0\in\varphi(F)\}$.

Extensivity and anti-extensivity of the operations are not always useful for practical situation of noise removal. Since median filter is not extensive or anti-extensive, but handles white objects and black backgrounds equivalently, morphological representation of median filter is very useful for de-noising. The cascades of opening and closing with an elementary (2×2) binary structuring element S is used for this purpose. Since the structuring element is symmetric, reflection of S is the S itself. The de-noised image F_{dn} is obtained from gray fabric image F by the following processing steps:

$$\text{open}(F)=(F\Theta S)\oplus S) \quad (14)$$

$$F_{dn}=\text{close}[\text{open}(F)]=([\text{open}(F)]\oplus S)\Theta S \quad (15)$$

By this process most of the noise present in the fabric image is removed while maintaining the shape and size of the defects almost intact.

5. Rank order operator: generalized morphological operations

The detection of small defects such as knots and broken picks in woven fabric is difficult because of the presence of confused background due to the interlaced periodic structure of the warp and weft, which result a large number of almost regular crossover points. Moreover, hairiness of the fabric introduces high dose of noise into the image pattern of the fabric.

In the Section 3, it has been shown that the morphological erosion operation can erode the grating structure of the fabric to a certain extent and might be suitable for the identification of defects. However, complete erosion of the background is not possible unless very judicious choice of the structuring element is made. For detection of defects of small size in woven fabric, simple morphological erosion operation with a comparatively large structuring element is not efficient. In such cases, there are always chances of complete erosion of the image. Dilation also may cause the entire cross point background of the image to fill in resulting in a completely dark search space (background). In many cases, there is a need to erode first by a small structuring element and then apply a larger one. This may somewhat solve the problem, however this is not always possible, particularly for the detection of defects in fabric where the failure rate of such a solution is frustratingly high.

It is shown in this section, that the binary rank order operation (Heygster, 1980; Arce & Foster, 1989) gives satisfactory result when the expected size of defect is considered in selecting of structuring element. Moreover, the rank order operator has unique property of image smoothing and presenting edges and at the same time it is very efficient for noise removal. In a woven fabric, repetition of the interlaced grating structure is not very accurate and therefore rank order operation proves to be an efficient tool for defect detection.

It may be noted that the rank-order operator is a generalization of the morphological erosion and dilation operations. Erosions and dilations are convolutions with maximum and minimum threshold value respectively. Therefore rank order operations require a structuring element as well. Rank order operator rotates in planar geometric structure, which is altered by probing with a structuring element called reference image. Each operation uses the reference image to determine the geometrical filtering process. The reference image or the structuring element of the operator is therefore a function defined in the domain of the spatial pattern of the operator.

5.1 Binary convolution operation

The rank order operation can be performed by a linear convolution between input image and reference image (structuring element image) followed by thresholding of the convolution output. Simplest image convolution operation of image is a process that copies one image at each pixel location of another while allowing for the effects of all pixel values in the area where the copy takes place. A multiplying, adding and shifting operation accomplish this. A convolution c(p) of a binary fabric image F(u) by structuring element S(u) is given by,

$$c(p) = \sum_{u} F(p-u)S(u) = F*S \tag{16}$$

where, p and u are two-dimensional spatial vectors, $u,p \in Z^2$, and the summation is over the Z domain of the image.

5.2 Binary nonlinear rank order filter

In recent years, several rank-order structuring filters are proposed. These filters are roughly divided into two categories. The set of adaptive rank-order filters fall under the first category (Lee & Fam, 1987), while the second class encompasses structure preserving rank-order filters (Niemnien et al, 1987). Adaptive rank-order filters are used on images with low signal-to-noise ratios and some a priory knowledge or local statistics are required for processing (Stevenson & Arce, 1987). The main advantage of the non-adaptive type lies in the fact that the extensive use of local statistics is not necessary.

The rank order filter while applied on binary image is also called an order statistic filter or a Ξ filter (Preston, 1983). A binary rank order operations can be replaced by operations of counting the number of image points that contacts the points of the structuring element probe and marks the reference point in the output image if at least a given percentage of points contact the probe. The percent point or the threshold is a variable and is adjusted for good performance in a particular application.

Widely employed nonlinear and locally adoptive filters are rank order filters. The threshold operation X_t is applied to eq.16, at threshold t, to get the output result after rank order filter operation and is given by,

$$R_t(F,S)(p) = X_t\left(\sum F(p-u)S(u)\right) = X_t(F*S) \tag{17}$$

For a rank order operator R_t expressed in this from, threshold t is the rank. The threshold operation X_t is performed under the conditions that,

$$\begin{aligned} X_t &= 1 \quad \text{if} \quad c(p) \geq t \\ &= 0 \quad \text{otherwise} \end{aligned} \tag{18}$$

5.3 Choice of the proper rank of the rank order operator

Two problems are encountered while applying the rank order operator for defect detection in fabric. They are related to the (a) choice of reference image (i.e. structuring element) and (b) the choice of the rank. The value of each pixel of the domain is the weight or coefficient employed by the pixel position. The selection of the size and shape of the reference image is hence an important step for defect detection. When the size of the reference image is almost same to the size of the defect to be detected the most efficient and optimum detection capability is expected. Evidently the size of the domain can be ascertained from priory knowledge of the likely defect in woven fabric.

The rank selection is based on the boundary characteristics. One of the most important aspects in selecting a rank is the capability of reliably identifying the mode peaks in a given image histogram. This is particularly important for automatic threshold selection in situations where image characteristics can change over a broad range of intensity distributions as in the case of woven fabric. Therefore the rank of the operator has to be decided also from a priori knowledge of the periodicity of the fabric structure and the diameters of the warp and weft yarns.

The rank of a particular operation is a function of the reference image or the structuring element S and the intensity values of the test fabric F. A simple relation is established between the rank R and the dimension D of the structuring element S and the maximum value M of the convoluted matrices for quick selection of a rank for a particular operation. The relation is given by,

$$R=255(k)(D)-M \tag{19}$$

where, k is a positive low value and M is obtained by dilation operation as $M=\max(F \oplus S)$. If F (the test fabric) is a $(m \times n)$ matrix and if S (the structuring element or the reference image) is a $(m_1 \times n_1)$ matrix, then the convoluted matrix has dimension $(m \times n) \times (m_1 \times n_1)$. So the dimension of the output matrix increases correspondingly. The conversion of $(m \times n) \times (m_1 \times n_1)$ matrix to $(m \times n)$ output matrix eliminates the data outside the region of interest.

5.4 Defect detection of fabric by rank order operator

Fig. 3 shows the presence of a knot, oil mark and thick yarns respectively, as defects in the test fabrics. A structuring element of size 15 X 15 is used for rank order filtering of the basic grating structure of the fabric at the value of k = 0.657.

It is seen that the knot is extracted efficiently (fig. 3, first column). In contrast to the erosion operation (Fig 2), the size of the knot is not reduced and is maintained in the detection process of the rank order filtering. The oilmark is also detected as defect by the same process (fig. 3, second column). The detection of thick yarns in the test image of fig. 3 is shown in third column. A scrutiny of fig. 3f will shows the existence of not only a thick weft at the bottom but also the existence of a slightly thicker yarn (weft) on the upper portion of the fabric. Also existence of the defect of slightly thicker yarn escapes visual inspection and the can not be detected by morphological operations described in sections 3 and 4. However, by selecting an elongated structuring element 4 X 132 and the value of k = 0.657 (which matches closely the diameter of the yarn) and the defect is detected by rank order filtering.

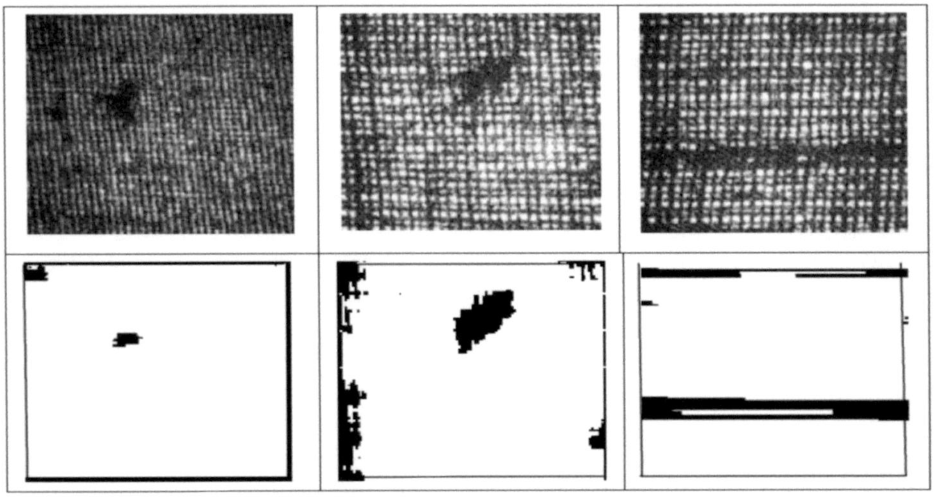

Fig. 3. Shows the presence of defects, knot, oilmark and thick yarns (top row). The bottom row shows the detection process by rank order filtering.

6. Selection of the size structuring element by artificial neural network

Selection of structuring element in all morphological operations is one of the major problems while detecting defects in woven fabric. The size and shape of structuring element has to be different for different textures of fabric and also has to be different for identification of different types of defects. In general, the selection of structuring element is done heuristically or by trial and error basis. In order to overcome the problem of selection of the size of the structuring element, decomposition of the structuring element is proposed (Park & Chin, 1995; Roerdink & Heijimans, 1988). One can also process (e.g., extract or eliminate) differently scaled objects of interest in the image by adapting the size of the structuring element(s) to the local intensity range (Masayasu, Masayoshi, & Akira, 2003). Attempt has been made to use optimally selected structuring element (Schonfeld, 1994) and optimal selection is applied for the defect detections in fabric (Mak, Peng, & Lau, 2005). Since biological neurons can adopt itself to a new situation and can be trained, the selection of structuring element can be done by training from the image data to be processed for a specific application. Any simple artificial neural network model may be useful for training (Lippman, 1989). However, particularly defect detection problems where association of a single class is required for many disjoint regions in the pattern space, a multi layer perceptron (MLP) model with a hidden layer of neurons is necessary (Widro, & Lehr, 1990). The signal flow in a MLP model is shown in Fig. 4, where w denotes the weight, I and O are the input and output signals with proper subscripts h and o for hidden and output layer. The hidden layer neurons generate hyper planes that are building blocks of decision regions. In MLP architecture, linear threshold unit is used for generating activation function form input layer neurons to the hidden layer neurons and sigmoidal threshold function S[.] is required for generating activation functions from hidden to output layer neurons. A real parameter λ determines the sigmoidal activation gain. Each neuron of the input layer is connected to each neuron of the hidden layer after being multiplied by the weight vectors and similarly each neuron of the hidden layer is also connected to each neuron of the output layer after being multiplied by the weight vectors. During the training process, up-gradation of the weight vectors takes place and continues till an error term ε goes below a certain predefined tolerance value. A learning rate coefficient η determines the size of the weight adjustments made at each iterations and hence influences the rate of convergence. The learning rate has to be kept small in order to maintain smooth trajectory in weight space. Large learning rate leads to oscillations during learning and it is necessary to introduce a momentum term α into weight update procedure. Many algorithms exist for training and testing in MLP model (Haykin, 2008).

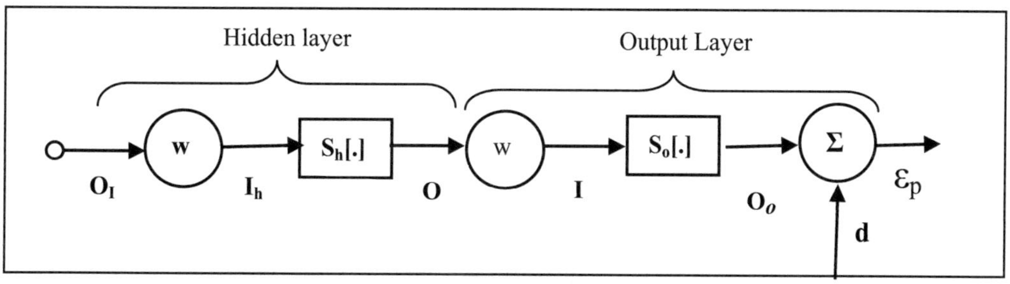

Fig. 4. Signal flow in artificial neural network model

If (u x v) is the size of the structuring element needed for the morphological operation, then this parameter should relate to the fabric parameters for training the ANN. For this purpose each sample of fabric is assigned a set of two normalized numbers Z_1 and Z_2 for their identification. Z_1 is a normalized number related to threads per centimeter of fabric along the weft direction (say *x*-axis) divided by an integer number (say N, where N>1), so that Z_1 can be expressed as a fraction. Similarly, Z_2 is a normalized number related to number of threads per centimeter of the fabric along warp direction (say y-axis). Z_2 is also expressed as a fraction by dividing by N. Given a training set vector pairs of fabric parameters Z_1 and Z_2 for P different types of fabrics, the network is trained for different types of defects to yield trained size of structuring element required for morphological operation.

7. Conclusion

It has been shown in this chapter that morphological operations can be exploited for detection of various types of defects in the woven fabric. However, the erosion operation on fabric image helps in detecting defects to a major extent. Binarization of the fabric image is necessary before applying the morphological opeartions. The detection capability improves if the image is spatially filtered and then thresholded to remove the interlaced grating structure of the fabric. Moreover, gray morphological operations is possible for the defect detection and more accurate detection is possible than the binary morphological processing. In this chapter it is alos shown that detection capability is greatly improved by rank-order filtering which is termed as generalized morphological operations. The method offers two

flexible controls. The structural element can be selected according to *a priori* knowledge of the defect. Secondly, the rank can be selected to suit a particular noisy situation. It has been established that very good detection is possible for small size defects (such as a knot) in all types of fabric by rank order filtering technique.
Proper selection of the structuring element in all morphological operations enhances the chance of detection of defects which otherwise may not be very distinct even during visual inspection. A method has been shown where the size of the structuring element can be selected as a result of training of the system by artificial neural network.

8. References

Arce, G. R.; & Foster, R. E. (1989). Detail preserving ranked order based filters for image processing, *IEEE, Trans, Acou. Speech & Sig. Processing,* Vol. 37, No 1, 83-89.

Botha, E.; Richards, J.; & Casasent, D. (1989). Optical laboratory morphological inspection processor, *Applied Optics.* Vol 24, No 11, 5342-6.

Camps, O.I.; Kanungo, T.; Haralick, R. L. (1996). Grey-scale structuring element decomposition. *IEEE Trans. Image Processing.,* 5, 111-120.

Casasent, D. (1990). Optical morphological processors, *Proc. SPIE,* Vol. 1350, p 380-384.

Chetverikov, D; Hanbury, A., (2002). Finding defects in texture using regularity and local orientation,. *Pattern Recognition,* 35,10-15.

Chu-Song, C; Ja-Ling, W.; Yi-Ping, H. (1999). Theoretical Aspects of Vertically Invariant Gray-Level Morphological Operators and Their Application on Adaptive Signal and Image Filtering, *IEEE Trans. on Signal Processing.,* 47, 4, 1049-53

Danielsson, P. E.; & Levialdi, S. (1981) Computer architectures for pictorial image information systems. *IEEE computer Magazine.* No 11, 53-58.

Drobino, R.; & Machnio, M. S. (2006). Application of image analysis technique for textile identification. *AUTEX Research J.* 6,1, 40-48.

Gracia, J.; Szoplik, T. & Ferreira, C. (1993). Optoelectronic morphological image processor. Optics Letter. Vol 17, No 10, 1952-60.

Haykin, S.,(2008) *Neural Networks,* Delhi, Pearson Prentice Hall.

Haralick, R.; Sternberg, S. & Zhuang, X. (1987). Image analysis using mathematical morphology. (1987). *IEEE Trans Pattern Analysis and Machine Intelligence.* Vol 9, no 4, 532.

Heijmans, H. J. A. M.; (1994) *Morphological image operators*" Academic , Boston.

Heijmans, H. J. A. M.; & Ronse, C. (1990). The algebraic basis of mathematical morphology - I: Dilations and erosions, *Comput. Vis., Graph., Image Process.: Image Understand.,* 50, 245–295.

Heijmans, H. J. A. M. (1999). Connected morphological operators for binary images, *Comp. Vis. and Image Understan.,* 73, 99–120.

Heygester, H. (1980). Rank filters in digital image processing. *Proc. 5th Int. Conf. on pattern Recognition, Florida,* 1165.

Lee. Y. H. & Fam, A. T. (1987). An edge gradient enhancing adaptive order statistic filter. *IEEE, Trans, Acou. Speech & Sig. Processing,* Vol. 35, No 5, 83-9.

Liu, L. (1989). Optoelectronic implementation of mathematical morphology. *Optics Letter.* Vol. 14, 482-88.

Lippmann.,R.P.,(1989). Pattern classification using neural networks. *IEEE Communication Magagine*. 2747-64.

Mak, K.L., Peng, P., & Lau, H.Y.K., (2005). Optimal morphological filter design for fabric defect detection, *IEEE International Conference on Industrial Technology*. 14-17 Dec. 799 - 804.

Masayasu, I., Masayoshi, T., & Akira, N. (2003). Morphological operations by locally variable structuring elements and their applications to region extraction. *Ultrasound images, System and Computation*, ,34, 98-103

Mallik-Goswami, B. & Datta A. K. (2000). Detecting Defects in Fabric With Laser- Based Morphological Image Processing, *Tex. Res.J.*, 70, 758-762.

Maragos, P. & Schafer, R. W. (1987). Morphological Filters- Part I: Their Set- Theoretic Analysis and Relations to Linear Shift-Invariant Filters, *IEEE Trans. Acoust., Speech, Signal Processing*. 35, 8, 1153-1169.

Maragos, P. & Schafer, R. W. (1987). Morphological Filters- Part II: Their Relations to Matheron, G. (1987). *Random sets and integral geometry*, Wiley, N. Y. 1975Median, Order-Statistic, and Stack Filters, *IEEE Trans. Acoust., Speech, Signal Processing*, 35, 8, 35. 8. 1170-1184.

Maragos, P. (1987) Tutorial on advances in morphological image processing, Opt. Engg. vol26, no3, , 623-28.

Nieminen, P. Heinonen and Y. Neuvo, (1987). A new class of detail-preserving filters for image processing, *IEEE Trans Pattern Anal. Machine Intelli.*, Vol 9, No. 1.

Park, H., & Chin. R.T., (1995), Decomposition of arbitrarily shaped morphological structuring elements. *IEEE Trans. Pattern Analysis & Machine Intelligence*, 17, 2-15.

Podaru, E.; & Stanomir, D. (2003) Noise suppression using morphological filters, *Proc. Int. Symp. Signals, Circuits, Systems*, 1, 61–64.

Preston, K. (1983). Ξ Filters. *IEEE, Trans, Acou. Speech & Sig. Processing*, Vol. 31, No 4.

Rumelhart, D.E., Hinton, G.E., & Williams., R.J. (1986). Learning representation by back propagating errors. *Nature*. 323, 533-36.

Roerdink, B.T.M., & Heijmans. H.J.A.M, (1988), Mathematical morphology for structuring elements without translation symmetry, *Signal Processing.*, 15, 271–277.

Schonfeld, D. (1994), Optimal structuring elements for the morphological pattern restoration of binary images, *IEEE Trans. Pattern Analysis & Machine Intelligence* 16, 6, 589–601.

Sakaguchi., Hua-Wen Guang. Matsumoto, Y. Toriumi., K & Hyungsup, K. (2001). Image Analysis of Woven Fabric Surface Irregularity , *Tex. Res. J.* 71, 666-67.

Serra, J. & Verchery, G. (1973). Mathematical morphology applied to fiber composite material, *Film. Sc. Tech.* Vol 6, No 1, 141-49.

Serra, J. (1988). *Image analysis and mathematical morphology*, Academic press, London.

Stevenson, R. L. & Arce, (1987). G. R. Morphological filters: Statistics and further syntactical properties. *IEEE Trans. Circuits & system*, Vol 34. No 11.

Sternberg, S. R. (1986). Grayscale morphology. *Comp. Vision, Graphics and Image Processing*, Vol. 29 no3.

Vincent, L. (1993). Morphological Grayscale Reconstruction in Image Analysis: Applications and Efficient Algorithms, *IEEE Trans. on Image Processimg.*, 2, 2, 176-187.

Wermen, M. & Peleg, S. (1985). Min-max operators in texture analysis, *IEEE Trans. Pattern anal. Mach. Intelli.* Vol 7, No6, 730-36.

Widrow, B. and Lehr, M.A., (1990) 30 years of adaptive neural networks: perceptron, Madaline and backpropagation, *Proc. IEEE*, 78, 9 1415-1442.

Xie, X. (2008). A review of recent advances in surface defect detection using texture analysis techniques. *Elect. Letters on Computer Vision and Image Analysis*, 7, 3. 1-22.

Xu, B. Pourdeyhemi, S & Spivak, M. (1991). Texture evaluation of carpets using image analysis, *Textile Res. J.* Vol 60, No 3, 407-16.

Zhang. Y. F. & Bresee, R. R. (1995). Fabric defect detection and classification using image analysis. *Textile Res. J.* Vol 65, No 1, 1-10.

12

Investigation of Wear and Surface Roughness of Different Woven Glass Fabrics and Aramid Fibre-Reinforced Composites

Haşim Pihtili
Firat University, Engineering Faculty, Department of Mechanical Engineering, 23279 Elazığ Turkey

1. Introduction

Composite Materials with Plastic Matrix has attained a broad utilization area due to their several qualities such as lightness, rigidity, heat resistance, high endurance limit, good resistance to wear in space and aircraft industry, automobile industry (piston, parts of engine, fan and compressor), sports and sea materials. Although manufacturing industry has developed good methods of manufacturing with modern machining, the traditional hole drilling method is still the most commonly used operation process due to its being economical and simple applicability. Drilling process is quite commonly used in defense, aviation and automobile industries [1]. In the production industry, non-metal composite materials are just beginning to be used alternatively instead of steel and other metals. Polymers and their composites are being increasingly employed in view of their high strengths and low densities. Many studies reported that the wear resistance with polymer sliding against steel improved when the polymers are reinforced with glass or aramid fibres. However, the behaviour is affected by factors, such as the type, amount, size, shape and orientation of the fibres, the matrix composition and the test conditions, such as load, speed and temperature [2.3.4–5]. The wear resistance of materials is determined by experiments that carefully performed in the laboratory.

Drilling process is quite commonly used in defence, aviation and automobile industries. When studies regarding processing composites are examined, we encounter commenly the composites with metal matrixes. In general in drilling tests made on metal matrix composites (MMC), it is seen that team wear is more and surface roughness is weak when classical high speed steel drills are used [6].

In this study, the effects of progress speed, rotation number, drill type and drill diameter on surface roughness of work sample are investigated while composite material is being machined by drilling. Besides, the wear behaviour of two woven glass fabric, namely the 300 and 500 grades, as well as aramid fibre-reinforced composite materials are investigated under different loads, speeds and sliding distances.

HSS, TIN and Carbide drills with 5, 10 and 15mm diameter were used in drilling processes. Experiments were performed with dry drilling by using values of cutting speeds of 125, 250 and 315 rpm and progresses of 0.056, 0.112 and 0.16 mm/rev.

2. Materials and specimens

In this study, woven glass fabric-reinforced composites, made of 300 and 500 gm^{-2} plain weave glass fabric, contain E glass fibres of diameter 10–24 µm were used. The woven glass fabrics are woven in two perpendicular directions. The woven glass fabrics composites consist of fibre, V_f = 30 vol.%, and matrix, V_m = 70 vol.%. Polyester resin was used as the matrix material for both composites. Another used specimen was the aramid fibre-reinforced composite. It has aramid fibres with a diameter of 10–24 µm. The aramid fibre-reinforced composite consists of fibre, V_f = 60 vol.% and matrix V_m = 40 vol.%. Epoxy resin was used as matrix material. The aramid fibre reinforced composite has been made by using filament winding process. There are three winds in this composite. The wind angle is equal to 45°. This method produces very strong composites; therefore extremely large cylindrical and spherical vessels can be built [7]. Automotive drive shafts, helicopter blades, oxygen tanks are among the applications of filament winding [8]. All composite specimens were cut in size of 47mm × 27mm for wear tests. The woven glass fabrics and aramid fibre-reinforced composites have 4 and 1.5mm thickness, respectively.

3. Wear tests

The wear tests were performed by using a block-on-shaft test method (Fig. 1). 15mm diameter ground shaft, SAE 1030 (DIN 22) steel, was used as the counterbody. The hardness and surface roughness (*R*a) of the shaft were 150 HB30 and 1.25 µm, respectively. The wear tests were performed on a specially prepared experimental set-up by using a lathe. The actual loads were placed on the pan of the load arm of this apparatus. The schematic view of wear set-up is shown in Fig. 1. All the experiments were conducted at room temperature. The temperature of the specimens was measured with a Cole-Palmer H–08406–46 infrared temperature measurement device. The woven 300 and 500 gm^{-2} glass fabrics and the aramid fibre-reinforced composite materials specimens were tested under different experimental conditions. Tests were conducted for speeds of 500 and 710 rpm and two different loads of 500 and 1000 g. Wear rate was determined by noting the weight loss (to 10^{-3} g) that was measured after the different sliding distances. One specimen was used for each experiment. The microstructures of the worn surfaces were examined by using a scanning electron microscope (SEM).

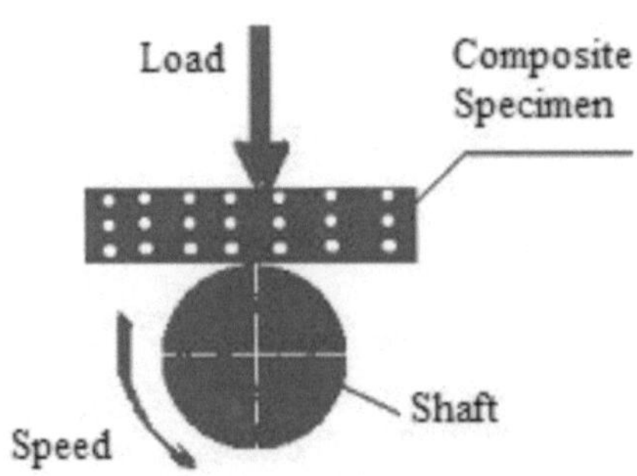

Fig. 1. Schematic showing a block-on-shaft wear test.

4. Experimental results and discussion

The effects of applied load and sliding speed on the weight loss of woven 300 and 500 glass fabrics and aramid fibre-reinforced composite specimens are shown in Figs. 2-3, respectively. The weight loss of the woven 300 glass fabric-reinforced composite specimens is lower than that exhibited by the woven 500 glass fabric-reinforced composite under the condition of 500 rpm speed and 500 g load (Fig. 2).

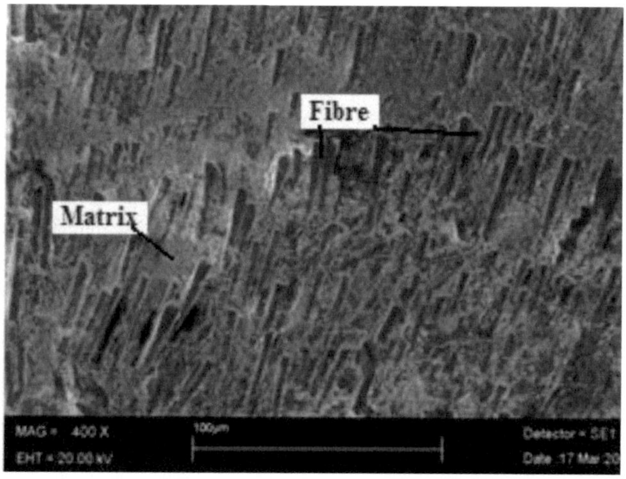

Fig. 2. SEM photograph showing the worn surface of the woven 500 glass fabric composite at 500 rpm speed and 500 g load.

The amount of matrix mass between fabrics in the woven 500 glass fabric-reinforced composite is more than that in woven 300 glass fabric-reinforced composite. For this reason, the reinforcement weight loss was very small, and weight loss was observed mainly as matrix loss. Epoxy-based composite exhibit lower wear loss than polyester-based composite [4,9]. The results also show that the wear did not occur in aramid fibre-reinforced composite specimens at 500 rpm speed and 500 g load, as shown in Fig. 2. Frequently, the wear and friction behaviour of polymeric composites have to be considered as a function of load, sliding speed and distance or temperature. The surface temperature plays an important role in the friction and wear of polymers. An increase in wear intensity can occur due to thermal softening. A low thickness of the softening layer normally results in low wear intensity [4]. From Figs. 2–3, it is seen that the wear increased with the increasing sliding distance. The weight loss of all the composite specimens generally increased depending on the sliding distance at the constant sliding speed 500 rpm when the applied load was increased from 500 to 1000 g (compare Fig. 3 with Fig. 2). The thickness of the softening polyester matrix layer on the specimen surface has increased proportionally with the temperature, so this softening layer has been detached from the surface of the specimens. The wear increased with an enhanced load, due to thermal softening [4]. As the earlier mentioned, epoxy-based composites exhibit lower wear loss than polyester-based composites.

In addition, aramid fibres usually exhibit much higher wear resistance than glass fibres [4,10] and aramid fibres exhibit lower friction than glass fibres. The low friction coefficient is prevented by the temperature increase. For this reason, the weight loss of the aramid fibre reinforced composite material is lower than that exhibited by the woven glass fabric-reinforced composites. The weight loss of the woven glass fabrics and the aramid fibre reinforced composites increased by a small amount at the constant applied load (500 g) when the sliding speed was increased from 500 to 710 rpm (compare Fig. 4 with Fig. 2). The surface temperature is normally not independent of sliding speed [4]. The applied load on the specimens has more effect on the wear than the sliding speed according to the datas given in Figs. 2–4. The woven glass fabric-reinforced composites was subjected to a larger weight loss depending on sliding distance, when both the sliding speed and the applied load are increased together (Fig. 4). The surface temperature plays an important role in the friction and wear of polymers and increases at higher sliding speeds and loads [4]. The weight loss of the aramid fibre-reinforced composite had no change, as shown in Fig. 4.

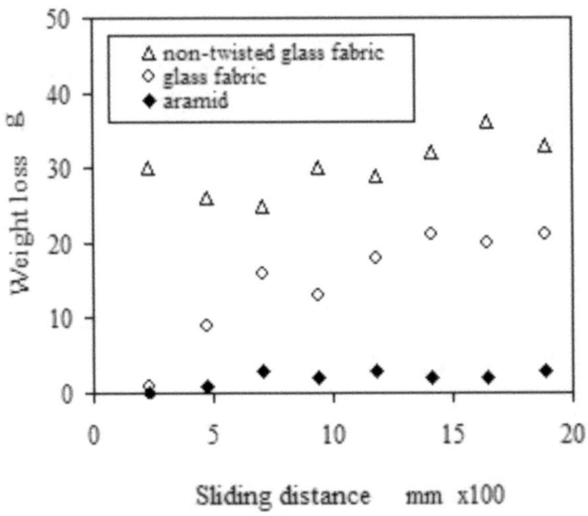

Fig. 3. Variation of weight lost with the sliding distances for 500 rpm speed and 1000 g load.

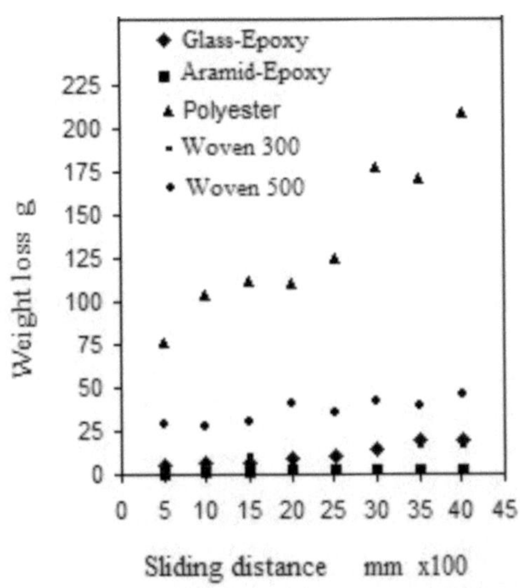

Fig. 4. Variation of weight lost with the sliding distances for 710 rpm speed and 1000 g load.

5. Drilling processes

In order to attain optimum results in drilling process, drilling length was chosed to be 15mm for the purpose of meeting the condition that drilling length should be three times or less than three times the hole diameter. Cooling liquid wasn't used in order to decrease possible heat shock in cutting tool during the experiment [11]. The experiments were done on the radial drill bench with a brand mane of Stanke Import with a progress interval of 0.056 mm/rev and 2, 5 mm/rev working between 20 and 2000 rpm. In drilling process, the effects of progress speed and number of rotations, drill type and drill peak angle, which are among the cutting parameters, on micro structure and surface roughness of work sample were investigated. In the experiments made, plastic matrixed composites manufactured in different ways were used. Some of the materials were supplied from TAI as semi-finished product. Various processes have been applied to these materials in order to use them in the experiments as products. Some other materials that were used in experiments by from (TEMSAN Turkish Elektromecanic Industry) as product. The composite materials used are shown in Table 1.

ISOVAL11- B	Epoxy / E-glass fiber	0.2 g/cm^3	3.5 mm
ISOVAL11- C	Epoxy / E-glass fiber	0.2 g/cm^3	5 mm
ISOVAL11- D	Epoxy / E-glass fiber	0.2 g/cm^3	7 mm
ISOVAL11- E	Epoxy / E-glass fiber	0.2 g/cm^3	10 mm
ISOVAL11- F	Epoxy / E-glass fiber	0.2 g/cm^3	14 mm

Table 1. The Composite Materials used in Experiments

5.1 ISOVAL11 composites

Final product with ISOVAL11 (Seen in Figure 5) series type and manufactured by TEMSAN has 0.2 g/cm^3 texture density. In Figure 6 the change in ISOVAL11 Epoxy/E-glass fiber composite and the average surface roughness is given for different drill materials. The biggest Ra value is found out for HSS drill material. In Figure 7 change in drill material

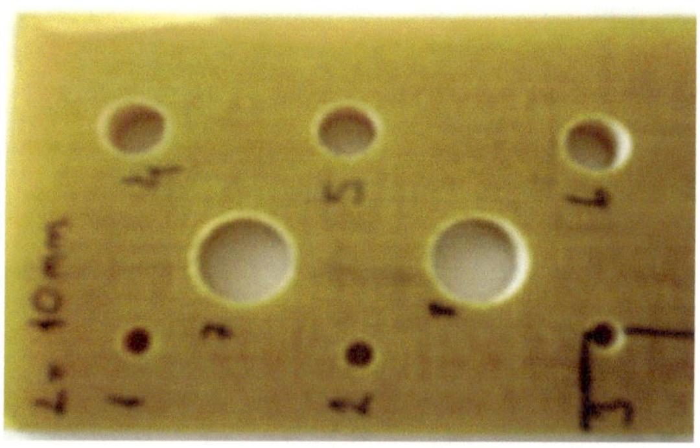

Fig. 5. ISOVAL 11 Composite materials used in experiments.

and the average surface roughness is given for ISOVAL11 Epoxy/E-glass fiber composite for different progress, rotation and drill diameters. In drills made at the same rotation and progress more sensitive surface is obtained in case Carbide drill is used and the same condition is applied for drills with 10 mm and 15 mm diameters. There is no heating on the cutting tool where swarfs extracted occures as dust. It is seen that Ry value for the holes with 15 mm diameter is lower than the other. As material thickness increases, the linearity of the results are observed and the surface roughness is measured accurately.

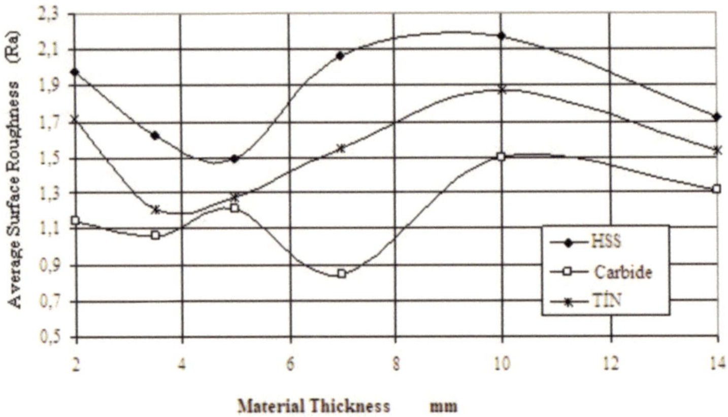

Fig. 6. The change in average surface roughness and ISOVAL11 Epoxy/E-glass fiber composite thickness. (s=0.112 mm/rot, n=250 d/min, d=10 mm).

In Figure 8 the SEM photos of middle point of hole section for ISOVAL11 Epoxy/E-glass composite drilled with HSS drill are shown. In these photos, the material texture denstity is 0.2 g/cm^3 and composite thickness is 5 mm, s=0.16 mm/rot, n=315 d/min, d=15 mm. In Kevlar materials, it is seen that the surface roughness is good in middle layers and roughness decreases in materials based on the texture condition. As material thickness increases extracting the swarf out during drilling become harder. As no cooling liquid is used during experimental studies, some heating occur on the cutting tool. Moreover as the material gets thicker it has difficulty in cutting the matrix material due to the fact that cutting tool is plastered over this region as matrix material per unit volıume increases and it is observed that heating occurs more in this region. The increased progress speeds and increased surface roughness values appear also as indicator of changes in texture condition of the material.

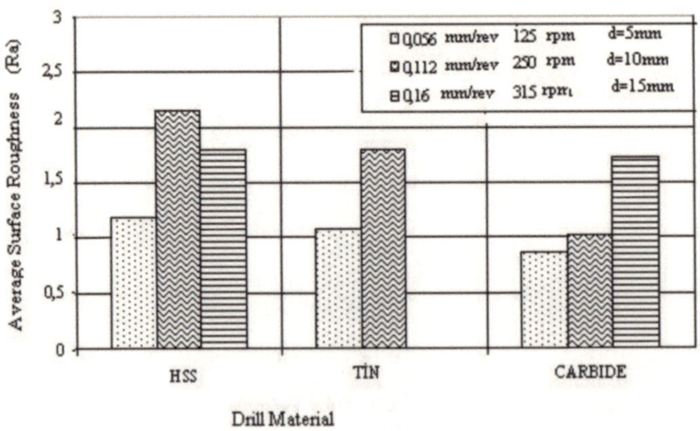

Fig. 7. The change in ISOVAL11 Epoxy/E-glass fiber composite and the average surface roughness (Texture density=0.2 g/cm^3, Composite material thickness=3.5 mm).

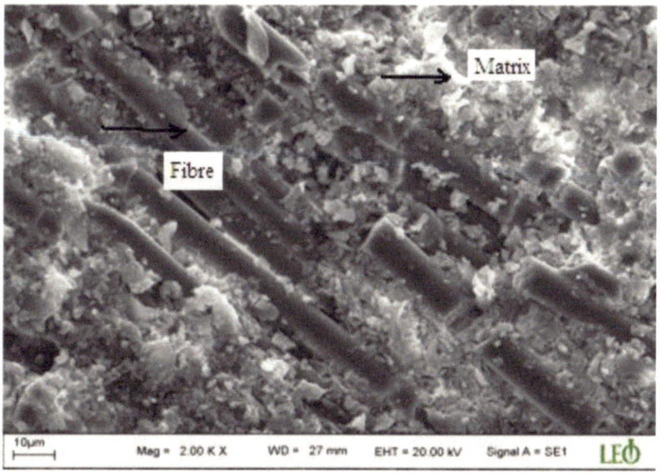

Fig. 8. 2000 times enlarged SEM photo of middlepoint of hole section for ISOVAL11 Epoxy/E-glass composite drilled with HSS drill (Texture density=0.2 g/cm^3, composite material thickness=5 mm, s=0.16 mm/rot, n=315 d/min, d=15 mm).

6. Conclusion

An experimental study of the wear behaviour of woven glass fabric and aramid fibre reinforced composites at various sliding speed, applied load and sliding distance reveals the following characteristics.

Wear of the composites depends on the experimental test parameters, such as applied load, sliding distance and speed. The applied load on the specimens has more effect on the wear than the sliding speed. The amount of matrix between fabrics in the woven 500 glass fabric-reinforced composite is more than in the woven 300 glass fabric. The wear occurs in the matrix rather than the reinforcement. Therefore, the wear in the woven 300 glass fabric-reinforced composite is lower than the woven 500 glass fabric-reinforced composite keeping all test parameters constant. Due to the aramid fibres having a low friction coefficient and high wear strength, and epoxy-based composite exhibit lower wear loss than polyester-based composite, the wear of the aramid fibre-reinforced composite is lower than the woven glass fabric-reinforced composites.

The surface roughness of the composite material is investigated experimentally depending on the effect of drill type, drill diameter, number of rotation and progress speed on machinability of different composite materials in drilling.

Since no cooling liquid is used, as a result of the heating occured around the hole matrix material gets softer. Due to this, in drilling processes with all drills there is matrix concentration on the edges of hole surfaces. Sudden ruptures occur between the textures on composite materials as progress speed increases and the value of surface roughness is seen to increase in all study in general.

Considering the value of surfaceroughness depending on the number of rotation, in all drilling processes made with drills it is observed that the value of surface roughness increases as the number of rotation increases. As material thickness increases in general, the surface roughness increases; naturally drilling becomes difficult and tool heat increases as well. It was previously stated that when an investigation is made among the materials, the rupture endurance of material with Kevlar support is higher than that of fiber glass material. Due to this it is seen that drilling Kevlar supported material is more difficult. Considering all results obtained as a result of the drilling processes; it is found out that using Carbide drills as the most suitable drill type in drilling operations would create better results. Besides this, considering the cost, as the price of Carbide drills is twice as expensive as TIN covered drills which have a similar performance as Carbide drills, it can be reccommendable that TIN covered drills to be used.

7. References

[1] Pihtili,H.,Canpolat,N., Investigating of Surface Roughness and Capability of Being Drilled of Different Reinforced Composite Materials, Journal of Composite Materials,Volume 43 Issue 19, 2070-2080, 2009.
[2] Pihtili, H.,N.Tosun, Investigation of the wear behaviour of a glass fibre-reinforced composite and plain polyester resin, Composite Sci. Technol. 62 (3) 367–370, 2002.
[3] N.S. El-Tayep, R.M. Gadelrap, Friction and wear properties of E-glass fiber reinforced epoksi composites under different sliding contact conditions, Wear 192 112–117, 1996.
[4] K.H. Zum Gahr, Microstructure and wear of materials, Tribology Series, Vol. 10, Elsevier, Amsterdam, 1987.
[5] B. Vishwanath, A.P. Verma, C.V.S. Kameswara Rao, Composite Sci. Technol. 44 (1) 77–86, 1992.
[6] Schaefer W.H., Christion J.L., "Evaluation of the structural behaviour of filament reinforced metal matrix composites", Technical report No. AFML-TR-69-36, vol. 3, air Force Materials Laboratory, 1969
[7] M.R. Piggot, Load-Bearing Fibre Composite, Pergamon Press, Oxford, 1980.
[8] P.K. Mallick, Fiber-Reinforced Composites: Materials, Manufacturing, and Design, Marcel Dekker, New York, 1988.
[9] N. Chand, M. Fahim, S.G. Hussain, Effect of 60Co gamma-irradiation on interface and abrasive wear of glass-reinforced polyester composite, J. Mater. Sci. Lett. 12 (1993) 1603- 1605.
[10] S.N. Kukureka, C.J. Hooke, M. Rao, P. Liao, Y.K. Chen, The effect of fibre-reinforcement on the friction and wear af polyamide 66 under dry rolling–sliding contact, Tribol. Int. 32, 107–116, 1999.
[11]Durante, S., Rutefli, G., Rabezzana, F., Aluminum-based MMC machining with diamond-coated cutting tools, Surface & Coatings Technology, 94-95 (1-3), 632-640, 1997.

d. Textile Production Engineering

13

Coated Textile Materials

Stana Kovačević, Darko Ujević and Snježana Brnada
University of Zagreb
Faculty of Textile Technology
Department of Textile Design and Management
**Department of Clothing Technology*
Baruna Filipovića 28a, 10000 Zagreb,
Croatia

1. Introduction

Top finishes began to be used already in the 18th century when fabrics were coated with linseed oil to produce oilcloth. This was the first procedure of coating several agents to the textile substrate and can be considered as the predecessor of multi-layered materials. Textile surface materials coated with chemical l structures have been developed continuously for several last decades. The basic substrate of the surface material is mostly textile fabric coated on one or both sides with one or more polymer layers. This kind of products with the basic textile material has many improved properties and multiple advantages over the classic textile material [1 - 3].

Polymer layers can be polyurethane, polyvinylchloride or polyacrylate layers. To improve their properties, appropriate additives are added: softeners, porosity-generating agents, filling materials, binders, fungicides etc. Coated polymers are applied to the textile material directly, and indirectly using paper or coagulation procedure. The constant development of the coating technique resulted in the newest achievements the result of which is the application of nanoporous polymers to the textile substrate. The use of these products is increasing and they are gaining greater importance in the clothing industry. They are especially widespread in the protective clothing where they meet all the market requirements. The design of a multilayered material is based on a target product so that it is very easy to obtain a material with desirable properties. As it is not always possible to satisfy the market with classic textile materials made of textile fibers and yarn, especially the requirements of the protective clothing, after-treatments of textile materials were introduced, either by applying polymers or some other agent or by thermal joining several different laminates. This was the introduction of the so-called composite materials which according to their structure and properties were in line with the requirements for specific purposes. With their especially good properties such as strength, durability, protection against UV radiation, wind, rain and other necessary properties they did not lose comfort, airiness, design and easy care. Multilayered materials, especially in the textile industry, contain at least one layer of the textile material which may be a woven fabric, knitted fabric or nonwoven fabric, whose only function is primarily comfort (comfort is the emotion of pleasant feeling related to tactile sensation, while pleasant is a broader term, a person can be pleasant), elasticity and airiness. Multilayered materials can be produced in different ways:

- by laminating a polymer layer to the textile surface material,
- by direct applying a polymer to the textile material,
- by indirect applying a polymer to the textile material.

The chemical compositions of polymer coatings are constantly developed and new types of polymer additives are increasingly introduced. The influence of industry globalization, the requirements of suppliers and consumers as well as new technologies change the market and expand the use of polymer coated textile materials.

The key factor for the success of polymer coating is its versatility and long durability. This chapter will deal with polyurethane (PU) polymer coats with different nanopores for coating and laminating the textile. Abrasion resistance and strength are by far higher in polyurethane in comparison to other polymers. Polyurethane has the property of good adhesion which can be strengthened by addition of cross-linking agents. Hardness or softness can be achieved by variation of polymer structures without using a plasticizer. It is also possible to reduce fragility by the impact of light [1, 4].

2. Lamination of fabrics with nanopur coating

Laminated fabric with polyurethane coating with nanopores is a multilayered composite material. Textile composite materials are composed of two or more different materials with at least one textile layer (woven fabric, knitted fabric or nonwoven material). All components composing the final product affect the properties of multilayered composites. The portion of individual components can be different which enables obtaining a composite with target properties for the predetermined purpose. Nowadays the material with woven fabric on the front side and polyurethane with nanopores on the back side is mostly used for military or police outerwear as well as for civil uses. For military purposes camouflage fabric in different shades and designs or less frequently single colored is used, while single-colored fabrics in blue shades are used for police purposes. This kind of composites have multiple advantages over the classic fabric since they are more durable and stronger, their body protection against meteorological effects (rain, wind, UV radiation), they did not lose their comfort (they are airy and have good sweat permeability), they are more resistant to abrasion and load, and they have less anisotropic properties in contrast to the classic fabrics.

Properties of composites with the woven fabric as the basis depend to a great extent on weave type, warp and weft density, yarn count and the angle of the straight line under which the load acts in relation to the warp and weft direction. The highest breaking strength is expected in the warp direction and then in the weft direction. According to the previous investigations the stress of the composite material with the woven fabric outside the warp and weft direction considerably reduces fabric breaking force. Through the action of the external force on the composite material the internal cohesion forces resist more strongly to the warp and weft direction in relation to other directions. The relaxation of the internal forces in the state of stress begins earlier if the force acts under a certain angle in relation to the warp and weft direction. This phenomenon defines fabric anisotropy which reflects on the composite material with one or more fabric layers. Load is expressed as the ratio of the internal forces acting on the area unit of the sample [5-7].

Deformations of the materials used for making clothing occur in joint areas such as elbows, knees and sitting trouser part due to multiaxial loading. After longer loading in the places mentioned deformations occur expressed as irreversible elongation and baggy appearance of the garment at this place. Using composite materials reduces this phenomenon in relation to standard fabrics, especially by use of nanopur coating on the back side of the fabric because it increases durability and strength of the composite material, while on the other hand it decreases anisotropy.

To join the fabric with other materials a binder for better adhesion or only thermal joining is used. Figure 1 shows examples of laminating a fabric to a fabric or fabrics with other materials.

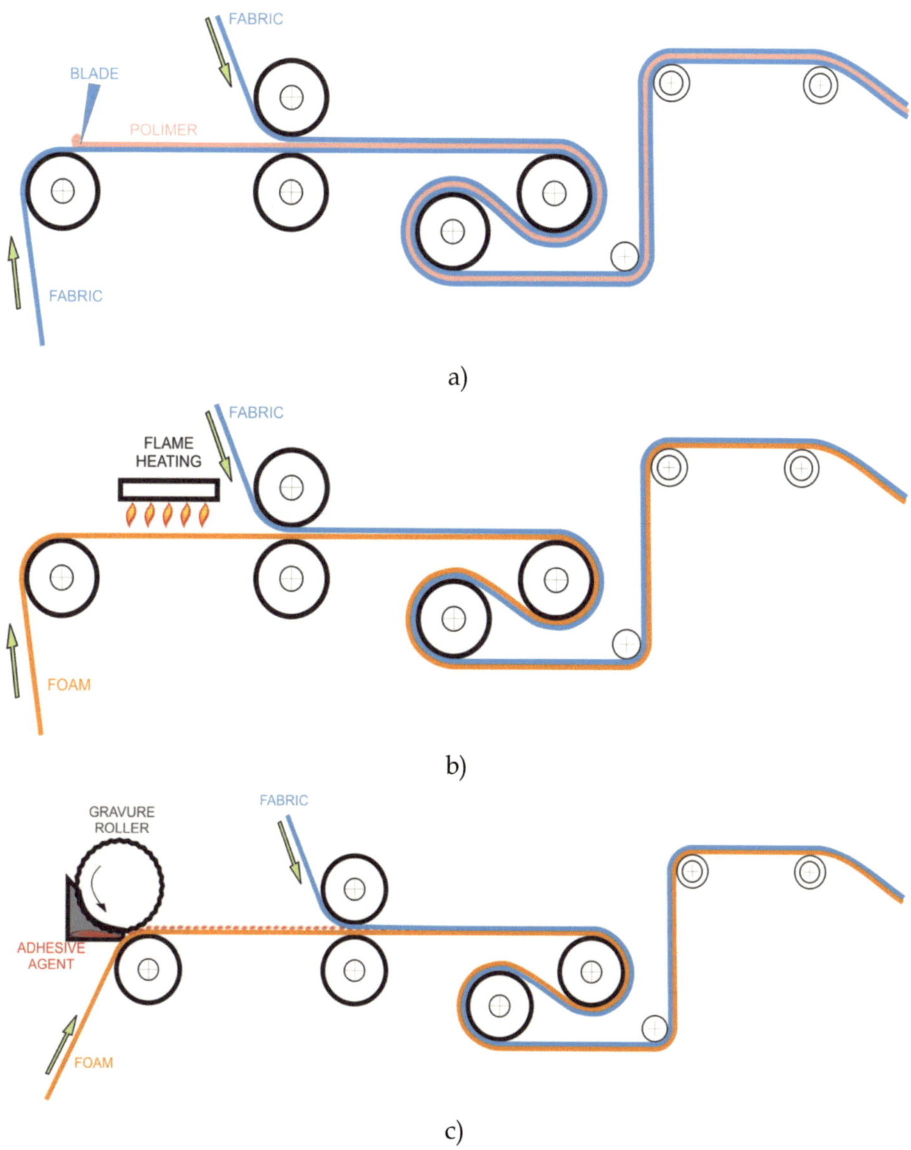

Fig. 1. Fabric lamination
a) lamination of two fabrics using a binder, b) foam lamination of the fabric using thermal joining, c) foam lamination of the fabric using a binder

3. Polymer coating on the textile material to produce the so-called artificial leather

Polymer coating on the textile material provides new properties of the fabric. Polymer-coated textile materials have a wide range of application, from the textile industry to technical textiles. The advantage of this kind of textiles in the clothing industry is their water impermeability, while on the other hand they are air and water vapor permeable, resulting in good properties of concurrent protection and comfort. Depending on the use, nonwoven, woven or knitted fabric is used as coating substrate. As far as their properties are concerned, new man-made fibers or materials can be similar to natural or even they can surpass the properties of natural fibers, resulting in an increasing application for coated materials. Nowadays companies are adapting to new challenges, they are trying to expand the domestic and world market; thus, it is understandable that great efforts are being made to develop new products and to improve the properties of the available ones. As the substrate for coated materials synthetic (perhaps better to say artificial than synthetic) materials are frequently used due to their relatively high strength and good abrasion resistance. Artificial materials possessing high elasticity, airiness and appropriate strength are used as the substrate for coated materials. They are mostly used for protective and sports clothing for children and adults.

It is important that the properties of the substrate meet the requirements of the final product. The following substrate parameters are of special importance:

- good mechanical properties such as elasticity, elongation at break, strength, frictional resistance,
- type of yarn: filament and texturized yarn, where spun yarn exhibits good adhesion because of protruding fibers which excellently join with the coat, but when making thin polymer materials these short hairs or cut filaments can penetrate the surface, causing water permeability of the fabric,
- dimensional stability,
- adhesion, absorption – the substrate must have good binding properties so that the coating could penetrate the substrate to a sufficient extent, and binding characteristics could improve by addition of the binding agent either in the pretreatment of the substrate or in the PU coat,
- pretreatment – agents such as softeners and dyes can negatively affect the subsequent production procedure; several treatment types, such as water repellent and antibacterial treatment can improve the properties of the final product,
- thermal stability – PU coating requires high temperatures to form the film, and thus the substrate have to endure high temperatures,
- uniformity of the substrate – uniform substrate thickness is a particularly important feature for subsequent treatments [4, 8, 9].

Polyurethane is mostly used for coating the textile material. The coating procedure can be direct or using siliconized paper. When polyurethane is coated directly, the polymer is coated using special coating blades indirectly to the textile material (Fig. 2a). When coating is indirect, the polyurethane polymer is coated first to the paper, and then is laminated with the substrate and the textile material respectively (Fig. 2b). When it is first coated to the paper, it can be in several layers. After each coating, the polymer is dried and cooled down. Upon completion of the coating procedure, the paper is separated from the finished material. The paper returns to the machine entry and can be used for further coatings, approximately from 8 to 10 times. This method is applied for low density fabrics so that the coated polyurethane remains on the back of the textile material and cannot penetrate to the front side. By adhesion to the fabric and by partial penetration into the fabric structure the PU coat remains permanently bonded and fused into a compact material. The composition of the layers in case of indirect PU coating does not need to be always the same, neither in the composition nor in the coating thickness. Both the material composition and the coating thickness and the number of coating procedures depend on the application of the final product.

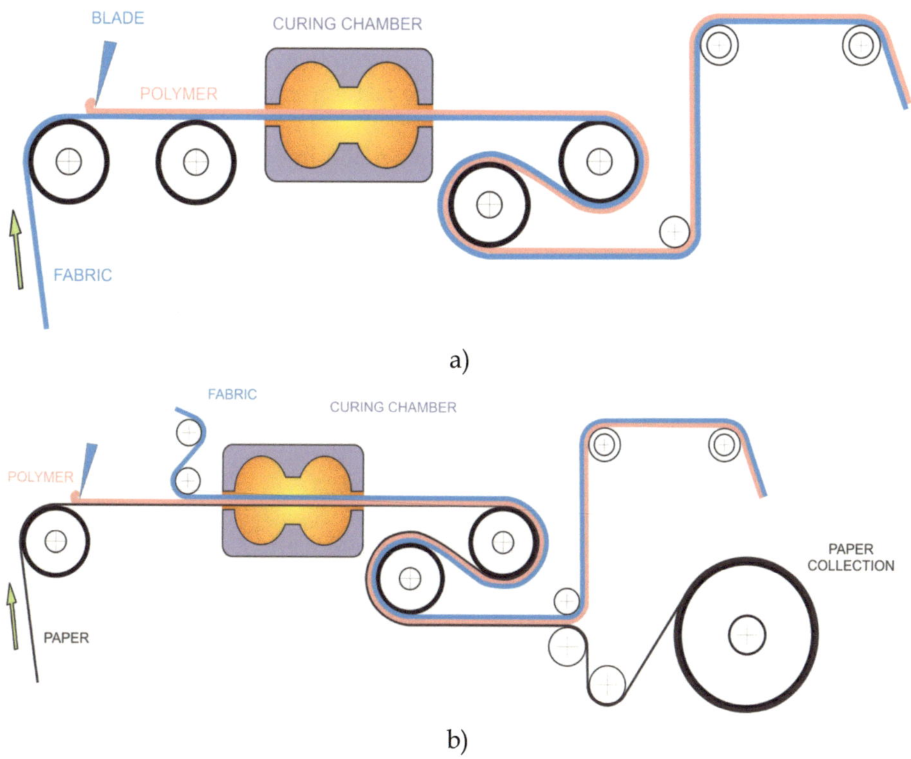

Fig. 2. Schematic of the process of polyurethane coating
a) Direct coating with one passage, b) Indirect coating using silicone paper with one coating passage

3.1 Polyurethane (PU) coating

The selection of polymers is very important to obtain desirable properties of the finished product, and the coating composition is determined according to the application of the finished product.

The coating consists of the basic polymer and additives. In the selection of the basic polymer the properties are as follows: thermo plasticity, mechanical properties of polymers, possibility of film formation, stiffness, good adhesion, abrasion resistance, heat, water and air conductivity, resistance to solvents and hydrolysis, resistance to UV radiation, melting point etc.

The basic polymer is mostly polyurethane that may be strong and rigid, soft and elastic. Polyurethanes belong to the group of very durable plastic materials. The main property of

polyurethane is its wide application. It can be coated to textiles, leather, in solution, dispersion, with a low solvent content or without it, as granules or powder. Softness or hardness can be obtained by varying polymer structures.
Polyurethane has good washproofness and cleaning resistance, good adhesion to the fabric, good durability at low temperatures, it is possible to use it without softeners, it has good viscosity and abrasion resistance, at the same time it has a pleasant and soft touch, a low specific mass, resistance to oils and fats. Polyurethane can be coated to textile materials in more ways:

- as a two-component polyurethane with isocyanate cross-linking,
- as one-component aromatic or aliphatic polyurethanes with chemical reactions performed in the production, and during the coating and drying process it is linked between evaporation chains and solvents,
- as a one-component product that enables dispersion in water and is environmentally friendly
- as a solid product with possible coating of greater quantities in each coating passage,.

3.2 Additives

Additives improve properties of coating polymers such as:

- softeners imparting better flexibility and softness of the finished multilayered product, and they enable a more uniform distribution of the polyurethane paste (Vithane),
- cross-linking agents and binders that improve the bond between the textile material and the coated polymer (Larithane CL 1, Larithane MA 80, Toulen),
- antimicrobial agents (Sanitized),
- light fastness agents (Tinuvin),
- various pigments for dyeing the polymer coating (pigments).

To achieve a good material quality, it is very important to dose the solvent regularly in the binding coating. Too small a quantity of the solvent in the binding coating causes the swelling of the binder instead of its dissolving, resulting in poor bonding of the material to the substrate. On the other hand, too great a quantity of the solvent in the binding coating causes too rapid dissolving of the binder, resulting in too great penetration of the PU coating into the substrate. The final result is too great material rigidity [5].

4. Properties of coated textile materials

Properties of coated textile materials primarily depend on their application. Nowadays modern technologies, optimization of the conditions of the production process and use of certain agents and recipes enable making a target product which will meet all the requirements. Since it is not possible to use the classic textile for many technical purposes, excellent properties are obtained by combination with other substances which are coated in the form of paste or laminated to the material. By use of the value-added material, nanoproducts and modern technology the use of textile materials has been enhanced several times. For the purposes of this study samples of the woven fabric with nanopur coating on the back side were chosen to test basic properties. They are used for police and military uniforms. The fabrics have the same construction parameters in different colors and different properties of the nanopur coating (Tables 1 and 2).
Likewise, samples of the artificial leather with polyurethane coating on the knitted fabric, namely with different properties of polyurethane in two colors (Table 3), will be considered.

Woven fabric parameters	Yarn/woven fabric			Laminated fabric		
	Green	Blue	Camouflage (printed)	Green	Blue	Camouflage (printed)
Fabric weave	Plain weave + Rep weave 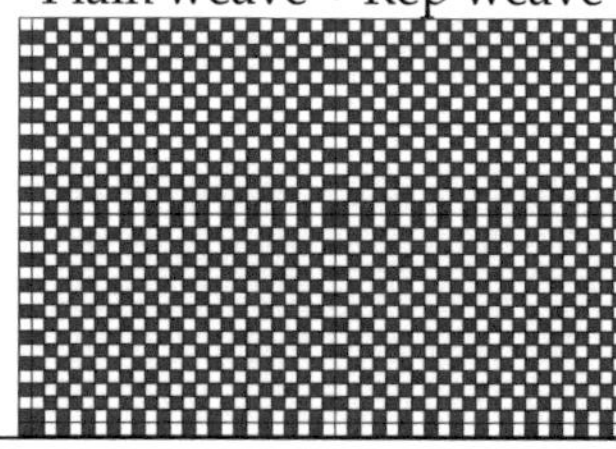					
Thread density/10 cm warp / weft	360 / 207					
Count of warp/weft (tex)	17x2 / 40					
Raw material composition of warp/weft	(PA 6.6/ cotton 50/50) / (PA 6.6/ cotton 50/50)					
Warp yarn twist single/plied/weft (turns/m)	938/622/673					
Fabric thickness + nanopur coating (mm)	0.39	0.39	0.39	0.41	0.41	0.41
Mass (g/m^2)	217	220	211	266	271	259

Table 1. Basic parameters of fabrics and yarns in fabrics

Laminate	Color	Nanopur coating	Properties of laminated fabric
1	Green	Nanopur FR 30 - 30 g/m^2	Not made water-repellent
2	Dark blue		Made water-repellent
3	Camouflage		Made water-repellent

Table 2. Properties of the laminated fabrics and polyurethane nanopur coating

4.1 Anisotropy of multilayered materials

Coated textile products either as artificial leather or as laminates assume properties of the materials they are made of. Since they are partially made of the textile that is in its properties mostly anisotropic, coated material as a whole is also anisotropic, meaning that the coated material behaves differently in different direction when stressed.

Two features are differentiated related to material load, namely: determination of the dependence of strength on the direction of force application in relation to the directions of the body structure known as anisotropy and assessment of the strength of anisotropic bodies in the case of complex states of stress [10-14].

Anisotropy of woven and knitted fabric is outstanding. Anisotropy is reduced by addition of a coating agent, but it is not eliminated. Thus, the properties of coated materials do not only depend on the components, but also on the direction of load.

In woven fabrics laminated with nanopur coating outstanding anisotropy is observable in all three samples (Figs. 3 and 4). The differences in the samples are in fabric color and hydrophobicity (Table 2). An exceptionally high anisotropy is present in the fabrics without nanopur coating. The highest anisotropy of breaking forces is visible in the camouflage fabric, and the lowest one in the single-colored, blue fabric. The highest sensitivity of all the fabrics is when they are loaded at an angle of 15°, then at an angle of 75° followed at angles of 30° and 60°. The essential point is that the fabric at an angle of 45° is even more strongly than in the weft direction. This means that the threads in the binding points are strongly bound and they do not allow shearing at an angle of 45°. Since the fabrics were woven in the combination of plain and rep weave with a relatively high density in the warp and weft direction, their compactness is outstandingly high. The thermal joining of the nanopur foil

Sample	Color	Mass (g/m^2)	Thickness (mm)	1st coating	2nd coating	3rd coating	Knitted fabric	Property of artificial leather
I	White	185	0.58	Larithane AB 4228	Larithane AB 4228	Ucecoat ID 9229	Plain jersey, mass: 90 g/m^2, raw material composition 100% PA 6.6, thickness: 0.45 mm	Standard recipe
Ia	Blue	175	0,56					
II	White	188	0.55	Larithane AB 4228 + Lomaflam TDCP	Larithane AB 4228 + Lomaflam TDCP	Ucecoat ID 9229		Additional flameproof treatment in the 1st and 2nd coating
IIa	Blue	174	0.55					
III	White	187	0.50	Larithane AB 4228 Larithane AB 4228	Larithane AB 4228 Larithane AB 4228	Larithane BTH 146 + Larithane CL 16		In the 3rd coating polyurethane was used, it has greater water vapor permeability
IIIa	Blue	163	0.48					

Table 3. Basic parameters of polyurethane coating, knitted fabric and artificial leather

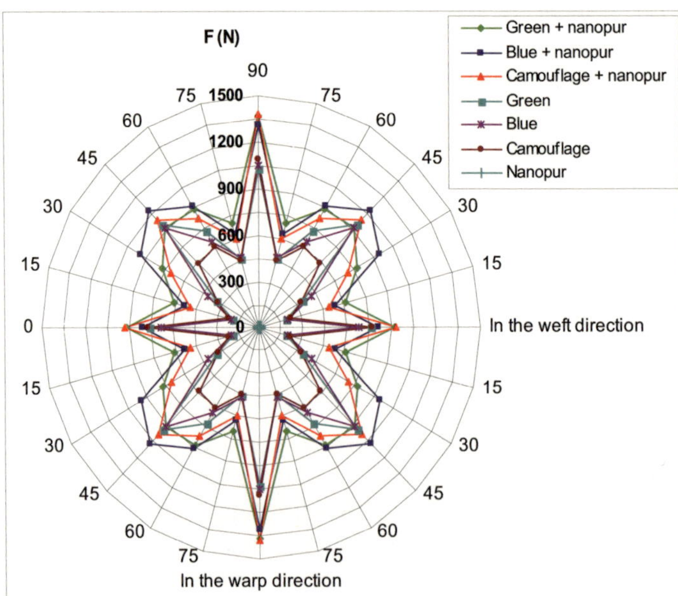

Fig. 3. Breaking force in different directions

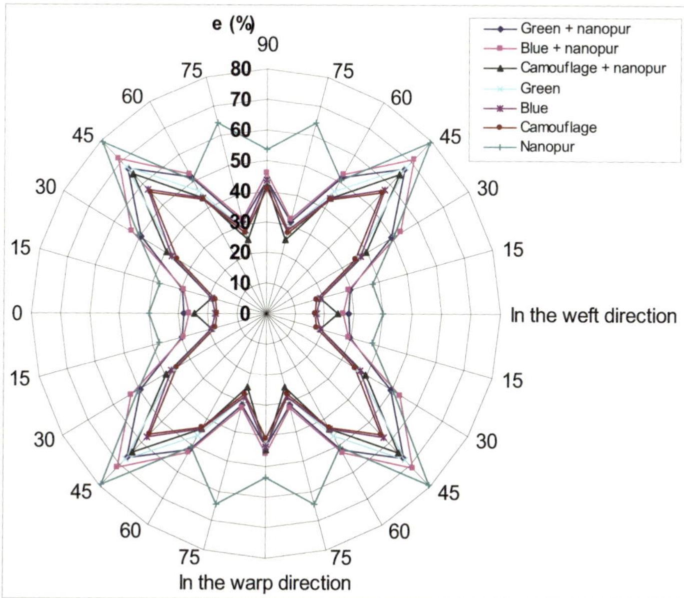

Fig. 4. Elongation at break in different directions of the fabric samples with nanopur coating

to the back side of the fabric increased material strength and reduced fabric anisotropy. However, anisotropy is present and follows the course of fabric anisotropy. The nanopur foil does not have outstanding strength and is almost invisible in the polar diagram. The breaking forces of the samples containing the fabric mostly followed the course of breaking forces, except in the weft direction where elongation at break is the lowest. The nanopur foil has an outstandingly high elongation at break which is in most cases higher than in the samples containing the fabric and its anisotropy is lower. The highest elongation at break occurs in all samples at an angle of 45°, and the lowest in the weft direction.

In the case of artificial leather the material strength is changed by altering the composition of coating (Fig. 5). Anisotropy of all materials in all testing directions is noticeable. Nevertheless, the course of the curve of breaking forces is almost identical in all three samples, and there is no difference among the samples. The greatest sensitivity of all materials is at angles of 15° and 30°. The highest breaking forces are in the warp direction or in the sample length, then at angles of 75° and 60°. In the weft direction or in the sample width breaking forces are similar as in the direction of 45°. The knitted fabric has a little lower breaking force than the artificial leather in all testing directions, meaning that the breaking force did not increase substantially by coating the polyurethane layer to the knitted fabric. Elongation at break is also different according to testing levels, and there is no particular difference among the samples (Fig. 6). The highest elongation at break is observable in the weft direction in all samples, and it is the lowest at angles of 60° and 75° and in the warp direction. The anisotropy of the artificial leather assumed the anisotropy of the knitted fabric, when observing breaking forces and elongation at break. The samples of artificial leather and knitted fabric for testing breaking force and elongation at break were prepared in dimensions 200x50 mm and tested on the Statimat M tensile tester made by Company Textechno in accordance with the standards ISO 13934-1:1999; EN ISO 13934-1:1999.

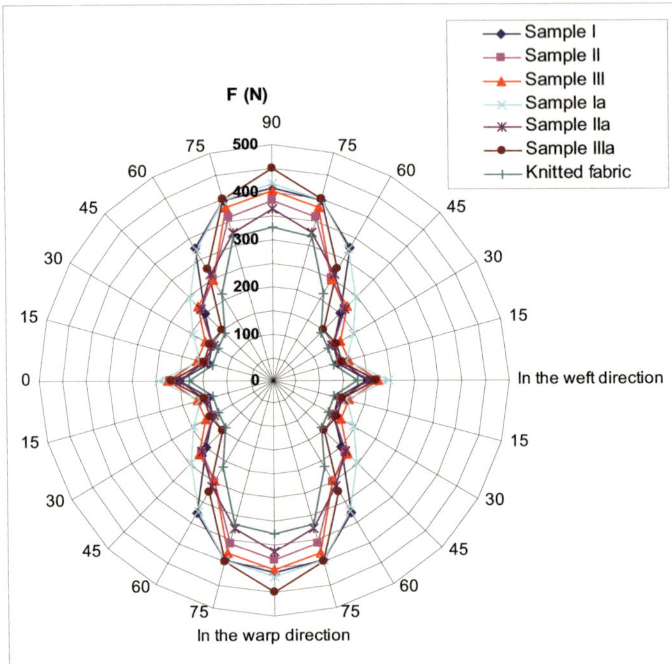

Fig. 5. Breaking force in different directions of the artificial leather (polyurethane coating + knitted fabric) and the knitted fabric before PU coating sample I, II, III - white samples defined according to Table 3; Ia, IIa, IIIa - blue samples defined according to Table 3

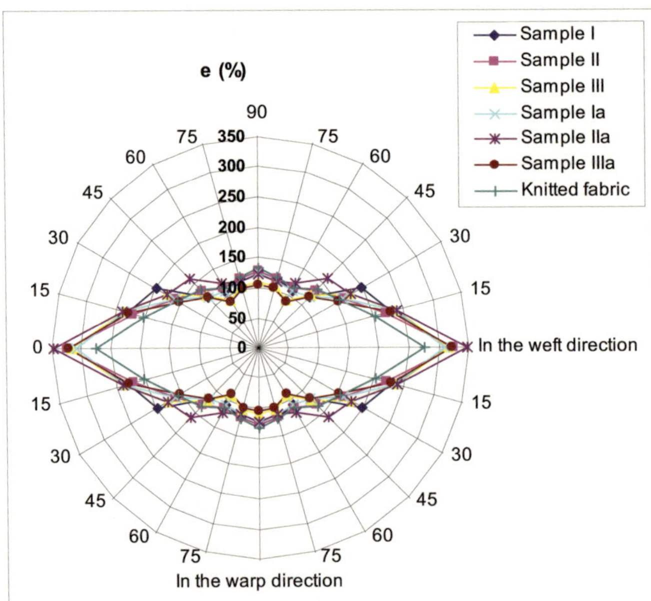

Fig. 6. Elongation at break in different directions of the artificial leather (polyurethane coating + knitted fabric) and the knitted fabric before PU coating sample I, II, III - white samples defined according to Table 3; Ia, IIa, IIIa - blue samples defined according to Table 3

4.2 Abrasion resistance

To test the abrasion resistance of laminated fabrics with nanopur coating, the determination of mass loss by the Martindale method after 5,000 and 10,000 cycles according to the standard ISO 12947-3:1998+Cor 1:2002; EN ISO 12947-3:1998+AC 2006 was used. According to the results obtained, a certain difference between the samples of the laminated fabrics and the artificial leather is noticeable. The lowest loss of mass records the blue sample followed by the green sample, while the printed or camouflage sample records the highest difference (Fig. 7).
In the artificial leather with knitted fabric on the back the loss of mass is also different (Fig. 8). The first white sample records a slightly lower loss of mass than the blue sample, while the other two samples in blue color record a noticeably higher loss of mass. This means that the pigments applied in the artificial leather affect the coating in such a way that they reduce abrasion resistance. The white coating has a lower loss of mass than the coating dyed with blue pigments.

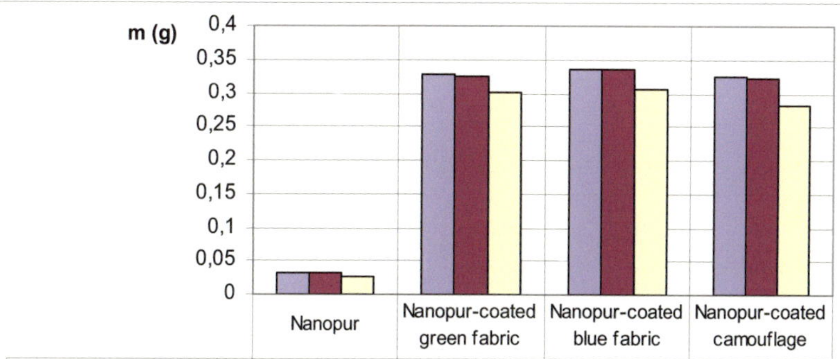

	Nanopur	Nanopur-coated green fabric	Nanopur-coated blue fabric	Nanopur-coated camouflage
Mass before abrasion	0,0332	0,327	0,337	0,325
Mass after 5,000 cycles	0,032	0,326	0,336	0,323
Mass after 10,000 cycles	0,028	0,301	0,306	0,282

Fig. 7. Mass loss of the nanopur-coated laminated fabrics

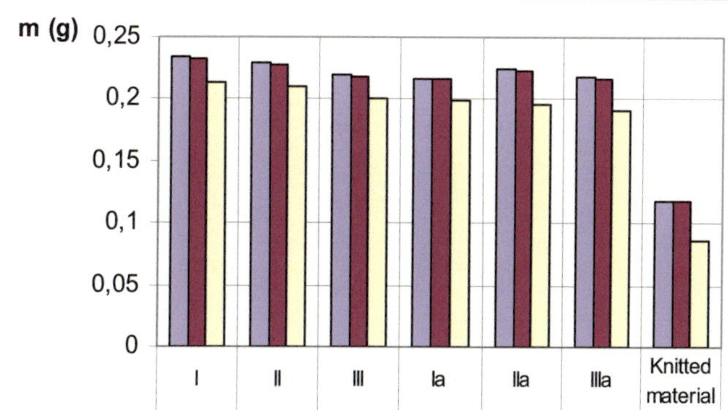

	I	II	III	Ia	IIa	IIIa	Knitted material
Mass before abrasion	0,2335	0,23	0,219	0,21655	0,22385	0,2174	0,11825
Mass after 5,000 cycles	0,23325	0,22845	0,2186	0,216	0,22315	0,2164	0,1171
Mass after 10,000 cycles	0,213	0,21	0,201	0,1983	0,1955	0,191	0,0855

Fig. 8. Mass loss of the coated textile materials

4.3 Bursting strength

The determination of bursting strength with a steel ball was carried out in accordance with the standard AN 12332 1:1998, ASTM 3787 using a strength tester made by Apparecchi Branca S.A., Italy. On the basis of the results obtained there is a difference in bursting strength and elongation at break among the tested fabric samples. The nanopur-coated blue fabric has the highest bursting strength, while the camouflage fabric has the lowest values (Fig. 9). The nanopur-coated blue fabric having the highest bursting strength has the lowest anisotropy (Fig. 10).
Differences in bursting strengths are also visible in the artificial leather (Fig. 10). Samples III and IIIa have the highest bursting strength, while samples I and Ia have the lowest values. It is essential to emphasize that white samples (I, II and III) have higher bursting strength and elongation at break than the blue ones (Ia, IIa, IIIa), which is not the case in testing bursting strength using strip test method (Fig. 5).

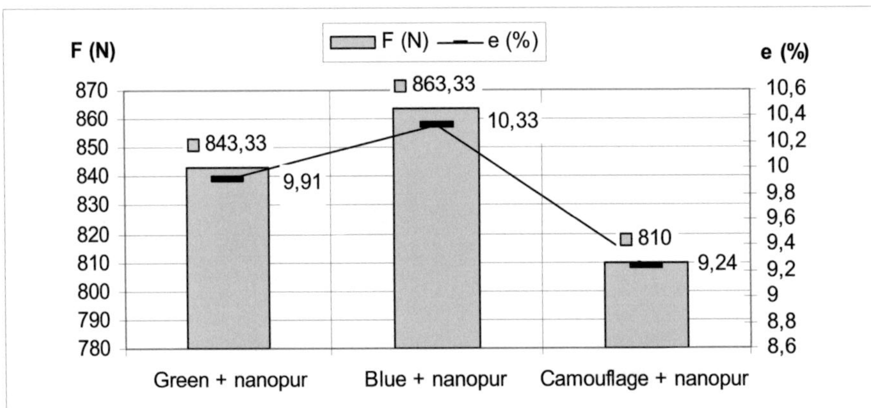

Fig. 9. Bursting strength of the artificial leather

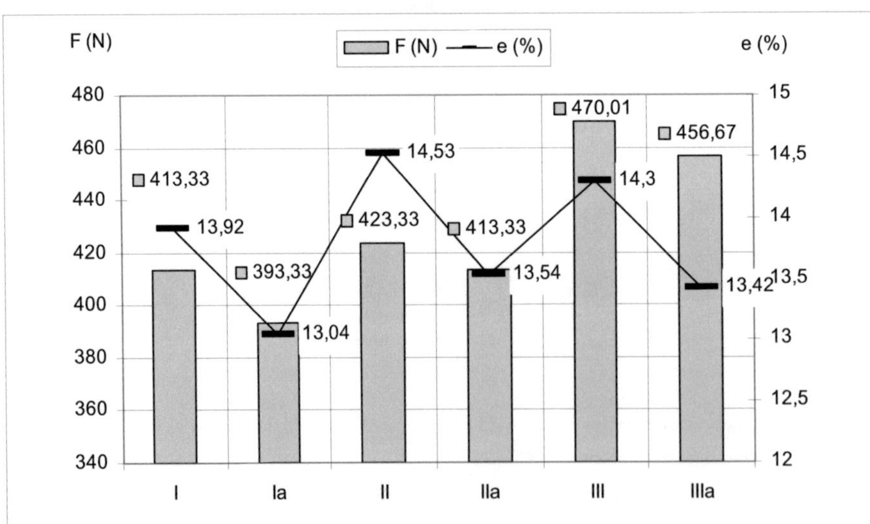

Fig. 10. Bursting strength of the nanopur-coated laminated fabrics

4.4 Thermal resistance

The determination of thermal resistance was performed in accordance to the standard ISO 11092 on the Sweating Guarded Hotplate made by MTNW, USA. According to the results obtained for the laminated fabric samples there is a certain difference (Tab. 4). The white fabric exhibited the highest thermal resistance before and after lamination, while the camouflage fabric exhibited the lowest thermal resistance. In the case of the artificial leather there is also a certain difference in thermal resistance among the samples. Flameproof samples (III white and IIIa blue) have the highest thermal resistance, while the samples with higher water-vapor resistance have the lowest thermal resistance.

No.	Sample designation	Measured value Rct (R_{ct} - m^2KW^{-1})	
		$\overline{X}$	CV (%)
1	Knitted fabric	0,0053	6,71
2	Green	0,0111	5,43
3	Blue	0,0127	7,36
4	Camouflage	0,0115	6,88
5	Green + nanopur	0,0101	4,31
6	Blue + nanopur	0,0113	4.06
7	Camouflage + nanopur	0,0103	5,62
8	Nanopur	0.0091	3,55
9	I	0,0134	4,71
10	Ia	0,0200	3,20
11	II	0,0143	3,70
12	IIa	0,0220	3,75
13	III	0,0105	4,88
14	IIIa	0,0113	3,80

Table 4. Thermal resistance using the sweating guarded hotplate

5. Conclusion

On the basis of the performed theoretical considerations, design of the coated textile products and corresponding properties, it is possible to make a target product which will meet all requirements. In the case of the observed multi-layered textile composites, it is necessary to define material anisotropy in the weakest directions, which are also the most responsible for deformation. In these places deformations in form of changes in material dimensions per unit of length are created and the so-called baggy shape results.

By use of woven fabrics as the basic layer of textile structured multi-layered composites in laminating a relatively high anisotropy occurs which can be reduced by polymer coating. However, due to an exceptionally good strength in the warp and weft direction and its abrasion resistance, breaking and good physiological properties its presence will be relatively widespread in relation to knitted and nonwoven fabrics. The use of the fabric on the composite front side provides great design possibilities such as printed fabric for camouflage military clothing etc.

By coating polyurethane paste to textile materials, materials known as artificial leather is obtained. They occupy an important place on the market. Artificial leather is unthinkable without the textile substrate. In most cases these are woven or knitted fabrics which transfer their properties to the final properties of artificial leather. Since they are materials mostly used as outerwear or upholstery fabrics, their physiological properties are essential. Air, water and water vapor permeability, their strength and durability depend on the properties of individual properties of coated materials and final products. Since structured multilayered materials consist of different materials and various binders, besides material comfort it is important to pay great attention to their compatibility in different conditions. The target product to meet market requirements can be produced by appropriate selection of recipes for polymer coating, and by determination of construction parameters of the textile fabric as well as raw materials and production conditions.

Subsequent investigations should include multiaxial testing of a series of models with different woven and knitted fabrics in order to reduce anisotropy, especially of the materials being less strong and having higher elongation. A change in polymer coatings and their properties related to textile materials affect final properties of multilayered materials. Likewise, adding a target polyurethane coating and after treatment, even the selection of color can provide a target product with appropriate properties.

6. References

[1] I. Soljačić: Textile Coating, Tekstil 42 (12) 673-686 (1993.)

[2] Y.E.El Mogahzy: Engineering textiles, Integrating the design and manufacture of textile products, The Textile Institute, Woodhead Publishing Limited, Cambridge England, 2009.

[3] W. Fung and M. Hardcastle: Textiles Automotive engineering, The Textile Institute, Woodhead Publishing Limited, Cambridge England, 2001.

[4] M. Skoko: Investigations of Properties and Multiaxial Strength and Deformations of Coated Textile Fabrics, Tekstil, 47 (7) 339-344 (1998.)

[5] P. Durst: PU Transfer Coating of Fabrics for Leather like Fashion Products, Journal of Coated Fabrics, 14, 227-241 (1985.),

[6] V. Lasić, M. Srdjak, V. Mandekić-Botteri: The Impact of Testing Angle on the Assessment of Mechanical Properties of Weft-knitted and Maliwatt stitch-Bonded Fabrics, Tekstil 50 (11) 549-557 (2001.)

[7] D. Jakšić: Possibilities of Determining Porosity of Textile Fabrics, Tekstilec, 37 (7-8) 221-228 (1994.)

[8] I. Frontczak-Wasiak: Measuring Method of Multidirectional Force Distribution in a Woven Fabric, Fibres & Textiles in Eastern Europe, 12 (3-5) 48-51 (2004.)

[9] M. Skoko: Investigation of the Properties with Multiaxial Strengths and Deformations of Coated Fabrics, Tekstil, 47 (7) 345-349 (1998.)

[10] M. Skoko: Contribution to Investigations of Stresses and Deformations of Particularly Loaded Textiles for Particular Purposes, Tekstil, 35 (6) 403-410 (1986.)

[11] I. Frontczak-Wasiak, M. Snycerski, M. Cybulski: Isotropy of Mechanical Properties of Multiaxial Woven Fabrics, 5th World Textile Conference Autex, 27-29 June 2005. Portorož, Slovenia

[12] A.K. Sengupta, D.De, and B.P. Sarkar: Anisotropy in Some Mechanical Properties of Woven Fabrics, Textile Research Journal, 42 (5) 268-271 (1972.)

[13] A.K. Sengupta, D.De, and B.P. Sarkar: Anisotropy of Breaking Load of Woven Fabric, Textile Research Journal, 41 (5) 277-278 (1971.)

[14] W. Schröer: Polyurethane Coating of Textile Materials, Tekstil, 38 (3) 147-154 (1989.)

The results shown in the paper resulted from the scientific program (Advanced Technical Textiles and Processes, code: 117-0000000-1376; Anthropometric Measurements and Adaptation of Garment Size System, code: 117-1171879-1887) conducted with the support of the Ministry of Science, Education and Sports of the Republic of Croatia.

14

Porosity of the Flat Textiles

Danilo Jakšić[1] and Nikola Jakšić[2]
[1]*University of Ljubljana*
[2]*Turboinštitut*
Slovenia

1. Introduction

Flat textiles play an important role in clothing and as component of composites. Besides of that, it would be difficult to imagine the processes of filtration without the flat textiles. They can be divided into three main groups: woven fabrics, knitted fabrics and non-woven textiles if their design is disregarded. The quality of the flat textiles can be defined by many parameters. In this chapter, we will focus on one of them - the porosity.

What is porosity? How could we define it? Way and where is it important? Usually we have more questions than answers. The porosity in flat textiles is defined as a void part of the textile's full volume. The full textile's volume is usually occupied by a mixture of three components: fibres, air and water. The part of the volume that is occupied by fibres is constant. On the contrary the portion of the volume that is occupied by water may vary considerably. For instance, there is no water in the absolutely dry flat textile, and in the absolutely wet condition, air is replaced by water. The water content in the textiles plays very important role in the clothing insulation due its effect on the clothes' thermal resistance. The coefficient of the thermal resistance of air is much larger than the coefficient of the fibres or water. Hence, it is extremely important to keep the clothing dry in a cold weather.

The coefficient of the thermal conductivity scales in the inverse manner with the coefficient of the thermal resistance and both are frequently used in the literature. The coefficient of the thermal conductivity is not solely influenced by the porosity in terms of its water content in the still weather, but also by the moving air - windy weather - that can penetrate the pores in the flat textiles.

The porosity can be defined by several parameters. The pore distribution is an important parameter and it is seldom well known. Even the average pore size is difficult to estimate. Yet, our aim was to describe the pore distribution by it attributes: average pore diameter, number of pores and distribution of pore diameters in a histogram form. The method that is capable of providing us with all these data is described in this chapter. The surface of the flat textile open to the flow of the fluid is of the interest as well. Additionally, the velocity of the fluid flow through the flat textile, driven by the pressure difference between textiles surfaces, is important when analysing the process of filtration or the properties of clothing. In the latter case the fluid is air.

Clothing has certainly a specific role in our life. It protects us against cold, wind, rain and sun radiation. Clothing must be suitable in the dry and wet, cold and hot weather and in the windy weather. Only one set of clothing can't be enough for all these situations – we do not have the universal clothing. Instead, we use clothing composed in layers. Problem may arise as energy or heat is produced by our metabolism. Heat production depends on the intensity and sort of the activity. Sweating is the body response on its own temperature rise and it is wetting the clothes. The thermal resistance coefficient of the wet clothing is smaller that the dry one. The similar effect can be observed when wind velocity increases. The influence of the temperature, water and wind velocity on the thermal coefficient is show in equation (1) for a flat surface (Jakšić, 2004).

$$R_s = \frac{d_c}{\lambda_0(1 + k_T \Delta T_c + k_w \Delta G_c) + b c_a \gamma_c V_a d_c} + \frac{0.0429}{0.4 + 2.0\sqrt{v}} \qquad (1)$$

where R_s stands for the clothing's coefficient of thermal resistance, d_c for the thickness of the clothing, λ_0 for the coefficient of thermal conductivity of clothing in the standard environment, k_T for the coefficient of direction curve temperature - the thermal resistance of the clothing, ΔT_c for the difference of the clothing temperature regarding the temperature in the standard environment, k_w for the coefficient of direction curve for the content of water in clothing - the thermal resistance of the clothing, ΔG_c for the change of the water content in the clothing, b for the coefficient, which describes the tightness of the clothing (if the value is 1, the air flows through the surface of clothing layers and not through the holes in the clothing, c_a for the specific heat of the air, γ_c for the specific mass of the air, V_a for the volume of the air which penetrate through the clothing due to the velocity v of the air flow.

The use the flat textile in the composites and as the geo textiles, the diameters of pores are also very important. For example, in a composite structure the diameters of pores must allow resin a good connection between the layers of the flat textile. The pores must simply be large enough to allow resin penetration. On the other hand, the diameters of pores in woven fabrics used as geo textiles must be small enough to effectively filtrate earth particles. Pores in the woven fabrics are voids between threads of the warp and weft and the light can go directly through. This sort of material is not suitable for use in the masks destined for protection against viruses. Viruses are extremely small and we can't get pores in textiles to be smaller. Hence, non-woven fabrics are used for the masks design in spite of the fact that the pores are many times larger than the viruses. The walls of pores are defined by fibres, and not by treads, in the non-woven fabrics. Pores change direction many times from one surface of the non-woven fabrics to another. The probability for the aerosol flowing through such a pore to deposit fine solid or liquid particles including bacteria and viruses on the fibres is extremely high, even 100% for some limited time. The micro fibres, which diameter is about 1 to 2 micrometers, must be used for this purpose. The porosity of the non-woven fabrics is high enough to enable us to breathe normally. The protection against microbes and viruses are tested in the special laboratories. However, if we could measure the composition and the porosity of masks, the number of those tests would be reduced. It would be enough to estimate porosity only, but it is not so strait forward without a suitable method.

We have developed a method for the assessment of the parameters of the porosity in all flat textiles. The method is relatively simple and efficient at the same time. The apparatus for measuring the airflow through a flat textile sample due to the pressure difference is needed. The application software has been developed on a basis of the method's algorithm.

2. Methods for estimating the porosity of the flat textiles

There are several different methods available for the assessment of the parameters of porosity, such as: geometrical methods (Matteson & Orr, 1987), (Piekaar & Clarenburg, 1967) and (Dubrovski & Brezocnik, 2002), liquid intrusion methods (Dosmar et al., 1993), (Rucinski et al., 1986) and (Rebenfeld & Miller, 1995), liquid extrusion methods (Miller & Tyomkin, 1986a), (Miller & Tyomkin, 1986b) and (Rushton & Green, 1968), liquid through methods (Hssenboehler, 1984), etc. Some of them can only give truly very approximate values, which may not be accurate enough. On the other hand some of them are not capable of estimating all the relevant porosity parameters.

A lot of work has been done over the years to overcome the mentioned shortcomings. We have developed a method for estimation of the parameters defining the textile's porosity. The method is suitable for all types of flat textiles: woven fabrics, knitted fabrics and non-woven fabrics (Jakšić, 2007). We have named it J-method after the first letter of authors' surname.

Main feature that set J-method apart of the other methods is that J-method is also suitable for the non-woven fabrics.

3. Theoretical bases for J-method

A flat textile product gets wet and the fluid pushes the air out of the product - especially from voids, if the product is immersed into a fluid. These voids are formed out of pores between fibres in the non-woven fabrics, as well as out of pores between threads in the woven and knitted fabrics. The pores between the threads of the warp and weft in the woven fabrics, figure 10a, are the most interest from the practical point of view. The pores between the threads of the warp and weft are well defined in textile fabrics made of monofilament and of some multifilament yarns. The pores can be counted on a defined area in such cases. This is not the case with the fabrics made of wool yarn where some fibres jut out of the yarn and thus cover the pores. A pore is thus divided into several smaller pores. It is thus impossible to ascertain the exact number of pores in the non-woven fabrics.

The porosity parameters that are needed in most of the cases are: the pore size distribution, the average hydraulic pore diameter, the open area for fluid flow and the air volume velocity as a function of the air pressure. The method under consideration is able to provide mentioned parameters with sufficient accuracy.

The method is based on selectively squeezing the fluid in the pores out of the wet fabrics by air pressure and on the presumption that a pore is approximated with a cylinder. The selectivity is assured by the fact that the fluid is squeezed out of the pores with a certain hydraulic diameter providing that the precise value of the air pressure is applied. The air pressure is inversely proportional to the hydraulic diameter of the pores (see equation (3)). Latter is important, while the process of squeezing out the fluid contained in the pores of the wet fabrics is under examination. There is always a small amount of the fluid that remains at the edges of pores if such edges exist.

The pore cross-section is approximated by a circle of the diameter d. The parameter d is the hydraulic diameter of the pore. It is defined by equation (2) where f denotes the surface of the cross-section of the pore, o the circumference of the cross-section of the pore w, the width of the pore cross-section and l denotes the length of the pore cross-section.

$$d = \frac{4f}{o} = \frac{2wl}{w+l} \tag{2}$$

The pressure difference p_i between the opposite surfaces of the flat textile, equation (3) and (4), results in squeezing the fluid out of the pores, which diameter is equal or larger than d_i. The fluid is characterised by the surface stress α.

$$d_i \geq \frac{4\alpha}{p_i} \tag{3}$$

$$p_i = \rho g h_i \; ; d_i \geq \frac{4\alpha}{\rho g h_i} \tag{4}$$

The fluid is first squeezed out from pores, which have the largest hydraulic diameter. The flow of air will establish itself through these pores that are now empty. The volume flow rate of air through the flat textile can be described by equation (5)

$$V_i = Ap_i^b = Pap_i^b = Pv_i \tag{5}$$

where V_i stands for the air volume flow rate through the sample at the air pressure p_i, A for a regression coefficient when fitting equation (5) to the measured dry data, P for the open surface, v_i for the linear air flow velocity, a for the coefficient and b for the exponent. The parameters a and P are unknown and they have to be estimated as well. The solution of the problem is enabled by equation (6) by putting the velocity v_i in the relationship with the air pressure p_i. The value for the exponent b is bounded between 0.5 and 1.0. The air volume flow rate depends on the degree of porosity of the flat textile fabrics and the air pressure difference between the two surfaces of the fabrics. Larger porosity means larger air volume flow rate through the fabrics at the constant pressure. The last part of equation (6) holds in the ideal circumstances, when all of the energy dissipation mechanisms are neglected.

$$v_i = a_0 p_i^b = 1.28 p_i^{0.5} \tag{6}$$

Suppose that the fluid is squeezed out from the largest n_1 pores with hydraulic diameter of d_1 at the pressure difference p_1. The volume flow rate of V_1 is thus established through empty pores, equation (7).

$$V_1 = \frac{\pi d_1^2}{4} n_1 v_1 = \frac{\pi}{4} a p_1^b n_1 d_1^2 \tag{7}$$

Additional n_2 pores will open at p_2, $p_2 > p_1$, and the volume flow will rise to value V_2, equation (8).

$$V_2 = \frac{\pi}{4} a p_2^b (n_1 d_1^2 + n_2 d_2^2) \tag{8}$$

The pressure value can be increased incrementally till all pores are opened. Hence at the ith incremental step the volume flow rate is V_i, equation (9).

$$V_i = \frac{\pi}{4} a p_i^b \sum_{j=1}^{i} n_j d_j^2 \tag{9}$$

The selective squeezing out the fluid from pores as described in equations from (3) to (9) enables us to compute the number of pores at each interval defined by the incremental pressure growth. The number of pores of the first interval n_1 can be estimated as

$$n_1 = \frac{4V_1}{\pi a p_1^b d_1^2}, \tag{10}$$

for the second interval as

$$n_2 = \frac{4}{\pi d_2^2}\left[\frac{V_2}{a p_2^b} - \frac{\pi d_1^2}{4} n_1\right], \tag{11}$$

and for the ith interval as

$$n_i = \frac{4}{\pi d_i^2}\left[\frac{V_i}{a p_i^b} - \frac{\pi}{4}\sum_{j=1}^{i-1} d_j^2 n_j\right]. \tag{12}$$

It is clear from the equation (9) that

$$\frac{\pi}{4}\sum_{j=1}^{i-1} d_j^2 n_j = \frac{V_{i-1}}{a p_{i-1}^b} \tag{13}$$

and hence, equation (12), which defines the number of pores in the ith interval, can be rewritten as

$$n_i = \frac{4}{\pi a d_i^2}\left[\frac{V_i}{p_i^b} - \frac{V_{i-1}}{p_{i-1}^b}\right] \tag{14}$$

and by taking into account equation (3), the final form of the equation for the number of pores in the ith interval can be derived as

$$n_i = \frac{p_i^2}{4\pi a \alpha^2}\left[\frac{V_i}{p_i^b} - \frac{V_{i-1}}{p_{i-1}^b}\right] \tag{15}$$

The air volume velocity through the wet sample depends on the air pressure and on the open surface of the sample. As the pressure increases, the open surface increases as well due to the squeezing the fluid out of pores with smaller hydraulic diameter. Hence, the rise of the air volume flow rate is consequence of the open surface and the pressure growth. As a consequence the sequential pore opening of the wet sample is achieved by increasing the air pressure gradually when testing. When the pressure is increased then the open surface and the linear velocity of the airflow is also increased. This enables us to calculate the portion of air volume flowing through the empty pores and to calculate the number of pores in i^{th} pore's diameter interval by starting from the first interval with the pores with the largest hydraulic diameter, equation (7), where p_1 and V_1 stand for the air pressure and the volume flow rate respectively when the first air bubble is spotted during the testing of the wet sample.

The presumption of the equal regime of the airflow through the wet sample's open area and the dry one at the same pressure is taken into account. Small values of the Reynolds number,

Re < 50, in the extreme causes (maximal hydraulic diameter of pore), support that presumption. The airflow is either laminar through open pores in the wet sample and through all pores in the dry sample, or the type of the airflow is the same. This is the criterion for using the exponent b, which is estimated when equation (5) is fitted to the measured dry data, in the process of determining the pore distribution from the measured wet data.

The method's algorithm can be presented in step-by-step scheme:

1. The measurements of the air volume velocity flowing through a dry sample as a function of the air pressure at several distinct air pressures produce the "dry data".
2. The measurements of the air volume velocity flowing through a wet sample as a function of the air pressure at several distinct air pressures produce the "wet data".
3. The weighted power approximation is fitted to the dry data, and thus the exponent b is estimated, see equation (5).
4. The approximating cubic splines are fitted to the wet data thus smoothing it.
5. The porosity parameters are computed with the help of b, estimated in the step 3, and with the help of smoothed wet data together with equations (2) – (4) and (6) – (15).
6. The procedure is repeated at step 3 on the portion of measurements (at the pressure interval) where pores were identified in the first algorithm sweep.

When the dry and wet data are measured (steps 1 and 2) the numerical data processing can start. A computer application was built for that purpose to enable one to interactively carry out the porosity parameters numerical computation. A user interaction with the application is needed at steps 3 and 4 when choosing weights to the approximations used to fit the dry and wet data and at the step 5 where a user chooses between two procedures for computing porosity parameters and defines the length of the base interval of the pore diameter distribution (histogram). At step 6 the algorithm is repeated from the step 3 on. The exponent b is computed on the portion of the dry data measurements (pressure interval) where pores were identified in the first algorithm sweep. The upper limit is the pressure, which squeezes the fluid from the smallest hydraulic pore detected by the first algorithm sweep.

Two different procedures are foreseen depending on the type of the flat textile under consideration. The first procedure is suitable for the flat textiles where the number of pores between threads of the warp and weft is known e.g. very thick monofilament woven fabric (sample d). The second procedure is used in other cases e.g. cotton fabric woven out of cotton yarn (sample a).

The corresponding coefficient a_j are also determined by equation (16)

$$a_j = 1.28\frac{n_{cj}}{n_t} = \frac{1.28}{a^*} \; ; a^* = \frac{n_t}{n_{cj}} \tag{16}$$

where n_t stands for the true number of pores, n_{cj} for the computed number of pores and a_j for the corrected a in equation (11). The values of theoretical limits, for exponent b (b_0 = 0.5) and coefficient a (a_0 = 1.28), that are used in the second procedure are shown in the last part of equation (6).

The first procedure is totally valid for the monofilament woven fabrics, which have the same or similar density of the warp and weft and have threads of the yarn of the similar size (yarn count) and quality. It can be used for monofilament and multifilament fabrics, which have similar density of the warp and weft and if the coefficient a_0, equation (6), is smaller then 1.28 (theoretical maximum). A single pore between threads of the warp and the weft can be counted for several hydraulic pores if pores are of rectangular shape (sample b) due to the differences in the densities of the threads of the warp and the weft or due to differences in fineness of the yarn and possibly due to the binding. The value of the coefficient a_0 is greater than theoretical maximum and the computation of the porosity, is continued by using the second procedure.

As a rule, the second procedure should be used if the number of pores is unknown or a_0 is larger than 1.28 or the type of fabrics unsuitable for the first procedure is used. The number of pores in intervals are computed first by using equations (10) and (15) and using maximal value of the coefficient a ($a_1 = 1.28$). The computed number of pores is minimal and so is the corresponding estimated open surface. If the computed coefficient a_0 is larger than 1.28 then the true value of the coefficient a, is computed as quotient between a_1^* and a_0'. Whole procedure is repeated with newly computed a. For example $a_1' = 1.28$; $a_0' = 8$; $a_0'/1.28 = 6.25$; $a_1/6.25 = 0.2048 = a$; $a_1^*/a_0' = 0.2048 = a$; $a_0 = 1.28$.

4. Experiment

Four different samples were used for the method's testing, which practically encompasses all the fabric types that the method is suitable for. The basic design parameters of the woven fabrics are presented in table 1. They are made of monofilament, multifilament and cotton yarn. The measured average pore's hydraulic diameters of the textiles are in the interval of 18 up to 200 micrometers. The wide assortment of textiles is thus covered.

Sample	Description	Interval of measurement [μm]	Numbers of pores per cm^2	Warp/weft, threads per cm
(a)	Cotton woven fabric	160 - 20	452	22/21
(b)	Thick monofilament fabric	80 - 10	2200	55/40
(c)	Multifilament woven fabric	270 - 140	960	32/30
(d)	Very thick monofilament woven fabric	24 - 12	32400	180/180

Table 1. Samples used in the testing of J-method

The results of the textile's porosity tests are presented in table 2 and in figures 1 - 8. The first procedure is used for all four samples. The second procedure was used for porosity parameters estimation of samples (a) and (b) due to large value of the parameter a_0.

We worked under two presumptions:

- The regime of the airflow through the dry and the wet sample is the same at same pressure difference regardless of the size of the open area of the wet sample.
- The number of the hydraulic pores is not the same as number of pores between threads of the warp and weft if the ratio of the rectangular sides, which represents real pore's cross-section, is at least 3:1.

The first presumption applies that the airflow regime through all pore's should be the same regardless of their diameter. This is certainly true for sample (d) due to the fact that the 90% of all pores are in the interval between 18 and 20 micrometers. If the regression parameters of the air flow through dry samples are obtained on the measurement's interval of pressures where pores actually exist then the values of the pore's average diameter obtained by the microscope and the scanning-electron microscope are in good agreement with those obtained with the method presented here indicates justification of the presumption of the same regime of the air flow through dry and wet sample at the same pressure difference. This holds for all tested samples due to low Reynolds number. Reynolds numbers have values 12 and 39 for flow through sample (d) and sample (c) respectively, if we take into account the average hydraulic diameters of 18.78 μm for sample (d) and 199 μm for sample (c). Hence, the flow through all samples is laminar and the exponent b, which is estimated by equation (5), can be used in equations (7) – (11).
The nomenclature in table 2 – b stands for the exponent in equation (5), h [μm] for the width of the interval of the pore distribution, m for the number of the distribution intervals, n_t for the true number of pores between the threads of the warp and the weft per cm², n for the computed number of hydraulic pores between the threads of the warp and the weft per cm², when the true number of pores (or number of hydraulic pores) is unknown (second procedure), d for the average hydraulic diameter of pores, d_t for the optically measured average hydraulic pore diameter – for samples (b), (c) and (d); the pores are ill-defined in sample (a), P [%] for the average open hydraulic flow area, P_t [%] for the average open flow hydraulic area computed on the bases of the optical experiment, a_0 for the coefficient a, equation (5), at presumption that exponent b has minimal value (b = 0.5).

Porosity test procedure	Parameter	Samples			
		(a)	(b)	(c)	(d)
Porosity parameters when the number of pores is known (first procedure)	b	0.5794	0.6647	0.8329	0.7174
	h [μm]	14	10	13	2
	m	10	7	10	7
	n_t	452*	2200	960	32400
	d [μm]	45.04	31.37	200.45	18.84
	d_t [μm]	53	30	199	18.78
	P [%]	0.98	2.07	31.32	9.06
	P_t [%]	1.00**	3.43	29.84	8.98
	a_0	9.4074	2.3872	0.28	0.8791
Porosity parameters when the number of pores is unknown (second procedure)	d [μm]	45.00	31.35		
	P [%]	6.96	3.87		
	n	3314	4115		
	a_0	1.28	1.28		

Table 2. Parameters of porosity estimated with J-method for all four samples. * – the number corresponds to the product of the warp and weft. ** – corresponds to the 452 measured pores – between the threads of warp and weft only one typical pore was measured in each void between the threads of warp and weft.

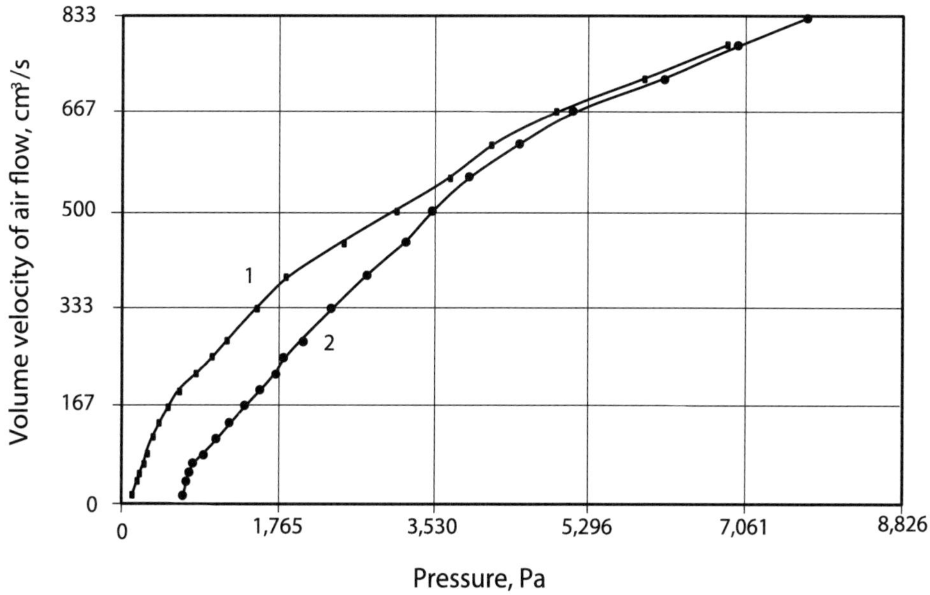

Fig. 1. Diagram of the volume velocity flow through open area of the sample (a) as a function of the pressure difference; 1 - velocity of the air through the dry sample; 2 - velocity of the air flow through the wet sample

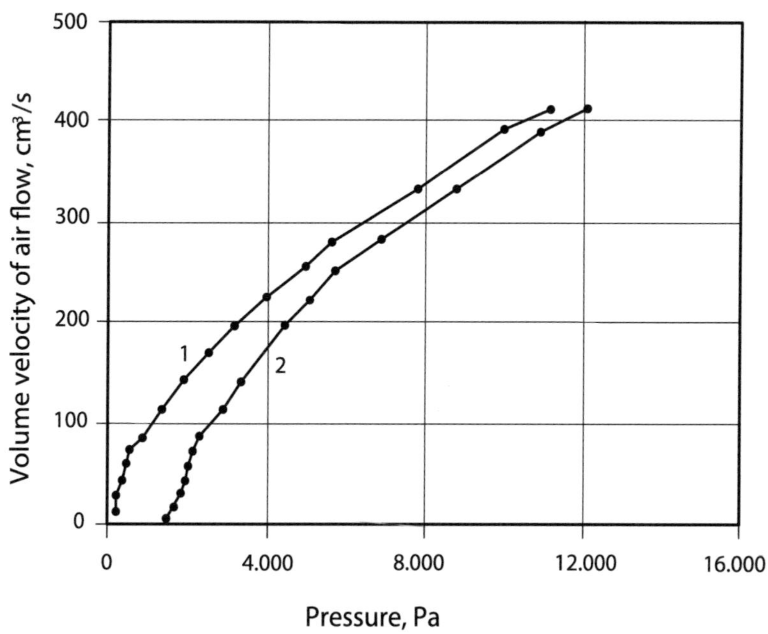

Fig. 2. Diagram of the volume velocity flow through open area of the sample (b) as a function of the pressure difference; 1 - velocity of flow air through the dry sample; 2 - velocity of the air flow through the wet sample

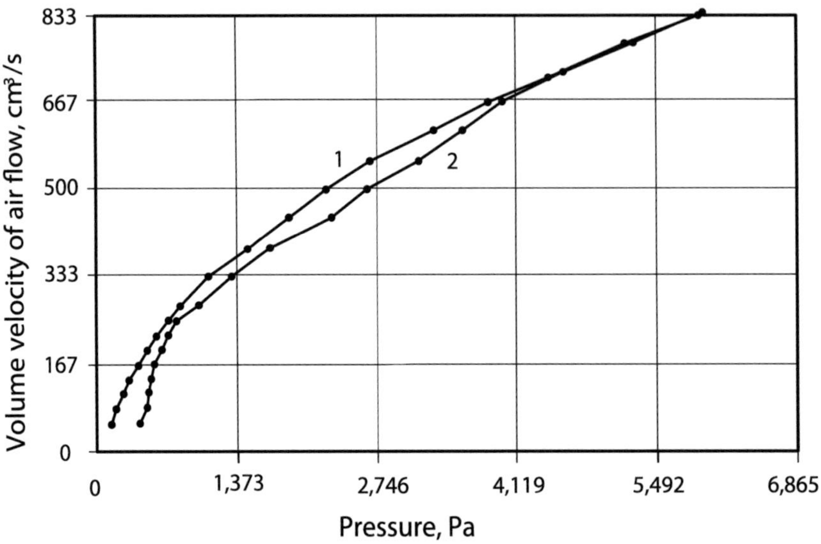

Fig. 3. Diagram of the volume velocity flow through open area of the sample (c) as a function of the pressure difference; 1 - velocity of the air flow through the dry sample; 2 - velocity of the air flow through the wet sample

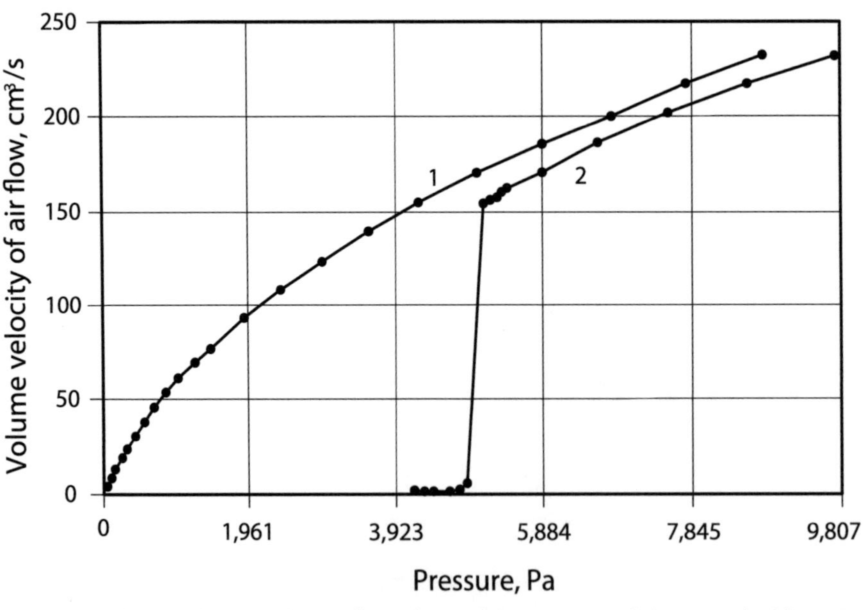

Fig. 4. Diagram of the volume velocity flow through open area of the sample (d) as a function of the pressure difference; 1 - velocity of the air flow through the dry sample; 2 - velocity of the air flow through the wet sample

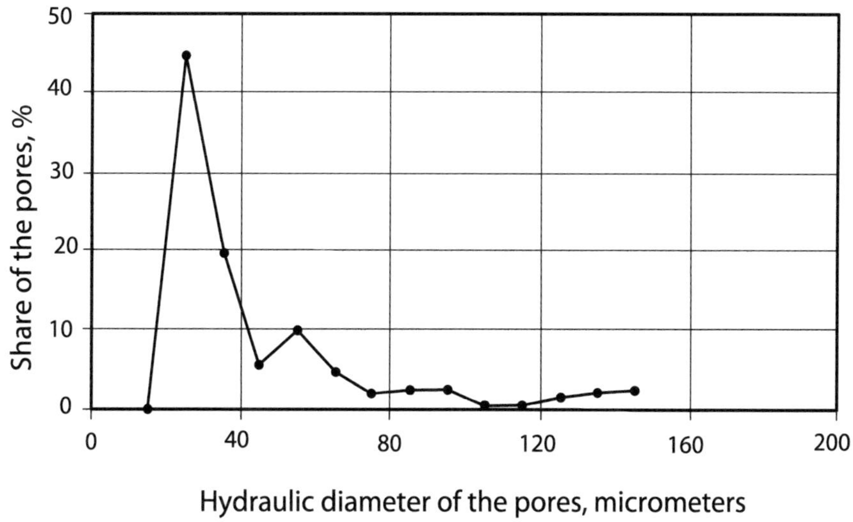

Fig. 5. Diagram of pore's distribution in sample (a)

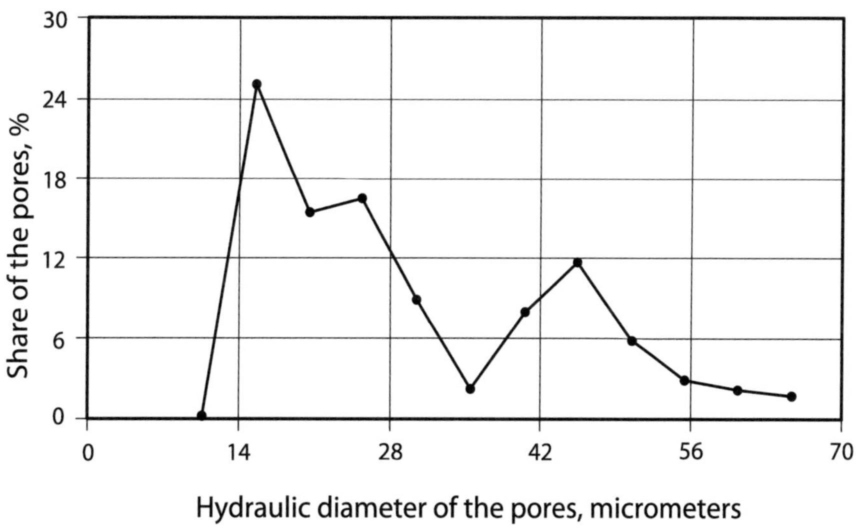

Fig. 6. Diagram of pore's distribution in sample (b)

Statistic parameters	w [µm]	l [µm]	d_t [µm]	$P_{real} = w^*l$ [µm²]	P_{hydr} [µm²]	l/d_t
Mean	20.13	66.06	30.00	1364.67	786.29	2.50
Standard deviation	7.77	13.75	10.19	664.13	461.38	1.19
Minimum	4.76	23.81	8.82	287.12	61.07	1.04
Maximum	34.92	87.3	46.91	2519.68	1727.43	6.86
Total				68233.43	39314.56	

Table 3. Results of the scanning-electron microscope pore's shape and open area measured on 50 pores of the sample (b)

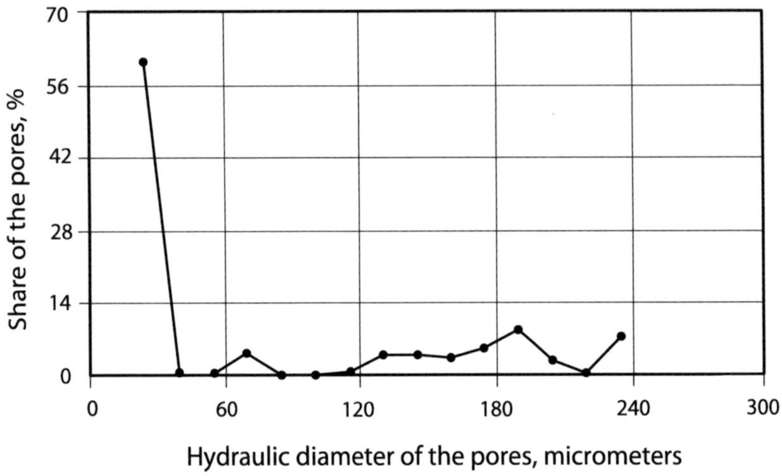

Fig. 7. Diagram of pore's distribution in sample (c)

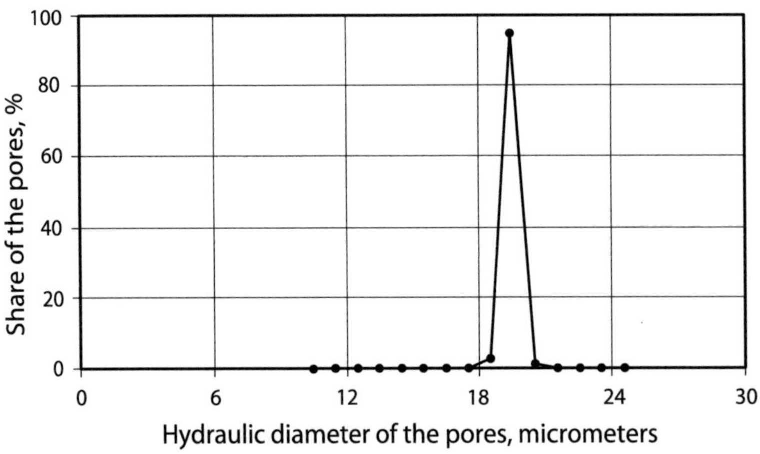

Fig. 8. Diagram of pore's distribution in sample (d)

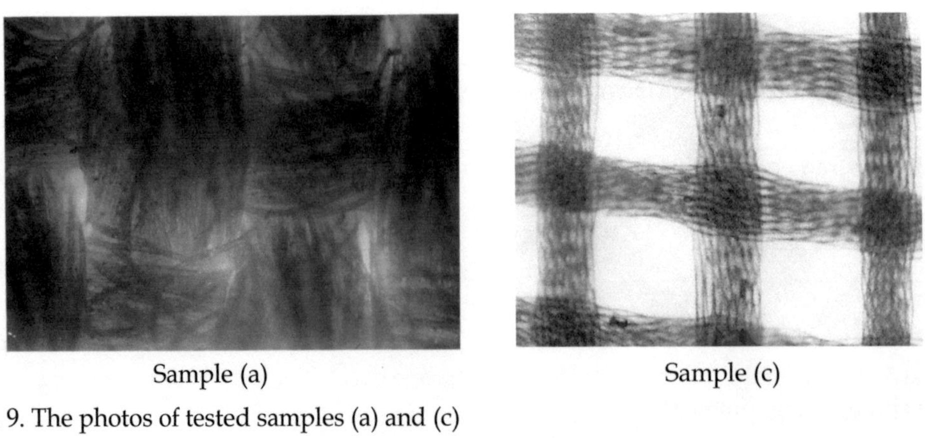

Sample (a) Sample (c)

Fig. 9. The photos of tested samples (a) and (c)

When dealing with the sample (b), the average ratio l/w, equation (2), in table 3 is 66.06/20.13 = 3.23, see also figures 9 - 11. Hence, the criterion of having more than one hydraulic pore in a pore between threads of the warp and weft, figure 12, is thus met. The maximum value of the ratio between the value of the longer rectangular side l and the hydraulic diameter belonging the same pore is 6.84.
This results in almost doubling in number of hydraulic pores - from the 50 pores measured by the scanning-electron microscope with magnification of 630, a detail can be seen in figure 11b, to estimated 99 hydraulic pores. The real number of hydraulic pores is thus 4356, if the results in table 3 are extrapolated to the test area of 1 cm^2, which is in good agreement, with 4112 hydraulic pores estimated by the J-method by using the second procedure, see table 2, sample (b). The difference in number of hydraulic pores is only 4.5 %. The true open area of pores extrapolated from results in table 3 to the test area of 1 cm^2 is 3.01 % and the true hydraulic open area 3.34 %. The estimated open area obtained by the method is 3.87 %, table 2, or 12.83 % more than true hydraulic open area, and 28.57 % more than true open area.

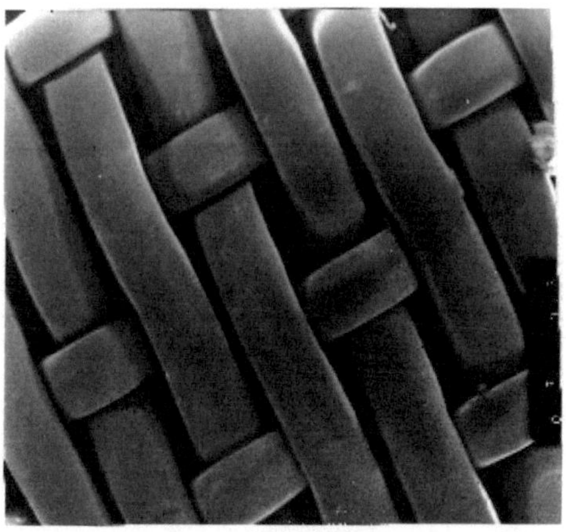

a)

b)

Fig. 10. Sample (b): a) magnification 63x, b) magnification 190x

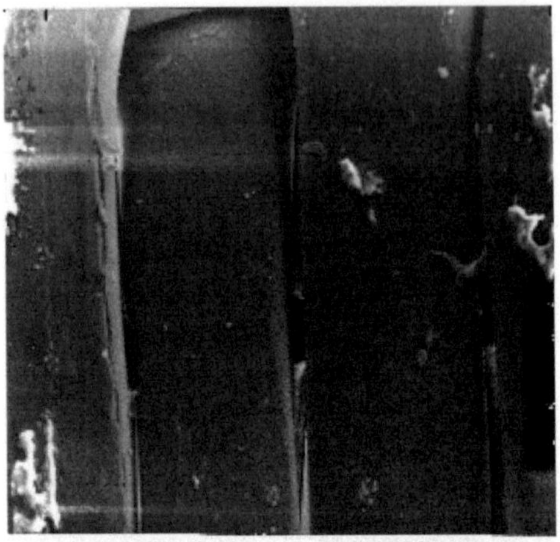

a)

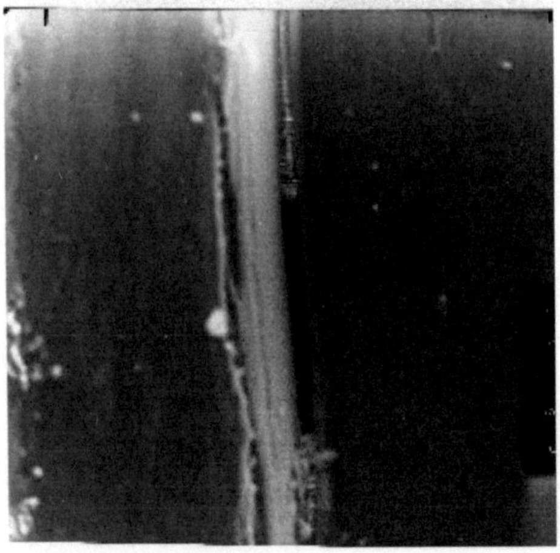

b)

Fig. 11. Sample (b): a) magnification 190x, b) magnification 630x

The fibres that jut out of the yarn enmesh the pores between the threads of the warp and weft and thus dividing them into smaller pores with no regular geometrical shape if the textile is made of spinning yarn e.g. cotton woven fabric such as sample (a). The values of the porosity parameters of the cotton fabric obtained by the first procedure are shown in table 2. The most distinctive pore in voids between the threads of the warp and weft is taken into account when inspecting the fabric with a microscope and thus obtaining the number of pores of 452 and the average pore's hydraulic diameter of 53 micrometers. The lower value of the pore's hydraulic diameters set to 20 micrometers when computing the porosity parameters in this case. The value of a_0, is high as well as the value of a_1 (a_1 = 1.7416), see equation (6). Both values are higher than theoretical maximum at b = 0.5. The number of pores is inversely proportional to the value of the coefficient a, equation (5), and the maximal value of the coefficient a, is 1.28. Hence, the porosity parameters of the samples (a) and also (b) are computed with the second procedure.

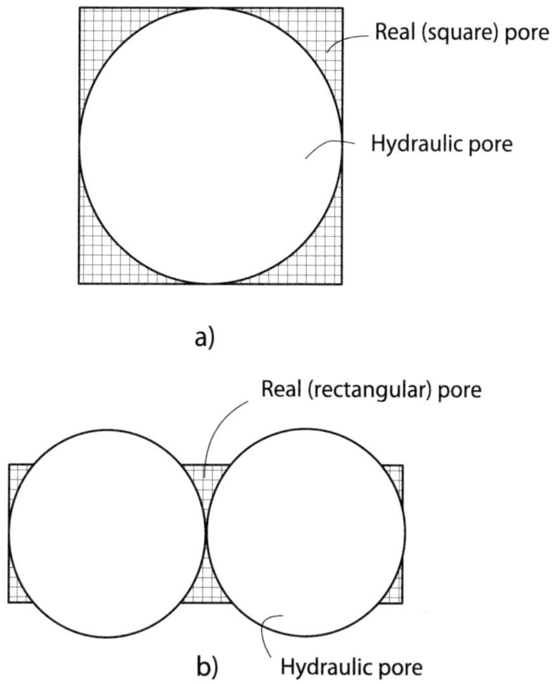

Fig. 12. Influence of the form of pores on the number of hydraulic pores in one pore between threads of the weft and warp in the woven fabric: a) square real pore, b) rectangular real pore

Fig. 13. The inner non-woven layer in the medical mask

The non-woven flat fabrics are extremely difficult to characterise in terms of porosity due to their irregular structure. The structure makes them better, more effective filtration media in comparison to the woven fabrics. Hence, the challenge is to estimate their porosity parameters. The experimental results presented in table 1, especially sample (b), proved that the porosity of the non-woven flat textiles can be estimated by J-method.

We have applied J-method in order to characterise the porosity of a medical mask. The mask fabric, figure 13, is composed of three layers. The data of layers is presented in table 4. The outer layer and the layer suite on the face of subject are composed from fibres which diameter is 18 μm. The inner layer is composed from microfibers which diameter is 2 μm. In this layer the fibres are arranged in 37 layers.

The walls of pores are defining by fibres. In contrast to a woven fabric, where pores are straight from one surface to the opposite one and where the length of pores is equal to thickness of the fabric, the pores in the non-woven fabric changes its direction and are thus much longer than the fabric's thickness. It is this property that makes them an excellent filtration media and at the same time, very difficult to characterise. Even though the viruses are much smaller than the hydraulic diameter of pores, the configuration of pores allows for 100% filtration efficiency.

The schematic airflow through the medical mask is shown in figure 14. The air flows through pores in a complex pattern. It is fairly difficult to developing a real theory of filtration due to that fact.

The number of the pores on 1 cm² is estimated as 63970, the maximal diameter of pores is 30.19 μm and the open area (free for air flow) is 8.42%.

The results for porosity parameters estimated by J-method are presented in table 5 and the pore distribution in table 6. The coefficient a (regression equation (5)) - flow air through dry sample is 0.0890 and the exponent b (regression equation (5) - flow air through dry sample) is 0.7521. The mean hydraulic diameter of pores is estimated to the value of 12.46 μm, table 5. The nomenclature of table 5 is as following: d_{max} stands for the average pore diameter of the first interval (the largest pores), d_{min} stands for the average pore diameter of the last interval (the smallest pores), d_p stands for the average pore diameter of the sample and P stands for the average open hydraulic flow area.

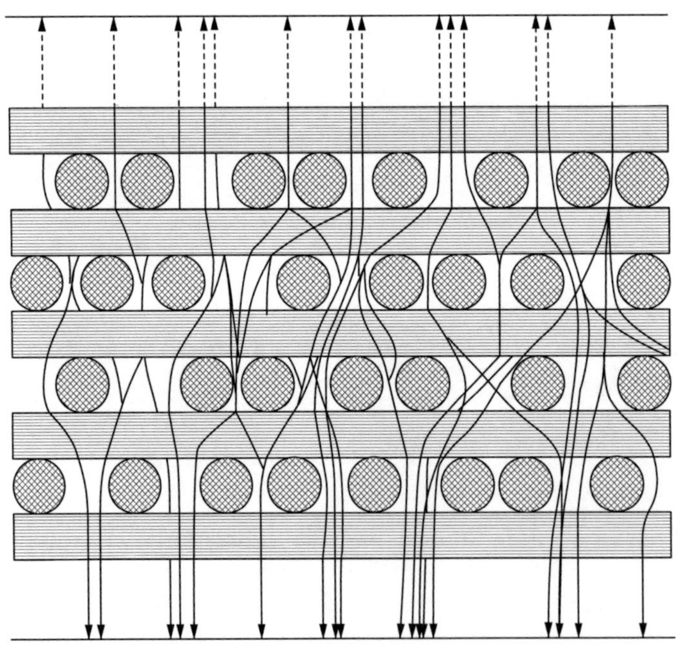

Fig. 14. Stream of air through the non-woven mask due to the respiration

Non-woven layer	Mass [g/m²]	Thickness [µm]	Numbers of the fibres layers	Thickness of the fibres [µm]	Active surface (not blocked whit resin) [%]	All surface of non-woven [cm²]
Outer	17.6	92	5	18	87	160
Inner	20.4	74	37	2	100	160
Outer on the subject face	19.1	120	7	18	75	160

Table 4. Comparison of chosen physical parameters of the fibres from different layers, that the mask is made of.

Parameters of porosity	Outer non - woven layer	Inner non - woven layer	Outer on the subject face	Mask (all three non-woven layers)
The biggest pore [µm]	305	38	211	30
dmax [µm]	275	28	195	28
dmin [µm]	15	8	15	8
dp [µm]	83.2	12.46	76.3	12.61
b	0.6183	0.7313	0.6143	0.7521
a	0.249	0.0889	0.2925	0.089
P [%]	27.71	8.43	25.32	8.42
Width of the classes [µm]	13	2	18	2
Number of the classes	13	10	10	10
Number of pores/cm²	3745	64016	4506	63970

Table 5. Parameters of porosity for all three non-woven layers of mask; the surface of samples: 1 cm2; liquid in the pores: n-butanol

Meas. num.	Limits of the gradual classes [µm]	Hydraulic diameter of pores [µm]	Pressure [kPa]	Volumes flow [cm³/s]	Number of pores	Portion of pores [%]
1	26-28	27	0.360	0.924	136	0.36
2	24-26	25	0.389	2.115	184	0.19
3	22-24	23	0.423	4.412	388	0.57
4	20-22	21	0.463	9.833	1029	1.61
5	18-20	19	0.512	21.504	2488	3.97
6	16-18	17	0.572	41.916	4859	7.75
7	14-16	15	0.649	74.346	8770	13.82
8	12-14	13	0.748	113.669	11324	17.70
9	10-12	11	0.884	157.730	10611	16.59
10	8-10	9	1.081	219.088	24281	37.96

Table 6. Parameters of porosity for mask (for all three non-woven layers)

5. Conclusions

J-method of porosity assessment of the flat textiles is presented here. It enables us to compute maximal and average hydraulic diameter of pores and relative distribution of pore's diameters regardless of the type of the flat textile by using both procedures. It is also possible to compute distribution of pore's diameters and true value of the hydraulic open surface if the number of pores is known and first procedure is used. If the number of pores is unknown the second procedure should be used. In that case the distribution of pore diameters and true value of the hydraulic open surface are determined approximately but well enough to meet most requirements.

The results are in good agreement with those obtained by the microscope and scanning electron microscope. Considering the results obtained when testing woven fabric we have concluded that the method could be used to determine the porosity parameters of knitted fabrics and thinner non-woven fabrics.

Method is suitable for assessment parameters of porosity in textiles filters, if the average hydraulic diameters are in interval 5 to 200 μm (Jakšić, 2007).

6. References

Matteson, M.J. & Orr, C. (Editors) (1987). *Filtration - Principles and Practices*, Marcel Dekker Inc., ISBN 0-8247-7582-1, New York

Dosmar, M.; Wolber, P.; Bracht, K. & Weibel, P. (1993). A new in - place integrity test for hydrofobic membrane filters. *Filtration & Separation*, Vol. 30, (Jun 1993), 305–308, ISSN 0015-1882

Dubrovski, D. P & Brezocnik, M. (2002). Using Genetic Programming to Predict the Macroporosity of Woven cotton Fabrics. *Textile Research Journal*, Vol. 72, (2002), 187–194, ISSN 0040-5175

Hssenboehler, C. B. Jr. (1984). A New Method for Pore Structure Analysis Using Air Flow. *Textile Research Journal*, Vol. 54, (1984), 252-261, ISSN 0040-5175

Jakšić, D. (2007). *Projektiranje in konstrukcija tekstilij*, p. 204-219, University of Ljubljana, ISBN 978-961-6045-32-2, Ljubljana

Jakšić, D. (2004). Effect of the parameters of environment and shape of clothing surfaces on their thermal resistance. *J. Text. Inst.*, Vol. 95, No. 1/6, (2004), 183–186, ISSN 0040-5000

Miller, B. & Tyomkin, I. (1986). An Extended Range Liquid Extrusion Method of Determining Pore Size Distributions. *Textile Research Journal*, Vol. 56, (Jan 1986), 35–40, ISSN 0040-5175

Miller, B. & Tyimkin, I. (1986). Methods for Determining Pore Size Distributions In Non-woven Materials, *Proceedings of Tappi 1986 Nonwovens Conference.*

Piekaar, H. W. & Clarenburg, L. A. (1967). Aerosol filters - Pore size distribution in fibrous filters. *Chemical Engineering Science*, Vol. 22, (1967), 1399–1408, ISSN 0009-2509

Rebenfeld, L. & Miller B. (1995). Research Approach to Quantification of Flow Behavior in Non-wovens. *J. Text. Inst.*, Vol. 86, No. 2, (1995), 241-251, ISSN 0040-5000

Rucinski, K.; Caronia, A. & McNeil, R.(1986). Filter media characterization by mercury intrusion. *Tappi Journal*, Vol. 23, (Nov 1986), 121-123, ISSN 0734-1415

Rushton, A. & Green, D. J. (1968) The Analysis of Textile Filter Media. *Filtration & Separation*, Vol. 6, (Nov/Dec 1968), 516-522, ISSN 0015-1882

15

Woven Fabrics and Ultraviolet Protection

Polona Dobnik Dubrovski
University of Maribor, Faculty of Mechanical Engineering
Slovenia

1. Introduction

The increasing awareness of negative effects of ultraviolet radiation and regular, effective protection are actual themes in general public in many countries. In professional journals, daily papers and internet sides a lot of different subscriptions can be noticed, where dermatologists, meteorologist, biologist and other professionals warn us about UV radiation, ozone depletion and give us some recommendations for effective protection. The problem of UV radiation is interdisciplinary - it is also the subject of textile scientists. Behaviour outdoors can significantly affect exposure to solar UVR and use of items of personal protection can provide a substantial reduction in the UVR dose received. Clothing made from woven fabrics can provide convenient personal protection however not all fabrics offer sufficient UV protection. This chapter gives the short overview of the role of UV radiation on human health, protection against UV radiation with the emphasis on woven fabric construction and other factors influencing the UV protection properties of woven fabrics.

2. Ultraviolet radiation

Ultraviolet radiation (UVR) is electromagnetic radiation, which we can not see or feel, and it is emitted by the natural or artificial sources. The natural source of UV radiation is the Sun, which emits different types of electromagnetic radiation with different wavelengths and energies. UV radiation has wavelengths shorter than that of visible region, but longer than that of soft X-rays, in the range of 10 nm to 400 nm, and energies from 3 eV to 124 eV. The UVR spectrum can be subdivided into near UV (400 - 300 nm), middle UV (300 - 200 nm) and vacuum UV regions (200 - 10 nm) by physicists, or into UVA (400 - 315 nm), UVB (315 - 280), UVC (280 - 100 nm) and UVD (100 - 10 nm) regions by biologists (Williams & Williams, 2002).

The artificial sources of UV radiation are different types of lamps for phototherapy, solariums, industrial/work place lightening, industrial arc welding, hardening plastics, resins and inks, sterilisations, authentication of banknotes and documents, advertising, medical care, etc. UV lasers are also manufactured to emit light in the ultraviolet range for different applications in industry (laser engraving), medicine (dermatology, keratectomy) and computing (optical storage). Lamps and lasers emit UVA radiation, but some of them can be modified to produce also a UVB radiation.

2.1 Effects of UV radiation on human health

There are big differences between the UVA, UVB and UVC (or UVD) radiation regarding their effects on human health. UVA radiation is also known as glass transmission region while ordinary glass blocks over 90% of the radiation below 300 nm and passes the radiation abut 350 nm. UVA radiation is thought to contribute to premature ageing and wrinkling of the skin while it damages collagen fibres and destroys vitamin A in the skin. It penetrates deeply under the skin but does not cause sunburn, only sun tanning. Sun tan is a defence mechanism of the skin. Brown pigment melanin namely absorbs UVA radiation and dissipates the energy as harmless heat, blocking the UV from damaging skin tissue. Today it is also known that UVA radiation can generate highly reactive chemical intermediates which indirectly damage the DNA and in this way induces the skin cancer. UVA is the main cause of immune-suppression against a variety of infectious diseases (tuberculosis, leprosy, malaria, measles, chicken pot, herpes and fungal disease) rather than UVB, but the effects are also positive (type 1 diabetes, multiple sclerosis, rheumatoid arthritis). UVB radiation is known as sunburn region and has been implicated as the major cause of skin cancers, sun burning and cataracts (Yalambie, 2003). It damages the fundamental building element - DNA directly at molecular level as well as collagen fibres and vitamin A in the skin.

UVA radiation	UVB radiation	UVC radiation
λ = 400-315	λ = 315-280	λ = 280-100
Energy: 3.10-3.94 eV	Energy: 3.94-4.43 eV	Energy: 4.43-12.4 eV
Mean energy: 340 kJ/mol	Mean energy: 400 kJ/mol	Mean energy: 810 kJ/mol
Intensity: 27 W/m^2	Intensity: 5 W/m^2	Intensity: -
It has 1.7 times bigger mean energy than visible radiation[1].	It has 2 times bigger mean energy than visible radiation[1].	It has 4.1 times bigger mean energy than visible radiation[1].
Its intensity represents the 7,9% of solar radiation[2].	Its intensity represents the 1,5% of solar radiation[2].	-
Damages collagen fibres and accelerates skin ageing.	Damages collagen fibres and accelerates skin ageing.	Damages collagen fibres and accelerates skin ageing.
Destroys vitamin A.	Destroys vitamin A. Initiates vitamin D-production.	Destroys vitamin A.
Responsible for tan.	Responsible for deeper tan of longer duration. Responsible for sunburn.	Responsible for sunburn.
Indirectly destroys DNA and contribute to skin cancer.	Directly destroy DNA and causes skin cancer.	Directly destroy DNA and causes skin cancer.
Suppresses immune system protection by some diseases or have positive effect by others.	Has negative or positive effect on immune system.	-
Penetrates under the skin.	Dangerous to the eyes.	Dangerous to the eyes.

[1] mean energy of visible radiation: 200 kJ/mol, [2] average solar radiation: 342 W/m^2 (Ron, 2005)
Table 1. Main differences between UVA , UVB and UVC radiation

UVB radiation increases the melanin production as a means of protection which leads to a long lasting tan with 2-day lag phase after irradiation. It is known that, biologically, sunburn corresponds to a real damage to the genome in the skin cells and that although the effects may be reversed by repair processes permanent genomic damage can occur (Cesarini, 2001). The cornea, the lens and the retina can be damaged if we are too much exposed to

UVB radiation. UVB radiation is also the component that initiates vitamin D production in the skin (Johnston, 2005). In this way it has a good effect on human health while the vitamin D is vital for normal functioning of the nervous system, bone growth, etc. UVC radiation is known as bacterial region and it is extremely dangerous while it has the highest energy. It destroys DNA directly. Table 1 shows the main differences between the UVA, UVB and UVC radiation (Zabetakis, 2002).

2.2 UVR transmission and ozone depletion

99% of the UV radiation that reaches the Earth's surface is UVA radiation and it is not absorbed by ozone. UVA radiation is most intense in early morning and afternoon and can pass through the window glass. UVB radiation is mostly absorbed by ozone, although some reaches the Earth. The amount of UVB radiation received by a location is strongly dependent on: latitude and elevation of the location (average UVB exposure at the poles is over a thousand times lower than at the equator), cloud cover (the reduction in UVB exposure depends on cover's thickness), and proximity to an industrial area (protection offered by photochemical smog, which absorbs UVB) (Sparling, 2001). UVB radiation does not penetrate through the window glass and is most pronounced midday. UVC radiation is completely absorbed by the atmosphere's ozone layer and normal oxygen before it reaches the ground. However, this is valid for the situation where there is no ozone depletion.

Ozone layer is a concentration of ozone, e.g. a naturally-occurring gas formed by three atoms of oxygen, in the stratosphere, which extends about 10-50 km above the Earth's surface (EPA, 2010). It filters the sun's ultraviolet radiation and protects plant, animal and human life on the planet. This natural shield has been gradually depleted by man-made chemicals like chlorofluorocarbons (CFCs), hydro fluorocarbons (HCFCs) and other ozone-depleting substances (ODS), which are used widely in refrigerators, food containers, plastic foam, home insulation etc. Ozone is degraded also by the absorption of UVB radiation (Cesarini, 2001). Global depletion of stratospheric ozone is nowadays one of the serious environmental and living problems. The reduction of stratospheric ozone has, as a direct consequence, increased the intensity of UVB radiation received at ground level. An increase in exposure to UV for the populations of New Zealand, Australia, Europe and North America has been recently reported. The 1% reduction in ozone will lead to 2% increase in solar UVB radiation at the Earth's surface (Sparling, 2001) and may eventually lead to a 2.3% increase in skin cancer (Roy et al., 1995). The potential biological consequences of this are modifications (of the genome) in the skin and the cornea - two interfaces between the environment and human body. Basal and squamous cell carcinoma (nonmelanoma skin cancers - NMSC) is the most common of all cancers associated with the increased exposure to UV radiation. Risk factors for NMSC include exposure to sun or radiation, fair skin, advancing age, and male sex. The most serious of all cancers e.g. malignant melanoma (MM) is increasing faster than any cancer except lung cancer. Melanoma risk is increased in men, fair-skinned individuals, those with the family history, those with multiple pigmented nevi, the immuno-suppressed and is more than double in those with one or more severe sunburns in childhood (Ferrini et al., 1998). We should be aware of that prolonged and repeated exposure to UV radiation in early childhood increases the risk of malignant tumours in adulthood.

2.3 Protection against solar UV radiation

It is obviously that protection against harmful UV radiation which includes recommendations on behaviour, environment, legislation and personal protection changes, is needed (Edlich et al., 2004; Gies et al., 1998; Turnbull & Parisi, 2005). Protection against solar UV radiation can reduce and individual's solar UV radiation exposure to between 1 and 10% of that without any protection (Gies et al., 1998). Sun avoidance and use of use of protective clothing have been associated with reduced risk of both melanoma and nonmelanoma skin cancers in multiple cohort, animal and case control studies; however, not all studies find effect (Ferrini et. al, 1998).

Behaviour. Avoiding the sun between the 10 and 14 hour is nowadays a well known recommendation on behaviour changes which can play a very important role in the reduction of individual's total UVR exposure. The intensity of solar UVR namely depends on the height of the sun in the sky, which varies according to the latitude, the hour of the day and the seasons (Cesarini, 2001). It is also worth to mention that reflections in the environment (snow cover, surface layer of the sea) increase the dose of UV received directly from the sky. For people dealing with outdoor activities who are not able to avoid the sun between 10 and 14 hour, the provision of shade and subsequent reduction in solar UVR exposure should be a high priority.

Environment and legislation changes. Creation and popularisation of a global UV index by the World Health Organisation was a step forward to reduce the amount of UV dose that people received. UV index represents the maximum effective radiance received on the skin surface, taking into account the cloud cover and all other variables of the environment (Table 2). It is obtained by multiplication of the effective radiance of the solar radiation by 40 and it takes values on a scale from 1 (low) to 11+ (extremely high) (Cesarini, 2001). Knowing the UV index, people can change the environment by provision of shade and other UV radiation protective structures or by adopting personal protection. People who are employed in the building trades, foresters, farmers, those who work on beaches and ski-slopes and other certain professions are much more exposed, by a factor of some 4-5 times, than those who live and work in towns, and in spite of the phenomenon of adaptation, are at higher risk from their exposure.

Personal protection. Sometimes it is even not possible to choose shaded position or to schedule the time spent outdoors. In this case the personal protection e.g. the protection of eye and skin is only option. Significant amounts of UVR personal protection is provided by application of sunscreen to all exposed areas of the skin as well as by wearing good quality sunglasses, a hat and clothing appropriate to the thermal conditions of the environment and the level of activity (Table 2). Sunscreen absorbs, diffuses or reflects incidental UVR thus reducing the fraction of solar UVR reaching the basal layer of the epidermis and dermis. Table 2 shows the recommendations for using the stated protection factors of sunscreens according to the UV index for people with sensitive and normal skin. Sunscreens should be used in conjunction with clothing to protect areas of the body not covered by clothing and should be reapply every two hours. However sunscreens may be associated with adverse effect. They allow users to increase time spent in the sun and to avoid sunburn, but users exposing themselves to harmful UVR, which may be carcinogenic or decrease immune function. Besides that some of the compounds in sunscreens are carcinogenic and are associated with reduced synthesis of vitamin D (Ferini et al., 1998). Sunglasses are equipped with lenses that filter both UV rays and visible light which arrive directly or indirectly through reflection. Eye protection is increased by wearing a hat or visor which extends the natural anatomical projection of the face, which include the nasal edge, eye-lashes, eyebrows and eyelids. A hat with 7 cm peak protects not only face but also the nape and sides of the neck. The best way to avoid the effects of UVR is to cover the skin as much as possible with clothing. Clothing is made by different types of fabrics, which provide a simple and convenient protection against UVR, however not all fabrics offer sufficient UVR protection. In general, personal protection items have been defined with different protection factor (PF) ratings in an attempt to quantify the UVR protection that such products can provide. The PF gives indication of the amount of UVR that is blocked by the protection items. Depending on the personal protection item there are different rating scales. For fabric UPF (Ultraviolet Protection Factor), for sunscreen SPF (Sun Protection Factor) and for sunglasses EPF (Eye Protection Factor) is used.

UV index 1-2	UV index 3-4	UV index 5-6	UV index 7-8	UV index 9 and above
duration of exposure to the sun without protection for sensitive skin / **for normal skin**				
weak sun continuous exposure of 1-2^{h}/3^{h} +	moderate sun continuous exposure of 40^{min}/1^{h} 30^{min}	strong sun continuous exposure of 25^{min}/50^{min}	very strong sun continuous exposure of 20^{min}/40^{min}	extremely strong sun continuous exposure of 15^{min} / 30^{min}
recommendations for UVR personal and environmental protection for sensitive skin / **for normal skin**				
sun glasses	sun glasses + sunscreen with SPF 15 / sun glasses	sun glasses + hat + T-shirt + sunscreen with SPF 25 + sunshade / sun glasses + hat + sunscreen with SPF 15	sun glasses + hat + T-shirt + sunscreen with SPF 40 + sunshade / sun glasses + hat + sunscreen with SPF 30	exposure strongly discouraged / sun glasses + hat + sunscreen with SPF 40 + sunshade

Table 2. Correspondence between solar rays and the universal UV index and recommendations for UVR protection (Cesarini, 2001)

3. Protection factor of fabrics

The ability of fabrics to protect against UV radiation can be tested by two major methods: in vitro method (or instrumental / spectrophotometric method) and in vivo method (or laboratory / human skin method). Both methods assess the amount/degree of sunburn protection provided by the fabrics with so called term UPF by in vitro method and SPF by in vivo method. Theoretically, the UPF and SPF value for any fabrics should be the same. However, some studies indicated that the results of UPF and SPF values are not statistically identical; never the less both values are in a good correlation (Hatch & Osterwalder, 2006).

3.1 Ultraviolet protection factor of fabrics – UPF and measurement techniques

To assign the degree of UVR protection of fabrics, the **u**ltraviolet **p**rotection **f**actor (UPF) is defined as the ratio of average effective UV radiation irradiance transmitted and calculated through the air (effective dose - ED) to the average effective UV radiation irradiance transmitted and calculated through the fabric (effective dose - ED_f) (EN 13758-1, 2002; Scott, 2005):

$$UPF = \frac{ED}{ED_f} = \frac{\sum_{\lambda=290}^{\lambda=400} E(\lambda)S(\lambda)\Delta\lambda}{\sum_{\lambda=290}^{\lambda=400} E(\lambda)T(\lambda)S(\lambda)\Delta\lambda} \tag{1}$$

where E(λ) is the relative erythemal spectral effectiveness, S(λ) is the solar spectral irradiance in $Wm^{-2}nm^{-1}$, Δλ is measured wavelength interval in nm, T(λ) is average spectral transmittance of the fabric specimen, and λ is the wavelength in nm. UPF indicates how much longer a person can stay in the sun when fabric covers the skin as compared with the length of time in the sun without fabric covering to obtain the same erythemal response. The higher the UPF of a fabric, the better is its ability to protect the skin it covers. To assign the degree of UVR protection of fabrics also the term penetration or erythema weighted transmittance - EWT is in use, which is the inverse value of UPF (Eq. (2)). The values of EWT lie between 0 and 1 (or 0% and 100%). The lower the percent of EWT is, the greater is the sunburn protection provided by fabric.

$$EWT = \frac{1}{UPF} \tag{2}$$

There are two in vitro quantitative measurement techniques to test UVR transmission through fabrics: radiometry, where the total transmission of UVR through a fabric is measured using a real or simulated solar spectrum, and spectrophotometry, where the transmission of UVR through a fabric is measured as a function of wavelength. Both methods include ultraviolet radiation source that emits both UVA and UVB radiation. The spectrophotometric technique relies on the collection of transmitted and scattered radiation with the aid of an integrating sphere positioned behind the fabric specimen. Suitable UV sources are Xenon arc lamps, Deuterium lamps and solar simulators. The procedure has two major steps: transmittance testing and calculations based on the transmittance data collected. The principle of this method is to direct a beam of monochromatic radiation in the UV light and of known quantity perpendicular to the surface of the fabric specimen and to measure the amount of radiation transmitted through and scattered by the fabric. The sending the beams of radiation continues until all wavelengths in the UV range (or wavelengths at 2 or 5 nm intervals) have been directed to the fabric face and transmittance data collected. It is also possible to direct a beam of polychromatic incident radiation. In that case the transmitted radiation is collected monochromatically. Then the transmittance data are used to calculate UVA and UVB percent transmittance values and a total percent transmittance value. The calculation of total UV percent transmittance for a fabric specimen is the ratio of the amount of radiation transmitted to the amount of radiation directed perpendicular to the fabric specimen surface. The calculation of a UPF is accomplished by combining the transmittance data with data collected that established the relative power of UV wavelengths to cause the skin to redden. These later data, data collected using human subjects, are given in the erythemal action spectra. In this way we may say that in vitro method also has an in vivo component to it. It is also worth to mention that percent transmittance data do not take into account that certain wavelengths in the UV range are more responsible for skin damage than others. On the other hand, the erythemal action spectra data in UPF calculation take into consideration that fabrics that allow a greater portion of the most harmful skin reddening rays to be transmitted will receive a numerical value lower than a fabric that allows less of the powerful skin reddening rays through, even when both fabrics transmit the same amount of radiation (Hatch et al., 2006).

The radiometric technique uses a broad band UV light source filtered for UVB or combined UVA and UVB bands to illuminate a fabric specimen. The total UV transmittance through a fabric is measured by a radiometer. The protection factor is determined by taking the ratio of the measured power in the absence of the fabric to the measured power in the presence of the fabric. Such measurements do not yield a definitive value for the protection factor of a fabric. This technique is more useful when a relative variation in UPF is needed, such as the variation in protection factor from site to site within a fabric or the effect of stretching the textile on the protection factor (Scott, 2005).

3.2 Sun protection factor of fabrics – SPF and measurement technique

SPF is defined as a ratio of radiation dose to produce minimal sunburn under fabric covered skin to the radiation dose to produce the same sunburn of uncovered skin:

$$SPF = \frac{MED_{ps}}{MED_{us}} \tag{3}$$

where MED_{ps} is minimum erythemal dose of protected skin in J, and MED_{us} is minimum erythemal dose of unprotected skin in J. The higher the SPF value, the better the fabric's protection ability against sunburn. MED is defined as the minimum quantity of radiant energy (using incremental UVB doses) required to produce first detectable reddening of the skin, 22 ± 2 hours after exposure. The measurement technique to estimate SPF of fabric is known as in vivo method. The procedure is to attach rectangular pieces of fabric to the back of human subject and determine the minimum erythemal dose of unprotected and protected skin.

3.3 Standards, classification and marking of UV protective fabrics/clothing

Several standards for measuring, classification and marking of fabrics' UV protection properties are in use. All of standards employ the Eq. (1) for measuring UPF of fabrics and differentiate regarding the scanning intervals, positioning of the fabrics in the instrument, the erythemal action spectrum designated, classification and marking.

European standard EN 13758-1:2002 specifies a method to asses UPF of fabrics, without those which offer protection at a distance (umbrellas, shade structures) or artificial UV radiation sources. The instrument records the transmittance between 290 nm and 400 nm by wavelength interval of at least 5 nm. The sample UPF is average UPF of sample minus standard error. When sample UPF is less than the lowest positive UPF measured for a particular sample, then the UPF of that specimen is reported. When the UPF of fabrics is greater than 50 only UPF>50 needs to be reported. By fabrics with areas of various shades and/or construction the lowest positive UPF value measured is reported as the sample UPF.

EN 13758-2:2003 specifies general clothing design requirements, marking and labelling. The clothing design which offers UV protection to the upper and/or lower body shall at least cover the upper and/or lower body completely. This standard classifies UV protective clothing only in one category for which the lowest UPF value is larger than 40 and the average UVA transmission is smaller than 5%. According to this standard UV protective clothing is marked with pictogram (Fig. 1), which includes the number of this standard and UPF 40+, and the wording: "Sun exposure causes skin damage" / "Only covered areas are protected" / "The protection offered by this item may be reduced with use or if it stretched or wet", or can be marked with the wording: "Provides UVA+UVB protection from the sun".

Fig. 1. Pictogram for UV protective clothing according to the EN standard 13758-2

Australian/New Zealand Standard AS/NZS 4399:1996 specifies requirements for determining UPF of sun protective (un-stretched and dry) textiles, garments and other items of personal apparel (hats) which are worn in close proximity to the skin, and appropriate detailed labelling. Standard is not valid for sunscreen products, fabrics for architectural or horticultural use (shadecloth) and items which offer protection at a distance from the skin. Also, it does not cover protection form UV radiation sources other than the sun. The rated UPF value is the mean UPF value of four testing samples reduced for the standard error in the mean UPF, calculated for the 99% confidence level, and finally rounded down to the nearest multiple of five. If the rated UPF is less than the lowest individual UPF sample measurement, the rated UPF is the lowest value of measured UPF rounded to the nearest multiple of five. According to this standard sun protective clothing is categorized to its rated UPF as given in Table 3.

UPF range	UVR protection category	Effective UVR transmission, %	UPF ratings
15 to 24	Good protection	6,7 to 4,2	15, 20
25 to 39	Very good protection	4,1 to 2,6	25, 30, 35
40 to 50, 50+	Excellent protection	< or = 2,5	40, 45, 50, 50+

Table 3. UPF classification system according to AS/NZS and ASTM standards

According to this standard UV protective clothing is accompanied by the following information: the manufacturer's name, trade name and mark; UPF rating; protection category; the wording: "This UPF rating is for the fabric and does not address the amount of protection which is afforded by the design of the article. The manipulations involved in garment manufacture such as stretching and sewing may lower the UPF of the material". For headwear following wording is used: "This item doses not provide protection against reflected or scattered solar UVR".

In the **United States** several standards or methods refer to the UV protective clothing. AATCC Test Method 183-2004 determines UPF and provides the procedure for measuring UPF for fabrics either in dry or wet states. The method prescribes a minimum of two specimens of tested fabric, which should be prior prepared according to the ASTM D 6544. Fabric samples should be exposed to the laundering (40 times), simulated sunlight and in the case of swimwear fabrics to the chlorinated pool water prior the UV transmission testing. ASTM D 6603-07 provides a uniform system of labelling on UV-protective textile products. The UPF value which is placed on a garment needs to be the lowest protection value expected during consumer use over a two-year period. The calculation of fabric UPF value and the protection classification (Table 3) are similar as described in AS/NZS standard. The fabric is not labelled as sun or UV-protective if the calculated UPF value is less than 15. If it is greater than 50, only 50+ is placed on the label. Label shall contain following elements: a UPF value; a classification category; a statement that the UV-protective textile product has been labelled according to the ASTM D 6603 standard guide and some other elements which are not obligatory as previous mentioned elements.

4. Woven fabric constructional parameters

Woven fabric is a flat product which consists of several interlaced thread systems oriented in different direction. Regarding the orientation of threads following woven fabrics are known: bi-axial, tri-axial and tetra-axial. Biaxial woven fabric has at least two orthogonal thread systems: lengthways - warp thread system (warp), and transversal - weft thread system (weft). In addition to the woven fabric classification by technology and type of weave, this chapter is focused on biaxial fabrics with a one warp and one weft thread system. In the phase of a new product development, woven fabrics are engineered to fit desired end-use properties with minimum production costs not only by real but also by trial production. End-use properties strongly depend on several woven fabric constructional parameters, which can be defined in general and particular manner (Dubrovski & Šujica, 1995). In general, woven fabric constructional parameters refer to: the **parameters of raw material** (type of fibre, *dimensional and physical properties* - length, specific density, cross-section shape, fineness, fibre crimp, etc., *mechanical properties* – stress/strain, elastic recovery, module, resilience, stiffness, flexibility, abrasion resistance, etc., *sorptive properties* – absorption of liquid water, vaporous water absorption, oil absorption, oil release, heat of wetting, etc., *thermal properties* - thermal conductivity, heat resistance, thermoplasticity, decomposition, combustibility, etc., *chemical properties* – chemical reactivity, chemical resistance, *miscellaneous properties* – electrical resistivity, resistance to UV radiation, resistance to biological organisms, etc.), the **parameters of yarns** (type of yarn, fibre composition, yarn linear density, number of strands, number of filaments, degree of twist, direction of twist, yarn flexibility, yarn packing factor, etc.), **parameters of woven fabric geometry**, **parameters of woven fabric patterning** (warp pattern, weft pattern, symmetry, colour composition, colour harmony, colour contrasts, etc.) and **technological parameters** of weaving and finishing processes (availability of dobby/jacquard, limitations regarding the fabric width, possibility for weft colour exchange, type of finishing processes, temperature, relative humidity, etc.). In particular, fabric constructional parameters relate to the geometrical structure of the fabric and are classified into **primary and secondary parameters** of fabric geometry. Primary parameters of fabric geometry are: yarn thickness, weave and thread density. Yarn thickness as a constructional parameter belongs first of all to the parameters of yarns, but because of its significant influence on fabric geometry it is classified as primary fabric constructional parameters. Yarn thickness and weave are independent variables, while the thread density is dependent variable. Via the defined selection of primary fabric constructional parameters, all other fabric structure parameters may be seen as constant and dependent on primary parameters. For this reason they are logically classified into the separate category, called secondary woven fabric constructional parameters (Fig. 2).

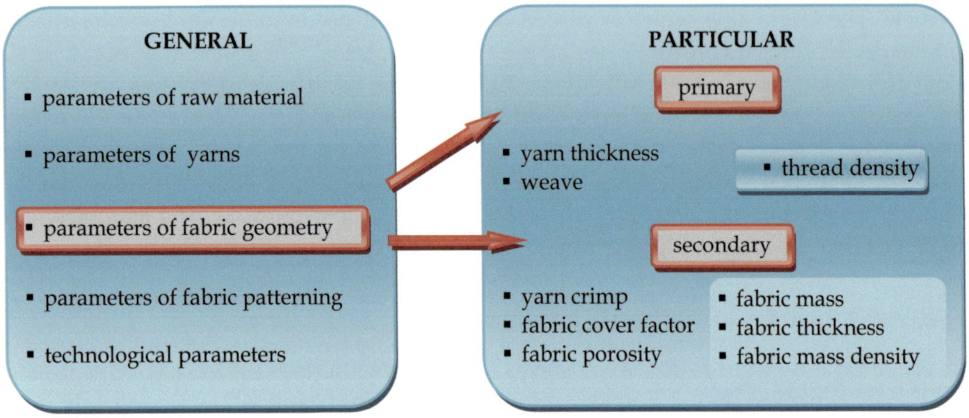

Fig. 2. Classification of woven fabric constructional parameters

4.1 Primary woven fabric constructional parameters

By dealing with woven fabric geometry we assume that yarn is cylinder with circular cross-section. The **yarn diameter or thickness** is then calculated on the basis of yarn linear density according to Eq. (4):

$$d = 3{,}568 \cdot 10^{-2} \cdot \sqrt{\frac{T}{\rho_y}} = 3{,}568 \cdot 10^{-2} \cdot \sqrt{\frac{T}{\rho_f \cdot i}} = 3{,}568 \cdot 10^{-2} \cdot v \tag{4}$$

where d is the yarn thickness in mm, T is the yarn linear density in tex, ρ_y is the yarn bulk density in g per cm^3, ρ_f is the fibre bulk density in g per cm^3, i is the yarn packing factor, and v is the yarn volume coefficient.

Weave is the pattern of interlacing of warp and weft in a woven fabric (Denton & Daniels, 2002). Several end-use properties of woven fabrics are influenced by the weave, which can be numerically expressed with **weave factor** in warp and weft direction (Kienbaum, 1990a):

$$V_1 = \frac{\sqrt{(v_1^2 + 2v_1v_2)} \cdot R_1}{v_1R_1 + \frac{a_2(2{,}6 - 0{,}6z_2)}{f_2}\left(\sqrt{(v_1^2 + 2v_1v_2)} - v_1\right)} \tag{5}$$

$$V_2 = \frac{\sqrt{(v_2^2 + 2v_1v_2)} \cdot R_2}{v_2R_2 + \frac{a_1(2{,}6 - 0{,}6z_1)}{f_1}\left(\sqrt{(v_2^2 + 2v_1v_2)} - v_2\right)} \tag{6}$$

where V is the weave factor, v is the yarn volume coefficient, R is the number of threads in weave repeat, a is the number of passages of yarn in weave repeat from face to back and vice versa, z is the smallest weave shift and f is yarn flexibility. Subscripts 1 and 2 denote warp and weft yarn, respectively.

Thread density is primary constructional parameter which is altered by weave and yarn thickness. It is usually defined as the number of threads per centimetre and expressed for warp (ends) and weft (picks) threads. While the fabric goes thought different technological phases, also thread densities are different and defined for particular production stage. Following densities are known: 1. thread density in finished woven fabric, 2. thread density in raw woven fabric, and 3. thread density by weaving (sett of warp in the reed and sett of weft by weaving). By developing a new fabric construction there is a need to know which densities can be reached by particular woven fabric geometry or yarn thickness and type of weave. The term **limit thread density** was introduced to calculate thread density by limit geometry. Limit geometry refers to the situation where the threads are not deformed and lie in one plane close to each other or there is a minimal space for thread passage. On the basis of limit thread density, woven fabric constructor can decide which **actual thread density** will be set for finished fabric. **Relative thread density** (or **thread tightness**) is the ratio between the actual and limit thread densities, expressed as per cent (Kienbaum, 1990a):

$$t_1 = \frac{G_1}{G_{\lim_1}} \cdot 100\% \qquad t_2 = \frac{G_2}{G_{\lim_2}} \cdot 100\% \tag{7}$$

where t is the thread tightness in per cent, G is the actual thread density in thread per cm, and G_{lim} it the limit thread density calculated according to the Kienbaum's setting theory. Subscripts 1 and 2 denote warp and weft yarn, respectively. **Woven fabric tightness – t** is then defined as the geometrical average of warp and weft tightness according to Eq. (8).

$$t = \sqrt{t_1 \cdot t_2} \tag{8}$$

According to the Kienbaum's setting theory (Kienbaum, 1990a, 1990b) there are four expressions for limit thread density calculation depending on the yarn construction and yarn fineness: 1. limit thread density calculation for the warp and weft systems with the same yarn construction and fineness (Eq. (9)), 2. limit thread density calculation for the

$$G_{\lim_{1,2}} = 5{,}117 \cdot \sqrt{\rho_{fib_{1,2}} \cdot i_{1,2} \cdot V_{1,2}} \cdot \sqrt{\frac{1000}{T_{1,2}}} \tag{9}$$

$$G_{\lim_1} = 8{,}8622 \cdot \sqrt{\frac{\rho_{fib_1} \cdot i_1}{1 + \frac{2v_2}{v_1}}} \cdot V_1 \cdot \sqrt{\frac{1000}{T_1}}$$
$$G_{\lim_2} = 8{,}8622 \cdot \sqrt{\frac{\rho_{fib_2} \cdot i_2}{1 + \frac{2v_1}{v_2}}} \cdot V_2 \cdot \sqrt{\frac{1000}{T_2}} \tag{10}$$

warp and weft systems with the same yarn construction and different yarn fineness (Eq. (10)), 3. limit thread density calculation for the warp and weft systems with different yarn

$$G_{\lim_1} = \frac{280{,}25}{\sqrt{v_1^2 + 2v_1v_2}} \cdot V_1 \qquad G_{\lim_2} = \frac{280{,}25}{\sqrt{v_2^2 + 2v_1v_2}} \cdot V_2 \tag{11}$$

construction and different yarn fineness (Eq. (11)), and 4. limit thread density calculation for the warp and weft pattern with different yarn construction and fineness (Eq. (11)). In the later case the average volume coefficient in warp/weft pattern is first calculated and then put into the Eq. (11). Yarn construction includes following constructional parameters: fibre bulk density, yarn packing factor, and yarn flexibility factor, with the exception of yarn fineness which is stated separately. In the Eq. (9) to Eq. (11), G_{lim} is the limit thread density in threads per cm, ρ_f is the fibre bulk density in g per cm^3, i is the yarn packing factor, v is the yarn volume coefficient, V is the weave factor, and T is the yarn fineness in tex. Subscripts 1 and 2 denote warp and weft yarn, respectively.

4.2 Secondary woven fabric constructional parameters

Yarn crimp is the consequence of yarn interlacing in woven fabrics. It is numerically expressed as percentage crimp, which is 100 divided by the fabric length and multiply by the difference between the yarn length and the fabric length (Denton & Daniels, 2002). Theoretically it can be calculated on the basis of fabric geometry with Eq. (12-14) (Kienbaum, 1990a):

$$\varepsilon_1 = \left[1 - \frac{p_2 R_2}{\left\{ m_1 \cdot \sqrt{\left(\frac{d_1 + d_2}{2}\right)^2 + p_2^2} \right\} + (R_2 - m_1) \cdot p_2} \right] \cdot 100\% \quad (12)$$

$$\varepsilon_2 = \left[1 - \frac{p_1 R_1}{\left\{ m_2 \cdot \sqrt{\left(\frac{d_1 + d_2}{2}\right)^2 + p_1^2} \right\} + (R_1 - m_2) \cdot p_1} \right] \cdot 100\% \quad (13)$$

$$p_1 = \frac{10}{G_1} \qquad p_2 = \frac{10}{G_2} \quad (14)$$

where ε is the yarn crimp in percentage, p is the distance between neighbourhood yarns in mm, R is the number of threads in weave repeat, m is the number of thread passages in weave repeat, d is the yarn thickness in mm, and G is the actual thread density in raw fabric in threads per cm. Subscripts 1 and 2 denote warp and weft yarn, respectively.

Fabric cover factor indicates the extent to which the area of a woven fabric is covered by one set of threads according to Eq. (15) (Fig. 3). It is calculated on the basis of warp/weft cover factor which indicates the ratio between the yarn thickness and the distance between neighbourhood yarns or the ratio between the actual thread density and maximal thread density. It is worth to mention that maximal thread density indicates the situation where threads lie close together without any distance among them. Of course, such situation does not occur in the real fabric, while there is always some space for thread passages. A lot of researches define fabric cover factor as a basic parameter which represents the woven fabric structure. However, the faultiness of fabric cover factor is the absence of weave influence.

$$K = \frac{ABGI + AEHD - AEFI}{ABCD} = K_1 + K_2 - K_1 K_2 \quad (15)$$

$$K_1 = \frac{d_1}{p_1} = \frac{G_1}{G_{\max_1}} = \frac{G_1 \cdot d_1}{10}$$
$$K_2 = \frac{d_2}{p_2} = \frac{G_2}{G_{\max_2}} = \frac{G_2 \cdot d_2}{10} \tag{16}$$

In Eq. (15) and Eq. (6), K is the fabric cover factor, K_1 is the warp cover factor, K_2 is the weft cover factor, d is the yarn thickness in mm, p is the distance between neighbourhood yarns in mm, and G_{max} is the maximal thread density in threads per cm. Subscripts 1 and 2 denote warp and weft yarn, respectively.

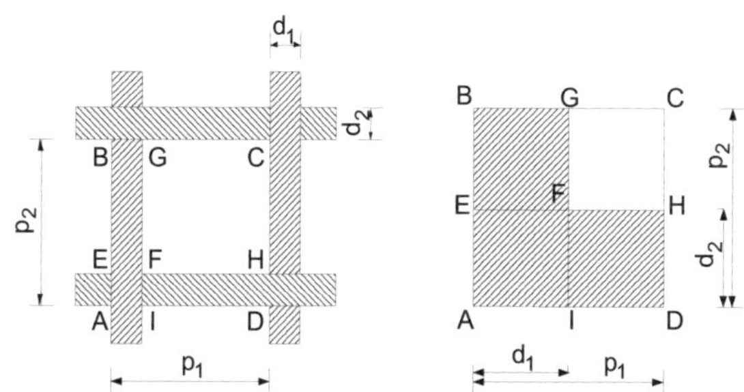

Fig. 3. Woven fabric geometry by fabric cover factor definition

Fabric porosity indicates the portion of pores in woven fabrics. While the woven fabric could be treated as two or three dimensional form, the terms open porosity and volume porosity are distinguished. **Open porosity** indicates the percentage of macropore's area in the fabric area unit. It is calculated on the basis of fabric cover factor or on the basis of the number of macropores and the area of macropore's cross section according to Eq. (17) and Eq. (18), respectively (Dubrovski & Brezocnik, 2005):

$$P_0 = (100 - K) \cdot 100\% \tag{17}$$

$$P_0 = N_p \cdot A_p \cdot 100\%$$
$$P_0 = G_1 G_2 (p_1 - d_1)(p_2 - d_2) \cdot 100\% \tag{18}$$
$$P_0 = (10 - d_1 G_1)(10 - d_2 G_2) \cdot 100\%$$

where P_o is the open porosity in percentage, K is the fabric cover factor in percentage, N_p is the number of pores in pores per cm^2, A_p is the area of macropore's cross section in cm^2, and G is the actual thread density in threads per cm. **Volume porosity** indicates the percentage of pore volume in the volume unit of woven fabric (Eq. (19)) with different types of pores (macro, mezzo, micro). It is calculated on the basis of fabric volume fraction which expresses the percentage part of yarn volume with regard to the fabric volume (Eq. (20)). While the fabric mass is actually the mass of yarns used ($m_{fab}=m_y$), the fabric volume fraction represents the ratio between the fabric mass density and yarn bulk density.

$$P_V = (100 - VF) \cdot 100\% \tag{19}$$

$$VF = \frac{V_y}{V_{fab}} \cdot 100\% = \frac{m_y \cdot \rho_{fab}}{\rho_y \cdot m_{fab}} \cdot 100\% = \frac{\rho_{fab}}{\rho_y} \cdot 100\% \tag{20}$$

In Eq. (19) and Eq. (20), P_V is the volume porosity in percentage, VF is the fabric volume fraction in percentage, V_y is the yarn volume in cm³, V_{fab} is the fabric volume in cm³, m_y is the yarn mass in g, m_{fab} is the fabric mass in g, ρ_{fab} is fabric mass density in g per cm³, and ρ_y is yarn bulk density in g per cm³. **Fabric thickness** is distance between the fabric face and back. Theoretically, it represents the sum of height of warp and weft arc according to Eq. (21) (Sokolovič, 1981):

$$h_1 = \frac{d_1 + d_2}{8}(F-1) \qquad h_2 = \frac{d_1 + d_2}{8}(9-F) \tag{21}$$

where h is the height of thread arc in mm, d is the yarn diameter in mm, and F is the number of Novik's fabric construction phase. While by a new fabric development there is no information in which Novik's fabric construction phase the woven fabric will appear, only minimal and maximal value of fabric thickness can be predicted. Minimal value of fabric thickness refers to V. Novik's phase, where warp and weft threads have equal yarn crimp. In this case the fabric thickness is the sum of warp and weft diameter (Fig. 4). Maximal value of fabric thickness refers to I. and IX. Novik's phases. In I. Novik's phase, where warp threads don't have any yarn crimp but weft threads maximal, fabric thickness is the sum of warp diameter and weft diameter multiple by two. In IX. Novik's phase, where warp threads have maximal yarn crimp and weft none, fabric thickness is the sum of warp diameter multiply by two and weft diameter.

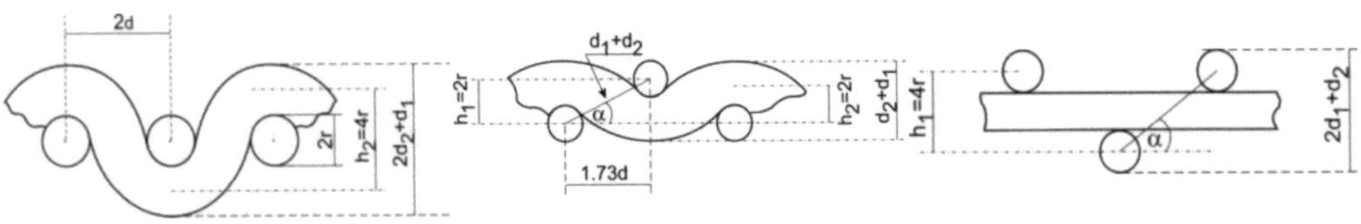

Fig. 4. I., V. and IX. Novik's fabric construction phases

Fabric mass can be expressed as mass per one meter or as mass per square meter. Later is more useful by comparing different types of woven fabrics. Theoretically it can be calculated according to Eq. (22) and Eq. (23) which refers to the unfinished and finished fabric state, respectively:

$$M_g = \frac{G_1 T_1}{100-\varepsilon_1} + \frac{G_2 T_2}{100-\varepsilon_2} \tag{22}$$

$$M_f = \frac{M_r(1-\frac{\Delta M}{100})}{(1-\frac{\varepsilon_{fin_1}}{100})\cdot(1-\frac{\varepsilon_{fin_2}}{100})} \tag{23}$$

where M_g is the raw fabric mass in g per m², G is the actual thread density in threads per cm, T is the yarn fineness in tex, ε is the yarn crimp in percentage, M_f is the fabric mass in finished state in g per m², ε_{fin} is the yarn shrinkage by finishing in percentage, and ΔM is mass change of raw fabric by finishing in pecentage. **Fabric mass density** expresses the mass of volume unit of woven fabric in grams per cm³. It can be calculated according to Eq. (24):

$$\rho_{fab} = \frac{M}{D \cdot 100} \tag{24}$$

where, ρ_{fab} is the fabric mass density in g per cm^3, M is the fabric mass in g per m^2 and D is the fabric thickness in mm. By fabric engineering it is possible only to predict minimal and maximal value of fabric mass density according to the minimal and maximal value of fabric thickness.

5. Effects of woven fabric construction, maintenance and usage on UV protection

5.1 Effects of yarn structure on UV protection

Woven fabrics are made from different types of yarns. Raw material of yarn or fibre composition is the initial yarn parameter which has an effect on UVR protection. Fibres have different ability to absorb UV radiation and to block most of the incident radiant energy and those prevent it from reaching the skin. There is a lack of studies dealing with the effect of fibre composition only. The reason is that yarn colour, additives and coatings have much more significant impact on UV transmission properties rather than fibre composition itself. Never less, Crews et al. (Hatch & Osterwalder, 2006) conducted a comparison of undyed woven fabrics and determined how fibre composition ranked relative in regard to UV absorbance. They established three distinct groups regarding the decreasing ability of fibre UVR absorbance: 1.group includes polyester, 2. group includes wool, silk and nylon and 3. group includes cotton and rayon fibres. Natural fibres have lower UV blocking properties regarding the synthetic ones, but from the thermo-physiology point of view there are more suitable in hot wearing conditions. Hustvedt et al. (2005) found that naturally-pigmented cotton fabrics have excellent sun protective properties, which are far superior to conventional, bleached or unbleached cotton fabrics. Stankovic et al. (2009) conducted a study of yarn twist effect on UPF of cotton knitted fabric and found that yarn twist to a great extent influenced the UV protection properties through the influence on yarn compactness and surface properties, which in turn influenced the open porosity of the fabric.

5.2 Effects of fabric geometry on UV protection

UV light passes direct through the macropores or fabric open area (direct UV transmittance) and also through the yarns, where changes the direction before leaving the fabric (scattered UV transmittance). Numerous studies focused on different fabric constructional parameters which represent the fabric structure the best and have direct and significant effect on UV protection. Such role has been given to fabric cover factor, fabric open porosity, fabric mass, fabric thickness etc. (Gies et al., 1998; Dimitrovski et al., 2009; Gabrijelčič et al., 2009; Hatch & Osterwalder, 2006).

5.2.1 Effect of cover factor or open porosity

To evaluate only the influence of fabric cover factor (or its complementary relationship - open porosity) on UPF and eliminate other significant factors such as colour and additives, the set of fabrics should be precisely prepared. Our experiment (Dubrovski & Golob, 2009) was focused on 100% cotton woven fabrics in a grey state with the same yarn fineness (14 tex) and different thread densities to achieve fabric cover factor between 59% and 87%. This was possible by introducing different types of weave (plain, twill, satin), while it is known that by plain weave lower densities are achieved due to the high number of thread passages regarding to the twill and satin weaves. Fabric cover factor and open porosity were calculated according to Eq. (15) to Eq. (18). While also cotton yarns absorb some of the incident UVR we could not focuse only on the UVR that goes through the macropores. To eliminate the influence of raw material, yarns with 100% absorption of UV light that strikes them should be used but this is not usually the case. From the Fig. 5 it can be seen that higher cover factor (or lower open porosity) means better UV protection and that cover factor should be at least 80% (or open porosity lower than 20%) to achieve good UV protection according to AS/NZ standard. This is possible only by higher thread densities and definitely not by plain weaves in our case. Even if the plain fabric would have the highest cover factor it would not reach the UPF 15. The results of mentioned study refer to the theoretical values of open porosity and cover factor. In real fabric open porosity is much lower, especially in the case of fabrics made from the staple-fibre yarns, where the phenomenon of latticed pores, the phenomenon of changing the position of warp threads according to the longitudinal axis and the phenomenon of thread spacing irregularity occur. In this case the correlation between the measured open porosity/cover factor (image analysis) and UPF is not so good and should be treated regarding the type of weave (Fig. 6).

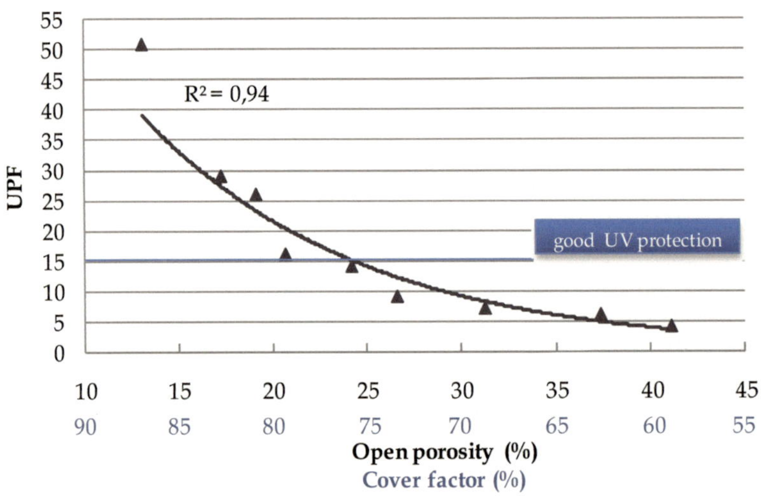

Fig. 5. The influence of theoretical values of open porosity or cover factor on UV protection of cotton fabrics in a grey state

The plain-weave fabric includes the maximum percentage of weave passages (67%) and it is reasonable to assume that all the threads are more or less equidistant and that the effect of fully latticed pores is reduced to its minimum, whereas by satin weave the effect of fully latticed pores is very high those reducing open area for UV transmission. If we observe measured values of open porosity, the limit values to reach good UV protection of fabrics is 12% or lower without taking into account the type of weave. Further observation regarding the type of weave shows that by plain and twill weave it is not possible to reach UPF 15 neither by 12% of open porosity, while by satin weaves this is possible. The results clearly indicate that theoretically defined open porosity/cover factor is not satisfactory parameter to asses its influence on UPF because of the absence of weave influence. In real fabrics, especially in fabrics with staple-fibre yarns, different types of pores regarding the type of weave and other phenomenon are involved, which all reduce the fabric open area in comparison with theoretically calculated values of open porosity. On the other hand, open porosity/cover factor could be a good parameter showing the influence on UPF if the set of fabrics with the same type of weave, raw material and yarn fineness is observed. In our previous research (Dubrovski & Brezocnik, 2002) we also proposed the predictive model of open porosity which is in better correlation with measured values than theoretical ones.

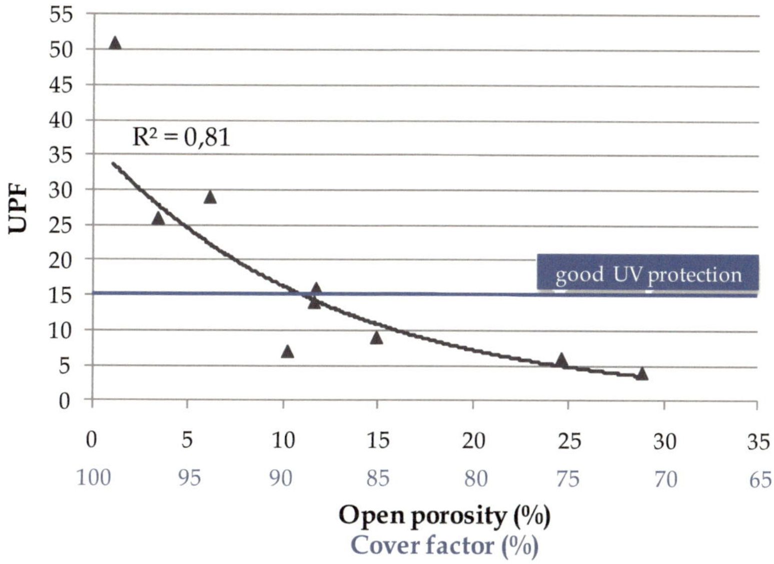

a

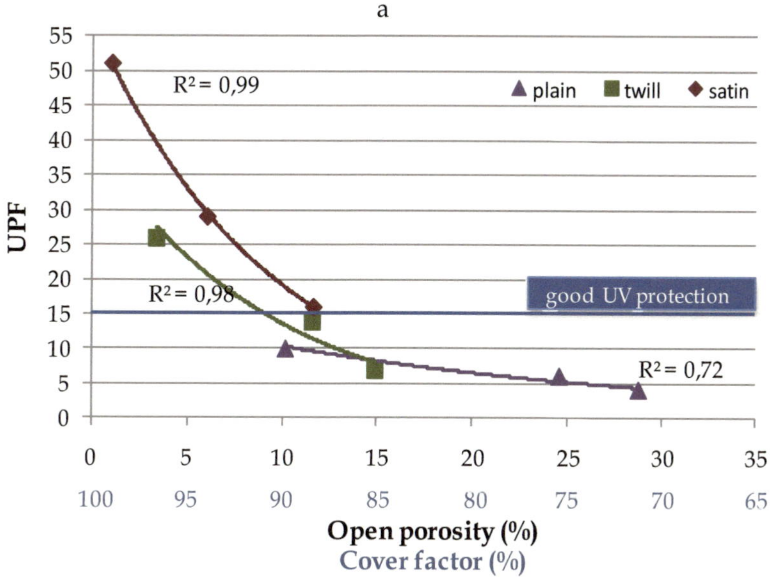

b

Fig. 6. The influence of measured values of open porosity and cover factor on UV protection of cotton fabrics in a grey state (a - without weave influence, b - with the weave influence)

5.2.2 Effect of fabric tightness

Fabric tightness or relative fabric density is another parameter which represents the fabric structure or how tight the fabric is woven, similar as cover factor. Advantage of fabric tightness is the consideration of weave by its calculation (Eq. (8) to Eq. (11)). It is known that by satin weave it is possible to achieve higher warp/weft density than with twill or plain weaves, so the limit density as well as actual density will be higher. Consequently, the macropores will be smaller and UV radiation will have less free space to pass through than in twill or plain weaves. The fabric tightness is relative term and according to previous mentioned experiment (Dubrovski & Golob, 2009), the following decreasing rate of UPF values could be seen within the same fabric tightness: satin - twill - plain (Fig. 7).

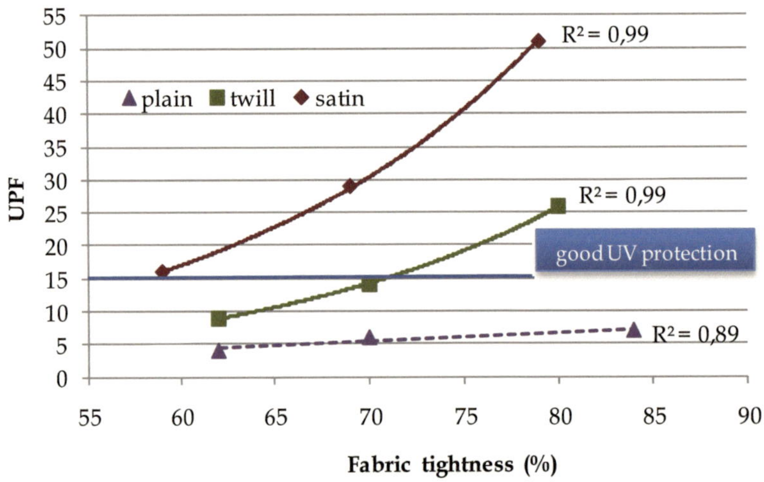

Fig. 7. The influence of fabric tightness on UV protection of cotton fabrics in a grey state

The macropores in plain fabrics have very stable and uniform form as a consequence of more thread passages. On the other hand, the pores in satin fabrics are not as stable due to few thread passages, and tend to group together which further reduces the free space area. By fabrics made from staple-fibre yarns macropores are further reduced because of the phenomenon of latticed pores. Nerveless higher actual warp/weft density by each weave means higher fabric tightness and consequently higher UV protection. Results for fabrics in a grey state show that none of the plain fabrics offered minimum UV protection, even if they were tightly woven. Twill fabrics had good UV protection if they were woven with tightness above 70%, while satin fabrics offered good UV protection already by 60% tightness.

5.2.3 Effect of volume porosity

Thicker and heavier fabrics minimize UVR transmission (Scott, 2005). While some of the researchers focused on fabric mass and thickness, we decided to include volume porosity as a parameter influencing UPF, while it includes fabric mass and thickness through the fabric volume fraction according to the Eq. (19), Eq. (20) and Eq. (25). Results for grey fabrics (Fig. 8) show that there is no direct correlation between volume porosity and UPF. Moreover, results indicate that volume porosity depends on the type of weave and affects UPF as well. This is in accordance with previous mentioned discussion about the macropores. The macropores as three-dimensional forms are bigger, more stable and uniform in plain fabrics compared with macropores in twill or satin fabrics at the same volume porosity. Lower volume porosity means higher UPF. Plain fabrics did not offer any UV protection, while twill and satin fabrics offered good UV protection when volume porosity was less than 64% and 66%, respectively.

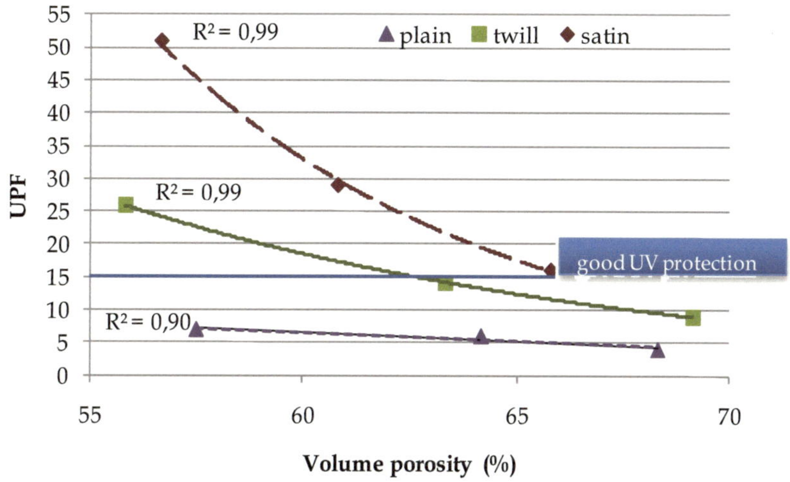

Fig. 8. The influence of fabric volume porosity on UV protection of cotton fabrics in a grey state

5.3 Effects of colour on UV protection

Undyed or bleached fabrics offer much lower protection against UV radiation if any in comparison with dyed fabrics. Dyes react like additives; they improve UV protection abilities, because they absorb UV radiation in the visible and UV radiation band. By bleaching process the naturally occurring pigments and lignin which act as UV absorbers are removed those affect UV absorber ability of cotton fabrics. Hypothesis that the hue of dye is responsible for UV protection of fabric is a matter to discuss.

Gatewood (Scott, 2005) noted that transmission/absorption characteristics of dyes in the UV band were a better predictor of UV protection than the colour of the dyestuff itself.

Srinivasaen et al. (Hatch & Osterwalder, 2006) who studied the effect of fourteen direct dyes on the UPF of cotton fabrics, concluded that colour (hue) is not related to UPF, while fabrics dyed with the dyestuff with the same hue (red 28, red 24, red 80) and identical concentration had different UPF values. Also the black fabric in this study did not have the highest UPF despite the common fact that darker colours (such as black, navy blue, dark green, red) of the same fabric type absorb UVR more strongly than light pastel shades (Yallambie, 2003; Wilson et al., 2008b). Neverless, the results of mentioned study indicate that higher dye concentration means higher UPF. Wilson et al. (2008a) concluded that black fabrics generally transmitted 20% less UVR than their matched white equivalent.

In another study Wilson et al. (2008b) examined the relationship between UV transmittance and colour and found that depth of colour, rather than colour per se is the principal aspect of colour affecting UV transmittance. The best description of the relationship between colour and UVR transmission was provided by the L*, and L* and b* components of the LAB system. He suggested that by developing fabrics for UV protection, selection of dyes that

generate colours with CIE Y or L* values of less than approximately 28 or 38, respectively is recommended.
Our study (Dubrovski & Brezocnik, 2009) focused on the effect of woven construction and colour of cotton woven fabrics dyed with the same concentration (1%) of reactive dyestuffs Cibacron LS (red, blue marine and black), bleached fabrics (white) and naturally pigmented fabrics (dirty white). The comparison of UPF of fabrics with the same construction but different colour was made for fabrics in plain, twill and satin weave and by three different levels of fabric tightness (55-65%, 65%-75%, 75%-85%). By satin and twill fabrics at third level of fabric tightness, where higher densities can be achieved and those the influence of open porosity is set to its minimum, the results show that all dyed fabrics posses excellent UV protection (UPF=1000), while naturally pigmented twill and satin fabric had UPF 25 and 50, respectively. UPF of bleached twill and satin fabric was 10 and 15, respectively. The L* component of fabric colour was around 93, 86, 44, 31 and 17 for white, dirty white, red, blue marine and black fabric, respectively. The previous mentioned recommendation that L* value of the dyed fabrics should be less than 38 to develop fabric with good UV protection, could not be generalized, while in our case also white satin fabric with L* of 93 showed good UV protection at third level of fabric tightness. Our results show that there were no big differences between red, blue and black coloured fabrics UPF at higher thread densities by twill and satin fabrics, but there was a huge difference between uncoloured and bleached fabrics UPF on one side and coloured fabrics UPF on another. The general conclusion of mentioned research was that UPF of cotton fabrics dyed with direct dyestuffs is influenced by the colour components (L*, a*, b*), fabric tightness and type of weave so we proposed a prediction model of UPF based on CIELAB colour components, weave factor, and warp/weft density.
Riva et al. (2009) analyzed the influence of the shade and colour intensity of the dyeing as well as their interaction with the initial UPF of the uncoloured cotton fabrics. They proposed UPF prediction model for cotton fabrics dyed with direct dyestuffs (yellow 98, blue 77, red 89) on the basis of the initial UPF of fabrics before dyeing, standard depth of colour, the corrected standard depth of colour and two categorical qualitative variables that define colour hue of dyestuffs.

5.4 Effects of additives on UV protection

During the fibre/yarn/fabric processes there is a possibility to include additives like a dye, pigment, delusterant, optical brighteners and UV absorbers, which have the ability to absorb UV radiation and those improve UV protection properties of fabrics with little UV protection like cotton, rayon, silk, wool, nylon and undyed fabrics. Besides dying, other techniques are known to incorporate additives in fabric structure: 1. addition of additives during fibre/yarn manufacturing, 2. addition of additives during fabric surface treatments or special treatments.
Pigments found in naturally-pigmented cotton are naturally UV absorbers and produce shades ranging from tan to green and brown. According to the study of Hustvedt & Crew (2005), fabrics from naturally pigmented cotton have excellent sun protection properties, which are far superior to conventional, bleached and unbleached cotton fabrics (green UPF=30-50+, tan UPF=20-45, brown UPF=40-50+, bleached conventional UPF=4, unbleached conventional UPF=8). Their UV protection properties remain high enough even after 80 AFUs light exposure.

TiO_2 or ZnO is delusterant pigment which is incorporated during fibre manufacturing and its effect is permanent. Optical brighteners convert a portion of incident UV radiation near 360 nm to the visible blue wavelengths about 430 nm and reflect it. UV absorbers are colourless additives having chromophore system that absorbs very effectively in the UV band. Optical brightness and UV absorbers are recently added to commercial laundry detergents (Yallambie, 2003).

Varga et al. (2009) introduced a nanoparticle coating on yarns. They applied nano ZnO finish on undyed and reactive dyed cotton yarns with the aim of studying the effect of the knitting operation on the durability of the coated nanoparticles and found that such yarns withstand the knitting process. They also performed sol-gel finishing of cotton fabrics, coated with TiO_2 nanoparticles and found that such fabrics are durable to domestic washing, and even there was a reduction in the load of nanoparticles on the fabric surface after washing, the UPF values were not affected.

Abidi et al. (2009) reported that titania or titania-silicia nonosol treatment in the form of thin film at cotton fabric surface offer excellent UV protection. Gorensek et al. (2007) treated cotton fabrics with nanosilver, which was in the form of nano powder added in the dyebath at two concentration (5 mg/L and 20 mg/L) and found that a noticeable increase of UPF was recorded by the 5% mock dyed sample with 20 mg/L nanosilver as well as by pale dyed fabrics in comparison with bleached and dyed cotton fabrics, respectively.

Grancaric et al. (2009) treated PET fabrics for summer clothing with ultrasound (US), ethylene-diamine (EDA), fluorescent whitening agents Uvitex ERN based on benzoxazole derivate (FWAs) and Tinofast PES UV absorbers based on triazine derivate and compared their UPF values. Untreated PET fabrics did not have any UV protection (UPF=5), while all other treatments lead to very good UV protection. EDA treated fabric resulted in better UV protection than US treated fabrics.

5.5 Effects of maintenance and usage on UV protection

When the fabrics for clothing are in use, their initial UPF of fabric is modified by laundering as well as by wearing conditions connected with the tension produced in contact with the body (fabric stretch) and with an exposure to the UV radiation in wet state (swimsuit). Stretching is more common in knitted rather than woven fabrics, with exception of elasticised woven fabrics. Most fabrics shrink when they are laundered which lead to significant improvement in the UPF of fabrics because of the open area reduction (Hatch & Osterwalder, 2006). Another reason of UPF improvement by laundering is optical whiteners which are added to laundry detergent.

Due to the effect of wetness and the effect of opening of the fabrics caused by the tension on tightened and/or elasticized garments, the initial UPF of unstretched and dry fabric does not have proper meaning. European standard EN 13758-1 in annex C considers measurements under stretched and wet conditions informatively, while ASTM D 6544 refers to the preparation of textiles prior to ultraviolet transmission testing which includes exposure conditions (laundering, simulated sunlight and chlorinated pool water).

Algaba et al. (2007) conducted a study on undyed woven fabrics made with three different cellulose fibres (cotton, modal and modal sun fibres that contain UV absorber in the spinning bath) which were exposed to the simulation of the wearing conditions of the clothing. Samples were stretched with a tension of 2, 4, and 6 N and the measurements were carried out after maintaining the samples (unstretched or stretched) in water until saturation. The UPF of fabrics decreased significantly when tension increased. The sign of the influence of the wetness on UPF depended on the fibre type. The UPF of wet cotton and modal fabrics was lower, while modal sun fabrics had higher UPF regarding the dry fabrics. Osterwalder et al. (Scott, 2005) concluded that UV absorbance is independent from environment and therefore treating cotton with UV absorber will afford complete protection when the fabric is wet. Wilson et al. (2008a) reported that by 10 x 20% extension UPF of cotton woven and knitted fabrics were decreased by -30% to -75%.

6. Conclusion

Woven fabrics can provide simple and convenient protection against harmful effects of UV radiation if the necessary attention is paid to their engineering in the phase of a new product development. There are several factors influencing UV protection properties of woven fabrics like yarn construction (fibre type, twist, yarn packing factor), fabric construction with its primary (type of weave, yarn fineness, warp/weft density, relative fabric density or fabric tightness) and secondary (cover factor, open porosity, mass, thickness, volume porosity) parameters of fabric geometry, additives (dye, pigment, delusterant, optical brighteners, UV absorbers), laundering and wearing conditions (stretch, wetness). The proper combination of mention factors allows production of passive woven fabrics with high UV protection properties, which may reduce risk associated with UV overexposure. For subject wearing garment made from UV protected fabrics the information about how long he/she could be exposed to the harmful UV rays before the serious skin damage occur, will be more useful, instead of knowing UPF value of garment. UV exposure time is affected by several factors like subject skin type, geographic position of subject, daily time or the sun position, the presence of clouds, altitude, portion of skin covered by fabric, etc. However, nowadays, there is a trend to develop smart textiles or active intelligent fabrics which, for example, could change their own colour in dependence on external stimulus like UV light (Vikova, 2004). Soon such smart textiles will be developed which will warn the subject how long he/she could be on the sun, what is the average UV index in a particular position, what is the UPF of wearing fabric in a particular moment, when subject should use the shadow, etc.

7. References

AATCC Test Method 183 (2004). Transmittance or Blocking of Erythemally Weighted Ultraviolet Radiation through Fabrics

Abidi, N.; Cabrales, L. & Hequet, E. (2009). Functionalization of a Cotton Fabric Surface with Titania Nanosols: Applications for Self-Cleaning and UV-Protection Properties. *ACS Applied Materials & Interfaces*, Vol.1, No.10, 2141-2146, ISSN 1944-8244

Algaba, I.; Riva, A. & Pepio, M. (2007). Modelization of the Influence of the Wearing Conditions of the Garments on the Ultraviolet Protection Factor. *Textile Research Journal*, Vol.77, No.11, 826-836, ISSN 0040-5175

ASTM D 6544-00 (2007). Standard Practice for Preparation of Textiles Prior to Ultraviolet (UV) Transmission Testing

ASTM D 6603-07 (2007). Standard Guide for Labelling of UV-Protective Textiles

AS/NZS 4399 (1996). Sun protective clothing - Evaluation and classification

Cesarini, J.P. (2001). Impact of Ultraviolet radiation on Humans. *Indoor and Built Environment*, Vol.10, No.5, 310-316, ISSN 1420326X

Denton, M.J. & Daniels, P.N. (2002). *Textile Terms and Definitions - 11th edition,* The Textile Institute, ISBN 1870372441, Manchester, England

Dimitrovski, K.; Sluga, F. & Urbas, R. (2009). Evaluation of the Structure of Monofilament PET Woven Fabrics and their UV Protection Properties. *Textile Research Journal on line first,* December 4, Vol.0, No.0, 1-11, ISSN 1746-7748

Dubrovski, D.P. & Golob, D. (2009). Effects of Woven Fabric Construction and Colour on Ultraviolet Protection. *Textile Research Journal,* Vol.79, No.4, 351-359, ISSN 0040-5175

Dubrovski, D.P. & Brezocnik, M. (2009). Prediction of the Ultraviolet Protection of Cotton Woven Fabrics Dyed with Reactive Dyestuffs. *Fibres & Textiles in Eastern Europe,* Vol.17, No.1, 55-59, ISSN 12303666

Dubrovski, D.P. & Brezocnik, M. (2002). Using Genetic Programming to Predict the Macroporosity of Woven Cotton Fabrics. *Textile Research Journal,* Vol.72, No.3, 187-194, ISSN 0040-5175

Dubrovski, D.P. & Šujica, M.Ž. (1995). The Connection Between Woven Fabric Construction Parameters and Air Permeability, *Fibres & Textiles in Eastern Europe,* Vol.11, No.4, 37-41, ISSN 12303666

Edlich, R.F.; Cox, M.J.; Becker, B.B.; Horowitz, J.H.; Nicher, L.S.; Britt, L.S.; Edlich, T.J. & Long, W.B. (2004). Revolutionary Advances in Sun-protective Clothing - an Essential Step in Eliminating Skin Cancer in our World, *Journal of Long-Term Effects of Medical Implants,* Vol.14, No.2, 95-105, ISSN 1050-6934

EN 13758-1 (2001). Textiles - Solar UV protective properties - Part 1: Method of test for apparel fabrics

EN 13758-2 (2003). Textiles - Solar UV protective properties - Part 2: Classification and marking of apparel

EPA (2010). *Ozone layer depletion - science,* http://www.epa.gov/ozone/science/q_a.html

Ferrini, R.I.; Perlam, M. & Hill, L. (1998). American College of Preventive Medicine, Practice Policy Statements: Skin Protection from Ultraviolet Light Exposure, *American Journal of Preventive Medicine,* Vol.14, No.1, 83-86, ISSN 0749-3797

Gabrijelčič, H.; Urbas, R.; Sluga, F. & Dimitrovski, K. (2009). Influence of Fabric Constructional Parameters and Thread Colour on UV Radiation Protection. *Fibres & Textiles in Eastern Europe,* Vol.17, No.1, 46-54, ISSN 12303666

Gies, P.H.; Roy, C.R.; Toomey, S. & McLennan, A. (1998). Protection against solar ultraviolet radiation. *Mutation Research - Fundamental and Molecular Mechanisms of Mutagenesis,* Vol.422, No.1, 15-22, ISSN 0027-5107

Gorensek, M. & Recelj, P. (2007). Nanosilver Functionalized Cotton Fabric. *Textile Research Journal,* Vol.77, No.3, 138-141, ISSN 0040-5175

Grancaric, A.M.; Tarbuk. A. & Majcen, A.M. (2009). The fluorescence of UV protected white polyester fabrics. *Proceedings of Autex 9th world textile conference,* 442-449, ISBN 978-975-483-787-2, Izmir, Turkey, May 2009, Ege University, Engineering Faculty, Department of Textile Engineering, Izmir

Hatch, K.L. & Osterwalder, U. (2006). Garments As Solar Ultraviolet Radiation Screening Materials. *Dermatolologic Clinics,* Vol.24, No.1, 85-100, ISSN 0733-8635

Hustvedt G. & Crews, P.C. (2005). Textile technology: The Ultraviolet Protection Factor of Naturally-pigmented Cotton. *Journal of Cotton Science,* Vol.9, No.1, 47-55, ISSN 15236919

Johnston, L. (2005). Sunlight, vitamin D & Health, accessible at http://www.healingtherapies.info/Sunlight&VitaminD.htm

Kienbaum, M. (1990a). Gewebegeometrie und Produktenwiklung. *Melliand Textilberichte,* Vol.71, No.10, 737-742, ISSN 0341-0781

Kienbaum, M. (1990b). Gewebegeometrie und Produktenwiklung. *Melliand Textilberichte,* Vol.71, No.11, 847-854, ISSN 0341-0781

Riva, A.; Algaba, I.; Pepio, M. & Prieto, R. (2009). Modeling the Effects of Colour on the Protection Provided by Cotton Woven Fabrics dyed with Azo Dystuffs. *Industrial & Engineering Chemistry Research,* Vol.48, No.22, 9817-9822, ISSN 0888-5885

Ron, N. (2005). Solar Radiation, http://home.iprimus.com.au/nielsens/solrad.html

Roy, C.R.; Gies, H.P. & Toomey, S. (1995). The Solar UV Radiation Environment; Measurement Techniques and Results, *Journal of Photochemistry and Photobiology B: Biology,* Vol. 31, No.1-2, 21-27, ISSN 1011-1344

Scott, R.A. (2005). *Textiles for protection,* Woodhead Publishing Limited, ISBN 13:978-1-85573-921-5, Cambridge, England

Sokolovič, R. (1981). Ispunjenost tkanina I uticaj na njena svojstva. *Tekstil,* Vol.30, No.4, 208-218, ISSN 0492-5882

Sparling, B. (2001). Ultraviolet radiation, http://www.nas.nasa.gov/about/education /ozone/radiation.html

Stankovic, S.B.; Popovic, D.; Poparic, G.B. & Bizjak, M. (2009). Ultraviolet Protection Factor of Gray-state Plain Cotton Knitted Fabrics, *Textile Research Journal,* Vol.79, No.11, 1034-1042, ISSN 0040-5175

Turnbull, D.J. & Parisi, A.V. (2005). Increasing the Ultraviolet Protection Provided by Shade Structures, *Journal of Photochemistry and Photobiology B: Biology,* Vol. 78, No.1, 61-67, ISSN 1011-1344

Varga, M.; Botet, J.M.; Puntes, E.S.V. & Marsal F. (2009). Nano-cotton Fabrics with High Ultraviolet Protection, *Textile Research Journal on line first,* October 13, Vol.0, No.0, 1-9, ISSN 1746-7748

Vikova , M. (2004). Visual assessment of UV radiation by colour changeable textile sensors. *Proceedings of AIC 2004 Color and Paints, Interim Meeting of the International Colour Association,* Porto Alegre, Brasil, November 2004, Brazilian Colour Association

Williams, R. & Williams, G. (2002). Reflected Ultraviolet Photography, *Medical and Scientific Photography - an online resource, http://msp.rmit.edu.au/Article_01/index.html*

Willson, C.A.; Bevin, N.K.; Laing, R.M. & Niven, B.E. (2008a). Solar Protection Effect of Selected Fabric and Use Characteristics on Ultraviolet Transmission, *Textile Research Journal,* Vol.78, No.2, 95-104, ISSN 0040-5175

Willson, C.A.; Gies, P.H.; Niven, B.E.; Mclennan, A. & Bevin, N.K. (2008b). The Relationship between UV Transmittance and Colour Visual Description and Instrumental Measurement, *Textile Research Journal,* Vol.78, No.2, 128-137, ISSN 0040-5175

Yallambie, V. (2003). *Resource Guide for UV Protective Products 2003 – Information Section,* Australian Radiation Protection and Nuclear Safety Agency, accessible at http://www.apparel.ca/_static/webupload/websiteDocuments/100000/UV%20reference.pdf

Zabetakis, A. (2002). Textiles for protection against solar UVR, weather conditions and fire, *Proceedings of 1st International Textile, Clothing & Design Conference,* 152-157, ISBN 953-96408-8-1, Dubrovnik, October 2002, University of Zagreb, Faculty of Textile Technology, Zagreb, Croatia

e. Textile Composite Engineering

16

Microwaves Solution for Improving Woven Fabric

Drago Katovic
University of Zagreb
Faculty of Textile Technology
Croatia

1. Introduction

According to well known physical definition, electromagnetic waves are oscillating electric and magnetic fields traveling together through space. In the electromagnetic radiation spectrum, shown in figure 1, microwaves (300 MHz - 300 GHz) lie between radio wave (RF) and infrared (IR) frequencies, with relatively large wavelength (1m-1mm) (Metaxas & Meredith)

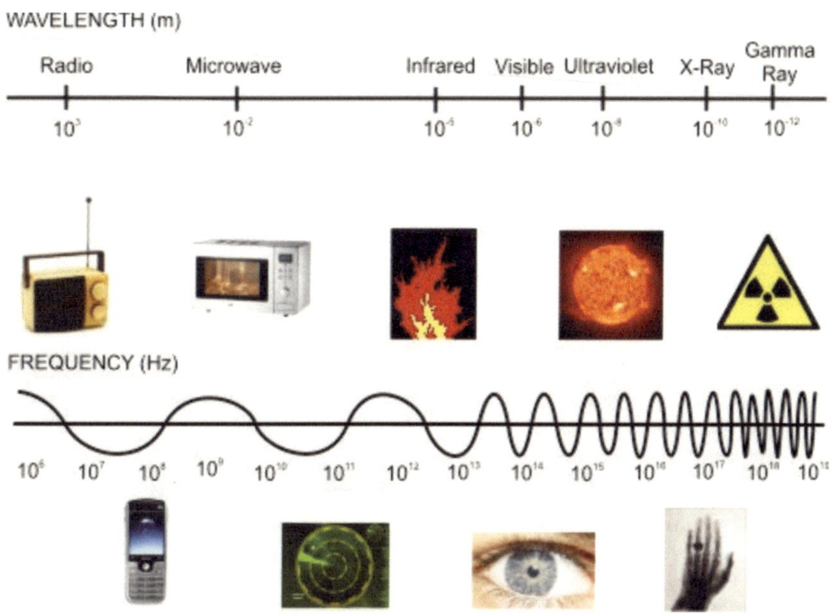

Fig. 1. Electromagnetic spectrum

Electromagnetic wave is formed by the electrical charge in the conductor that produces an electrical field in the spreading direction. The electrical field produces the magnetic filed. The so-formed magnetic field reproduces the electrical field in the space. The electrical field is perpendicular to the magnetic field, and both are perpendicular to the direction of the spreading wave.

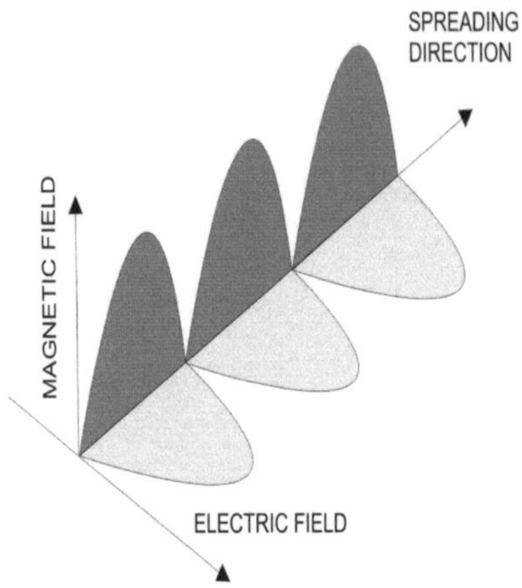

Fig. 2. Direction spread of electromagnetic wave

The energy of microwave photons is very low (0,125 kJ/mol) relative to the typical energies for chemical bonds (335-84 kJ/mol); thus microwave will not directly affect the molecular structure. They cannot change the electronic structure around atoms or among them, but they can interact with the electronic differences between atoms. However, chemical reactions can accelerate due to selective absorption of microwave energy by polar molecules, while non-polar molecules are inert to the microwave (MW) radiation (Varma 2001).
Different materials can be divided according to their response on microwave radiation:

- The materials that reflect MW radiation (stayed cold)
- The materials that are transparent to MW radiation (non-heated)
- The materials that absorb MW energy (being heated).

However, chemical reactions can be accelerated due to selective absorption of electromagnetic energy by polar molecules, while non-polar molecules are inert to the electromagnetic radiation. Besides influencing dipole water molecules, an alternating electromagnetic field also acts on partially polar molecules of textiles such as polyurethane (PU), polyacrylonitrile (PAN), or polyamide (PA)
A microwave electromagnetic field oscillating at 2.45 GHz, which is preferred frequency for heating applications, the charge changes polarity nearly 5 billion times per second. Microwave radiation is specially tuned to the natural frequency of water molecules to maximize the interactions.
Some important applications of microwaves come from their interaction with various types of material. The interaction of microwaves with dielectric materials causes a net polarization of the substance. There are several different mechanisms of polarization: electronic polarization, ionic, molecular (dipole) polarization and interfacial (space-charge) polarization. The overall net polarization creates a dipole moment. Dipole rotation is an interaction, in which polar molecules or species try to align themselves with the rapidly changing electric field of applied radiation. The motion of the molecule as it tries to orient to the field results in a transfer of energy. The second way to transfer energy is ionic conduction that occurs if there are free ions or ionic species present in the substance being heated.
The main difference between conventional heating with hot air and microwave heating is the heating mechanism. While conventional techniques heat a surface, the microwaves heat the whole volume of the treated object. During the conventional heating, the heat is

generated outside the treated product and conveyed by conduction or convection. Hence, the surface is heated at first and afterwards the heat flows toward the inside, which always remains colder than the surface. The required internal temperature can be reached only by sufficient increase of the surface temperature of the material above the temperature needed for particular treatment.

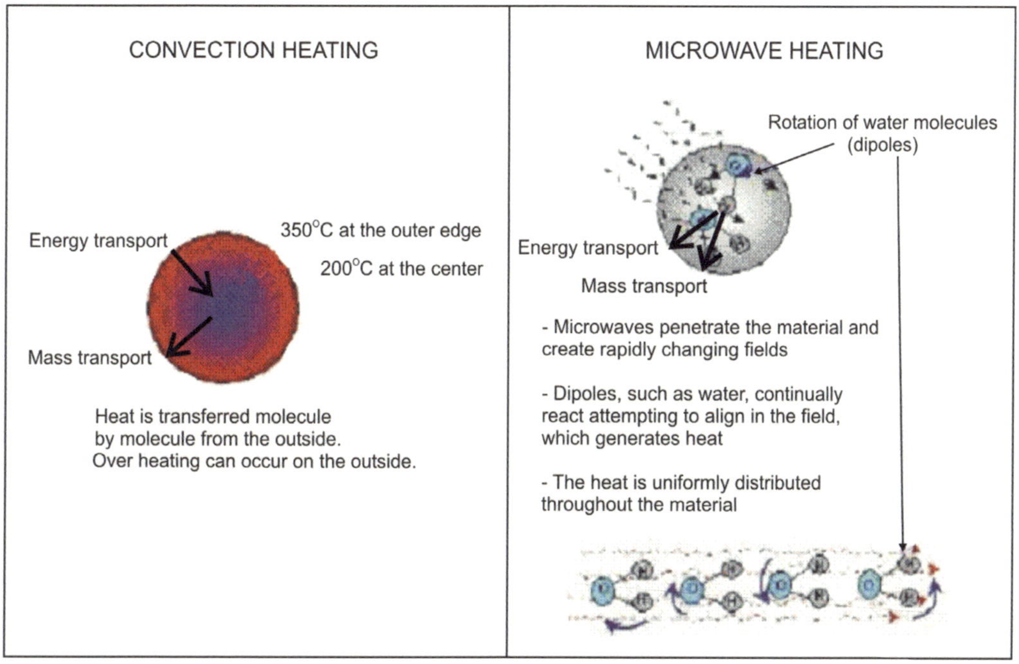

Fig. 3. Energy transfer comparison

On the contrary, in electromagnetic treatment, the heat is generated in a distributed manner inside of the material, allowing more uniform and faster heating. According to the literature (Metaxas & Meredith 1983) the energy consumption is 60-70 % lower in a case of electromagnetic treatment. For dielectric heating the generated power density per volume is calculated by

$$p = \omega \cdot \varepsilon_r'' \cdot \varepsilon_0 \cdot E^2 \tag{1}$$

where ω is the angular frequency, ε_r'' is the imaginary part of the complex relative permittivity, ε_0 is the permittivity of free space and E the electric field strength. The imaginary part of the complex relative permittivity is a measure for the ability of dielectric material to convert radio frequency electromagnetic field energy into heat.

What are the advantages microwave? Because volumetric heating is not dependent on heat transfer by conduction or convection, it is possible to use microwave heating for applications where conventional heat transfer is inadequate. One example is in heterogeneous fluids where the identical heating of solids and liquids is required to minimize over-processing. Another is for obtaining very low final moisture levels for product without over-drying. Other advantages include: Microwaves generate higher power densities, enabling increased production speeds and decreased production costs.

Microwave systems are more compact, requiring a smaller equipment space or footprint. Microwave energy is precisely controllable and can be turned on and off instantly, eliminating the need for warm-up and cool-down. Lack of high temperature heating surfaces reduces product fouling in cylindrical microwave heaters. This increases production run times and reduces both cleaning times and chemical costs. Microwaves are a non-contact drying technology. One example is the application of IMS planar dryers in the textile industry, which reduce material finish marring, decrease drying stresses, and improve product quality. Microwave energy is selectively absorbed by areas of greater moisture. This results in more uniform temperature and moisture profiles, improved yields and enhanced product performance. The use of industrial microwave systems avoids combustible gaseous by-products, eliminating the need for environmental permits and improving working conditions.

What are the disadvantages? Historically, the primary technological drawback to using microwave energy for industrial processing has been the inability to create uniform energy distribution. If uniform energy distribution is not present, wet regions of the target material are underexposed, and other regions are overexposed. This is analogous to the hot spots and cold spots generated in your microwave oven at home when heating or defrosting food like a potato or frozen chicken. Severe overexposure of non-uniform energy distribution may provide excessive focus of heat build up resulting in burnt material or a fire hazard. The uniformity of distribution designed into IMS microwave equipment overcomes this problem. Another disadvantage is the depth of penetration achievable using microwave energy. This is a function of microwave frequency, dielectric properties of the material being heated and its temperature. As a general rule, the higher the frequency, the lower the depth of penetration. 2,450 MHz versus 915 MHz? 915 MHz generators can provide up to 100 KW from a single magnetron. Although the cost is similar, the largest commercial 2,450 MHz units available use 30 KW magnetrons. 915 MHz generators lose about 15% efficiency in producing electromagnetic energy from electric power. However, the conversion of that energy into useful heating or drying is often greater than 95% so that the total system efficiency usually exceeds 80%. This compares with 55 to 70% total system efficiency obtainable from 2,450 MHz generators. The depth of penetration of microwave energy at 915 MHz is about three times as great as that at 2,450 MHz. With their higher total system efficiencies, 915 MHz heaters and dryers tend to have lower running costs than comparable 2,450 MHz units. One 100 KW 915 MHz generator will be about 50% cheaper than seven 15 KW 2,450 MHz units. The low power 2,450 MHz magnetrons developed from the proliferation of domestic microwave ovens are inexpensive and readily available. This makes them ideal for low flow capacity R & D applications. The size of magnetrons and wave-guides for a 2,450 MHz system is considerably smaller than those used in 915 MHz units. This makes them suitable for small-scale installations. 2,450 MHz is efficient where fast product expansion is required, such as dry frying of starch-based foods.

Today they are widely accepted and spread to mobile phones, television, wireless computer networks and some special applications such as rocket engines.

2. Microwave in textile finishing

The term "microwaves" was used for the first time in 1932nd, and its first usage was during the Second World War in radio communication and radar technology. The activity of electromagnetic field of high frequency was discovered accidentally during a radar-related

research project, while testing a new vacuum tube, called a magnetron. After more then 50 years of investigation and development, the microwave heating technology is nowadays widely used in number of fields. Studies in the last decade suggest that microwave energy may have a unique ability to influence chemical processes. These include chemical and materials syntheses as well as separations (Tompsett et al. 2006). Until now, MW have been used for food preparation, chemical sludge, medical waste, organic synthesis (Cablewski et al. 1994), analytics and curing (Saito et al. 2004) of hi-tech polymers (Zubizarreta L at all). There are a number of papers dealing with synthesis of organic compounds using microwave (Varma R. 2001). Numerous chemical reaction of textile materials are discussed and presented; e.g. (Barantsev et al. 2007) substitution, additions esterfication (Satge et al (2000), transesterfications, acetylization, amidation and decarboxylation (Hou & Wang 2008). One of the advantages of using microwave radiation is its influence on the reaction kinetics. Kaynak investigated the influence of polymerization time and dopant concentraction on the absorption of microwave radiation in conducting polypyrrole coated textiles (Kaynak et al. 2009). Chang investigated microwave heating for butyrylation of wood with aim of reducing the reaction time (Chang & Chang 2003).
The effect of the sintering temperature on the structural characteristies of nanosized zirconium dioxide particles treated by microwave radiation the during process was investigated by small-angle X-ray scattering and the BET method. It was shown that the specific surface area, particle size, polydispersvity index, and surface and mass fractal dimensionality of zirconium dioxide depend on heat treatment conditions (Strizhak et al).
Microwave moisture measurement is capable of measuring the moisture application behind the padder in continuous dyeing processes and of evaluating the measured values for the padder control. They can also used to determine the residual moisture content behind the stenter exit. A defined microwave emission is thereby beamed onto the damp fabric. The proportion of microwaves, not absorbed because of its density, is measured and relates to the humidity by calibration (Rouette 2002).

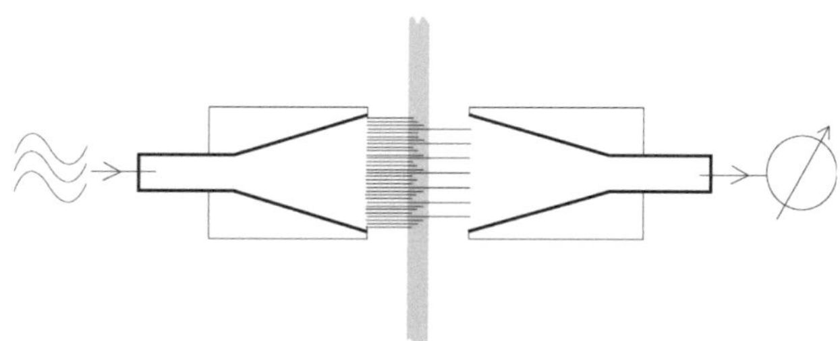

Fig. 4. Principle of moisture measurement using microwaves

The first idea of microwave application for textile finishing processes originated in 1970-es when cellulose fabrics were treated with Durable Press (DP) finishing agents and cured in microwave oven (Englert & Berriman 1974). Until now, microwave irradiation for textile finishing has been used (anonymous 1996) for the combined de-sizing, scouring and bleaching processes, dyeing (Nando & Patel 2002), printing (Neral et al. 2007), and drying processes, as well as for eradication of insects from wool textiles (Regan 1982). Microwave sterilization has many advantages in comparison with conventional methods. It is able to raise the temperature of a material in a short time and selectively heat the material. This results in the reduction of usage and the rapid completion of sterilization Bacillus subtilis (ATCC 9372) and Bacillus stearothermophilus (ATCC 7953) (Wang et al. 2005)

Although these first results microwave irradiation for textile finishing, were promising, the idea was abandoned until 1955, when Miller patented his Pre-set process without awareness of the earlier patent. Both cases involved garment microwave treatment, but they were abandoned because of efforts to control the process failed.

2.1 State of the art

The influence of three different drying methods, convection, contact and as a novelty – microwave one, on physical-mechanical parameters of yarn sizing was investigated by Katovic et al. The research was performed on 4 different types of 100% cotton yarn which had been sized on newly constructed laboratory sizing device. In this way the following parameters: sizing velocity, temperature of the sizing agent, tension and inlet moisture of the warp, outlet moisture of the warp, after drying and drying intensity were continuously controlled and regulated. The application of microwave drying method for wrap sizing showed to be good or even better in some cases, compared to the other drying methods (Katovic et al., 2008).

For microwave vacuum drying (Therdthai & Zhou 2009), three microwave intensities were applied with pressure controlled at 13.33 kPa. For hot air drying, two drying temperatures were examined. The microwave vacuum drying could reduce drying time by 85-90% compared with the hot air drying. In addition, colors change during drying was investigated. From scanning electron micrographs, the microwave vacuum dried mint leaves had a more porous and uniform structure than the hot air ones.

Microwave heating has been proved to be more rapid, uniform and efficient, and easy penetrate to particle inside. To investigate the effect of microwave irradiation on the physical property and morphological structure of cotton cellulose, cellulose fabric was treated with microwave irradiation at different condition. The morphological structures and thermal stabilities of the untreated and treated cellulose were investigated with differential structures and thermal stabilities of were investigated with differential scanning calorimetry and X-ray diffraction. The thermal stability of the treated cellulose was changed. The crystallinity and preferred orientation of the treated cotton cellulose increased (Hou et al. 2008).

The release of formaldehyde from plywood has been greatly reduced by treatment with microwave radiation. Microwave released formaldehyde from plywood samples more effectively compared to samples subjected to thermal energy from external heating. This suggests that microwaves directly activate free formaldehyde molecules, which have a polarity that is susceptible to microwaves. (Saito, Y. et al, 2004).

The influence of microwaves on the efficiency of polycarboxilic acid esterification was studied by FT-IR spectroscopy. Polycarboxilic acid is used as non-formaldehyde durable press finishing agents and maximum effects can be obtained with 1,2,3,4 butantetracaroxylic acid (BTCA) and citric acid (CA). Instead of the usual curing process performed at very high temperatures microwaves were used. Fabric resilience improved while the whiteness was not significantly lowered (Katovic & Bischof Vukusic, 2002).

The esterification involved in Durable Press (DP) finishing is one among several chemical reactions that can be improved by microwave radiation. Cotton material is usually esterified with modified 1, 3 dimethylol 4, 5 dihydroxyethylene urea (DMDHEU). In this study, a novel microwave planar device was used for simultaneous drying and curing processes. The experimental results showed that microwave-assisted textile finishing yields better results than conventional curing at tender frame. Noticeable improvements were obtained in wrinkle recovery resistance and tensile strength reduction. In addition, the influence of microwaves on formaldehyde release was investigated in order to decrease formaldehyde emission from textile material. Several different experimental methods were used in order to identify a mechanism of formaldehyde release (Katovic et al. 2002, 2005). An alternative approach to formaldehyde-releasing conventional N-methylol compounds is based on the use of non-formaldehyde durable press polycarboxlic acid (PCA) finishing agents. Another alternative approach, investigated, is using microwave energy to impart durable crease resistance to dyed cotton fabric. The bi-functional reactive dyes are used in the study, and the isocratic HPLC method is employed to quantify the PCA reacted with the cellulosic material for two different curing procedures. Shade change evaluation reveals that microwave curing has a greater influence on the dE values than conventional curing. In all other aspects, primarily wrinkle recovery and deformation resistance, microwave curing offers much better results (Katovic et al 2000) and (Bischof Vukusic et al, 2000).

A new microwave curing system was used to affect cross-linking of cotton fabric with non-formaldehyde finishes, namely, glyoxal, glutaraldehyde and BTCA along with water soluble chitosan in order to impart ease care and antibacterial properties to the fabrics (Fouda et al., 2009).

The esterification involved in Durable Press (DP) finishing is one among several chemical reactions that can be improved by microwave radiation. Cotton material is usually esterified with modified 1,3 dimethylol 4,5 dihydroxyethylene urea (DMDHEU). The experimental results obtained on a novel microwave planar device used for simultaneous drying and curing processes showed that MW-assisted textile finishing yields better results than conventional curing at stenter frame, especially for wrinkle recovery resistance and tensile strength reduction (Katovic et al., 2005).

Esterification of cellulose with fatty acids is relatively more recent than acylation of cellulose. The fatty acid esters of cellulose are potentially biodegradable plastics. Most of the undertaken studies using conventional heating resulted in long reaction times. Rapid homogeneous esterification of cellulose with long chain acyl chloride induced by microwave irradiation was studied by Satge et al. The use of microwave resulted in dramatic drop in reaction time: 1 min irradiation was sufficient, compared with 30 min to 2 days, when conventional heating is used. In this work, a systematic study of the effect of degree of substitution as the main parameter for estimating biodegradability was performed (Satge et al., 2002).

Temperature changes in conducting polypyrrole/para-toluene-2-sulfuric acid coated nylon textiles due to microwave absorption in the 8-9 GHz frequency ranges were obtained by a thermography station during simulataneous irradiation of the samples. The temperature values are compared and related to the amounts of reflection, transmission and absorption obtained with a non-contact free space transmission technique, indicating a relationship between microwave absorption and temperature increase. Non-conductive samples showed no temperature increase upon irradiation irrespective of frequency range. The maximum temperature difference around 4° C in the conducting fabrics relative to ambient temperature was observed in samples having 48 %absorption and 26.5 ± 4%reflection. Samples polymerized for 60 or 120 min with a dopand concentraction of 0.018 mol/l or polymerized for 180 min with a dopant concentraction of 0.009 mol/l yielded optimum absorption levels. As the surface resistivity decreased and the reflection levels increased, the temperature increase upon irradiation reduced (Kaynak et al., 2009).

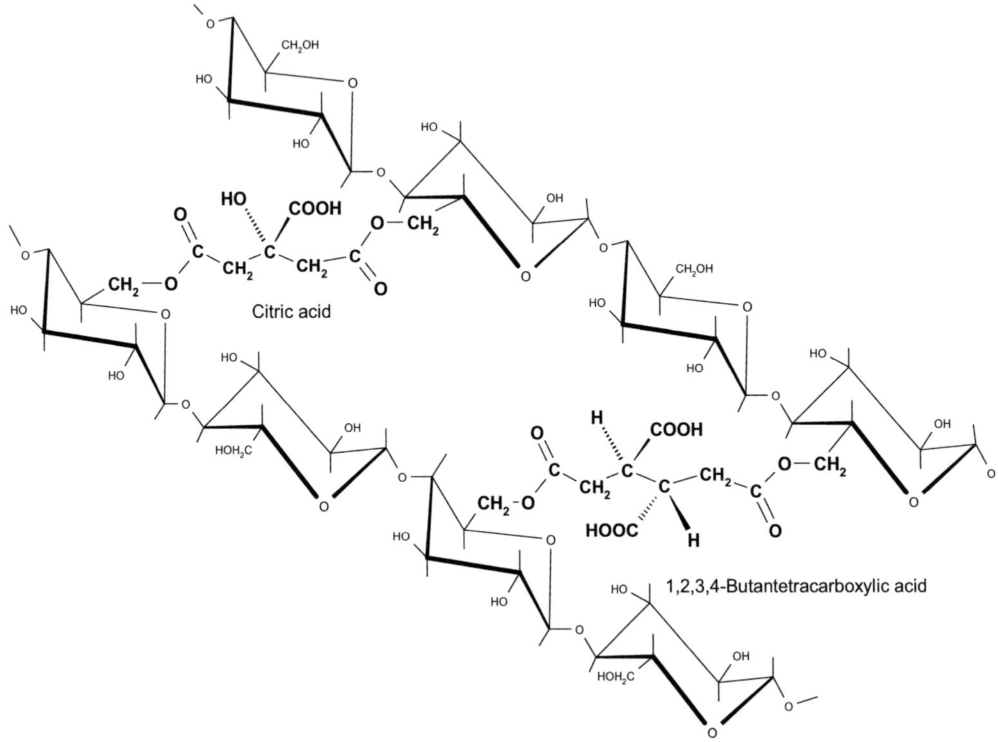

Fig. 5. Cross-linking via ester linkages of CA and BTCA with cellulosic chains

The important possibility of conducting sol-gel synthesis of oxide systems on the surface of para-aramid fibres under the effect of microwave radiation was demonstrated. Selection and control of the basic process parameters (duration and intensity of exposure to electromagnetic radiation, concentration of salts and carbamide) allow regulation of the effectiveness of interaction of reaction system with the microwave fields and eliminating degradation of the polymer (Barantsev et a., 2007).

Results of water- and oil-repellency obtained on planar MW apparatus have been compared with the ones obtained using conventional curing treatment. Simultaneous drying and curing processes have been conducted with MW at planar microwave device for the first time. Microwave technology offers better effects than conventional curing at stenter frame. Only in the case of durability to washing, cotton material treated with microwaves has shown a decrease. Lower effects, primarily caused by de-orientation of fluorocarbon chains, have been improved with thermal re-activation performed after washing and dry cleaning. Greatest advantages of the microwave device constructed are lower production costs and the elimination of separate drying procedure. In this way, conventional treatment, which might cause uneven effects, is eliminated (Bischof Vukusic et al., 2004). The influence of microwave pre-treatment on the UV protective properties of white polyester woven fabrics was investigated. The fabric samples for shade structures of various construction characteristics have been air dried and dried using laboratory microwave device. The impact of microwave pre-treatment has been verified after microwave untreated and treated samples examination and micro-structural changes of the fabrics treated influenced by intra- and inter- structural PES multifilament yarn changes attributed to specific character of the treatment applied. It was found that changes mentioned, confirmed with obtained results of fabric mass per unit area, fabric thickness, yarns diameter, percent cover, volume porosity and air permeability, have a strong influence on UVA and UVB transmission trough the fabrics. Synergistic influence on the UV protection effect obtained by unconventional pre-treatment and agents based on organic UV absorbers has also been evaluated (Tomljenovic & Katovic 2008).

The efficiency of microwave fixation of prints of the reactive dye applied to a cotton fabric using the digital print technology has been investigated. The results of the fixation of prints with saturated steam and hot air were compared with the characteristics of the microwave-fixed prints. The effects of time and microwave power on change in characteristics of impregnated textile substrates were teed. Based on the results obtained it may be concluded that the characteristic of microwave-fixed prints comparable with the characteristics of digital prints are of reactive dyes fixed by classic methods (Neral et al. 2007).

This paper deals with calibration and standardization of microwave oven, selection of energy level, configuration and placement of fabric swatch in the oven and fixation time to get optimum results viz., shade closest to that obtainable by the cold pad-batch method. Both vinyl sulphone as well asheterobifunctional dyes have been studied. The major finding has been that high energy and low exposure time in microwave oven gives comparable results to those by pad batch method in terms of K/S values, bleaching in post-dyeing wash-off and dry wet rub fastness. Several bulk trials have been taken successfully (Nanda & Patel 2002).

Thermosetting is an important part of the finishing of thermoplastic poly (ethylene terephthalate) (PET) fabrics and garments that confers stability in dimensions and shape as well as appropriate hand to the final product. Conventional thermosetting methods for PET include hot air and steaming treatments. In the present work we used solid state NMR as well DSC methods in order to investigate any differences in the behavior of PET chips when annealed with either a conventional or microwave technique. (D'Arrigo et al. 2002).

2.2 Electromagnetic devices in textile finishing

There are three types of devices for microwave processing of flexible materials. The device based on the resonant cavity principle can be used on discontinuing principle. Therefore it is suitable for lab research of small quantities of textile materials. The major part of the research was conducted on this type of a device. Devices based on the open resonator and waveguide applicator principle operate according to a continuing principle, and they are still being tested. These devices for microwave textile finishing are prevalently laboratory apparatus. Their main problem is reduced spreading of microwaves into the environment through gaps for flexible material. The only devices using electromagnetic waves that are used in textile industrial applications are radio-frequency dryers.

2.2.1 Resonant cavity

The frequencies used in microwave ovens were chosen based on two constraints. The first is that they should be in one of the industrial, scientific, and medical (ISM) frequency bands set aside for non-communication purposes. Three additional ISM bands exist in the microwave frequencies. Two of them are centered on 5.8 GHz and 24.125 GHz, but are not used for microwave cooking because of the very high cost of power generation at these frequencies. The third, centered on 433.92 MHz, is a narrow band that would require expensive equipment to generate sufficient power without creating interference outside the band, and is only available in some countries. For household purposes, 2.45 GHz has the advantage over 915 MHz in that 915 MHz is only an ISM band in the ITU Region while 2.45 GHz is available worldwide.

Most microwave ovens allow users to choose between several power levels. In most ovens, however, there is no change in the intensity of the microwave radiation; instead, the magnetron is turned on and off in duty cycles of several seconds at a time. This can actually be heard (a change in the humming sound from the oven), or observed when microwaving airy foods which may inflate during heating phases and deflate when the magnetron is turned off. For such an oven, the magnetron is driven by a linear transformer which can only feasibly be switched completely on or off. Newer models have inverter power supplies which use pulse width modulation to provide effectively-continuous heating at reduced power so that foods are heated more evenly at a given power level and can be heated more quickly without being damaged by uneven heating.

The cooking chamber itself is a Faraday cage which prevents the microwaves from escaping. The oven door usually has a window for easy viewing, but the window has a layer of conductive mesh some distance from the outer panel to maintain the shielding. Because the size of the perforations in the mesh are much less than the microwaves' wavelength, most of the microwave radiation cannot pass through the door, while visible light (with a much shorter wavelength) can.

This type of device has precisely determined dimensions depending on the characteristics of microwaves. Until now, the use of different types of resonant cavities has been tested for the purpose of microwave treatment and one of them is a domestic oven. A magnetron operating most often in the 2.45 GHz band (ISM) generates microwave power between a few hundred watts and few kilowatts, depending upon the application. It is connected by means of a waveguides to resonant cavity oven, which contains the materials to be heated or dried: food, wood, paper, plastics chemicals textiles, building materials. A mode stirrer distributes the microwave energy among the different resonant modes of the cavity, ensuring homogeneous heating. Main problems related to the use of such resonant cavities are the

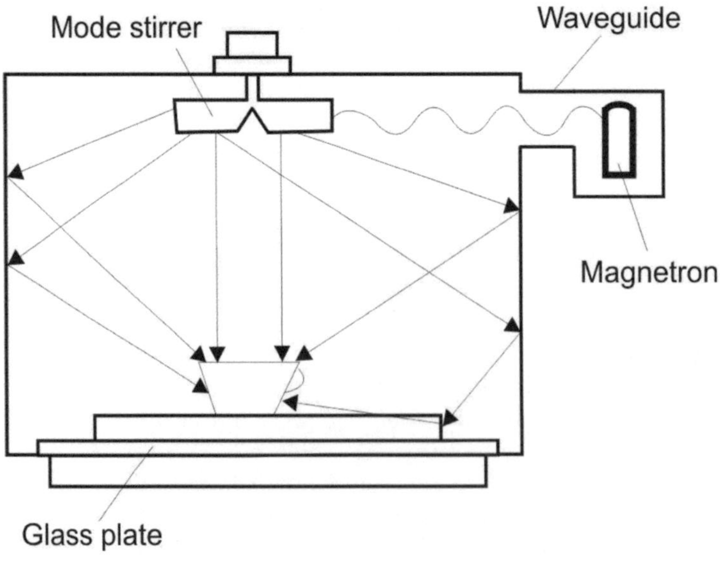

Fig. 6. Microwave oven

non-uniform energy distribution and possible MW leakage from the door seals in the case of inadequate chokes. The distribution of microwave energy within the cavity is always imperfect and the rotator (turntable) will cause the passage of the material through hotter and cooler spots, averaging out the exposure to microwaves. (Thewlis & Barnold 1999), (Hong & Thompson 1998), (Enderling 1988).

2.2.2 Open-resonator

The second reported microwave drying machine consist of many drying cells (17 in their prototype machine), which are positioned above the moving textile material. Each of the drying cells is based on the idea of an open resonator. These cells have their own magnetron placed in a waveguide holder. This applicator, which derives from the Fabry-Perot open resonator, has a magnetron as a source of high electromagnetic power. Dried textile material is located in the middle plane between the parallel conductive plates and the distance between these plates is equal to $3/2\ \lambda$. The use of this device is mostly for drying in the factory production of fabrics. This type of semi industrial dryer was developed at the Czech Technical University, Prague, Research Institute of Textile Machines Liberec and Technical University of Liberec (Pourova & Vrba 2006) (Vrba et al. 2005).

In their research of the open applicator they determined the position of the magnetron. In the same manner they also found the distribution of the electric field strength in drying textile materials. This applicator has a magnetron as a source of high electromagnetic power, placed in the waveguide holder. The power of the used magnetrons is 800 W and its working frequency is 2.45 GHz.

Their drying resonant system is optimized by criteria to create the maximum electric field strength in the plane of the drying textile. They described this structure by means of an oriented graph, which is represented in the Figure 8.

Drying resonant system is optimized by criteria to create the maximum electric field strength in the plane of the dying textile. We can describe this structure by means of an oriented graph, and we can also create a diagram of the electromagnetic waves inside this structure. By modifying the diagrams we can arrive at the resulting expression for calculating the E-field strength in the textile plane:

a) b)

Fig. 7. Open-resonator a) Interior the microwave drying machine, b) Prototype of semi-industrial microwave drying machine

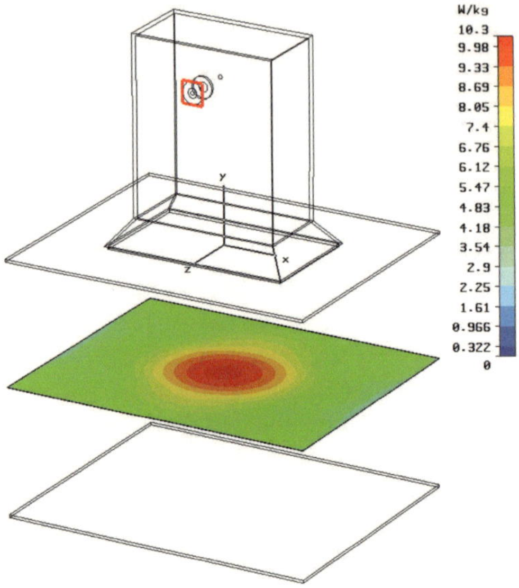

Fig. 8. Distribution of electric field strength in one applicator

$$E(l,p_2,\alpha_{tex}) = \sum_{n=0}^{\infty} p_1^n (p_2 + \delta_2 e^{-x})^n \cdot e^{-j\beta l(l+2n)} \tag{2}$$

parameters p_2 and α_{tex} are given by the dielectric properties of the textile, so we can write the electric field strength depended on relative permittivity ε_{tex} and loss factor tg δ_{tex} as follows

$$E(l,p_2,\alpha_{tex}) = \frac{e^{(j\beta l+\alpha_{tex}\cdot t)}}{e^{(j\beta l+\alpha_{tex}\cdot l)} - p_1 p_2 \cdot e^{(\alpha_{tex}\cdot t)} - p\sqrt{1-p_2^2}} \tag{3}$$

Electric field strength with respect to distance l and to relative permittivity ε_{tex} (tg δ = 0.566)
Were p_1 is reflection coefficient of metallic plate; p_2 is reflection coefficient of textile; α_{tex} is attenuation factor of textile; β is phase constant of free space; tg δ_{tex} - loss factor; l is distance between reflective plate and textile; t is thickness of textile; δ is $\sqrt{1-p_2^2}$ – transmission factor; e^{-} is $e^{(j-\alpha_{tex}\, t)}$ – absorption in textile; ε_{tex} – relative permittivity

2.2.3 Waveguide applicator

Waveguide are metallic tube, in the section-plane rectangle or circle. They transport electromagnetic energy from magnetron that runs along the waveguide. Waveguides work according to the principle of waves reflecting from the waveguide from one part to another. Fields in the waveguide can bee seen as a group of planar waves. They reflect from one part to another part, distributing in the direction of waveguide shown in the figure 9.
There are two characteristic wave lengths:

- one in the direction of vertical with the waveguide:

$$\lambda_n = \lambda / \cos\theta \tag{4}$$

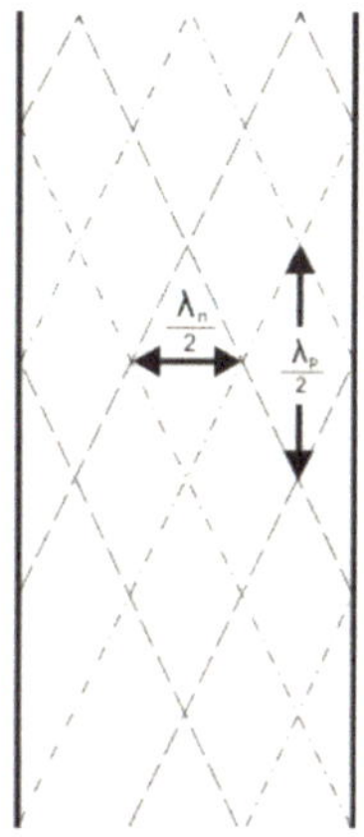

Fig. 9. Characteristic wave length in waveguide

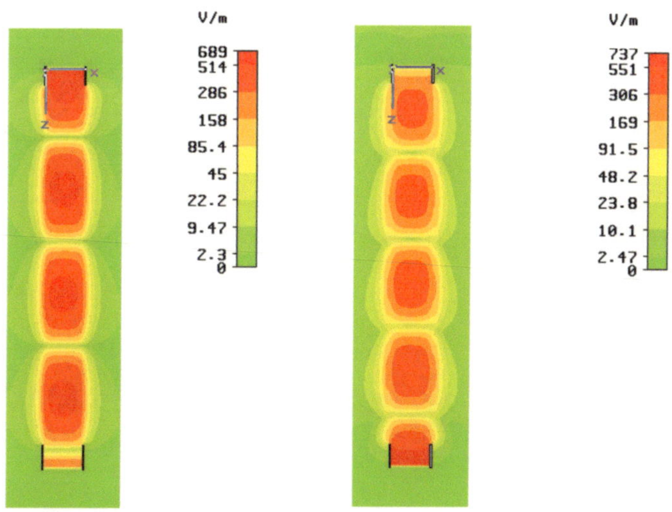

a) b)

Fig. 10. 2D distribution of electric field strength in one waveguide a) without the textile material, b) with textile material

- one in the direction of parallel with the waveguide:

$$\lambda_p = \lambda / \sin\theta \tag{5}$$

Were λ is wave length appointed signal; λ_n - is wave length in the direction vertical with the waveguide; λ_p I wavelength in the direction parallel with waveguide; θ is entrance angle (angle of incidence).

This drying system for the treatment of flexible textile material consists of rectangular waveguides centrally slotted in order to obtain planar passage of textile mater in wide state (Katovic et al. 2008). With proper design of the waveguides and supporting equipment, a specific environment (at the particular wavelength) can be created in order to provide controlled distribution of the microwave energy, making it possible to achieve uniform exposure to material passed through a channel. The leakage of microwave energy is inherently small due to the fact that waveguide slots are oriented along the waveguide line of symmetry, and therefore they cannot act as efficient slot antennas. Furthermore, in this way the material lies in the maximum of the electric field that assures effective coupling to the flowing microwave energy. In a case that request for slots symmetry is fulfilled, only the load (textile material) which passes through the waveguides has an influence on energy loss. The amount of microwave energy absorbed by the textile in each waveguide pass depends on the material thickness and moisture content. This laboratory drying system for the treatment of flexible textile material consists of 6 rectangular waveguides (4 x 8 cm) centrally slotted in order to obtain planar passage of textile material in a wide state.

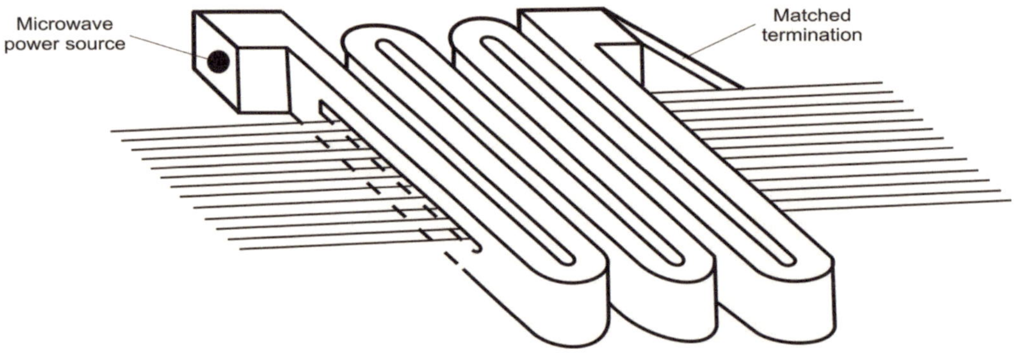

Fig. 11. Scheme of the textile material passing through the waveguides

Fig. 12. Laboratory microwave device for the treatment of textile materials

In a case of single pass applicator, exponential decay of electric field might cause non-uniform heat distribution.

To prevent this negative tendency, the material is passed through a number of waveguide passes. In order to

obtain a uniform absorption of microwave energy on the whole material an even number of waveguides must always be used. Number of waveguides used depends on the desired speed of the textile material passing and the amount of water on the material. Due to special design of waveguide slot for textile materials there is only minimal leakage of microwave energy into the environment. Namely, passing of the textile material through the waveguides leads to transition of the part of energy out of the waveguide together with the material. In order to reduce this energy transition as much as possible, waveguide slots are elongated and beveled which enables the return of microwave energy into the waveguide. Reduced energy is guided through the waveguide to the absorber of microwave energy (water) (Katovic et al. (2005).

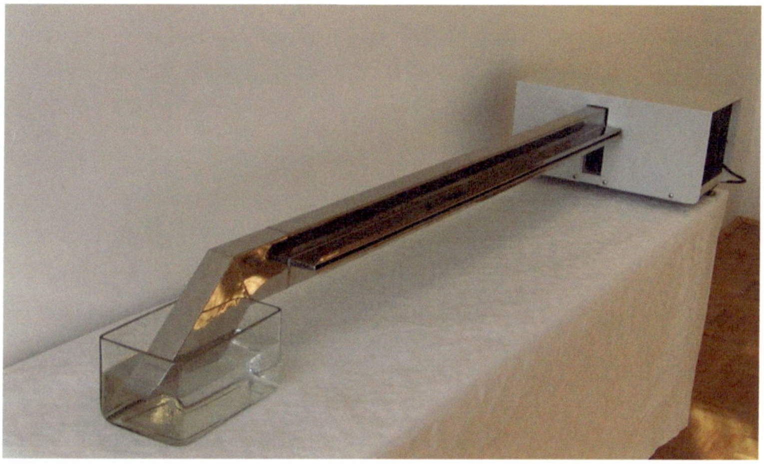

Fig. 13. The modular microwave unit

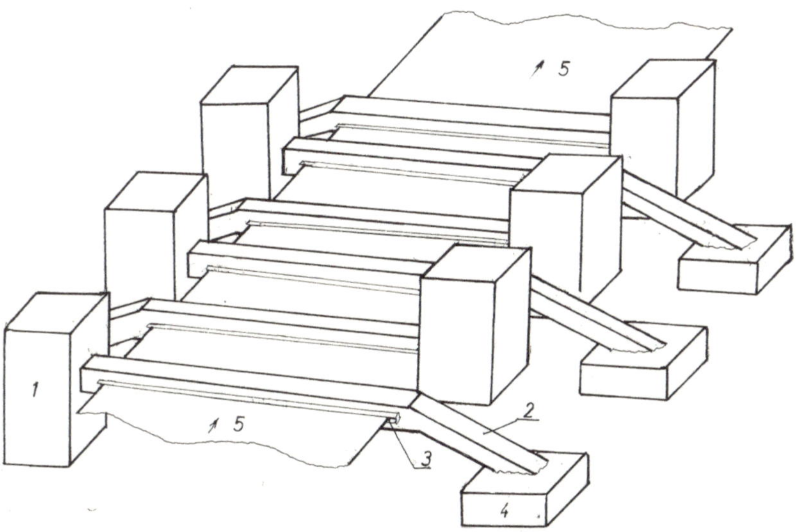

Fig. 14. Modular microwave units
1. Microwave unit box 2. Waveguides 3. Slots 4. Absorber of microwave energy (water) 5. Textile material.

For paper manufacturing, textiles, and other flat materials, American company Industrial Microwave System (IMS) offer an exceptional improvement over other drying alternatives.

A completely scalable configuration of slotted separated waveguides in combination with high power microwave generators can accommodate materials up to 5 cm in thickness and 10 m wide. Because of the efficiency of microwaves along with the uniform energy distribution, production speed can be dramatically increased and product quality improved.

Fig. 15. IMS Planar System (prospect of company *Industrial Microwave System*)

3. Radio frequency dryers

Radio frequency (RF) and microwaves (MW) are forms of electromagnetic energy but differ in operating frequency and wavelength. Both are allocated specific bands of operation by international governments. Industrial radio frequencies typically operate between 10 and 30 MHz with wavelengths of 30 to 10 meters. Radio frequency dryers are operating with power from 10 till 100 kW. Generally speaking, the efficiency of power utilization is far lower in a RF generator than a microwave unit, although the initial capital cost per KW of power output is higher. Selection of RF or microwave heating will depend on product physical properties and required process conditions for a particular application. Where penetration depth in excess of 15 cm is required and control of uniformity of heating is not a major issue, radio frequency offers a good solution. However, where uniformity of drying and moisture control is essential. For planar applications requiring belt widths in excess of 100 cm, where edge-to-edge uniformity is essential, control of microwave energy is superior to RF. Low moisture levels and high production belt speeds, such as those encountered in the textile industry, are far better suited to IMS microwave heating due to their characteristics of control and response time respectively. Electromagnetic waves have been used in the textile industry finishing the purpose of drying of thick materials, performed at radio frequency (RF) dryers, which are operating at different frequencies between 10 and 30 MHz. In textile processing, radio frequency waves are used in dryers for thick and multi-layered materials. In these machines, energy is transferred by means of two metal electrodes plates, between which the fabric is transported on a conveyer belt. An alternating electric field is created between the electrodes, with alternating voltage created by on RF generator.

Under the influence of the alternating electric field, dipole water molecules start vibrating, which causes them to heat up and be transformed into water vapor. A wet fabric submitted to a radiofrequency fields absorbs the electromagnetic energy, so that its internal temperature increases. If a sufficient amount of a energy is supplied, the water is converted into steam, which leaves the product; that is to say, the wet product is dried. Radiofrequency dyers have some specific design and construction features which allow their users to obtain the maximum benefits from the radio frequency technology in terms of quality of the dried products, reduced operating cots flexibility and reliability. The RF generators are of the „lumped components" type, having high efficiency (Q quality factor) and outstanding reliability. The cooling system of triodes is made up of a double water circuit; it is designed to allow the longest possible life of the triodes and does not require periodic maintenance operation. The RF power adjustment is accomplished by means of a semi-automatic circuit which controls the power supplied to the product being dried through a variable capacitor, located in the generator. The electrode is fixed or automatically positioned at pre-set heights. The range of power density for textile industry is from is 3 (nylon) to 18 kW/m^2 (cotton, viscose) of electrode surface.

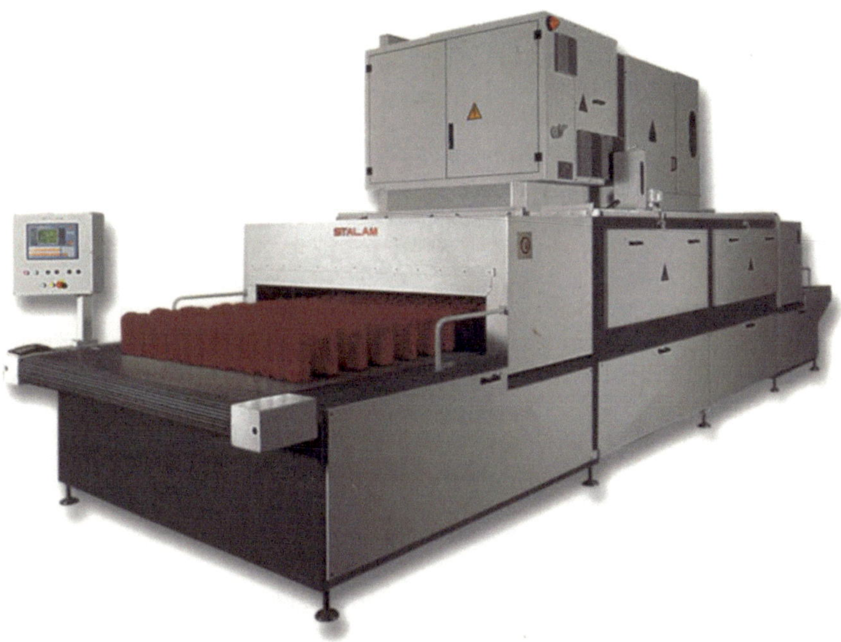

Fig. 16. Radio frequency dryer (Prospect of company *Stalam*)

4. Future development

The main advantage of the microwave energy application is that the energy consumption is 60-70 % lower respect to conventional heating treatments. Another advantage is its influence on the reaction kinetics: a reaction that takes place in two days under conventional treatment methods terminates after a few minutes applying MW energy.

Recent studies have documented a significantly reduced time for fabricating zeolites, mixed oxide and mesoporous molecular sieves by employing microwave energy. In many cases, microwave syntheses have proven to synthesize new nanoporous structures. By reducing the times by over an order of magnitude, continuous production would be possible to replace batch synthesis. This lowering of the cost would make more nanoporous materials readily available for many chemical, environmental, and biological applications. Further, microwave syntheses have often proven to create more uniform (defect-free) products than from conventional hydrothermal synthesis.

The main disadvantage of a wide application of microwave energy in textile finishing is the negative influence of electromagnetic irradiation on the environment. It means that preventive security measures are needed to be developed prior to microwave energy use on a larger scale. The exposure to an excessive level of radiation can produce hazards. The microwave radiation is non-ionizing, its main effect being of a thermal nature, commonly used in applications. The body absorbs radiation and automatically adapts to the resulting temperature increase, excess heat being removed by the blood flow. However, should the radiation become too intense, the thermal balance no longer could be restored by the body processes, and burns would then occur. As microwaves tend to heat deeply into the body, one might fear deep burns would occur while the surface temperature remained acceptable. There exists a certain radiation threshold, beyond which irreversible changes do occur. A considerable number of studies were carried out to determine this threshold. No permanent effect was observed for power level lower than $100 mW/cm^2$. Severe overexposure of non-uniform energy distribution may provide excessive focus of heat build up resulting in burnt material or a fire hazard. Another disadvantage is the depth of penetration achievable using microwave energy. This is a function of microwave frequency, dielectric properties of the material being heated and its temperature. As a general rule, the higher the frequency, the lower the depth of penetration.

5. References

Anonymus (1996). Microwave Processes for the Combined Desizing, Scouring and Bleaching of Grey Cotton Fabrics, *J.Text. Institute*, 3, *pp. 602-607,* ISSN 0400-5000

Barantsev, V.M., Larionov, O.S., Pavlov, N.N. (2007). Prospects for modification of para-aramid fibres with metal complex salts in conditions of microwave expositure, *Fibre Chemistry* 39, *pp.193-196,* ISSN 0018-3830

Bischof Vukusic, S., Schramm, C., Katovic, D. (2003). Influence of Microwaves on Nonformaldehyde DP Finished Dyed Cotton Fabrics, *Textile Research Journal, 73, pp.733-738,* ISSN 0040-5175

Bischof Vukusic, S., Katovic, D. (2004). Textile finishing treatments influenced with microwaves, *The Textile Institute 83rd World Conference,* Shangai, *China, pp.1165-1169,* ISBN 1-8703-7261-1

Bischof Vukusic, S., Katovi D., Flincec Grgac S. (2004). Effect of microwave treatment on fluorocarbon finishing, *Colourage Annual, 51, pp.1000 -1004,* ISSN 0010-1826

Cablewski, T. et al (1994). Development and Application of Continuous Microwave Reactor for Organic Synthesis, *J. Org.Chem 59 pp. 3408 – 3412,* ISSN022-3263

Chang, H-T., Chang S-T.: (2003) Improvements in dimensional stability and lighfastnedd of wood by butyrylation using microwave heating *J.Wood Sci (2003) 49 p.455-460* ISSN 1435-0211

D'Arrigo, Focher, B., Pellacani, G.C., Cosentino, C.Torri, G. (2002). Textiles Thermosetting by Microwaves, *Macromol. Symp. 180 pp. 223-239,* ISNN 1022-1360

Enderlig, R., (1988). US Patent 4,907,310

Englert, R.D., Berriman, L.P. (1974), Curing chemically treated cellulosic fabrics, *US Patent 3846845, 1974. 1112*

Fouda, M. El Shafei, A., Hebeish, A. (2009). Microwave curing for producing cotton fabrics with care and antibacterial properties, *Carbohydrate Polymers 77, pp. 651-655,* ISSN 0144-8617

Hong, S., Thompson, D. (1998), Canadian Patent CA 2 235 439

Hou, A., Wang, X., Wu, L. (2008). Effect of microwave irradiation on the physical properties and morphological structures of cotton cellulose, *Carbonate Polymers 74 pp. 934-937,* ISSN 0144-8617

Katovic, D., Bischof Vukusic, S., Soljacic, I., Stefanic, G. (2000). Application of Electromagnetic Waves in Durable Press Finishing with Polycarboxylic Acid, *AATCC International Conference & Exhibition, Winston-Salem, NC, USA, 17-20 September 2000, CD-ROM,*

Katovic, D., S. Bischof Vukusic, (2002), Application of Electromagnetic Waves in Durable Press Finishing with Polycarboxylic Acid, *AATCC Review 2 (2002) 4,pp. 39-42,* ISSN 1532-8813

Katovic, D. Bischof Vukusic S., Versec, J. (2002), The application of microwave energy in Durable Press Finishing, *International Textile Clothing & Design Conference Dubrovnik 6-9 October (2002) 283-287,* ISBN 953-96408-8-1

Katovic, D., Bischof Vukusic, S. Flincec Grgac, S. (2005). Application of Microwaves in Textile Finishing Processes, *Tekstil 54(7) 313-318,* ISSN 0492-5882

Katovic, D., Bischof Vukusic, S., Hrabar, S., Bartolic, J. (2005). *Microwaves in Chemical Finishing of Textiles* 18th International Conference on Applied Electromagnetics and Communications 12-14 October (2005), Dubrovnik, 255-25, ISBN 953-6037-44-0

Katovic, D., Kovacevic, S., Bischof Vukusic, S., Schwarz, I., Flincec Grgac, S. (2007), Influence of Drying on Psysico-mechanical Properties of Sized Yarn, *Tekstil 56,8, pp .479 - 486,* ISSN 0492-5882

Katović, D. Kovacevic, Bischof Vukusic, S., Schwarz I., Flincec Grgac, S. (2008). The Effect of Microwave on Warp Sizing, *Textile Research Journal 74, pp. 353-360,* ISSN 0040-5175

Kaynak A., Hakansson E., Amiet A. (2009) The influence of polymerization time and dopant concentration on the absorption of microwave radiation in conducting polypyrrole coated textiles, *Synthetic Metals 159 (2009) pp.1373-1380,* ISSN 0379-6779

Metaxas, A.C., Meredith, R.J. (1983). Industrial Microwave Heating, *Peter Peregrinus, pp. 111-150,* ISBN 0-90604-889-3, London

Nanda, R., Patel, G. (2002). *Microwave oven: A tool for quick response in shade development and lab-to bulk shade translation in reactive dyeing* 7th International & 58th All India Textile Conference, Mumbai 14 -15 Dec 2002 pp. 83-88

Nanda, R., Patel, G. (2002). Microwave Oven: A tool for guide response in shade translation in reactive dyeing, *Colourage* 49,12, *pp.83-88,* , ISSN 0010-1826

Neral, B., Sostar Turk, S., Schneider, R (2007). Efficiency of Microwave Fixation of Digital Prints of the Reactive Dyestuff, *Tekstil 56, 6, pp.358-367,* ISSN 0492-5882

Pourova, M., Vrba, J. (2006). Microwave Drying of Textile Materials and Optimization of Resonant Applicator *Acta polytechnica 46 5, pp. 3-7,* ISSN 0323-7648

Reagan, B.M. (1982), Eradication of insects from wool textiles, *Journal of the American Institute for Conservation 21, 2, pp. 1-34,* ISSN 0197-1360

Rouette, H.K. (2001). Encyclopedia of Textile Finishing, *Springer-Verlag, Berlin Heidelberg pp. 1399-1401,* ISBN 3-540-65031-8

Saito,Y., Nakano,K., Shida S., Soma,T., Arima, T. (2004). Microwave-enhanced release of formaldehyde from plywood *Holzforschung 58, pp. 548-551,* ISSN 1437-434

Satge,C., Verneuil, B., Brandland, P. Granet, R. Krausz P., Rozier, J., Petit, C. (2002). Rapid homogeneneous esterification of cellulose induced by microwave irradiation Carbonate *Polimers 49 pp. 373-376,* ISSN 1385-772

Strizhah, P.E., Tripol`shii A.I., Gurnik T.N., Tuzikov, F.V., Moroz, E.M. Konstandinova, T.E., Tuzikova,N.A., Kol`ko, V.P., Danilenko,I.A. Gorban, O.A. (2008). Effect of temperature on the structural characteristics ofd zirconium dioxide nanoparticles produced under conditions of microwaver treatment, *Theoretical and Experimental Chemistry,44, 3, p.144-148,* ISSN 0040-5760

Therdthai, N., Zhou,W., (2009). Characterzation of microwave vacuum drying and hot air drying of mint leaves (Mentha cordifolia Opiz ex Fresen), *Journal of Food Engineering 91 pp.482-489,* ISBN 0260-8774

Thewli, R., Barnoldswick (1999). European Patent EP 0 974 693 A1 (1999)

Thiry, M. (2000), The Magic of Microwave, *Textile Chemist and Colorist–American Dyestaf Reporter 32, 10, pp. 2-4,* ISSN 0040-490

Tomljenovic, A., Katovic, D. (2008). *Microwaves – solution for improving Polyester woven fabric UV protective properties* 4 rd International Textile, Cloting & Design Conference October 5th to 8th 2008; Dubrovnik, 898-903 ISBN 978-953-7105-26-6

Tompsett G., Conner W.C., Yngresson K.S. (2006). Microwave Synthesis of Nanoporous Materials *ChemPhysChem 7,296-319* ISSN 1439-764

Varma, R. (2001). Solvent- free accelerated organic syntheses using microwaves, *Pure Appl. Chem 73, pp.193- 198* ISSN 0033-4545

Vrba, J., Stejskal, M., Klepl, R., Richter, A., Pourova, M., Žak, O., Herza, J., Oppi, L. (2005). *Microwave Drying Machine for Textile Materials* European 35th Microwave Conference ISBN 2-9600551-2-8

Wang H., Takashima H., Miyakawa Y., Kanno Y. (2005) Development of catalyst materials being effective for microwave sterilization *Science and Technology of Advanced Materials 6 pp. 921-926* ISBN 1878-5514

Zubizarreta, L.,Arenillas,A, Menéndez,J.A., Pis,J.J., Pirard,J.P., Job,N. (2008). Microwave Drying as an effective method to obtain porous carbon xerogels, *Journal of Non-Crystalline Solids 354 pp. 4024-4026,* ISSN 0022-3093

17

Composites Based on Natural Fibre Fabrics

Giuseppe Cristaldi, Alberta Latteri, Giuseppe Recca and Gianluca Cicala
University of Catania – Department of Physical and Chemical Methodologies for Engineering, Catania
Italy

1. Introduction

In the latest years industry is attempting to decrease the dependence on petroleum based fuels and products due to the increased environmental consciousness. This is leading to the need to investigate environmentally friendly, sustainable materials to replace existing ones. The tremendous increase of production and use of plastics in every sector of our life lead to huge plastic wastes. Disposal problems, as well as strong regulations and criteria for cleaner and safer environment, have directed great part of the scientific research toward eco-composite materials. Among the different types of eco-composites those which contain natural fibers (NF) and natural polymers have a key role. Since few years polymeric biodegradable matrices have appeared as commercial products, however their high price represents the main restriction to wide usage. Currently the most viable way toward eco-friendly composites is the use of natural fibres as reinforcement. Natural fibres represent a traditional class of renewable materials which, nowadays, are experiencing a great revival. In the latest years there have been many researches developed in the field of natural fibre reinforced plastics (Bledzki & Gassan, 1999). Most of them are based on the study of the mechanical properties of composites reinforced with short fibers. The components obtained therefore are mostly used to produce non-structural parts for the automotive industry such as covers, car doors panels and car roofs (Magurno, 1999, John at al., 2008) (Fig.1,2).

Fig. 1. Mercedes-Benz A natural fibre composites components (source: DaimlerChrysler AG)

Few studies deal with structural composites based on natural reinforcements. These studies are mainly oriented to the housing applications where structural panels and sandwich beams are manufactured out of natural fibres and used as roofs (Saheb & Jog., 1999).

Considering the high performance standard of composite materials in terms of durability, maintenance and cost effectiveness, the application of natural fiber reinforced composites as construction material holds enormous potential and is critical for achieving sustainability. Due to their low density and their cellular structure, natural fiber posses very good acoustic and thermal insulation properties and demonstrate many advantageous properties over glass or rockwool fibre (e.g. handling and disposal).

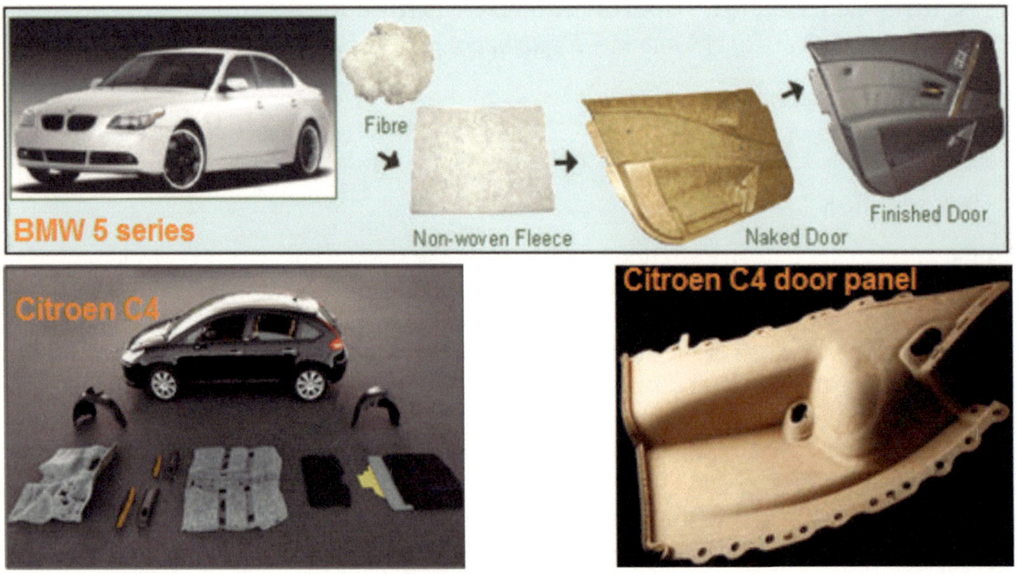

Fig. 2. Examples of applications of Natural Fibres in the automotive field

Nowadays natural fibre composites are not exploited only in structural and semi-structural applications of the automotive sector, but in other fields too (Fig.3).

Fig. 3. Examples of use of Natural Fibres in several applications

Natural fibres (Fig.4) can be divided, according to their origin, into: animal, vegetable and mineral. The most used are the vegetable ones due to their wide availability and renewability in short time respect to others, so when we say "natural fibres" We refer here to the vegetables ones. In the past, natural fibres were not taken into account as reinforcements for polymeric materials because of some problems associated with their use:

- Low thermal stability, in other terms the possibility of degradation at moderate temperature (230-250 ° C).
- Hydrophilic nature of fibre surface, due to the presence of pendant hydroxyl and polar groups in various constituents, which lead to poor adhesion between fibres and hydrophobic matrix polymers (John et al., 2008, Kalia et al., 2009). The hydrophilic nature can lead to swelling and maceration of the fibers. Furthermore, moisture content decreases significantly fibre's mechanical properties.
- Properties variability depending on the quality of the harvest, age and body of the plant from which they are extracted, the extraction techniques and the environmental conditions of the site.

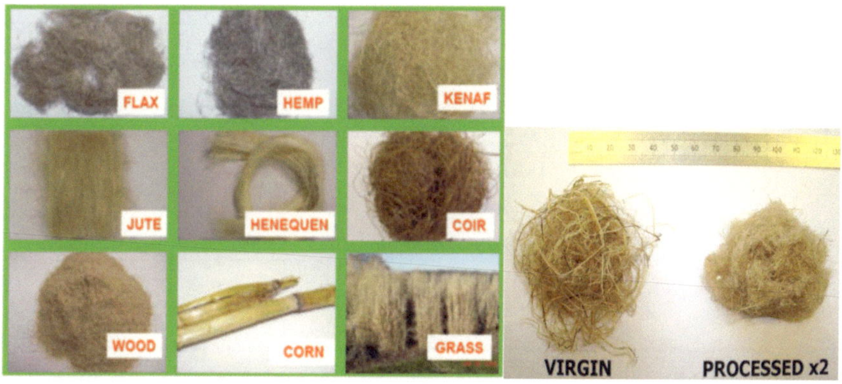

Fig. 4. (a) Some natural fibre, (b) Unprocessed and Processed hemp fibres (source: University of Exeter)

Lack of good interfacial adhesion, low degradation temperature, and poor resistance towards moisture make the use of natural fibre reinforced composites less attractive than synthetic fibre (glass, carbon, aramid, etc.) that have been up to now the only choice for reinforcing polymeric composites, due to their superior mechanical properties. However, the production of composites reinforced with synthetic fibres and matrices requires a large amount of energy which is only partially recovered with incineration of fibre reinforced composites. This has once again drawn the attention towards natural fibres due to their environmental advantages. It has been demonstrated that the energy needed for production of natural fibres is, on average, more than half of the amount needed for synthetic fibres (Fig.5). Thus, the renewed interest in the natural fibers, due to their lightweight, nonabrasive, non irritating, combustible, nontoxic, biodegradable properties (Saheb & Jog, 1999), low energy consumption for production, budget zero CO_2 emissions if burned, low cost (Table 1), main availability and renewability compared to synthetic fibres, has resulted in a large number of applications to bring it at par and even superior to synthetic fibers. Because of such properties natural fibers are fast emerging as a viable choice as reinforcing material in composites (kalia et al., 2009).

Even if natural fibre has a very low energy consumption for production compared to other synthetic fibre, such as glass or carbon, careful environmental impact evaluation must be take in consideration in order to make the right choice. In fact, the validity of "green" case for substitution of synthetic fibre by natural ones is dependent on the type of reinforcement and related production processes. A parameter which better describe the environmental impact is the *embodied energy* calculated with reference to all related agricultural operations (from ploughing to harvest), fibre extraction operations (retting and decortication), fibre preparation operations (hackling and carding), fibre processing operations (spinning or finishing) and materials used for these operations. The use of embodied energy parameter reveals that not any kind of natural fibre reinforcement is "greener" then synthetic ones. Fig. 6 shows that, even if adopting the most environmental friendly option (no-till and water retting) for flax fibre production, only mat fabrics are, in energetic terms, "greener" while flax yarns has a higher embodied energy respect to glass fibre continuous filament production.

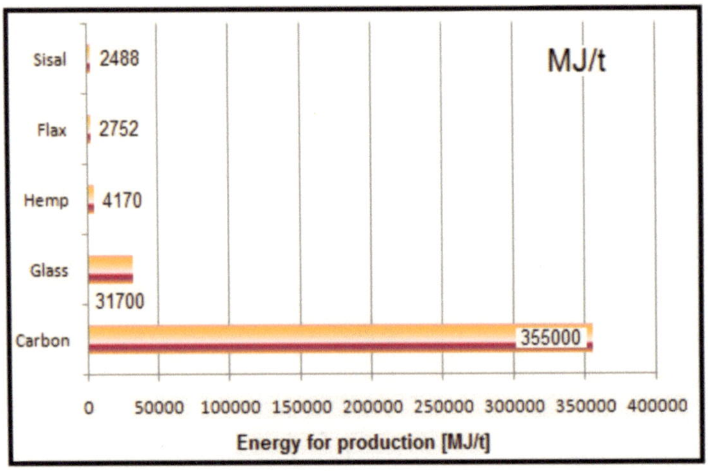

Fig. 5. Energy for production of some fibre *(sources: SachsenLeinen; Daimler 1999; BAFA; NOVA; AVB; CELC; REO)*

Fiber	Price	Specific Gravity	Price
	$ /m³	Kg/m³	$ /kg
Wood	420	1600	0,26
Flax	600	1500	0,40
Glass	4850	2600	1,87
PP	650	900	0,72

Table 1. Cost comparison between natural and synthetic fibre *(Source: Georgia Institute of Technology www.me.gatech.edu/jonathan.colton/me4793/natfiber.pdf)*

Natural fibres can be classified according to their origin and grouped into *leaf:* abaca, cantala, curaua, date palm, henequen, pineapple, sisal, banana; *seed*: cotton; *bast*: flax, hemp, jute, ramie; fruit: coir, kapok, oil palm. Among them flax, bamboo, sisal, hemp, ramie, jute, and wood fibres are of particular interest (Kalia et al., 2009). The most important physical and mechanical properties are summarized in Table 2.

Physical and mechanical properties depend on the single fibre chemical composition (Cellulose, hemicelluloses, lignin, pectin, waxes, water content and other minors) according to grooving (soil features, climate, aging conditions) and extraction/processing methods conditions. Grooving conditions is recognized as the most influent parameter for the variability of mechanical properties of the fibres. The chemical composition of several natural fibres is summarised in Table 3.

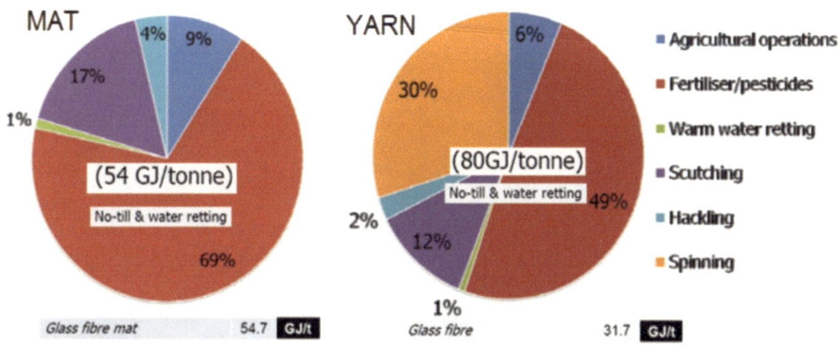

Fig. 6. Embodied energy of flax fibre mat and yarn (source: ACMC Advanced Composites Manufacturing Centre - University of Plymouth)

Plant fibre	Tensile strength (MPa)	Young's modulus (GPa)	Specific modulus (GPa)	Failure strain (%)	Length of ultimates, l (mm)	Diameter of ultimates, d (μm)	Aspect ratio, l/d	Microfib angle, θ (°)	Density (kg.m^{-3})	Moisture content (eq.) (%)
Cotton[s]	300-700	6-10	4-6.5	6-8	20-64	11.5-17	2752	20-30	1550	8.5
Kapok[s]	93.3	4	12.9	1.2	8-32	15-35	724	-	311-384	10.9
Bamboo[b]	575	27	18	-	2.7	10-40	9259	-	1500	-
Flax[b]	500-900	50-70	34-48	1.3-3.3	27-36	17.8-21.6	1258	5	1400-1500	12
Hemp[b]	310-750	30-60	20-41	2-4	8.3-14	17-23	549	6.2	1400-1500	12
Jute[b]	200-450	20-55	14-39	2-3	1.9-3.2	15.9-20.7	157	8.1	1300-1500	12
Kenaf[b]	295-1191	22-60	-	-	2-61	17.7-21.9	119	-	1220-1400	17
Ramie[b]	915	23	15	3.7	60-250	28.1-35	4639	-	1550	8.5
Abaca[l]	12	41	-	3.4	4.6-5.2	17-21.4	257	-	1500	14
Banana[l]	529-914	27-32	20-24	1-3	2-3.8	-	-	11-12	1300-1350	-
Pineapple[l]	413-1627	60-82	42-57	0-1.6	-	20-80	-	6-14	1440-1560	-
Sisal[l]	80-840	9-22	6-15	2-14	1.8-3.1	18.3-23.7	115	10-22	1300-1500	11
Coir[f]	106-175	6	5.2	15-40	0.9-1.2	16.2-19.5	64	39-49	1150-1250	13

Table 2. Natural fibre properties. Source: Natural fibre'09 Proceedings (University of Bath)

%	Jute	Flax	Hemp	Kenaf	Sisal	Cotton
Cellulose	61-71	71-75	70,2-74,4	53-57	67-78	82,7
Hemicellulose	13,6-20,4	18,6-20,6	17,9-22,4	15-19	10-14,2	5,7
Lignin	12-13	2,2	3,7-5,7	5,9-9,3	8-11	-
Pectin	0,2	2,2	0,9	-	10	-
Others	-	3,8	6,1	7,9	1	-
Waxes	0,5	1,7	0,8	-	2,0	0,6
Water	12,6	10,0	10,8	-	11,0	-

Table 3. Natural fibre composition (Williams et al., 2000; Bogoeva-Gaceva et al., 2007)

Natural fibre mechanical properties depends on the type of cellulose and the geometry of the elementary cell. The celluloses chains are arranged parallel to each other, forming bundles each containing forty or more cellulosic macromolecules linked by hydrogen bonds and through links with amorphous hemicelluloses and lignin which confer stiffness to fibre called microfibrils. More interwoven microfibrils form a rope-like structure (Rong at al., 2001) (Fig.7).

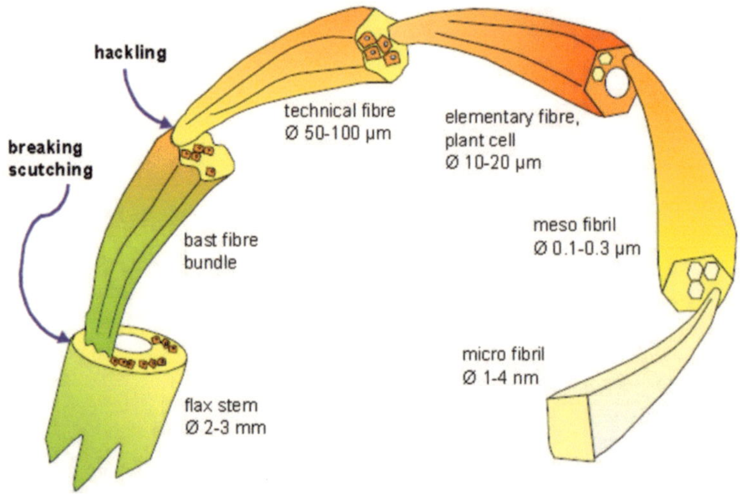

Fig. 7. Natural fibre hierarchal structure

Among natural fibres the bast fibres, extracted from the stems of plants such as jute, kenaf, flax, ramie and hemp are widely accepted as the best candidates for reinforcements of composites due to their good mechanical properties. Hemp was shown to have very promising tensile properties for applications where mechanical properties are a requisite (Nair et al., 2000)

As many authors agree, the two basic parameters that allow to characterize mechanical behavior of natural fibers are the cellulose content and the spiral angle. In general, the tensile strength of the fibers increases with increasing cellulose content and with decreasing angle of helix axis of the fibers.

The strength of natural fibre composites in on average lower compared to the synthetic fibre reinforced composites, even under optimised fibre-matrix interaction (Heijenrath & Peijs, 1996 , Berglund & Ericson, 1995), but their lower density and cost make them competitive in terms of specific and economic properties. This is basically due to the composite-like structure of natural fibres (Van den Oever et al., 1995); they are generally not single filaments as most manmade fibres but they can have several physical forms, which depend on the degree of fibre isolation. Composite strength depends also on fibre diameter (smallest diameter could achieve higher mechanical resistance due to larger specific contact surface with matrix) and fibre length.

2. Natural fibre fabric types

The possibility to have long or short fibres depends on the material under consideration, in fact, for synthetic fibre it is easy and common to have long continuous fibres out of production plant, while, for natural fibres, the fibre's length is an inherent limit for the material itself due to their natural origin which limits their length (for example the plant stem). This is a basic reason why natural fibres are usually found as short reinforcements which are used to produce mat fabrics. Discontinuous fibres (*chopped*) are generally used for a randomly oriented reinforcement (*mat*) when there is not any preferential stress direction and/or there is a low stress/strain level in the composite (Fig.8). As it will be shown in the case studies mats, due to the random fibre orientation, are non-optimised fabric for mechanical performances.

Fig. 8. Hemp mat

The alternative to the use of short fibres is the manufacture of long yarns. Yarn is a long continuous assembly of relatively short interlocked fibres, suitable for use in the production of textiles, sewing, crocheting, knitting, weaving, embroidery and ropemaking that are twisted with an angle to the yarn axis in order to provide axial strength to the yarn. Spun yarns are made by twisting or otherwise bonding staple fibres together to make a cohesive thread and may contain a single type of fibre or a blend of various types. Two or more spun yarns, if twisted together, form a thicker twisted yarn. Depending on the direction of this final twist, the yarn will be known as s-twist or z-twist (Fig.9). Two or more parallel spun yarns can form a *roving*. The main advantage of using natural yarns is the possibility to weave them into 2D and 3D fabrics with tailored yarn orientations.

A common measure unit used to classify fibres and yarns is the *denier* which corresponds to the linear mass density of the yarns. Denier is defined as the mass in grams per 9000 meters. In the International System of Units the tex is used instead, defined as the mass in grams per 1000 meters. The most commonly used unit is actually the decitex, abbreviated dtex, which is the mass in grams per 10000 meters. Similar to tex and denier, yield is a term that helps describe the linear density of a roving of fibres. However, unlike tex and denier, yield is the inverse of linear density and is usually expressed in yards/lb. Linear mass of twisted yarn is expressed by a fraction where the numerator is the yarn count and the denominator is simply the number of ends (e.g. 30/3).

Spun yarns obtained from natural fibres present usually some short fibres protruding out of the main yarn body (Fig.10). This short fibres are commonly referred to as yarn hairiness. Although not desirable in many cases, the hairiness can lead to better mechanical yarn/resin interlocking in composites. Another advantage of natural yarns is the increased surface roughness of yarns compared to fibres, which increases the interfacial strength due to mechanical interlocking, improving the transverse properties. In addition, twisting localizes the micro damages within the yarn leading to higher fracture strength.

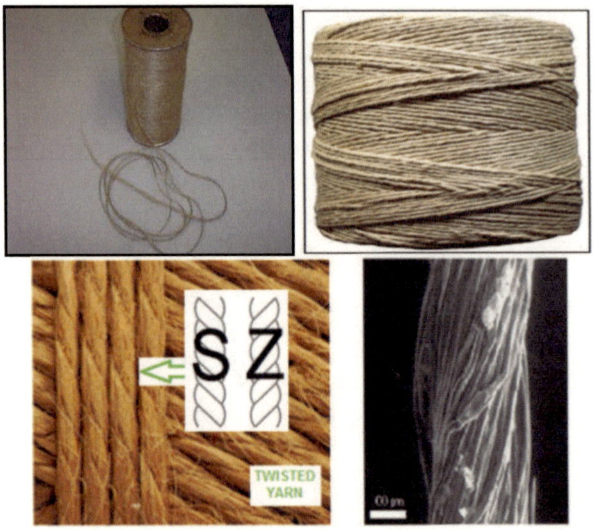

Fig. 9. Hemp twisted yarn and scanning electron microscope image of hemp twisted yarn

Fig. 10. Hemp and flax fibre rovings

An important control parameter for such natural yarns is the twist level. It has been shown (Goutianos & Peijs , 2003) that very low twisted yarns display a very low strength when tested in air and therefore they cannot be used in processes such as pultrusion or textile manufacturing routes like knitting or weaving.(Fig.11) where heavy loading is experienced by the yarns while processing. In the case of short staple (length) fibres, higher twist level is necessary to prevent fibre slippage and to develop sufficient strength.

Besides yarn strength, the amount of twist also affects the inter-yarn impregnation while fabricating reinforced composites. With increased twist level yarns become more compact making it difficult for the resin to penetrate into the yarn. Dry yarns lead to lower bonding between yarns and resin thus leading to delamination and lowering of the composite tensile properties. Several authors showed that when highly twisted yarns are impregnated in a polymer resin, their strength may decrease significantly with decreases similar to the drop in strength of an off-axis composite (Goutianos & Peijs , 2003; Baley, 2002). Thus, there is an optimum level of twist, which should be kept as low as possible for optimal composite mechanical properties to allow for proper yarn's wetting to be achieved.

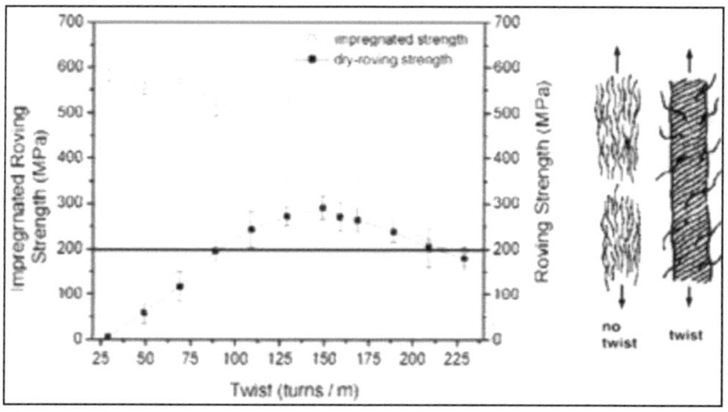

Fig. 11. Effect of twist level on mechanical properties (Goutianos at al., 2006)

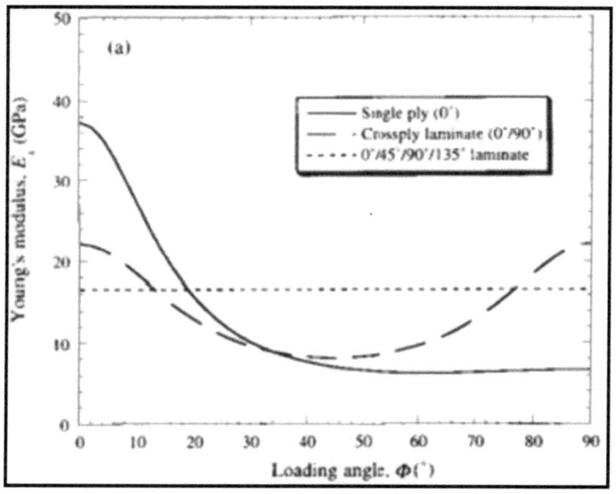

Fig. 12. Effect of fibre orientation on elastic modulus. Data for 50% fibre volume fraction of glass-epoxy laminate (source: Hull & Clyne)

The fibre contribution to composite mechanical properties improvement is emphasized when the stresses have components along the fibre direction (Fig. 12). However, most of the studies reported in literature are focused on the use of mat which are the cheapest alternative (Paiva et al., 2004) among technical fabrics. Several studies showed that the random orientation of the fibres in mat fabrics leads to lowering of the reinforcing efficiency (Baiardo et al., 2004).

Yarns offer a viable and interesting alternative to the use of short fibres as multiple filament yarns can be weaved into 2- or 3-Dimension textiles. Weaving is a textile production method which involves interlacing a set of longer threads, twisted yarn or roving, (called the warp) with a set of crossing threads (called the weft) (Fig.13). This is done on a frame or machine known as a loom, of which there are a number of types. Some weaving is still done by hand, but the vast majority is mechanised. The main advantage of using weaved fabrics is the possibility to pre-orient the filaments in the designed directions. Natural yarns differ from multifilament of synthetic fibres (ie.tow) because they are an assembly of short fibre instead of an assembly of aligned continuous fibres. However, the fibres which constitute the yarn have a preferential orientation along an helical trajectory which make the use of natural yarns attractive compared to short fibres because in such yarns fibres are mostly along the load direction.

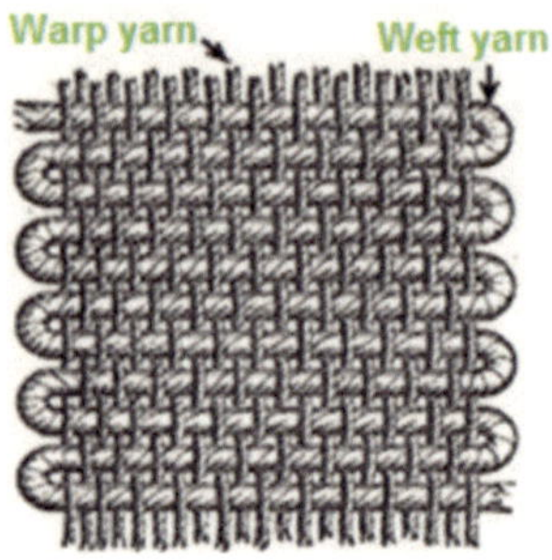

Fig. 13. Warp and weft in plain weaving

The manner in which the warp and weft threads are interlaced is known as the *weave style*. The three basic weaves styles or architectures are:

- plain weave
- satin weave
- twill weave

Plain weave is the most basic type of textile weaves.The warp and weft are aligned so they form a simple criss-cross pattern. Each weft thread crosses the warp threads by going over one, then under the next, and so on (Fig.14, 15). The next weft thread goes under the warp threads that its neighbour went over, and vice versa. In balanced plain weaves the warp and weft are made of threads of the same weight (size) and the same number of ends per inch. *http://en.wikipedia.org/wiki/Plain_weave - cite_note-1*

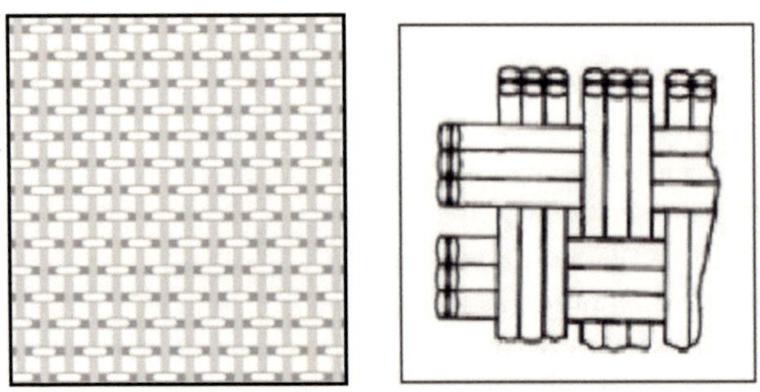

Fig. 14. Plain woven yarn and woven roving schemes (0°/90° reinforcement directions)

Fig. 15. Examples of plain woven flax yarns. H-181 100% Hemp Canvas weave 18oz/sq yd Wide 59" 5N/2 x 8N/2 x23x21. *Source: dongpinghemp.com*

The satin weave is characterized by four or more weft yarns floating over a warp yarn or vice versa, four warp yarns floating over a single weft yarn (Fig.16).
Twill is a type of fabric woven with a pattern of diagonal parallel ribs. It is made by passing the weft thread over one or more warp threads and then under two or more warp threads and so on, with a "step" or offset between rows to create the characteristic diagonal pattern (Fig.17,18). Because of this structure, twills generally drape well. In a twill weave, each weft or filling yarn floats across the warp yarns in a progression of interlacings to the right or left, forming a distinct diagonal line. This diagonal line is also known as a wale. A float is the portion of a yarn that crosses over two or more yarns from the opposite direction.

Fig. 16. Satin weave with 16 warp yarns floating over each weft yarn.

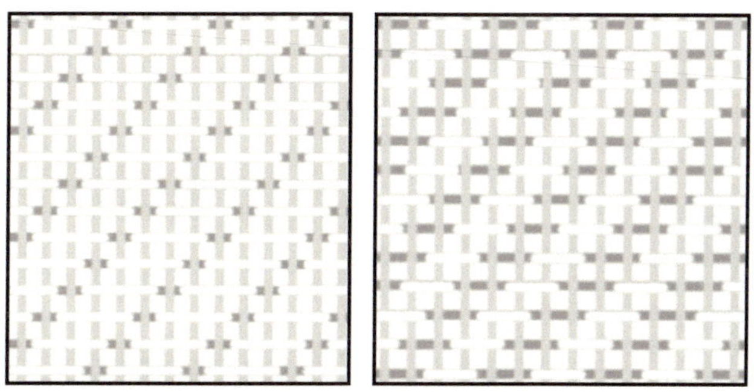

Fig. 17. Structure of a 3/1 and 2/2 twills

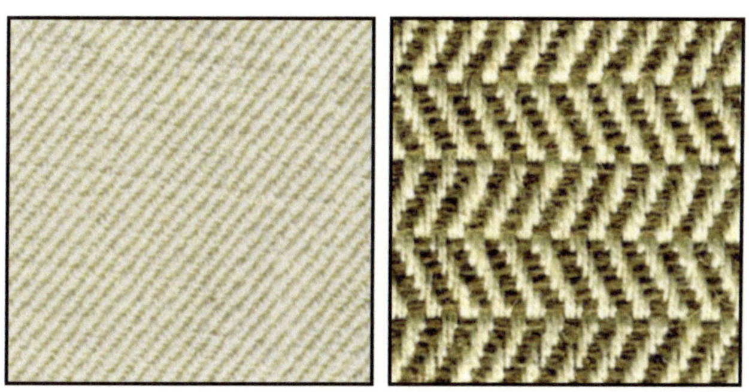

Fig. 18. Examples of plain woven flax yarns. (A) Natural Twill Weave 100% Hemp 12oz Width 57/58" (B) Natural Herringbone Weave 52% Hemp 48% Flax 20oz Width 57/58". *Source: EnviroTextile.com*

A twill weave can easily be identified by its diagonal lines. and is often designated as a fraction—such as 2/1—in which the numerator indicates the number of harnesses that are raised, in this example, two, and the denominator indicates the number of harnesses that are lowered when a filling yarn is inserted, in this example one. The fraction 2/2 would be read as "two up, two down." with two warp threads crossing every two weft threads. The offset at each row forms the diagonal pattern. The minimum number of harnesses needed to produce a twill can be determined by totalling the numbers in the fraction.

The fewer interlacings in twills allow the yarns to move more freely, and thus they are softer and more pliable, and drape better. Twills also recover better from wrinkles than plain-weave fabrics. When there are fewer interlacings, yarns can be packed closer together to produce high-count fabrics.

There is an increasing number of producers of natural fibre fabrics around the world which are tailoring their products for composites technology. Table 4 shows some costs for a selection of fabrics commercialized in U.S.A. by the company EnviroTextile LLC.

Fabric Descriptions and Specifications	Units = Yards					
	1-49	50-99	100-499	500-999	1K-2999	3K-4999
Natural 100% Hemp Canvas Plain Weave, 16n/2x7n, 41x28, Width 57/58" 12oz Semi-Bleached, Preshrunk, Cationic Softener, SBP® 100%	$14.60	$12.00	$11.10	$10.30	$10.00	$9.65
Natural 100% Hemp Canvas Basket Weave, 10n/3x 10nm/3, 30x20, Width 57/58" 18.5oz Semi-Bleached, Preshrunk, Cationic Softener, SBP® 100%	$16.45	$13.50	$12.55	$11.60	$11.30	$10.90
Black 100% Hemp Canvas Basket Weave, 10n/3x 10nm/3, 30x20, Width 57/58" 18.5oz, Preshrunk, No Softener, SBP® 100%	$17.60	$14.45	$13.40	$12.40	$12.05	$11.65
Dark Brown 100%Hemp Canvas Basket Weave, 10n/3x 10n/3, 30x20, Width 57/58" 18.5oz Semi-Bleached, Preshrunk, No Softener, SBP® 100%	$17.60	$14.45	$13.40	$12.40	$12.05	$11.65
Sand 100 % Hemp Canvas Basket Weave, 10n/3x 10n/3, 30x20, Width 57/58" 18.5oz Semi-Bleached, Preshrunk, No Softener, SBP® 100%	$17.60	$14.45	$13.40	$12.40	$12.05	$11.65
Natural 100% Hemp Herringbone Weave, 16nm/2x8.5nm, 41x27, Width 55/56" 10.5oz Soft Finish, SBP® 100%	$14.95	$12.30	$11.40	$10.55	$10.25	$9.90

Table 4. Costs of some fabrics sold by EnviroTextile

Other examples of commercial products available on the market are the flax fabric (Fig. 19) manufactured by Biotex (http://www.compositesevolution.com) which are also available as pre-impregnated fabric with PLA (polylacticacid) and PP (polypropylene). Other products available are the pre-impregnated fabrics (FLAXPLY©) produced by Lineo. The products sold by Lineo have pre-treated fibers for increased fiber-matrix adhesion. The FLAXPLY© are proposed to be used for internal layer of mixed carbon/flax design for improved vibration absortion (Fig.20).

As mentioned before, yarns and rovings can be weaved in 3-Dimension fabrics, even if they are not so widespread as plain ones. To date no commercial example of 3D weaved fabric based on natural yarns is available.

Fig. 19. Biotex Flax 3H Satin 420gsm

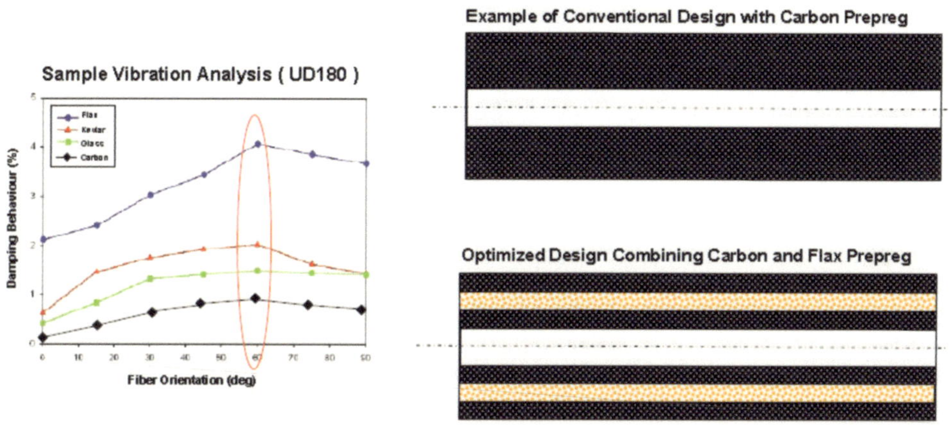

Fig. 20. Example of the use of FLAXPLY© for vibration absorption

3. Fiber surface treatments

The contribution of fibres to the final properties of the composite depends on:

- Mechanical properties of fibres;
- Type (continuos/discontinuos) and orientation of fibres in the composite (anisotropy).
- Volume fraction of fibres;
- Fibre-matrix interface;
- Processing technique used for composite manufacturing.

Shortcomings associated with natural fibres have to be overcome before using them in polymer composites. The most serious concern with natural fibres is their hydrophilic nature due to the presence of pendant hydroxyl and polar groups in various constituents, which can lead poor adhesion between fibres and hydrophobic matrix polymers (Rong et al., 2001, Bledzki & Gassan, 1996). The hydrophilic nature of the fibre surface lead also to high moisture up take for the natural fibres which can seriously lower the mechanical properties of the fibres themselves.

The natural fibres are inherently incompatible with nonpolar-hydrophobic thermoplastics, such as polyolefins. Moreover, difficulty in mixing because of poor wetting of the fibres with the matrix is another problem that leads to composites with weak interface (John & Anandjiwala, 2008).

There are some physical fibre treatments (e.g Plasma), but nowadays when we speak about surface treatments we almost mean chemical ones. These treatments can clean the fibre surface, modify the chemistry on the surface, lower the moisture up take and increase the surface roughness. As the natural fibres bear hydroxyl groups from cellulose and lignin they are amenable to chemical modification. The hydroxyl groups may be involved in the hydrogen bonding within the cellulose molecules thereby reducing the activity towards the matrix. Chemical modifications may activate these groups or can introduce new moieties that can effectively lead to chemical interlock with the matrix. Mercerization, isocyanate treatment, acrylation, permanganate treatment, acetylation, silane treatment and peroxide treatment with various coupling agents and other pretreatments of natural fibres have achieved various levels of success for improving fiber strength, fiber fitness and fiber-matrix adhesion. In the following section we report a review of the main preteatments techniques.

3.1 Alkali treatment

Alkali treatment of natural fibers, also called mercerization, is the common method to produce high-quality fibers. The scheme of the reaction is:

$$\text{FIBER-OH} + \text{NaOH} \rightarrow \text{FIBER-O}^{-}\text{Na}^{+} + H_2O.$$

Mercerization leads to fibrillation which causes the breaking down of the composite fibre bundle into smaller fibres. Mercerization reduces fibre diameter, thereby increases the aspect ratio which leads to the development of a rough surface topography that results in better fibre/matrix interface adhesion and an increase in mechanical properties (Kalia at al., 2009). Moreover, mercerization increases the number of possible reactive sites, allows better fibre wetting and gets an effect on the chemical composition of the hemp fibres, degree of polymerization and molecular orientation of the cellulose crystallites due to cementing substances like lignin and hemicelluloses which were removed during the mercerization process. As a result, mercerization had a long-lasting effect on the mechanical properties of hemp fibres, mainly on fibre strength and stiffness. If the treatment is done at high percentage of NaOH there could be an excessive extraction of lignin and hemicelluloses which can results in damage of the ultimate cells walls. Similar reduction of mechanical properties after alkali treatment have been reported in the literature (Rodriguez at al., 2007).Alkali treatment is recognized to hydrolyses the amorphous parts of cellulose present in fibres so that after treatment the material contains more crystalline cellulose (Le Troedec, 2008). Furthermore, it removes waxes and oils from the surfaces (Sgriccia, 2008).

3.2 Acetylation

Acetylation was originally applied to wood cellulose to stabilize the cell walls against moisture, improving dimensional stability and environmental degradation and to introduce plasticization to cellulosic fibers by esterification. Acetylation is based on the reaction of cell wall hydroxyl groups of lignocellulosic materials with acetic or propionic anhydride at elevated temperature (Fig.21). Pretreatment of fibers with acetic anhydride substitutes the polymer hydroxyl groups of the cell wall with acetyl groups, modifying the properties of these polymers so that they become hydrophobic (Andersson & Tillman, 1989; Murray, 1998; Rowell, 1991) Hydroxyl groups that react with the reagent are those of lignin and hemicelluloses (amorphous material), whereas the hydroxyl groups of cellulose (crystalline material) are being closely packed with hydrogen bonds, prevent the diffusion of reagent and thus result in very low extents of reaction (Rowell, 1998).

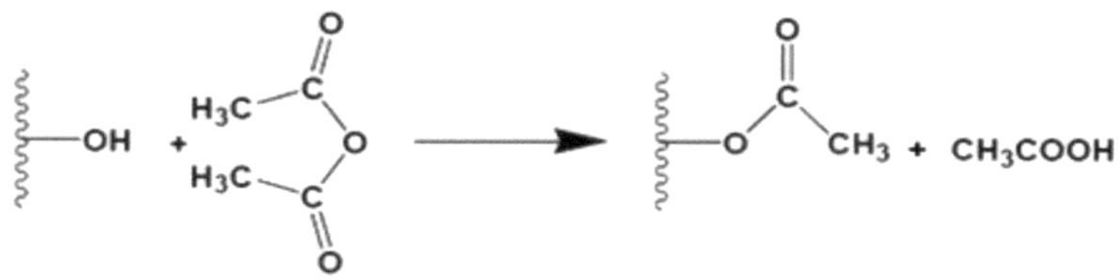

Fig. 21. Scheme of acetylation reaction

3.3 Peroxide treatment

Peroxide treatment of cellulose fibre has attracted the attention of various researchers due to easy processability and improvement in mechanical properties. Organic peroxides tend to decompose easily to free radicals, which further react with the hydrogen group of the matrix and cellulose fibers. In peroxide treatment, fibers are treated with 6% benzoyl peroxide or dicumyl peroxide in acetone solution for about 30 min after alkali pretreatment (Sreekala at al., 2002; Sreekala et al., 2002; Paul et al., 1997) conducted at a temperature of 70°C to support the decomposition of the peroxide.

3.4 Graft copolymerization

Synthesis of graft copolymers by creation of an active site, a freeradical or a chemical group which may get involved in an ionic polymerization or in a condensation process, on the preexisting polymeric backbone is one of the common methods. Polymerization of an appropriate monomer (e.g. benzoyl chloride, maleated polypropylene/maleic anhydride MAH-PP, acrylation, titanate) onto this activated back-bone polymer leads to the formation of a graft copolymer with a higher surface energy and wettability and adhesion interface by polymer matrix. It has been reported that maleic anhydride treatment reduced the water absorption to a great extent in hemp, banana and sisal fibers and their composites (Mysra et al.2000).

Modification of cellulosic fibers by etherification enhances certain new ranges of properties and makes it more useful and acceptable in diversified applications. Sodium hydroxide plays an important role in forming a charged intermediate species with the fiber, which allows the faster nucleophilic addition of epoxides, alkyl halides, benzyl chloride, acrylonitrile, and formaldehyde (Matsuda, 1996).

Benzoyl chloride is the most often used benzoylation pretreatment. Benzoyl ($C_6H_5C{=}O$) groups react with the cellulosic OH group of fiber decreasing hydrophilic nature of the treated fiber (Joseph et al., 2000) after a 30 min pre-soaking with NaOH solution to activate the hydroxyl groups of the cellulose and lignin in the fiber, followed by filtration and washing with water (Fig.22).

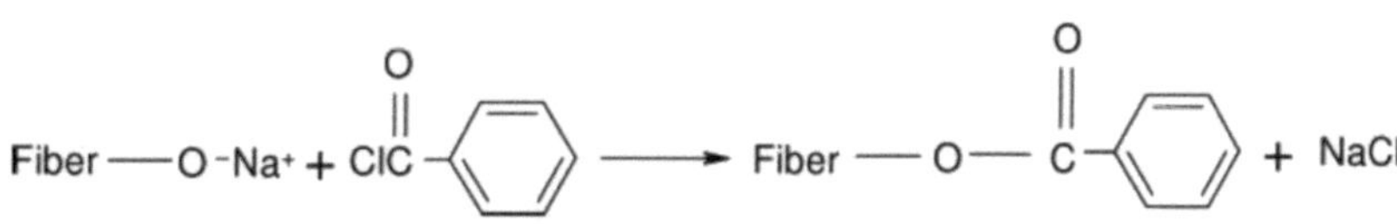

Fig. 22. Possible reaction between cellulosic-OH and benzoyl chloride (Joseph et al., 2000)

A number of methods can be used for the generation of active sites on the polymeric backbone and can be described as: physical, chemical, physicomechanical, radiation method and enzymatic grafting. The conventional techniques of grafting of natural fibers require significant time and energy. It has been found that grafting under microwave radiations is the best method in terms of time consumption and cost effectiveness. Microwave radiation technique reduces the extent of physicochemical stresses to which the fibers are exposed during the conventional techniques (Kaith & Kalia 2008).

3.5 Coupling agents

Coupling agents usually improve the degree of crosslinking in the interface region and offer a perfect bonding. Among the various coupling agents, silane coupling agents were found to be effective in modifying the natural fiber-matrix interface. Silane grafting is based on the use of reactancts that bear reactive end groups which, on one end, can react with the matrix and, on the other end, can react with the hydroxyl groups of the fiber (Fig.23). The alkoxy or ethoxy are the end groups which can form stable covalent bonds reacting with the hydroxyl groups of the fiber. The end groups which can react with the matrix vary according to the polymer matrix type. If unsaturated polyester is used silanes bearing methacryl-, amine- and vinyl- can be used (Soo-Jin et al., 2001; Li Hu at al., 2009). Efficiency of silane treatment was high for the alkaline treated fiber than for the untreated fiber because more reactive site can be generated for silane reaction. Therefore, fibers are pretreated with NaOH for about half an hour before its coupling with silane. Fibers are then washed many times in distilled water and finally dried. Silane coupling agents may reduce the number of cellulose hydroxyl groups in the fiber-matrix interface minimizing fibre sensitivity to humidity. In the presence of moisture, hydrolizable alkoxy group leads to the formation of silanols. The silanol then reacts with the hydroxyl group of the fiber, forming stable covalent bonds to the cell wall that are chemisorbed onto the fiber surface (Agrawal et al., 2000). Therefore, the hydrocarbon chains provided by the application of silane restrain the swelling of the fiber by creating a cross-linked network because of covalent bonding between the matrix and the fiber.

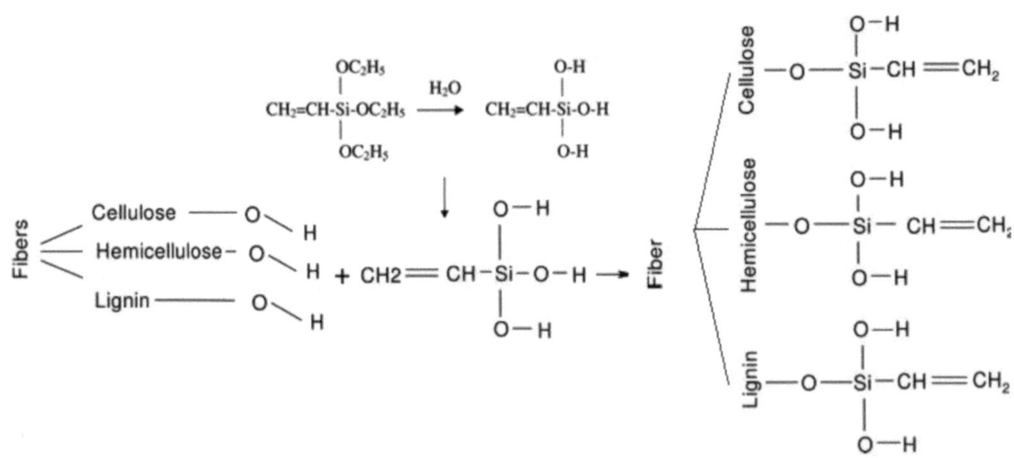

Fig. 23. Reaction of silane with OH groups of natural fiber

Silanes are effective in improving the interface properties (Coutinho at al., 1997; Gonzales et al., 1997). Alkoxy silanes are able to form bonds with hydroxyl groups. Fiber treatment with toluene dissocyanate and triethoxyvinyl silane could improve the interfacial properties.

Silanes after hydrolysis undergo condensation and bond formation stage and can form polysiloxane structures by reaction with hydroxyl group of the fibers. Silane grafting can modify the mechanical performances of fiber as a consequence of the use of acid solution for the treatment.

Isocyanate has -N=C=O functional group, which is very susceptible to reaction with the hydroxyl group of cellulose and lignin in the fibers and forms strong covalent bonds, thereby creating better compatibility with the binder resin in the composites (Kokta et al. 1990).

3.6 Permanganate treatment

Pretreatments with permanganate are conducted by using different concentration of potassium permanganate ($KMnO_4$) solution in acetone with soaking duration from 1 to 3 min after alkaline pretreatment. As a result of permanganate treatment, the hydrophilic tendency of the fibers is reduced, and thus, the water absorption of fiber-reinforced composite decreases with increase in $KMnO_4$ concentration (Sreekala et al., 2000; Paul et al., 1997). Permanganate treatment is indicated as one of the best method to improve the bonding at the fiber-polymer interface.

3.7 Physical plasma treatment

Plasma treatment is an effective method to modify the surface of natural polymers without changing their bulk properties. The plasma discharge can be generated by either corona treatment or cold plasma treatment. Both methods are considered as a plasma treatment when ionized gas has an equivalent number of positive and negative charged molecules that react with the surface of the present material. The distinguishing feature between the two categories of plasmas is the frequency of the electric discharge. High-frequency cold plasma can be produced by microwave energy, whereas a lower frequency alternating current discharge at atmospheric pressure produces corona plasma. The type of ionized gas and the length of exposure influenced the modification of the wood and synthetic polymer surfaces (Young et al., 1992; Goring & Bolam, 1976).

3.8 Chemical treatments on natural fibre: effect on mechanical properties

Chemically treated fibers can show a considerable decrease in tensile properties and this decrease is attributed to the substantial delignification and degradation of cellulosic chains during chemical treatment. The extension at break of these fibers does not change much. Most of the chemical treatments have been found to decrease the fiber strength due to breakage of the bond structure, and disintegration of the noncellulosic materials but silane and acrylation treatment leave to strong covalent bond formation and the stiffness is enhanced marginally due to the crystalline region (cellulosic) of the fiber.

The alkali treatment can produce a drop in both tensile strength and Young's modulus of the fibers if a very high percentage treatment is adopted. This result is attributed to the damage induced in the cell walls and the excessive extraction of lignin and hemicellulose, which play a cementing role in the structure of the fibers.

Morphological studies showed that the silane, benzoylation and peroxide pretreatment of flax fiber improved the surface properties. Silane and peroxide treatment of flax led to a higher tensile strength than that of untreated flax (Wang et al., 2007).

4. Case study: hybrid glass/natural fibre composites for curved pipes

4.1 Case study outline

The case study presented here refers to the analysis of the hybridization of glass fibres with natural fibres for applications in the piping industry (Cicala et al, 2009). The natural fibres studied were hemp, flax and kenaf. The pipe selected for the study was a curved fitting (90°) flanged at both ends designed to withstand an internal pressure of 10 bar and in the presence of acid aqueous solutions. This type of fitting is widely used in chemical plants which bear acid solution. The actual fittings are manufactured by hand layup with a complex sequence of glass mats and fabrics impregnated with epoxy vinyl ester resins. The problem was how to save cost without significant loss in mechanical properties and solvent resistance. Natural fibres mats were investigate as an alternative to glass mats.

4.2 Experimental

A commercial epoxy vinyl ester resin was used as thermoset matrix. Several glass fabrics were used varying from E-glass woven to E-glass random mat and C-glass liner (Table 5). The hemp mat was purchased by Hempcore Ltd., United Kingdom. Kenaf and Flax mats were kindly offered by Sachseinleinen Gmbh.

Component	Cost (€/m^2)	Areal weight (g/m^2)
E-glass woven	2.80	600
E-glass mat	2.13	600
C-glass liner	0.50	30
Hemp mat	0.31	600–650
Flax mat	0.37	750
Kenaf mat	0.33	630–650

Table 5. Technical data of the fabrics used

The lamina for mechanical testing were impregnated by hand lay-up and cured at room temperature for 48 h. The fittings were also manufactured by hand lay-up by wrapping the fabric onto a steel mandrel which is shown for reference in Fig. 24.

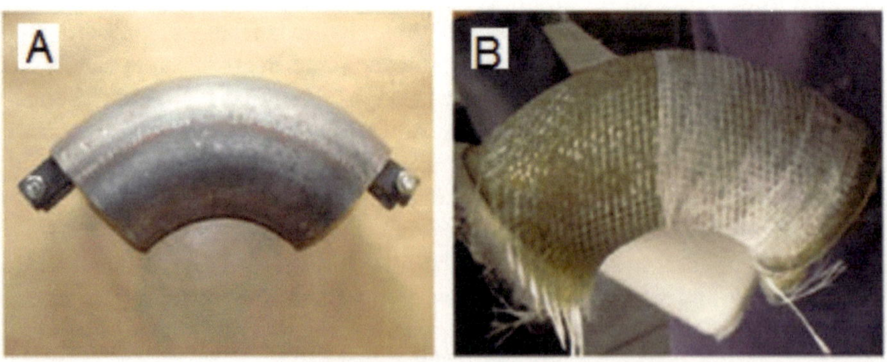

Fig. 24. (A) Steel mandrel used and (B) example of the fabric wrapping step

Tensile tests of single fibres (free fibre length was 15 mm), manually extracted from each mat, were carried out with a speed of 1 mm/min.

The cured laminas were tested accordingly to EN ISO 527 either on laminas obtained from a single fabric or on laminates obtained with a lay-up similar to those used for the fitting. Some laminate specimens were also conditioned in different HCl solutions with pH varying from 1 to 7. The specimens were immersed for 40 days and then tested to analyze the effect on mechanical properties. This test was designed to predict the mechanical behaviour of the specimens in real working conditions. All the specimens were wrapped with C-glass liner to simulate the real surface of the interior of the fittings which is usually exposed to acid solutions.

4.3 Results and discussion

The results of tensile testing on single ply lamina are summarized in Fig. 25 for the tensile strength and modulus respectively normalized with respect to the density of each lamina. Bending showed similar results. The lamina reinforced with glass woven fabric showed the best performances in terms of tensile strength and modulus. This result is the consequence of the presence of long and aligned continuous glass fibres. The glass mat showed better mechanical properties compared to the natural fibre mats. The decrease of tensile strength compared to neat resin was observed for the lamina obtained from natural fibre mats. However, slight improvements of tensile modulus were observed compared to neat resin for the same samples. This behaviour can be explained as a consequence of the low fibre volume fraction (V_f) achieved for the lamina reinforced with the natural fibres and of the scarce adhesion between fibre and matrix. The latter and matrix were due to the absence of surface treatment on the fibres used in the present study. The natural fibre surface was not treated because this choice avoids to increase the price of the natural fibre. Measurements of V_f were performed on the natural fibre mat samples and an average of 8-11% was obtained. The reason for such low V_f are twofold: the hand lay-up method does not allow to achieve high compaction pressure and poor control on resin quantity is obtained; the natural fibres have a porous structure that increase the amount of resin adsorbed when lamina are impregnated. Moreover, the architecture of the natural fibre mats is quite open and thus higher percentages of resin are allowed to impregnated the mat. If liquid molding techniques like RTM (Resin Transfer Moulding) were employed for the manufacturing a V_f of 30% could be achievable. Table 6 reports the mechanical data of Fig. 25 after normalization to a V_f of 30%. The data clearly show that natural fibres can compare to glass fibres also in terms of mechanical performances if higher volume fraction of natural fibres are achieved.

The laminate sequence leads to a thickness of 11.92 mm and a cost for the fittings of 15.74€ in terms of raw materials cost) with a weight of 2.97 kg.

The laminates for fittings which are currently manufactured present the following ply sequence: [C/C/M/W/M/W/M/M/W/M/W/M] where C stands for C-glass liner, M for E-glass mat and W for E-glass woven.

The resistance of the laminate sequence was verified accordingly to the Tsai-Hill criterion and to the maximum tension criterion using the data from single lamina testing for the calculations. The calculations were carried out for each single ply considering the relative position in the lay-up sequence (table 7).

Accordingly to this finding and taking into account the cured ply thickness of the hemp mat the following alternative design was proposed for the fittings in order to achieve a pipe thickness similar to the original pipe construction: [C/C/M_n/W/W/M_n] where M_n stands for the natural fibre mat.

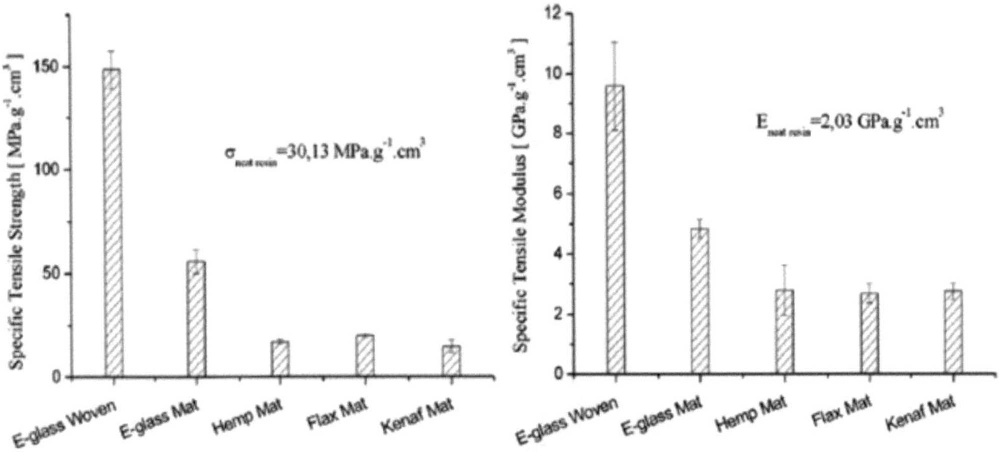

Fig. 25. Specific tensile strength and modulus on single lamina

Material	V_f (%)	Tensile strength ($MPa\,g^{-1}\,cm^3$)	Tensile modulus ($GPa\,g^{-1}\,cm^3$)
E-glass woven	35	127.64	8.22
E-glass mat	20	83.78	7.26
Hemp mat	12	42.95	7.00
Flax mat	10	60.24	8.04
Kenaf mat	9	49.50	9.17

Table. 6. Mechanical properties of single lamina after normalization to V_f of 30%

Ply	R (mm)	Maximum tension criterion			Tsai–Hill criterion			
		σ (MPa)	Res. (MPa)	Verify	σ (MPa)	Res (MPa)	Value R	Verify
Liner 1	50	–	–	–	–	–	–	–
L1	50	51.9	74.8	OK	51.9	74.8	0.36	OK
L2	51.54	123.1	240.2	OK	123.1	240.2	0.20	OK
L3	52.21	54.2	74.8	OK	54.2	74.8	0.39	OK
L4	53.75	128.4	240.2	OK	128.4	240.2	0.21	OK
L5	54.42	56.5	74.8	OK	56.5	74.3	0.43	OK
L6	55.96	58.1	74.8	OK	58.1	74.8	0.45	OK
L7	57.5	137.3	240.2	OK	137.3	240.2	0.25	OK
L8	58.17	60.4	74.8	OK	60.4	74.8	0.49	OK
L9	59.71	142.6	240.2	OK	142.6	240.2	0.26	OK
L10	60.38	62.7	74.8	OK	62.7	74.8	0.53	OK

Table. 7. Calculations according to the maximum tension and the Tsai-Hill criterion

The novel hybrid lay-up has been used to predict the cost (raw material) and the weight of the fittings produced using natural mat as replacement of glass mat. The results are summarized in Figs. 26 and 27 where the data for the original lay-up (named Glass) is reported for comparison purposes.

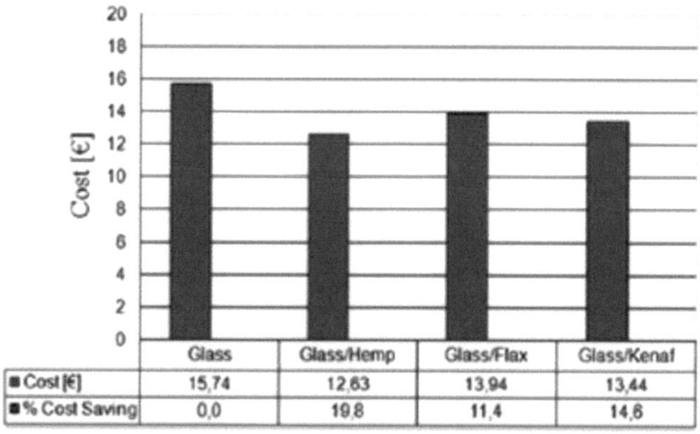

Fig. 26. Cost comparison for different lay-up solutions

The comparison shows that the novel lay-up allows for the reduction of cost and weight for all the types of the natural fibres selected. The best performances were obtained with the hemp mat. A prototype of the fitting was build with the proposed laminate sequence using the hemp mat and it was tested under pressure up to 16 bar without any significant deformation or fluid leakage.

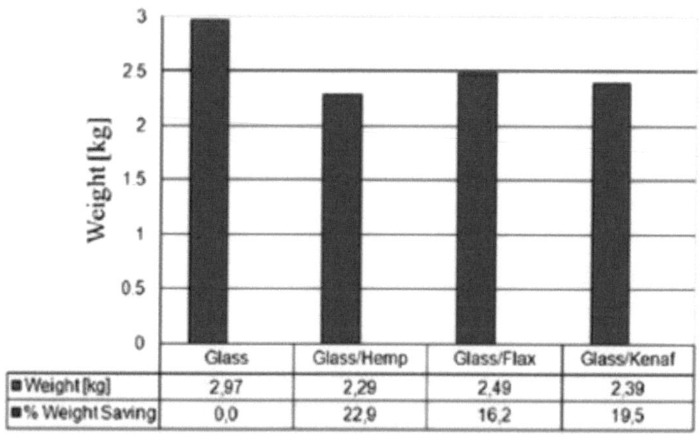

Fig. 27. Weight comparison for different lay-up solutions

Finally some laminates were tested after immersion in aqueous acid solutions for 40 days. In order to have significant data that laminates were wrapped with C-glass liner impregnated with the resin. This construction of the test lamina allows reproducing the conditions of the internal layer of the pipe which is usually exposed to the acid solution. The mechanical test showed that only small variations of the mechanical properties after immersion were obtained. The resistance to acid solution is a consequence of the barrier effect of the liner wrapping.

5. Case study: twisted hemp fabric versus hemp fabric

5.1 Case study outline

The objective of this case study is to compare the mechanical properties of twisted hemp fabric with hemp mats as viable reinforcement for composites. It has been mentioned previously that hemp mats do not represent a fabric with optimised properties for composites reinforcement due to their random fibre orientation. To overcome the limitation offered by mats the use of fabric made with aligned yarn has been investigated. Two fabric architectures were considered: unidirectional and twill 2x2.

5.2 Experimental

The general purpose unsaturated polyester resin ECMALON 4411, purchased by Ecmass Resins Pvt. Ltd, India, was used as thermoset matrix. Methyl ethyl ketone peroxide (MEKT) and cobalt naphthenate were purchased by Aldrich, Italy, and used as catalyst and accelerator respectively. 3-aminopropyltriethoxysilane (A1100) was purchased from Aldrich, Italy, and used without further purification.
Several hemp fabrics were used in this study, varying from random mat fabric, purchased by Hempcore Ltd., United Kingdom, to unidirectional [0°] and bidirectional [0°/90°] woven fabrics purchased by Canipificio Italiano, Italy. The woven fabrics were obtained weaving yarns of natural fibres made of stable filaments twisted together.
Methyl ethyl ketone peroxide (MEKT) and cobalt naphtenate were added at room temperature at percentages of 1.5 wt% and 0.07 wt% respectively. Hand layup was used to prepare the laminates for mechanical testing. Each composite was cured at room temperature for 48 h.
The cured laminas were tested accordingly to EN ISO 527 for tensile test. Five replicas for each specimen were tested. Tensile test was carried out with a Zwick universal testing machine (model Z050) equipped with a 50 kN load cell. The experiment was performed in displacement-control mode at a stroke rate (i.e. cross-head displacement rate) of 2 mm/. All output data (strain, displacement of cross-head, and load) were collected by an acquisition system and transferred to the PC.

5.3 Results and discussion

The mechanical properties of the laminates reinforced by mat are reduced by a factor of about 3/8 because of the random distribution of the fibres. To overcome this limitation the use of weaved fabrics made of twisted yarns has been considered here. Two architectures, namely, unidirectional (UD) and 0/90 were considered (Fig.28). The laminates were obtained by hand layup. The results of tensile testing obtained for laminates prepared with these fabrics are summarized in Fig. 29.
Fig.29 clearly shows that both modulus and strength are greatly enhanced when twisted yarns are used despite their low mechanical properties in dry form compared to single fibres extracted from hemp mats. This finding is the outcome of the impregnation of the yarns with the resin which, upon curing, stabilizes the yarn reducing the sliding effect of the filaments. The good properties measured for the composites reinforced with hemp is the results of the favourable orientation, along the loading direction, of the staple fibres of the yarns. As it can be expected the 0/90 fabrics present lower mechanical performances compared to unidirectional fabrics. This result is due to the presence in the 0/90 fabric of yarns directed transversely compared to loading tensile direction. The modulus and strength reported in Fig. are slightly lower than the values found in literature because of the manufacturing method (ie. hand layup) selected and of the low fibre volume fraction achieved.

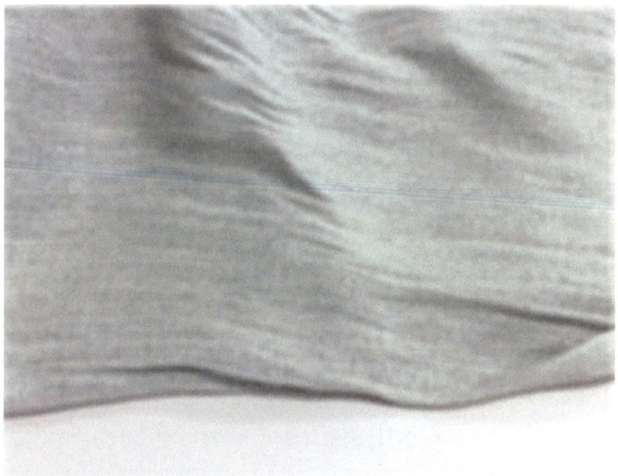

Fig. 28. Weaved fabric (0/90) with twisted yarns

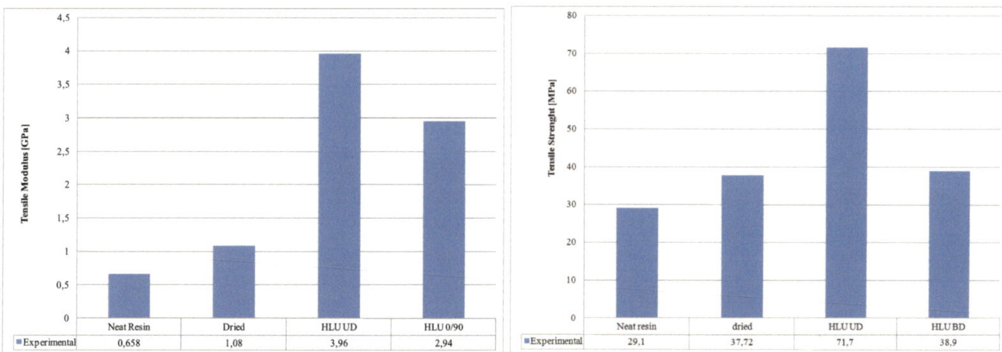

(a) (b)

Fig. 29. Tensile testing of weaved fabrics: Modulus (a), Strength (b)

6. Conclusions

The present chapter was focused on the use of natural fibre fabric as reinforcement for composite materials. The environmental and cost benefits connected with the use of natural fibre based fabrics are at the basis of their wide success. However, several limitations must be overcome in order to exploit the full potential of natural fibres. At first proper fibre surface treatment should be developed and implemented at industrial scale. Secondly, the use of mats should be investigated and the hybridization of mats with different textile further improved by analysing the effects of different layup and manufacturing techniques. Finally, the use of advanced textile based on twisted yarn should be developed further by optimising the yarn manufacturing and realising 3D architectures which are still missing from the market.

7. References

Bledzki AK & Gassan J. (1999). Composites reinforced with cellulose based fibres. *J Prog Polym Sci*; 24, 221-74.

Magurno A. (1999). Vegetable fibres in automotive interior components. Die Angew Makromol Chem; 272, 99–107.

John M.J., Francis B., Varughese K.T. & Thomas S. (2008), Effect of chemical modification on properties of hybrid fiber biocomposites. Composites: Part A - Applied Science and Manufacturing, 39 (2008) 352-363.

Saheb DN & Jog JP. (1999) Natural fiber polymer composites: A review. Adv Polym. Technol., 18, 351–63.

Kalia S., Kaith B.S. & Kaura I. (2009), Pretreatments of Natural Fibers and their Application as Reinforcing Material in Polymer Composites - A Review. Polymer Engineering and Science, 49, 1253-1272.

Williams G.I. & Wool R.P.(2000), Composites from Natural Fibers and Soy Oil Resins. Appl.Compos. Mater., 7, 421.

Bogoeva-Gaceva G., Avella M., Malinconico M., Buzarovska A., Grozdanov A., Gentile G. & Errico M.E. (2007), Natural Fiber Eco-Composites. Polymer Composites, 28, 98-107.

Rong. M.Z., Zhang M.Q., Liu Y., Yang G.C. & Zeng H.M. (2001), The effect of fiber treatment on the mechanical properties of sisal-reinforced epoxy composites. *Compos.Sci.Technolo.*, 61, 1437.

Nair KCM, Kumar RP, Thomas S, Schit SC & Ramamurthy K. (2000) Rheological behavior of short sisal fiber-reinforced polystyrene composites. Composites Part A. 31, 1231–40.

Heijenrath R. & Peijs T. (1996), Natural-fibre-mat-reinforced thermoplastic composites based on flax fibres and polypropylene, Adv. Comp. Let, 5, 81-85.

Berglund L.A. & Ericson M.L. (1995), Glass mat reinforced polypropylene in: Polypropylene: Structure, blends and composites, Vol 3, J. Karger-Kocsis (ed.), 202-227, Chapman & Hall, London.

Van den Oever M.J.A, Bos H.L. & van Kemenade M.J.J.M. (1995), Influence of the physical structure of flax fibres on the mechanical properties of flax fibre reinforced polypropylene composites, Appl. Comp. Mat. 7, 387-402.

Paiva MC, Cunha AM, Ammar I & Ben Cheikh R. (2004), Alfa fibres: mechanical, morphological, and interfacial characterisation, In: Proceedings of ICCE-11, pag. 8–14 USA, August 2004.

Baiardo M, Zini E & Mariastella S. (2004), Flax fibre-polyester composites. Composites: Part A ; 35, 703–10.

Goutianos, S. & Peijs, T. (2003) The optimisation of flax fibre yarns for the development of high performance natural fibre composites. Adv. Compos. Lett. 12, 237–241.

Baley, C. (2002) Analysis of the flax fibres tensile behaviour and analysis of the tensile stiffness increase. Composites A, 33, 939–948.

Goutianos S., Peijs T. & Nystrom B. (2006), Development of Flax Fibre based Textile Reinforcements for Comnposite Applications, Appl. Compos. Mater., 13, 199-215.

John M.J. & Anandjiwala R.D. (2008), Chemical modification of flax reinforced polypropylene composites, Polym. Compos., 29, 187.

Bledzki AK & Gassan J. (1996), Natural fiber reinforced plastics. Kassel, Germany: University of Kassel; 1996.

Rodriguez E.S., Stefani P.M. & Vazquez A. (2007), Effects of Fibers'Alkali Treatment on the Resin Transfer Moulding Processing and Mechanical Properties of Jute-Vinylester Composites, Journal of Composite Materials, Vol. 41, No. 14.

Le Troedec M., Sedan D., Peyratout C., Bonnet J.P., Smith A., Guinebretiere R., Gloaguen V. & Krausz P. (2008), Influence of various chemical treatments on the composition and structure of hemp fibres, Composites- Part A: applied science and manufacturing, 39, 514-522.

Sgriccia N., Hawley M.C. & Misra M. (2008), Characterization of natural fiber surfaces and natural fiber composites, Composites- Part A: applied science and manufacturing, 39, 1632-1637.

Andersson M. & Tillman A.M. (1989), Acetylation of jute: Effects on strength, rot resistance, and hydrophobicity, *J. Appl. Polym. Sci.*, 37, 3437.

Murray J.E. (1998), Acetylated Natural Fibers and Composite Reinforcement, *21st International BPF Composites Congress*, Publication Number 293/12, British Plastics Federation, London.

Rowell R.M. (1991), Natural Composites, Fiber Modification, in *International Encyclopedia of composites*, 4, S.M. Lee, Ed., VHC, New York,.

Rowell R.M. (1998), Property Enhanced Natural Fiber Composite Material based on Chemical Modification, in *Science and Technology of Polymers and Advanced Materials*, Prasad P.N., Mark J.E., Kendil S.H. & Kafafi Z.H Eds., pag. 717-732, Plenum Press, New York.

Matsuda H. (1996), Chemical Modification of Solid Wood in Chemical Modification of Lignocellulosic Materials, D. Hon Ed., pag. 159, Marcel Dekker, New York.

Sreekala M.S., Kumaran M.G., Joseph S., Jacob M & Thomas S. (2000), Appl. Compos. Mater., 7, 295.

A. Paul, K. Joseph, and S. Thomas, Compos. Sci. Technol., 57, 67 (1997).

M.S. Sreekala, M.G. Kumaran, and S. Thomas (2002), Compos. Part A: Appl. Sci. Manuf., 33, 763.

Joseph K., Mattoso L.H.C., Toledo R.D., Thomas S., de Carvalho L.H., Pothen L., Kala S. & James B. (2000), Natural Fiber Reinforced Thermoplastic Composites in *Natural Polymers and Agrofibers Composites*, Frollini E., Leao A.L. & Mattoso L.H.C. Eds., 159, San Carlos, Brazil, Embrapa, USP-IQSC, UNESP.

Kaith B.S. & Kalia S. (2008), *Polym. Compos.*, 29, 791.

Soo-Jin Park & Joong-Seong Jin (2001), Effect of Silane Coupling Agent on Interphase and Performance of Glass Fibers/unsaturated Polyester Composites, *Journal of Colloid and Interface Science*, 242, 174-179.

Li Hu, Yizao Wana, Fang He, H.L. Luo, Hui Liang, Xiaolei Li & Jiehua Wang (2009), Effect of coupling treatment on mechanical properties of bacterial cellulose nanofibre-reinforced UPR ecocomposites, *Materials Letters*, 63: 1952–195.

Mishra S.,. Naik J.B &. Patil Y.P (2000), *Compos. Sci. Technol.*, 60, 1729.

Agrawal R., Saxena N.S., Sharma K.B. (2000), Thomas S. &. Sreekala M.S, *Mater. Sci. Eng. A*, 277, 77.

Coutinho F.M.B., Costa T.H.S. & Carvalho D.L. (1997), *J. Appl. Polym. Sci.*, 65, 1227.

Gonzalez L., Rodriguez A., de Benito J.L.& Marcos-Fernandez A. (1997), *J. Appl. Polym. Sci.*, 63, 1353.

Sreekala M.S., Kumaran M.G., Joseph S., Jacob M. & Thomas S. (2000), *Appl. Compos. Mater.*, 7, 295.

Kokta B.V., Maldas D., Daneault C. & Beland P. (1990), *Polym.-Plast. Technol. Eng.*, 29, 87.
Wang B., Panigrahi S., Tabil L. & Crerar W. (2007), *J. Reinf. Plast. Compos.*, 26, 447.
Young R., Rowell R., Shulz T.P. & Narayan R. (1992), Activation and Characterization of Fiber Surfaces for Composites in *Emerging Technologies for Materials and Chemicals from Biomass*, Eds., American Chemical Society, pag.115 Washington D.C., 115.
Goring D. & Bolam F. (1976), Plasma-Induced Adhesion in Cellulose and Synthetic Polymers in *The Fundamental Properties of Paper Related to its uses*, Ed., Ernest Benn Limited, pag.172, London.
Cicala G., Cristaldi G., Recca G., Ziegmann G., ElSabbagh A. & M.Dickert (2009). Properties and performances of various hybrid glass/natural fibre composites for curved pipes, *Materials & Design*, 30, 2538-2542.

18

Crashworthiness Investigation and Optimization of Empty and Foam Filled Composite Crash Box

Dr. Hamidreza Zarei[1] and Prof. Dr.-Ing. Matthias Kröger[2]
[1]Aeronautical University, Tehran,
[2]Institute of Machine Elements, Design and Manufacturing, University of Technology Freiberg,
[1]Iran
[2]Germany

1. Introduction

Metallic and composite columns are used in a broad range of automotive and aerospace applications and especially as crash absorber elements. In automotive application, crashworthy structures absorb impact energy in a controlled manner. Thereby, they bring the passenger compartment to rest without subjecting the occupant to high decelerations. Energy absorption in metallic crash absorbers normally takes place by progressive buckling and local bending collapse of columns wall. A distinctive feature of such a deformation mechanism is that the rate of energy dissipation is concentrated over relatively narrow zones, while the other part of the structure undergoes a rigid body motion. In comparison to metals, most composite columns crush in a brittle manner and they fail through a sequence of fracture mechanism involving fiber fracture, matrix crazing and cracking, fiber-matrix debonding, delamination and internal ply separation. The high strength to weight and stiffness to weight ratios of composite materials motivated the automobile industry to gradual replacement of the metallic structures by composite ones. The implementation of composite materials in the vehicles not only increases the energy absorption per unit of weight (Ramakrishna, 1997) but also reduces the noise and vibrations, in comparison with steel or aluminum structures (Shin et al., 2002). The crashworthiness of a crash box is expressed in terms of its energy absorption E and specific energy absorption SEA. The energy absorption performance of a composite crash box can be tailored by controlling various parameters like fiber type, matrix type, fiber architecture, specimen geometry, process condition, fiber volume fraction and impact velocity. A comprehensive review of the various research activities have been conducted by Jacob et al. (Jacob et al., 2002) to understand the effect of particular parameter on energy absorption capability of composite crash boxes.

The response of composite tubes under axial compression has been investigated by Hull (Hull, 1982). He tried to achieve optimum deceleration under crush conditions. He showed that the fiber arrangement appeared to have the greatest effect on the specific energy absorption. Farley (Farley, 1983 and 1991) conducted quasi-static compression and impact tests to investigate the energy absorption characteristics of the composite tubes. Through his experimental work, he showed that the energy absorption capabilities of Thornel 300-fiberite and Kevlar-49-fiberite 934 composites are a function of crushing speed. He concluded that strain rate sensibility of these composite materials depends on the relationship between the mechanical response of the dominant crushing mechanism and the strain rate. Hamada and Ramakrishna (Hamada & Ramakrishna, 1997) also investigate the crush behavior of composite tubes under axial compression. Carbon polyether etherketone (PEEK) composite tubes were tested quasi-statically and dynamically showing progressive crushing initiated at a chamfered end. The quasi-staticlly tested tubes display higher specific energy absorption as a result of different crushing mechanisms attributed to different crushing speeds. Mamalis et al. (Mamalis et al., 1997 and 2005) investigated the crush behavior of square composite tubes subjected to static and dynamic axial compression. They reported that three different crush modes for the composite tubes are included, stable progressive collapse mode associated with large amounts of crush energy absorption, mid-length collapse mode characterized by brittle fracture and catastrophic failure that absorbed the lowest energy. The load-displacement curves for the static testing exhibited typical peaks and valleys with a narrow fluctuation amplitude, while the curves for the dynamically tested specimens were far more erratic. Later Mamalis et al. (Mamalis et al., 2006) investigated the crushing characteristics of thin walled carbon fiber reinforced plastic CFRP tubular components. They made a comparison between the quasi-static and dynamic energy absorption capability of square CFRP.

The high cost of the experimental test and also the development of new finite element codes make the design by means of numerical methods very attractive. Mamalis et al. (Mamalis et al., 2006) used the explicit finite element code LS-DYNA to simulate the crush response of square CFRP composite tubes. They used their experimental results to validate the simulations. Results of experimental investigations and finite element analysis of some composite structures of a Formula One racing car are presented by Bisagni et al.(Bisagni et al., 2005) . Hoermann and Wacker (Hoermann & Wacker, 2005) used LS-DYNA explicit code to simulate modular composite thermoplastic crash boxes. El-Hage et al. (El-Hage et al., 2004) used finite element method to study the quasi-static axial crush behavior of aluminum/composite hybrid tubes. The hybrid tubes contain filament wound E glass-fiber reinforced epoxy over-wrap around an aluminum tube.

Although there is several published work to determine the crash characteristics of metallic and composite columns, only few attempts have been made to optimize those behaviors. Yamazaki and Han (Yamazaki & Han, 1998) used crashworthiness maximization techniques for tubular structures. Based on numerical analyzes, the crash responses of tubes were determined and a response surface approximation method RSM was applied to construct an approximative design sub-problems. The optimization technique was used to maximize the absorbed energy of cylindrical and square tubes subjected to impact crash load. For a given impact velocity and material, the dimensions of the tube such as thickness and radius were optimized under the constraints of tube mass as well as the allowable limit of the axial impact force. Zarei and Kroeger (Zarei & Kroeger, 2006) used Multi design objective MDO crashworthiness optimization method to optimize circular aluminum tubes. Here the MDO procedure was used to find the optimum aluminum tube that absorbs the most energy while has minimum weight.

This study deals with experimental and numerical crashworthiness investigations of square and hexagonal composite crash boxes. Drop weight impact tests are conducted on composite crash boxes and the finite element method is used to reveal more details about crash process. Thin shell elements are used to model the tube walls. The crash experiments show that tubes crush in a progressive manner, i.e. the crushing starts from triggered end of the tubes, exhibit delamination between the layers. Two finite element models, namely single layer and multi layers, are developed.

In the single layer model, the delamination behavior could not be modeled and the predicted energy absorption is highly underestimated. Therefore, to properly consider the delamination between the composite layers, the tube walls are modeled as multi layer shells and an adequate contact algorithm is implemented to model the adhesion between them. Numerical results show that in comparison to the one layer method, the multi layer method yield more meaningful and accurate experimental results. Finally the multi design optimization MDO technique is implemented to identify optimum tube geometry that has maximum energy absorption and specific energy absorption characteristics.

The length, thickness (number of layers) and width of the tubes are optimized while the mean crash load is not allowed to exceed allowable limits. The D-optimal design of experiment and the response surface method are used to construct sub-problems in the sequentially optimization procedure. The optimum tube is determined that has maximum reachable energy absorption with minimum tube weight. Finally the optimum composite crash box is compared with the optimum aluminum crash box. Also the crash behaviour of foam filled composite crash boxes are investigated and compared with empty ones.

2. Experimental and numerical results

Axial impact tests were conducted on square and hexagonal composite crash boxes. The nominal wall thicknesses of the composite tubes are 2 mm, 2.4 mm and 2.7 mm. Square tubes with length of 150 mm and hexagonal tube with the length of 91 mm are used, see Figure 1. The specimens are made from woven glass-fiber in a polyamide matrix, approximately 50% volume fiber. Equal amount of fibers are in the two perpendicular main orientations. They are produced by Jacob Composite GmbH. Similar tubes are used in the bumper system of the BMW M3 E46 as well as E92 and E93 model as crash boxes.

A 45 degree trigger was created at the top end of the specimens. Generally injection moulding can be used to produce complex reinforced thermoplastics parts with low fiber length/fiber diameter aspect ratio. With increasing aspect ratio the crush performance increases but the flow ability of the material decreases. For this reason continuous reinforced thermoplastic have to be thermoformed. In this way and by using other post processing technologies like welding, complex composite parts with an excellent crush performance can be realized (Hoermann & Wacker, 2005). Here, the crash boxes are produced from thermoplastic plates by using thermoforming technique. The square specimens have overlap in one side and the overlaps have been glued by using a structural adhesive. The hexagonal crash boxes consist of two parts that are welded to each other.

The experimental tests have been conducted on the drop test rig, see Fig. 2, which is installed in the Institute of Dynamics and Vibrations at the Leibniz University of Hannover. This test rig has an impact mass which can be varied from 20 to 300 kg. The maximum drop height is 8 m and maximum impact speed is 12.5 m/s. The force and the displacement are recorded with a PC using an AD-converter. The force is measured using strain gauges and laser displacement sensors provide the axial deformation distance of the tubes. Here an impact mass of 92 kg was selected. The interest in this study is the mean crashing load P_m and the energy absorption E. The mean crash load is defined by

$$P_m = 1/\delta \int_0^{\delta} P(\delta)d\delta \qquad (1)$$

where P(δ) is the instantaneous crash load corresponding to the instantaneous crash displacement d. The area under the crash load–displacement curve gives the absorbed energy. The ratio of the absorbed energy to the crush mass of the structure is the specific energy absorption. High values indicate a lightweight absorber. Figure 1 shows the geometry of the specimens.

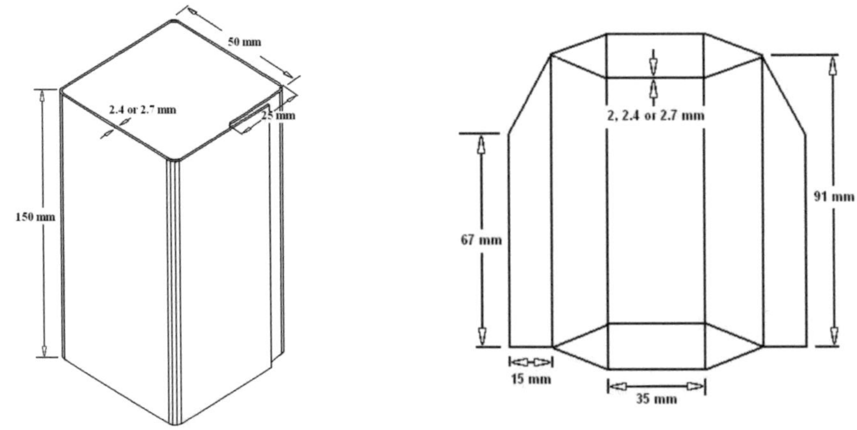

Fig. 1. (a) Square crash box (b) hexagonal crash box

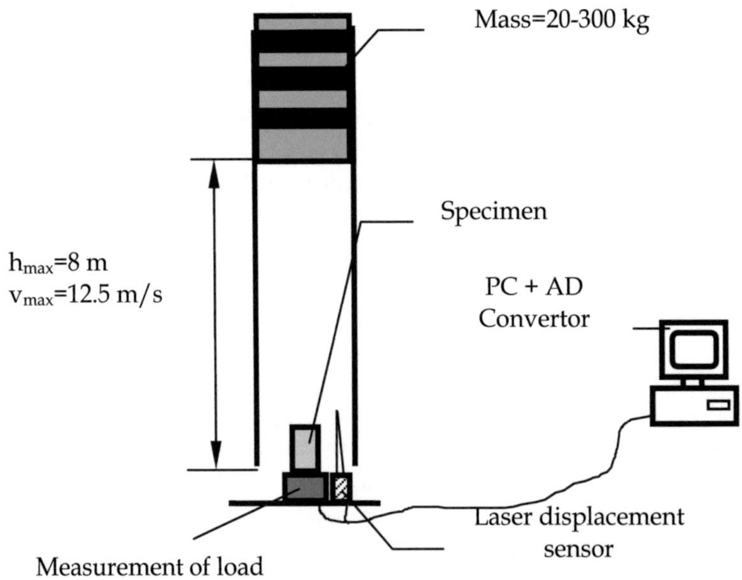

Fig. 2. Test rig

Numerical simulations of crash tests are performed to obtain local information from the crush process. The modeling and analysis is done with the use of explicit finite element code, LS-DYNA. The column walls are built with the Belytschko-Tsay thin shell elements and solid elements are used to model the impactor. The contact between the rigid body and the specimen is modeled using a node to surface algorithm with a friction coefficient of $\mu=0.2$. To take into account the self contact between the tube walls during the deformation, a single surface contact algorithm is used. The impactor has been modeled with the rigid material. The composite walls have been modeled with the use of material model #54 in LS-DYNA. This model has the option of using either the Tsai-Wu failure criterion or the Chang-Chang failure criterion for lamina failure. The Tsai-Wu failure criterion is a quadratic stress-based global failure prediction equation and is relatively simple to use; however, it does not specifically consider the failure modes observed in composite materials (Mallick, 1990). Chang-Chang failure criterion (Mallick, 1990) is a modified version of the Hashin failure criterion (Hashin, 1980) in which the tensile fiber failure, compressive fiber failure, tensile matrix failure and compressive matrix failure are separately considered. Chang and Chang modified the Hashin equations to include the non-linear shear stress-strain behavior of a composite lamina. They also defined a post-failure degradation rule so that the behavior of the laminate can be analyzed after each successive lamina fails. According to this rule, if fiber breakage and/or matrix shear failure occurs in a lamina, both transverse modulus and minor Poisson's ratio are reduced to zero, but the change in longitudinal modulus and shear modulus follows a Weibull distribution. On the other hand, if matrix tensile or compressive failure occurs first, the transverse modulus and minor Poisson's ratio are reduced to zero, while the longitudinal modulus and shear modulus remain unchanged. The failure equations selected for this study are based on the Chang-Chang failure criterion. However, in material model #54, the post-failure conditions are slightly modified from the Chang-Chang conditions. For computational purposes, four indicator functions e_f, e_c, e_m, e_d corresponding to four failure modes are introduced. These failure indicators are based on total failure hypothesis for the laminas, where both the strength and the stiffness are set equal to zero after failure is encountered,

(a) Tensile fiber mode (fiber rupture),

$$\sigma_{aa} > 0, \text{ and } e_f^2 = (\sigma_{aa}/x_t)^2 + \zeta(\sigma_{ab}/S_c)^2 - 1 \begin{cases} \geq 0 \Rightarrow \text{faild} \\ > 0 \Rightarrow \text{elastic} \end{cases} \tag{2}$$

Where ζ is a weighting factor for the shear term in tensile fiber mode and $0<\zeta<1$. $E_a=E_b=G_{ab}=\upsilon_{ab}=\upsilon_{ba}=0$ after lamina failure by fiber rupture.

(b) Compressive fiber mode (fiber buckling or kinking),

$$\sigma_{aa} > 0, \text{ and } e_c^2 = (\sigma_{aa}/x_c)^2 - 1 \begin{cases} \geq 0 \Rightarrow \text{faild} \\ > 0 \Rightarrow \text{elastic} \end{cases} \tag{3}$$

$E_a=\upsilon_{ab}=\upsilon_{ba}=0$ after lamina failure by fiber buckling or kinking.

(c) Tensile matrix mode (matrix cracking under transverse tension and in-plane shear),

$$\sigma_{bb} > 0, \text{ and } e_m^2 = (\sigma_{bb}/y_t)^2 + \zeta(\sigma_{ab}/S_c)^2 - 1 \begin{cases} \geq 0 \Rightarrow \text{faild} \\ > 0 \Rightarrow \text{elastic} \end{cases} \tag{4}$$

$E_a=G_{ab}=\upsilon_{ab}=0$ after lamina failure by matrix cracking

(d) Compressive matrix mode (matrix cracking under transverse compression and in-plane shear),

$$\sigma_{bb} > 0, \text{ and } e_d^2=(\sigma_{bb}/2S_c)^2 + \left[(y_c/2S_c)^2 - 1\right]\sigma_{bb}/y_c + (\sigma_{bb}/2S_c)^2 - 1 \begin{cases} \geq 0 \Rightarrow \text{faild} \\ > 0 \Rightarrow \text{elastic} \end{cases} \quad (5)$$

$E_b=\upsilon_{ab}=\upsilon_{ba}=0 \rightarrow G_{ab}=0$ after lamina failure by matrix cracking

In Equations (2)–(5), σ_{aa} is the stress in the fiber direction, σ_{bb} is the stress in the transverse direction (normal to the fiber direction) and σ_{ab} is the shear stress in the lamina plane aa-bb. The other lamina-level notations in Equations (2)–(5) are as follows: x_t and x_c are tensile and compressive strengths in the fiber direction, respectively. Y_t and y_c are tensile and compressive strengths in the matrix direction, respectively. S_c is shear strength; E_a and E_b are Young's moduli in the longitudinal and transverse directions, respectively. Here, to model the trigger, two elements with progressively reduced thicknesses were placed in the triggers zone. The tied surface to surface contact algorithm has been used to glue the overlapping walls.

Tables 1 and 2 show the test results of the square and hexagonal composite tubes . Here, the area under crush load-displacement curve is considered as energy absorption E. The maximum crush load P_{max} is a single peak at the end of the initial linear part of the load curve. The mean crush load P_m has been determined with the use of Equation (1). The maximum crush displacement S_{max} is the total displacement of the impactor after contact with the crash box. The values of specific energy absorption SEA, which is the energy absorption per crush weight, and the crush load efficiency η, which is the ratio of the mean crush load and maximum crush load, are also presented in these tables.

Figure 3 shows the specimen (S-67) and (S-75) after crush, respectively. Relatively ductile crush mode can be recognized. The tubes are split at their corners. This splitting effect is initiated at the end of the linear elastic loading phase, when the applied load attains its peak value P_{max}. The splitting of the corners of the tube is followed by an immediate drop of the crush load, and propagation parallel to the tube axis results in splitting of the tube in several parts. Simultaneous of splitting, some of these parts are completely splayed into two fronds which spread outwards and inwards and some parts are split only partially. Subsequent to splitting, the external and internal fronds are bended and curled downwards and some additional transverse and longitudinal fracture happened.

Photographs from high speed camera for different impact moments are presented in Figures 4 and 5. Here it can be seen that local matrix and fiber rupture results in a formation of pulverized ingredients material just after initial contact between impactor and crash boxes. As compressive loading proceeds, further fragments are detached from the crash box. Furthermore, the crush performance of tests has been simulated with the use of LS-DYNA explicit code. Figure 6 shows the experimental and simulated crush load-displacement and energy absorption-displacement curves of tests (S-67) to (S-69).

The same results for hexagonal crash boxes, tests (S-75) to (S-77), are presented in Figure 7. The crush-load displacement curves indicate that the mean crush load of simulation is obviously lower than experimental results. The numerical simulation can not cover the experiments very good.

Test No.	V [m/s]	t [mm]	P_{max} [kN]	P_m [kN]	S_{max} [mm]	E [J]	SEA [J/kg]	η [%]
S-67	10.3	2.4	77.2	40.6	126.9	4956	41844	53
S-68	10.4	2.4	75.3	46.03	118.9	5053	45533	61
S-69	10.2	2.4	83.7	43.3	117.3	4923	44967	52
S-70	10.4	2.7	82.2	58.7	86.2	5075	55542	71
S-71	10.4	2.7	92.3	59.3	84.7	5024	55957	64

Table 1. Experimental dynamic test on square composite tube

Test No.	V [m/s]	t [mm]	P_{max} [kN]	P_m [kN]	S_{max} [mm]	E [J]	SEA [J/kg]	η [%]
S-72	7.3	2.0	51	42.6	72.8	3103	35681	83
S-73	7.3	2.0	55	45.5	68.3	3109	35750	83
S-74	7.3	2.0	46	37.9	78.2	2964	34083	82
S-75	8.4	2.4	72	53.7	76.95	4133	39604	75
S-76	8.4	2.4	81	69.4	61.03	4235	40582	86
S-77	8.9	2.4	72	65.6	71.4	4683	44875	91
S-78	8.3	2.7	83	66.9	59.96	4012	34173	81
S-79	8.3	2.7	80	68.4	58.6	4008	34139	86
S-80	8.8	2.7	84	58.8	75.5	4442	37836	70

Table 2. Experimental dynamic test on hexagonal composite tube

Fig. 3. Crush pattern of square tube S-67 (left) and hexagonal tube S-75 (right)

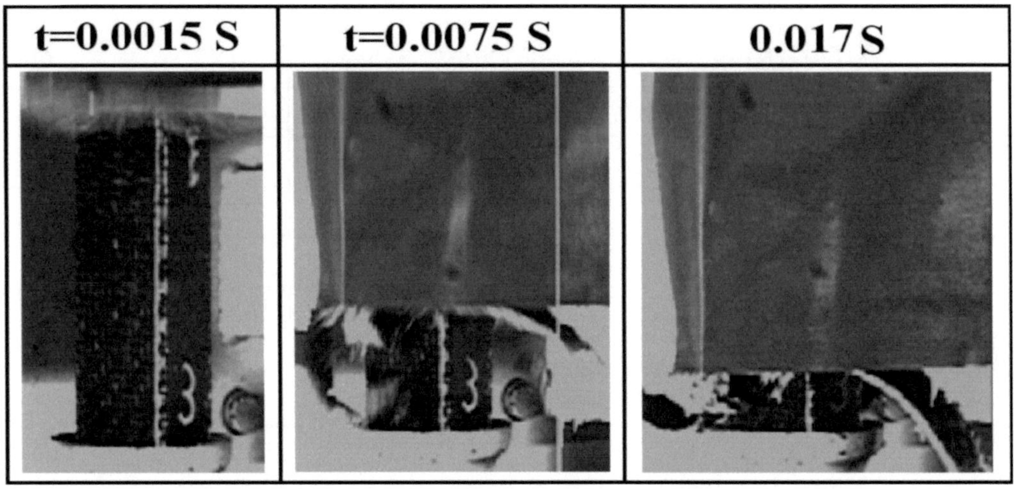

Fig. 4. Crush pattern of a square composite tube (S-67) for different crush moments

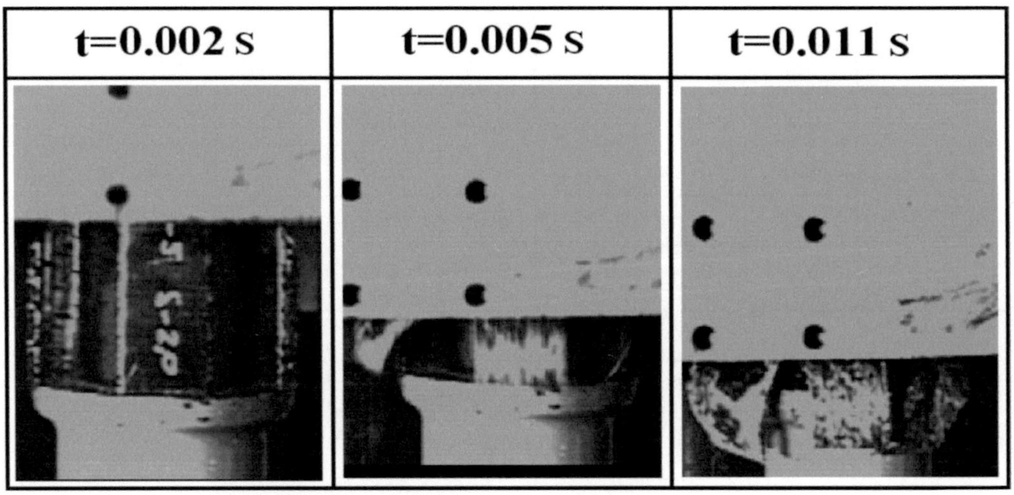

Fig. 5. Crush pattern of a hexagonal composite tube (S-75) for different crush moments

The energy absorption E and specific energy absorption SEA of the experiments and simulations at the same crush length (80 mm for square tubes and 60 mm for hexagonal ones) are presented in Table 3. Here, index S indicates simulation results. Again, it can be seen that the numerical simulations highly underestimate the tube crush behavior. The numerical crush patterns show the tube experiences the progressive crushing with some damages in tube walls instead of splitting and spreading, see Figure 8 and 9. It is evident that the total energy absorption of the composite tube is the sum of the energy needed for splitting of the tube corners, delamination and spreading of tube walls into two inwards and outwards fronds, bending and curling of each fronds, fracture and damage created in fronds during bending, fragmentations of tube walls and friction between the impactor and inwards and outwards fronds. The single layer finite element model does not have the capability to consider all aspects of crushing damages observed experimentally. Therefore, a new finite element model has to be developed to overcome this problem.

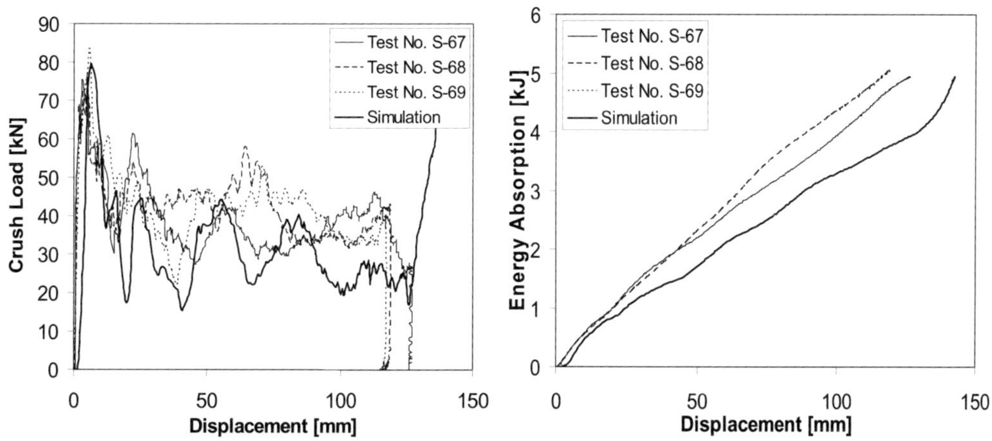

Fig. 6. Comparison between experimental and numerical (single layer method) crush load-displacement curves (left) and energy absorption-displacement curves (right) of square composite tubes

Test No.	E [J]	SEA [J/kg]	Es [J]	SEAS [J/kg]	Difference [%]
S-67	3259	43647	2686	35973	-17.6
S-68	3682	49313	-	-	-27.1
S-69	3520	47143	-	-	-23.7
S-75	3718	54035	2890	42002	-22.3
S-76	4170	60604	-	-	-30.7
S-77	3930	57116	-	-	-26.5

Table 3. Comparison between experimental and numerical (single layer method) energy absorption and specific energy absorption of the square and hexagonal tubes

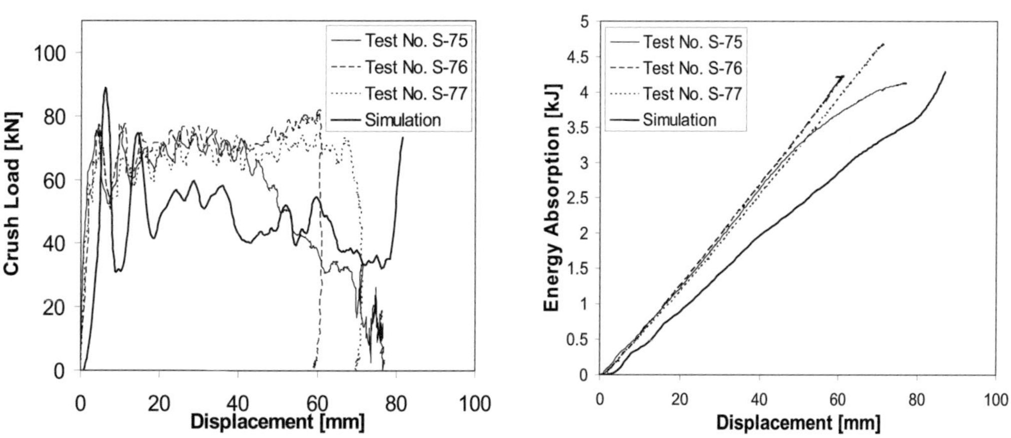

Fig. 7. Comparison between experimental and numerical (single layer method) crush load-displacement curves (left) and energy absorption-displacement curves (right) of hexagonal composite tubes

3. Advanced finite element model

The numerical crush behavior of the composite crash box are shown above for tube walls modeled with only one layer of shell elements, simulated crush pattern are quite different from experiment. The delamination, a main energy absorption source of composite crash boxes, can not be modeled and, therefore, the predicted energy absorption by the simulation is highly underestimated. Several methods have been used by the researchers to model the delamination growth in composite materials, including the virtual crack extension technique (Farley & Jones, 1992), stress intensity factor calculations (Hamada & Ramakrishna, 1997), stresses in a resin layer (Kindervater, 1995), and, the virtual crack closure technique (Fleming & Vizzini, 1996).

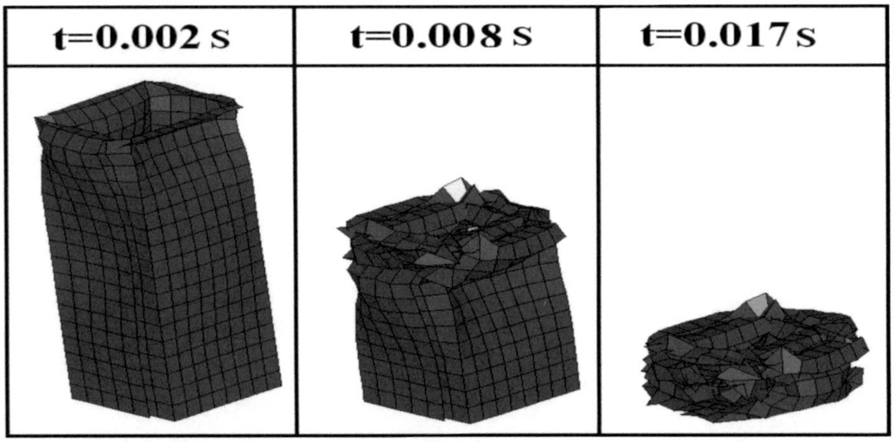

Fig. 8. Crush pattern of single layer finite element model of square composite tube

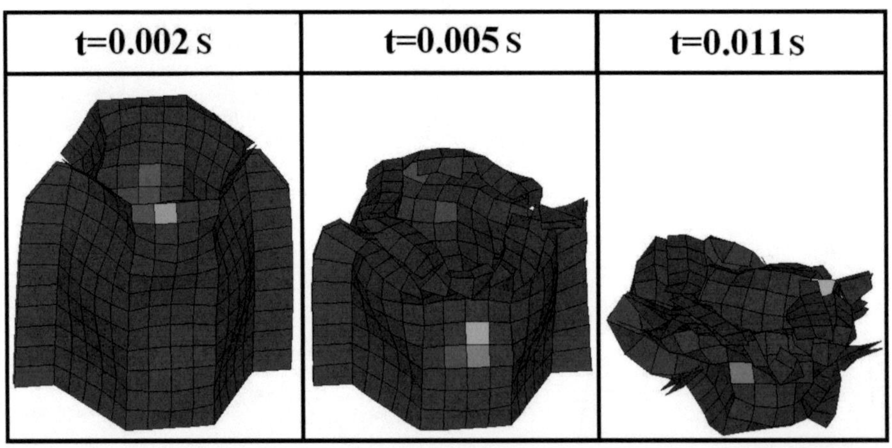

Fig. 9. Crush pattern of single layer finite element model of hexagonal composite tube

However, choices for modeling delamination using conventional finite element crush codes are more limited. Good correlations are obtained in many cases using models that do not fully capture all aspects of crushing damage observed experimentally. They only provide sufficient attention to the aspects of crushing that mostly influence the response. Models of composite structures using in-plane damaging failure models to represent crushing behavior are used in (Haug et al., 1991), (Johnson et al., 1996 and 1997), (Feillard, 1999) and (Kohlgrueber & Kamoulakos, 1998). These models appear to be effective for structures whose failure modes are governed by large-scale laminate failure and local instability. However, crushing behavior in which wholesale destruction of the laminate contributes significantly to the overall energy absorption cannot be accurately modeled by this approach (Fleming, 2001). Further, if delamination or debonding forms a significant part of the behavior, specialized procedures must be introduced into the model to address this failure mechanism . Kohlgrueber and Kamoulakos (Kohlgrueber & Kamoulakos, 1998)and Kerth et al. (Kerth et al., 1996) used tied connections with a force-based failure method to model the delamination in composite materials. By this method, nodes on opposite sides of an interface where delamination is expected are tied together using any of a variety of methods including spring elements or rigid rods. If the forces produced by these elements exceed some criterion, the constraint is released. The primary disadvantage of this method is that there is no strong physical basis for determining the failure forces. Reedy et al. (Reedy et al., 1997) applied cohesive fracture model for the same reason. This method is similar to the previous method. However, instead of relying on simple spring properties the force-displacement response of the interfacial elements is based on classical cohesive failure behavior. Virtual crack closure technique is often used by researchers in the area of fracture mechanics. Energy release rates are calculated from nodal forces and displacements in the vicinity of a crack front. Although the method is sensitive to mesh refinement, but not so sensitive like the other fracture modelling techniques, those requiring accurate calculation of stresses in the singular region near a crack front. Further, the use of conventional force and displacement variables obviates the need for special element types that are not available in conventional crash codes.

In this study for the delamination, tube walls are modeled with two layers of shell elements. The thickness of each layer is equal to the half of the tube wall thickness [130]. To avoid tremendous increase of the required simulation time, a larger number of layers is avoided. The surface to surface tiebreak contact is used to model the bonding between the bundles of plies of the tube walls. In this contact algorithm the tiebreak is active for nodes which are initially in contact. Stress is limited by the perfectly plastic yield condition. For ties in tension, the yield condition is

$$[\sqrt{(\sigma_n{}^2 + 3|\sigma_s|^2)} / \varepsilon_p] \leq 1 \tag{6}$$

Where ε_p is the plastic yield stress and σ_n and σ_s are normal and shear stresses, respectively. For ties in compression, the yield condition is

$$[\sqrt{(3|\sigma_s|^2)} / \varepsilon_p] \leq 1 \tag{7}$$

The stress is also scaled by a damage function. The damage function is defined by a load curve with starts at unity for crack width of zero and decays in some way to zero at a given value of the crack opening (Hallquist, 1998)], see Figure 10. The surface to surface tied contact is implemented between the overlapped walls and single surface contact is used for each layer. The node to surface contact is applied between rigid impactor and composite layers. To model the rupture at the corners of the tube, the vertical sides of the tube have offset 0.5 mm and deformable spot-welds are used to connect the nodes of the vertical sides.

The spot-welds are defined by the use of material number #100 in LS-DYNA (MAT_SPOTWELD). Based on this material model, beam elements, based on Hughes-Liu beam formulation, are placed between the tube walls and contact-spotweld algorithm ties the beam elements to the tube shell elements. The normal strength of spot-welds is calculated from the transverse tensile strength of the composite material.

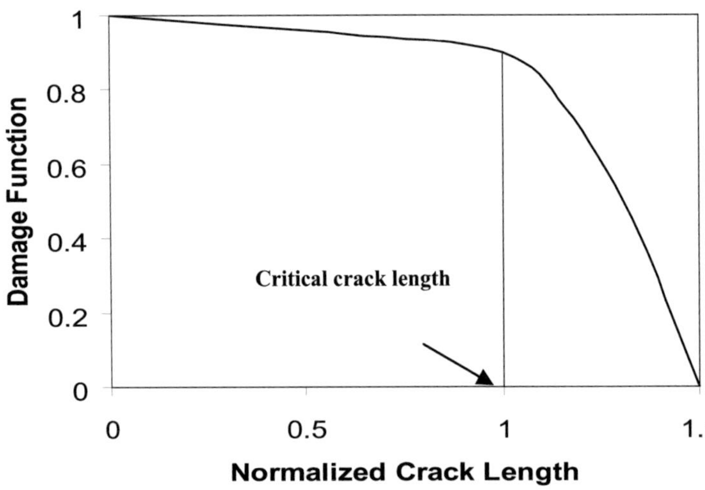

Fig. 10. Variation of damage function

To account for the reduced strength of the composite material at the corners, material strength is reduced by 50%. The shear strength is considered as half of the normal strength. In order to model the trigger, the length of the outer layer of the composite tube is a little bit smaller than the inner layer. The crush patterns of the multi layer square and hexagonal crash boxes are presented in Figures 11 and 12. Here it is possible to see the delamination which starts in some tube walls and propagates during the crush process.

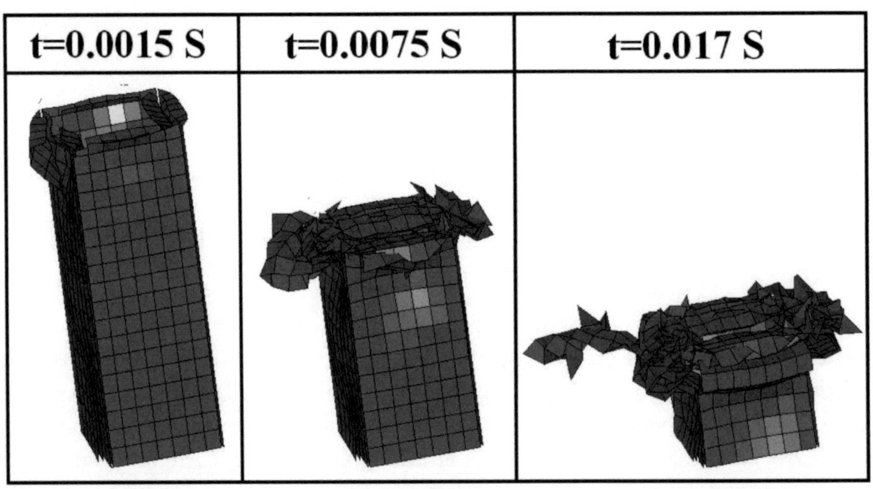

Fig. 11. Crush pattern of multi layer finite element model of square composite tube

The Figures 13 and 14 left compare the crush load-displacement curves of experimental and numerical impact on square and hexagonal crash boxes, respectively. Acceptable correlations are reached between experiments and simulations. In addition the experimental and numerical energy absorption is presented in Figure 13 and Figure 14 right. The multi layers method can predict the energy absorption of the crash box very well.

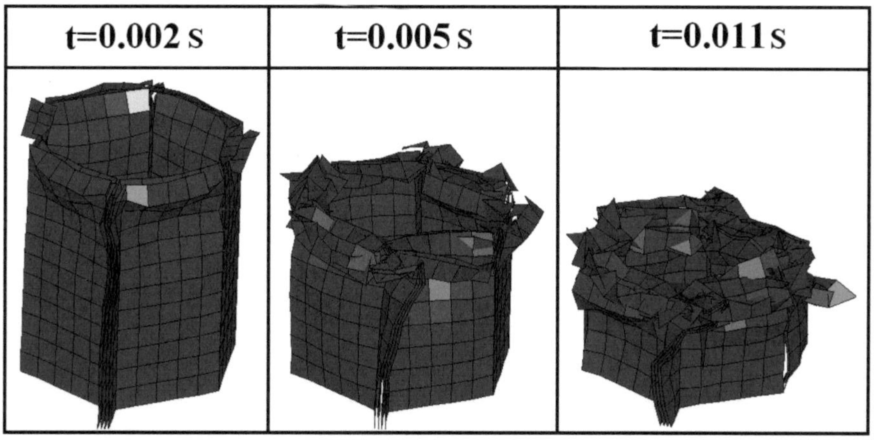

Fig. 12. Crush pattern of multi layer finite element model of hexagonal composite tube

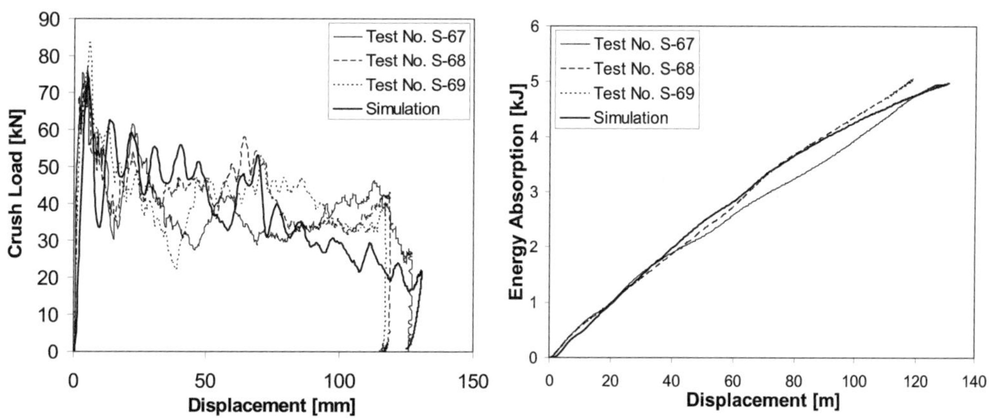

Fig. 13. Comparison between experimental and numerical (multi layers method) crush load-displacement curves (left) and energy absorption-displacement curves (right) of square composite tubes

4. Multi design optimization of crush behavior of square composite crash box

There are high interests to find the effect of composite tube geometry on its energy absorption capability. Generally, variation in tube geometry influences the fracture mechanisms and, therefore, the energy absorption capability. Thornton and Edwards (Thornton and Edwards, 1982) investigated the crush performance of square, rectangular and circular composite tubes. They concluded that for a given fiber lay up and tube geometry, circular tubes have the highest specific energy absorption followed by square and rectangular tubes. Farley (Farley, 1986) investigated the effect of geometry on the energy absorption capability of the composite tubes. He conducted a series of quasi-static crash tests of Graphite/Epoxy and Kevlar/Epoxy composite tubes with the ply orientation of ±45 degree. He found that the tube diameter to wall thickness ratio d/t has significant effects on the energy absorption capability. The energy absorption was found to be a decreasing nonlinear function of tube d/t ratio. A reduction in d/t ratio increases the specific energy absorption of the tube. Similar result has been reported by Farley and Jones (Farley & Jones, 1992) for elliptical composite tubes.

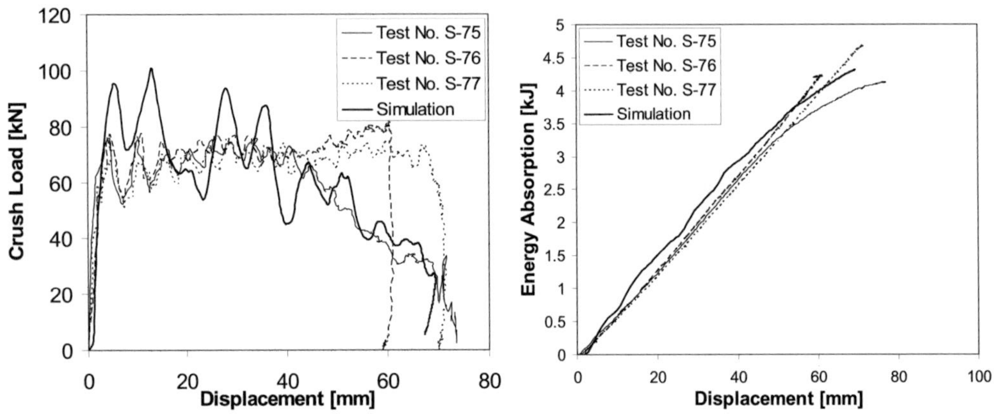

Fig. 14. Comparison between experimental and numerical (multi layers method) crush load-displacement curves (left) and energy absorption-displacement curves (right) of hexagonal composite tubes

Zarei and Kroeger (Zarei & Kroeger, 2006) used Multi design objective MDO crashworthiness optimization method to optimize circular aluminum tubes. Here, the same optimization procedure is used to find optimum composite crash box. The finite element method is used to calculate the absorbed energy and specific absorbed energy of the tubes. The design variables are the tube thickness (number of layers), width and length of the composite tubes. The composite tubes with the thickness between 1 mm and 4 mm are selected while the tube width is varied between 70 mm and 120 mm and the tube length between 100 mm and 350 mm. Here 0.5 mm thickness is considered for each layer of composite tube. To have acceptable crush performance in oblique crash conditions, the tube width lower than 70 mm is not considered. An impact force constraint is usually required to reduce the occupant injury when passenger vehicles are considered. Therefore, in the optimization process, the mean crush load P_m should not exceed the allowable limit P_{ma} i.e.:

$$g = Pm/Pma - 1 \leq 0. \quad (8)$$

Where Pma=68.5 kN is selected in this research. The optimization problem can be rewritten as follows

Maximize energy absorption E and specific energy absorption SEA of tube

Subjected to

$$0.5\ \text{mm} \leq t \leq 3.0\ \text{mm},$$
$$100\ \text{mm} \leq l \leq 350\ \text{mm},$$
$$50\ \text{mm} \leq d \leq 120\ \text{mm},$$
$$P_m \leq 68.5\ \text{kN}.$$

The optimization procedure which is presented in Figure 15 is applied to the maximization of absorbed energy and specific absorbed energy of the composite tube under axial impact load. Since the interest is to find the crush behavior of tubes up to the final effective crush length, all tubes are encountered with a large amount of impact energy. Here 75 percent of tube length is considered as effective crush length. In order to reduce the optimization time, the single layer finite element models are used to find the energy absorption of composite tubes in every subproblem and the final optimum tube is modeled as a multi layer composite tube.

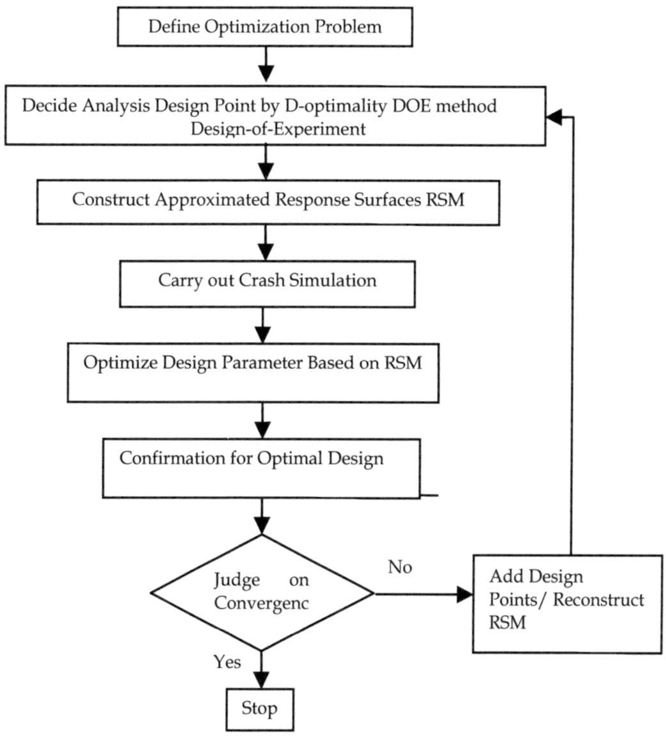

Fig. 15. Flowchart of the optimization process

Table 4 shows the final optimum composite tube that absorbs maximum energy with minimum weight. Here it can be seen that the optimum tube thickness t is 3 mm (N_l=6 layers). The thicker tube will have mean crush load higher than allowable limit. The variable d coincides with the lower bound which shows an increase of the crashworthiness efficiency by reduction of tube width. But here values lower that 70 mm are not allowed to guarantee enough bending resistance of the composite crash box in oblique crash conditions. The tube length coincides with the upper bound but in order to avoid global buckling, longer tubes are not considered. Previously the MDO procedure was used to find optimum aluminum tubes. There, to avoid global buckling in the aluminum tubes the maximum allowed tube length to width ratio is set to $l/d \leq 3$ based on experimental observations (Mamalis et al., 2005) and (Hanssen et al., 1999 and 2000). In order to compare crashworthiness behavior of the optimum composite and aluminum crash boxes, this new optimization constraint is considered for composite crash tube. Table 5 shows the results of optimum composite and aluminum crash boxes. It can be seen that the composite tube absorbs about 17 percent more energy than aluminum crash box while it has about 27 percent lower weight.

Tube type	T; N_l [mm; -]	d [mm]	l [mm]	E [J]	SEA [J/kg]
Square composite	3; 6	70	350	15316	35580

Table 4. Optimum square composite tube

Tube Type	t [mm]	d [mm]	l [mm]	E [J]	Increase [%]	SEA [J/kg]	Increase [%]
Square aluminum	2.1	70	210	7602	-	26124	-
Square composite	3	70	210	9198	17.4	35716	26.9

Table 5. Comparison between optimum composite and optimum aluminum crash boxes

5. Crush performance investigation of foam-filled composite crash box

Here, Alporas aluminum foam with a relative density of 0.085 is used to produce foam filled square composite crash box. Dynamic compression tests were conducted on them. The composite square tubes with the dimensions which previously presented in Figure 1 are used. The nominal wall thickness of the composite tubes is 2.4 mm. Dynamic tests were done in drop weight test rig. Simply support boundary conditions were applied for the tubes. Table 6 shows the results of experimental tests. The crush pattern of test number (F-37) is shown in the Figure 16. Here, similar to empty composite tubes, the tube is split from its corners. In comparison to the empty composite tubes, lower delamination area can be seen. The tube is ruptured from its corners and the foam filler is crushed progressively. Numerical simulations of crash tests are performed using the explicit finite element code LS-DYNA. The new developed finite element model in this study is used to describe the composite square tubes, see section 4. The foam filler is modeled with solid elements and rigid body elements are used to model the rigid impactor. The contact between the rigid body and the specimen is modeled using a node to surface algorithm with a friction coefficient of $\mu=0.2$. To account for self contact between the tube walls during deformation, a single surface contact algorithm is used. The node to surface contact is implemented between tube walls and foam filler. The composite walls are modeled with the use of material model #54 in LS-DYNA The aluminum foam was modeled with the foam model of Dehspande and Fleck (2000) [19] material number #154 in LS-DYNA. Figure 17 shows that the predicted energy absorption by the simulation is in good agreement with the experimental one.

Table 7 shows a comparison between energy absorption E and specific energy absorption SEA of the empty and foam-filled composite square tubes at the 80 mm crash length. Here, it can be seen that the foam insertion of the composite tube results in higher energy absorption but unlike the aluminum foam-filled tubes, the specific energy absorption in the composite filled tubes is decreased in comparison with empty one. As mentioned in the chapter four, the benefit of using foam inside the crash absorbers is the interaction between foam and crash absorber walls during crush process. But as one can see in the Figure 16, in the foam-filled composite tubes, the composite tube is split into four parts and the tube and foam crushed independently. Here no interaction between tube and foam is taken place. From

Figure 3 it can be seen that the empty composite tubes are split into several parts and each part is splayed into two fronds which spread outwards and inwards. From Figure 16 it is clear that the foam filler forced the tube parts outward during the crush process and prevent from splaying of the parts. Therefore no frond is created and delamination between the composite layers, which is one of the main energy absorption sources of the composite, is not taken placed. Therefore, the specific energy absorption of the filled composite tube is lower than empty tubes.

Another interesting result which is extracted from experimental results of dynamic tests on simple foam filler is that the energy absorption of foam filler is about 4950 J at 80 mm crash length. That means the some of the energy absorption of the empty composite tube alone and foam filler alone is higher than energy absorption of the foam-filled composite tube. In other word not only inserted foam plays no positive roll in the crush process of the filled composite crash box but also it has destructive effect.

Test No.	V [m/s]	t [mm]	P_{max} [kN]	P_m [kN]	S_{max} [mm]	E [J]	SEA [J/kg]	η [%]
F-37	10.4	2.4	85.1	46.9	105.4	4994	34006	55.1
F-38	10.3	2.4	95.1	47.3	97.7	4890	35922	49.7
F-39	10.3	2.4	87.8	46.2	108.5	4954	32770	42.6

Table 6. Experimental dynamic test on foam filed square composite tube

Fig. 16. Crush pattern of foam-filled composite crash box

Test No.	Filler type	E [J]	Increase [%]	SEA [J/kg]	Increase [%]
Average of S-67, S-68, S-69	-	3487	-	46701	-
Average of F-37, F-38-F-39	Foam	3832	9.0	34233	-26.7

Table 7. Comparison between empty and foam-filled composite tubes

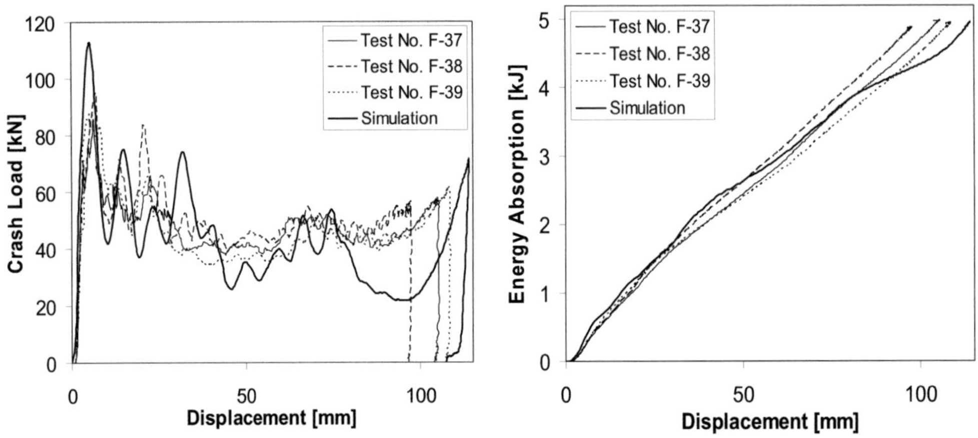

Fig. 17. Comparison between experimental and numerical (multi layers method) crush load-displacement curves (left) and energy absorption-displacement curves (right) of square composite foam-filled tubes

6. Conclusion

Experimental crash tests on square and hexagonal composite crash boxes showed that unlike metallic crash boxes which are crushed in a progressive buckling manner, the composite tubes are crushed in a progressive damaging manner.
A new multi layer finite element model was developed to simulate the crush process of the composite crash box.
The MDO procedure was used to find an optimum design of the composite crash box. The comparison between crashworthiness behavior of the optimum composite and aluminum crash boxes showed that the composite crash box absorbs about 17 percent more energy than the aluminum crash box while it has about 27 percent higher SEA.
For light weight crash box or bumper beam designs, low density metal fillers, such as aluminum honeycomb or foam, are superior to tubes and beams with thicker walls in terms of achieving the same energy absorption. The crush performance of foam-filled square composite crash box was investigated experimentally and numerically. The results showed that the foam insertion results in higher energy absorption but unlike the aluminium foam-filled tubes, the specific energy absorption of the composite filled tubes is decreased in comparison with empty one.

7. References

Bisagni, C.; Pietro, GD.; Fraschini, L. & Terletti, D. (2005). Progressive crushing of fiber-reinforced composite structural component of a formula one racing car. *Compos. Struct.*, Vol. 68, 491–503

Chang, FK. & Chang, KY. (1987). Post-failure analysis of bolted composite joints in tension and shear-out mode failure. *J. Compos. Mater.*, Vol. 21, 809–33

El-Hagel, H.; Mallick, PK. & Zamani, N. (2004). Numerical modeling of quasistatic axial crash of square aluminum-composite hybrid tubes. *Int. J. Crashworthiness*, Vol. 9, No. 6, 653–64

Farley, G.L. (1991). The effects of crushing speed on the energy-absorption capability of composite tubes. *J. Compos. Mater.*, Vol. 25, No. 10, 1314–29.

Farley, G.L. (1983). Energy absorption of composite materials. *J. Compos. Mater.*, Vol. 17, No. 3, 267–79

Farley, GL. (1986). Effect of specimen geometry on the energy absorption capability of composite tubes. *J. Compos. Mater.*, Vol. 20, 390–400

.

Farley, GL. & Jones, RM. (1992). Prediction of the energy-absorption capability of composite tubes. *J. Compos. Mater.*, Vol. 26, No. 3, 388–404

Fleming, DC. & Vizzini, AJ. (1996). Off-axis energy absorption characteristics of composites for crashworthy rotorcraft design. *J. American Helicopter Soc.*, Vol. 41, No. 3, 239–46

Fleming, DC. (2001). Delamination modeling of composite for improved crash analysis. *J. Compos. Mater.*, Vol. 35, No. 19, 1777–92

Feillard, P. (1999). Crash modeling of automotive structural parts made of composite materials, *Proceedings of the SAE international congress and exposition*, March 1-4, Detroit, MI

Jacob, GC.; Simunovic, JFS. & Starbruk, JM. (2002). Energy absorption in polymer composite for automotive crashworthiness. *J. Compos. Mater.*, Vol. 36, No.7, 813–50

Johnson, AF.; Kindervater, CM.; Kohlgrueber, D. & Luetzenburger, M. (1996). Predictive methodologies for the crashworthiness of aircraft structures, *Proceedings of the 52nd american helicopter society annual forum*, pp. 1340–52, June 4–6, Washington DC

Johnson, AF. & Kohlgrueber, D. (1997). Modeling the crash response of composite structures. *J. Phys. IV France, Colloque C3, Supple ́ment au Journal de Physique III*, Vol. 7, C3-981-6 (in English).

Hallquist, JO. (1998). *LS-DYNA theoritical manual.* Livermore Software Technology Corporation

Hamada, H.; Ramakrishna, SA. (1997). FEM method for prediction of energy absorption capability of crashworthy polymer composite materials. *J. Reinf. Plast. Compos.*, Vol. 16, No. 3, 226–42

Hanssen, AG.; Langseth, M. & Hopperstad, OS. (1999). Static crushing of square aluminum extrusions with aluminum foam filler. *Int. J. Mech. Engng.*, Vol. 41, 967-993

Hashin, Z. (1980). Failure criteria for unidirectional fiber composites. *J. Appl. Mech.*, Vol. 47, 329–34

Haug, E.; Fort, O.; Tramecon, A.; Watanabe, M. & Nakada, I. (1991). Numerical crashworthiness simulation of automotive structures and components made of continuous fiber reinforced composite and sandwich assemblies. *SAE technical paper series 910152*

Hoermann, M. & Wacker, M, (2005). Simulation of the crash performance of crash boxes based on advanced thermoplastic composite, *Proceedings of the 5th European LS-DYNA users conference*, pp. 25-6, UK, May ,Birmingham

Hull, D. (1982). Energy absorption of composite materials under crash displacement variables obviates the need for special element types that are not available in crash codes, *Proceeding of the 4th international conference on composite materials: progress in science and engineering of composites*, pp. 861–87, Japan, Tokyo

Kerth, S.; Dehn, A.; Ostgathe, M. & Maier M. (1996) Experimental investigation and numerical simulation of the crush behavior of composite structural parts, *Proceedings of the 41st international SAMPE symposium and exhibition*, pp. 1397–408

Kindervater, CM. (1995). Crash resistant composite helicopter structural concepts thermoset and thermoplastic corrugated web designs, *Proceedings of the AHS national technical specialists meeting on advanced rotorcraft structures*, Williamsburg, VA

Kohlgrueber, D. & Kamoulakos, A. (1998). Validation of numerical simulation of composite helicopter sub-floor structures under crash loading, *Proceedings of the 54th American helicopter society annual forum*, May 20–22 Washington DC

Mamalis, AG.; Manolakos, DE.; Demosthenous, GA. & Ioannidis, MB. (1997). The static and dynamic axial crumbling of thin-walled fiberglass composite square tubes. *Composites Part B*, Vol.28B, No. 4, 439–51

Mamalis, AG.; Manolakos, DE.; Ioannidis, MB. & Papapostolou, DP. (2005). On the response of thin-walled composite tubular components subjected to static and dynamic axial compressive loading: experimental. *Compos. Struct.*, Vol. 69, 407–20

Mamalis, AG.; Manolakos, DE.; Ioannidis, MB. & Papapostolou, DP. (2006).The static and dynamic axial collapse of CFRP square tubes: finite element modeling. *Compos. Struct.*, Vol. 74, 2213–50

Mallick, PK. (1990) *Fiber reinforced composites*. 2nd ed. NY, Marcel Dekker

Ramakrishna, S. (1997). Microstructural design of composite materials for crashworthy applications. *Mater. Des.*, Vol.18, 167–73

Reedy, ED.; Mello FJ. & Guess, TR. (1997). Modeling the initiation and growth of delaminations in composite structures. *J. Compos. Mater.*, Vol. 31, No. 8, 812–31

Shin, K.C.; Lee, JJ.; Kim, KH.; Song, MC. & Huh, JS. (2002). Axial crash and bending collapse of an aluminum/GFRP hybrid square tube and its energy absorption capability. *Compos. Struct.*, Vol. 57, 279–87

Thornton, PH. & Edwards, PJ. (1982). Energy absorption in composite tubes. *J. Compos. Mater.*, Vol. 16, 21–45

Yamazaki, K. & Han, J. (1998). Maximization of the crushing energy absorption of tubes. *Struct. Optim.*, Vol. 16, 37–49

Zarei, HR. & Kroeger, M. (2006). Multiobjective crashworthiness optimization of circular aluminum tubes. *Thin-Walled Struct. J.*, Vol. 44, 301–8

Zarei, HR.; Kröger, M. & Albertsen, H. (2007). Crashworthiness investigation of the composite thermoplastic crash box, *Proceeding of the Sixth Canadian-International Composites Conference*, pp. 1-14, August , Winnipeg

Zarei, HR.; Kröger, M. & Albertsen, H. (2008). An experimental and numerical crashworthiness investigation of the thermoplastic composite crash boxes. *Comp. struc. J.*, Vol. 85, 245-258

19

Effects of the Long-Time Immersion on the Mechanical Behaviour in Case of Some E-glass / Resin Composite Materials

Assoc.prof.dr.eng. Camelia CERBU
„Transilvania" University of Braşov, Faculty of Mechanical Engineering, Romania

1. Introduction

The chapter deals with the actual and difficult problem of analysing the mechanical structures from the perspective of using composite materials in aggressive environment. Optimising the mechanical structures, made by composite materials is a great actual and important problem that includes two of the most modern, difficult and demanded aspects in mechanical engineering. If we point out this subject, meaning the aggressive environment, we already have the complete image of an extreme actual, important and special complexity subject.

The major studies in the field of structural optimising of the components made of composite materials, followed to obtain structures of components having higher strength and rigidity, lower weight, under conditions of a lower cost. There have been analysed composite material components, for which have been varied the material structure for fibre and matrix, the orientation of the fibres in layers, the shape of component, etc. The present study proposes an objective and supplementary criterion: the conservation of the mechanical characteristics of strength and rigidity under the long time action of the aggressive environment factors.

The results presented within this chapter address to the researchers and specialists in the field of the composite materials, to the ph.d. students and students from master, etc. Concurrently, reading of this working, may establish a point of start in the researching activity in this direction because it notes some important remarks regarding the effects of the aggressive environment (humidity, basic and acid solutions, temperature, thermal cycles, electrons radiation, UV rays etc) on the degradation of the mechanical characteristics of some composite materials.

The specialists interested in the field of composite materials will find a rich source of information by establishing a method of testing the specimens made of composite materials, subjected to statically forces after maintaining in aggressive environment; recommendations concerning the polymeric composite structure, having long durability under the action of the humidity and variations of temperature.

When an organic matrix composite is exposed to humid air or to a liquid, both the moisture content and temperature of the composite material may change with time. These changes affect the mechanical characteristics (Corum et al., 2001; Pomies et al., 1995; Cerbu, 2007; Takeshige et al., 2007).

Glass fibre reinforced resins are used widely in the building and chemical industry (wall panel, window frames, tanks, bathroom units, pipes, ducts, boat hulls, storage tanks, process vessels), automotive industry, aerospace industry. These structure elements may also be exposed to the environmental conditions (moisture, temperature etc).
In the last years, it was published many scientific papers concerning to the mechanical behaviour in wet environment of the composite materials and of the structure made of composite materials. For example, it was shown that the long-time immersion of the polymeric composite materials in water, seawater or detergent solutions, lead to the degradation of the mechanical characteristics (Corum et al., 2001; Pomies et al., 1995 ; Cerbu, 2005; Cerbu, 2007). Experimental results (Cerbu s.a. 2009) also demonstrated the influence of the immersion time on the degradation of the mechanical characteristics of some polymeric composite materials.
A recent paper (Takeshige et al. 2007) investigated some interactions between mechanical and chemical fatigue in case of some resin composite materials. Therefore, in that research was remarked that the fatigue crack propagation is retarded under humid conditions but accelerated after water immersion.
Some recent works shown the new tendencies from the manufacture field of the composite materials by recycling of the wood wastes (Adhikary s.a. 2008), plastic wastes, polyetilena waste, paper, CDs / DVDs (Cerbu s.a. 2009) and so forth. Consequently, some of the new composite materials were studied from moisture absorption point of view. For example, Adhikary s.a. (2007) analysed the long-term moisture absorption and thickness swelling in case of some specimens made of recycled thermoplastics reinforced with pinus wood flour.
An interesting book (Klyosov 2007) focused on wood-plastic composites regarding the paticularities of their fabrication; the effects of cellulose fillers, mineral fillers and coupling agents on their mechanical properties. It also showed the effects of the moisture absorption on changes of some mechanical characteristics. Thus, it was experimentally demonstrated that the flexural strength and flexural modulus E increased in case of an water-saturated board made of a commercial wood-plastic composite material.
The first of all, the present chapter proposes the analysing of the efffects of moisture absorbed concerning the changing of the mechanical characteristics of four kinds of polymeric composite materials randomly reinforge with chopped glass fibres. Both the nature of resin and the immersion environment, were analysed regarding their effects on the changing of the mechanical characteristics.
On the other hand, it focus on the using of the wood flour obtained by recycling of the wood wastes from industry to manufacture of hybrid composite materials. The new polymeric composite material described within this section, is reinforced with both glass woven fabric EWR145 (145g/m^2) layed in six layers and wood flour.
It will be comparatively shown the results concerning to the mechanical characteristics (tensile strength, flexural modulus E, flexural maximum stress σ_{max}) determinated by tensile tests and flexural tests (method of the three points), before and after immersion in different environments (water, natural seawater, detergent solution). The results will be compared with the ones obtained in case of four kinds of composite materials randomly reinforced only with chopped glass fibres, free of admixture of wood flour. Moreover, it will be comparatively analysed the data concerning the quantity of the moisture absorbed and the its effect on the change of the mechanical characteristics.

2. Materials. Work method

The first of all, it was manufactured six laminated composite plates whose material structures are different. In this work, E-glass fibres (50 mm length) were used to randomly reinforce four kinds of resins: two polyester resins (Heliopol 8431 ATX and Polylite 440-M880); an epoxy resin (LY554); a vinyl-ester resin (Atlac 582). These composite materials had the average volume fibre ratio equal to 26%, while the weight fibre ratio was 40%. Another two plates having the dimensions 350 x 250mm^2 and 8mm in thickness, were manufactured by using a polyester resin reinforced with both glass woven fabrics EWR145 (six layers) and wood flour (oak wood flour or fir wood flour).

Then, the plates were cut to obtain the specimens according to the european standards concerning of determination of the both tensile properties (ISO 527) and flexural properties (SR EN 63). A total number of 164 specimens were manufactured for tensile tests (Table 1) while 194 specimens were prepared for the flexural tests (Table 2).

It may be noted that all sides of a half of the total number of specimens reinforced only with E-glass fibres, were coated using the resins used for the matrix while the others were not coated.

No.	Composite material	Number of specimens			
		Dry	Environment		
			Water	Detergent solution	Seawater (Black Sea)
1	E-glass / polyester Heliopol 8431 ATX	5	12	12	12
2	E-glass / polyester Polylite 440-M880	5	12	12	12
3	E-glass / epoxy LY 554	5	12	12	12
4	E-glass / vinyl-ester Atlac 582	5	12	12	12

Table 1. Specimens for tensile test

No.	Composite material	Number of specimens			
		Dry	Environment		
			Water	Detergent solution	Seawater (Black Sea)
1	E-glass / polyester Heliopol 8431 ATX	5	12	12	12
2	E-glass / polyester Polylite 440-M880	5	12	12	12
3	E-glass / epoxy LY 554	5	12	12	12
4	E-glass / vinyl-ester Atlac 582 composite	5	12	12	12
5	E-glass EWR145 / fir wood flour / polyester Colpoly 7233	5	5	-	5
6	E-glass EWR145 / oak wood flour / polyester Colpoly 7233	5	5	-	5

Table 2. Specimens for flexural test

Then, the specimens were kept at the room temperature and dried environment for three weeks.

Additionally, some specimens were stored in an oven at 30 ± 1°C and weighted to ensure that they were dried prior to the immersion in water (SR EN ISO 62, 2008). Water (Fig. 1, a), detergent solution (Fig. 1, b) and fresh natural seawater from Black Sea (Fig. 1,c) at room temperature (20 °C) were used as wet environments. Stands were used to maximise the contact surface between specimens and water (see the detail presented in the Fig. 1, c). The salinity of the natural seawater was approximately of 1.6%. The water tanks were covered to minimise evaporation and the water was changed every month to keep conditions constant.

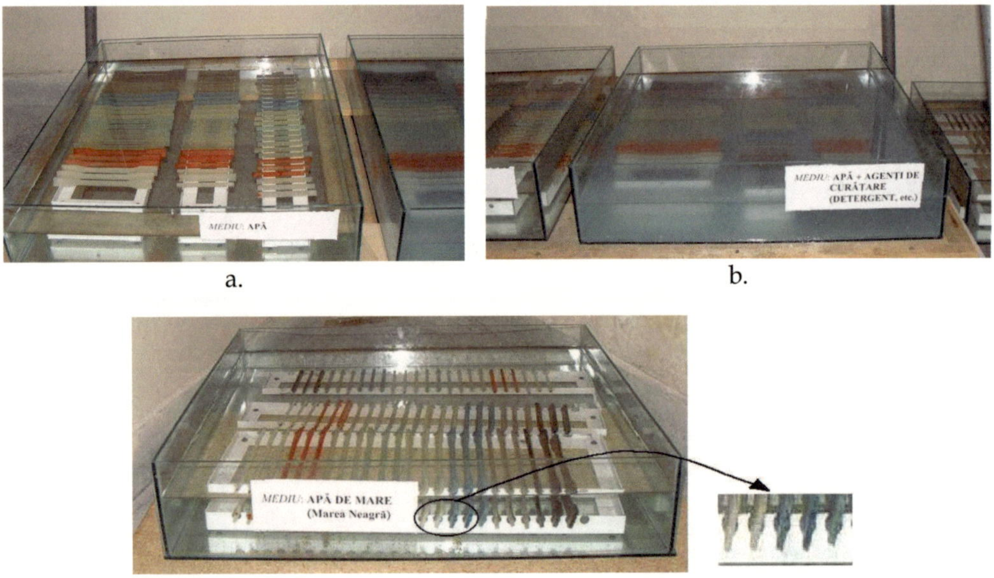

Fig. 1. Environments a. Water ; b. Water / detergent mix c. Seawater (Black Sea);

No.	Composite material	Specimen type	Immersion time *t* (hours)		
			Environment		
			Water	Detergent solution	Seawater (Black Sea)
1.	E - glass / polyester Heliopol 8431 ATX	Uncoated	7197	7134	6987
		Coated	7197	7134	6987
2.	E - glass / polyester Polylite 440-M880	Uncoated	7197	7134	6987
		Coated	1803	1732	2762
3.	E - glass / epoxy LY 554	Uncoated	7197	7134	6987
		Coated	7197	7134	6987
4.	E-glass / vinyl-ester Atlac 582 composite	Uncoated	2975	2865	6987
		Coated	1821	1732	2762
5.	E-glass EWR145 / fir wood flour / polyester Colpoly 7233	Uncoated	5612		5612
6.	E-glass EWR145 / oak wood flour / polyester Colpoly 7233	Uncoated	5853	-	5853

Table 3. Immersion times for the composite materials tested

The times of immersion in water were different in case of the composite materials tested (Table 3). To monitor the uptake of water, quantified by the moisture content *m*, the specimens were periodically removed from tanks, superficially dried with absorbing paper and weighted by using an electronic balance (maximum mass 250 g) accurate within ± 0.0001g.

After long-time immersion, the specimens were subjected to both tensile test and flexural test by the method of the three points (SR EN ISO 178, 2001). LR5K Plus machine manufactured by LLOYD Instruments, was used for mechanical testing of the composite specimens involved. The maximum force provided by the testing machine is $F_{max} = 5kN$.

The shape and the dimmensions of the tensile specimens are shown in the figure 2. Diagrams having the coordinates force - elongation $(F - \Delta l)$ coordinates were directly recorded from tensile machine.

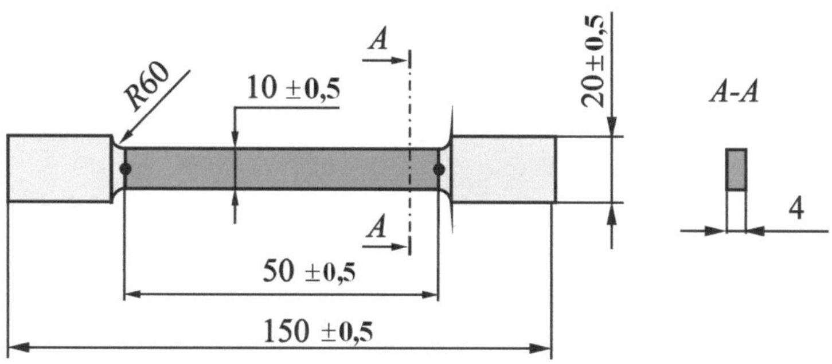

Fig. 2. Specimen for tensile test

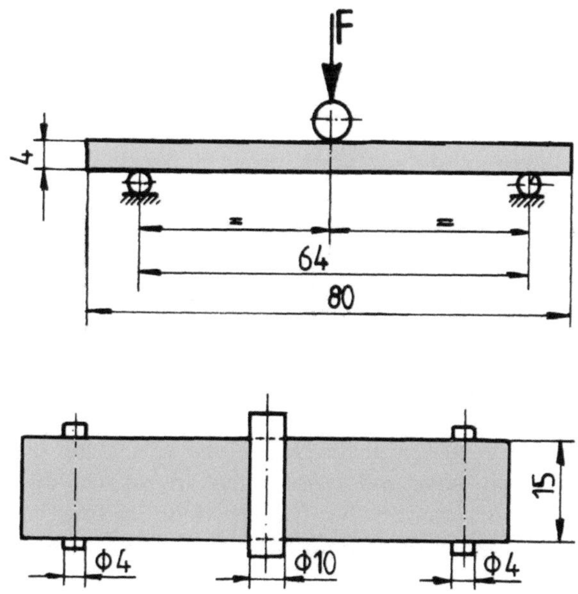

Fig. 3. Specimen used in flexural test

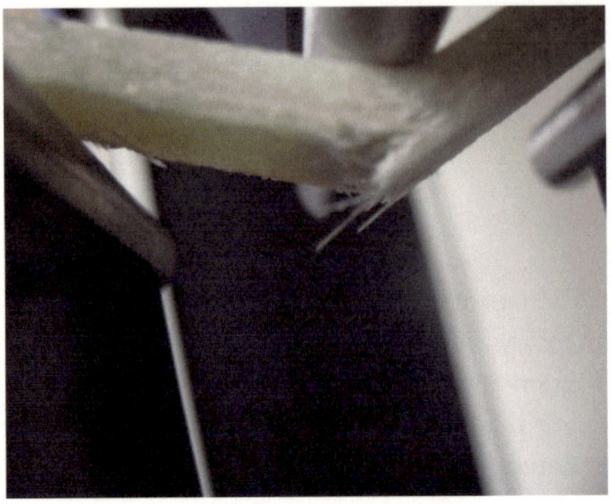

Fig. 4. Specimen made of E-glass / polyester Heliopol 8431 ATX during flexural test

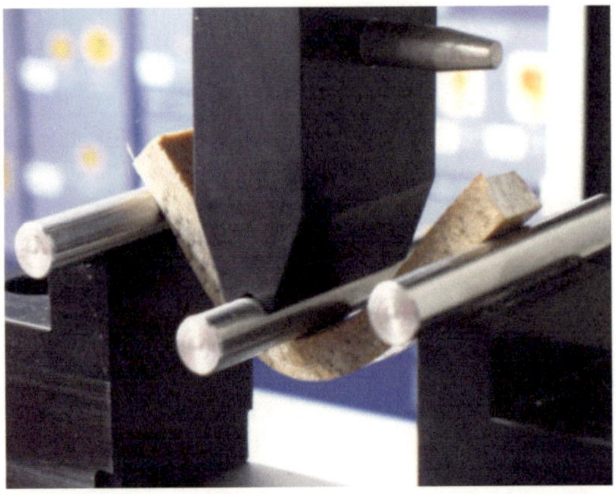

Fig. 5. Specimen made of E-glass EWR145 / oak wood flour / polyester Colpoly 7233 during flexural test

The loading scheme (three-point method) is shown in the figure 3. The dimensions of the specimens are valid only for specimens reinforced only with E-glass fibres because these are functions of the specimen thickness. In case of the flexural tests, the span between the supports was accorded with the specimen thickness (SR EN ISO 62, 2008).

Figures 4 and 5 show two photos of specimens during the flexural test. The specimens shown are made of E-glass / polyester Heliopol 8431 ATX composite material (Fig. 4) and E-glass EWR145 / oak wood flour / polyester Colpoly 7233 (Fig. 5), respectively. It may observe that the deformations are much more greater in case of the specimens made with wood flour (Fig. 5) than in case of the specimens reinforced only with glass fibres (Fig. 4).

The speed of loading was 3mm/min. in case of the specimens made of the composite materials filled with wood flour and 1.5mm/min. in case of the other four kinds of composite materials tested, respectively. The speed of loading was greater in case of additionaly reinforcing with oak wood flour because this composite material is much more flexible. Consequently, the time of the flexural test was approximately 40min. when the speed of loading was equal to 1.5mm/min., in case of this composite material. It was remarked that the the time of the flexural test halved by doubling of the loading speed while

the shape of the curve obtained was approximately the same. It may be mentioned that the time of the flexural test was approximatelly equal to 10min. when the speed of loading was 1.5mm/min., in case of the specimens reinforced only with the E-glass fibres.
Before each mechanical test of a specimen, the dimensions of the cross-section were accurately measured (0.1mm) and then, they were considered as input data in the software program of the machine.
In case of the flexural testing, the testing equipment allowed to record pairs of values (force F and deflection v at midpoint of the specimens) in form of files having up to 3000 recordings. The testing machine also gave the results of a statistical calculus for the set of the specimens tested. Experimental results recorded during the flexural tests, were graphically drawn using F – v coordinates and finally, the following quantities were computed:
- flexural modulus E of the composite material

$$E = \frac{1}{48} \cdot \frac{l^3}{I_z} \cdot \frac{\Delta F}{\Delta v} \qquad (1)$$

- flexural strength σ of the composite material:

$$\sigma = \frac{M_{bz\,\max}}{W_z}, \qquad (2)$$

where $l = 64\ mm$ represents the span of the specimen between simple supports (Fig. 3), I_z - moment of inertia, W_z - elastic cross-section modulus, $M_{bz\,max} = Fl/4$ - maximum value of the bending moment. Formula used for the flexural modulus E is a good approximation because $\frac{l}{h} = 16$, where h represents the thickness of the specimen and one can neglect the effect of the shearing force.

3. Results

3.1 Water absorption

The first, moisture behaviour was analysed. The absorption data were shown in the figures 6 – 9 for all composite materials reinforced only with glass fibres. Important remarks are noted by analysing these results.

- Moisture absorption in composite materials depends on the resin used for matrix and type of the wet environment. The absorption process is a long-term process in case of the composite materials tested.
- E - glass / Heliopol 8431 ATX and E-glass / Polylite 440-M880 composites closed the saturation point after 7000 hours of immersion time while the moisture content was approximately the same.
- E-glass / epoxy LY 554 composite does not reach the saturation point after 7000 hours of immersion (Fig. 8) and moisture content is much more greater than in case of the others three composite materials (Fig. 6, 7 and 9). E-glass / epoxy LY 554 composite material absorbs more water than seawater or detergent solution.
- Glass-reinforced polymers absorb more water than seawater. Rate of diffusion of the water through composite materials analysed is greater than that of the seawater.
- Sodium chloride molecules contained in seawater (as well as sulphate) appear to be limiting the diffusion of water into the matrix material.

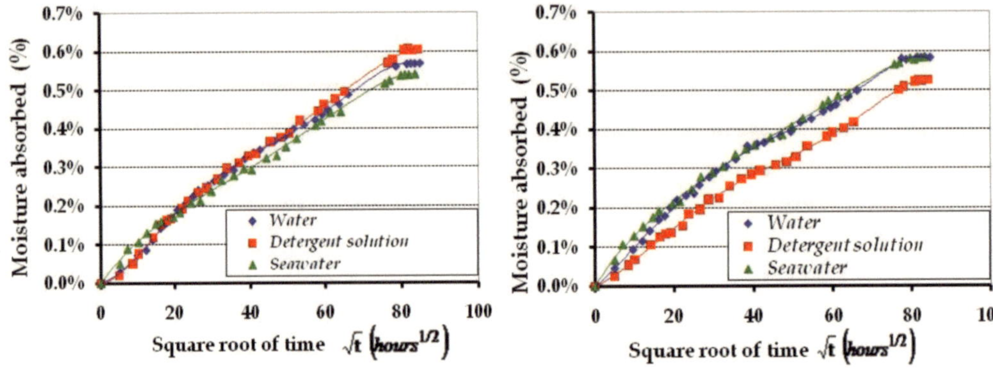

a. Coated specimens

b. Uncoated specimens

Fig. 6. Absorption data in case of E-glass / Heliopol 8431 ATX composite material

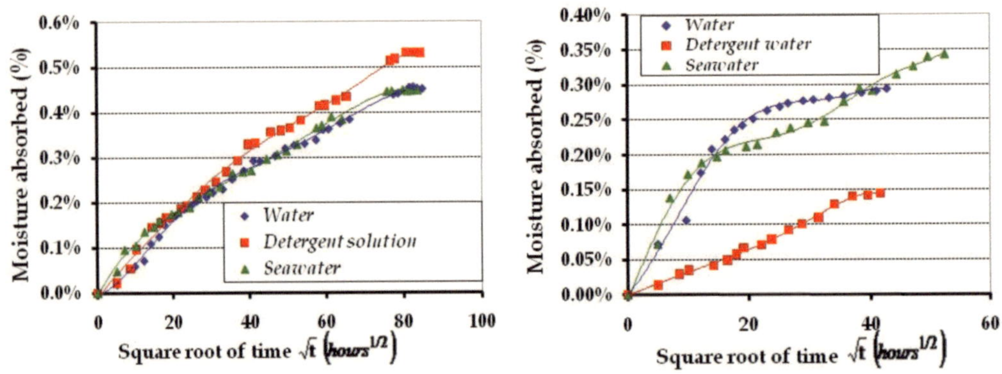

a. Coated specimens

b. Uncoated specimens

Fig. 7. Absorption data in case of E-glass / Polylite 440-M880 composite material

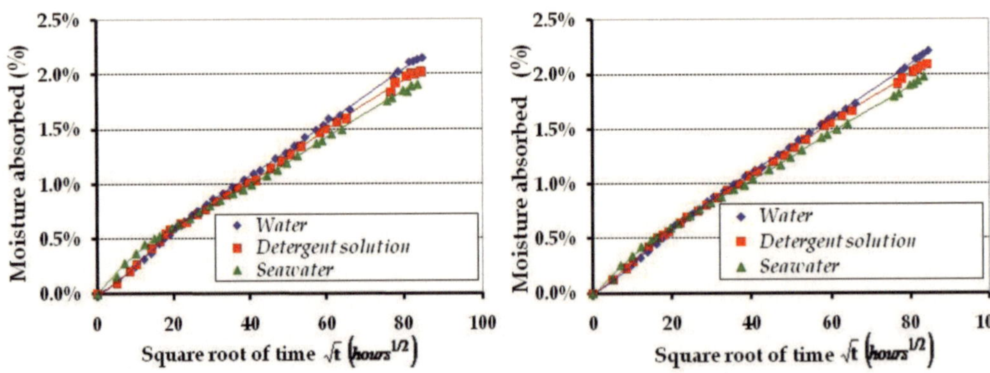

a. Coated specimens

b. Uncoated specimens

Fig. 8. Absorption data in case of E-glass / epoxy LY 554 composite material

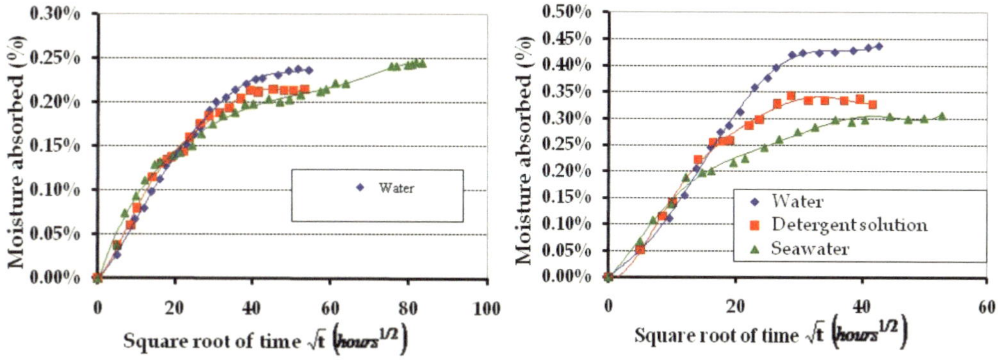

a. Coated specimens b. Uncoated specimens

Fig. 9. Absorption data in case of E-glass / vinyl-ester Atlac 582 composite

The absorption curves recorded in case of the two hybride composites are drawn in the figure 10 in case of the immersion in water and in the figure 11 in case of the immersion in seawater.

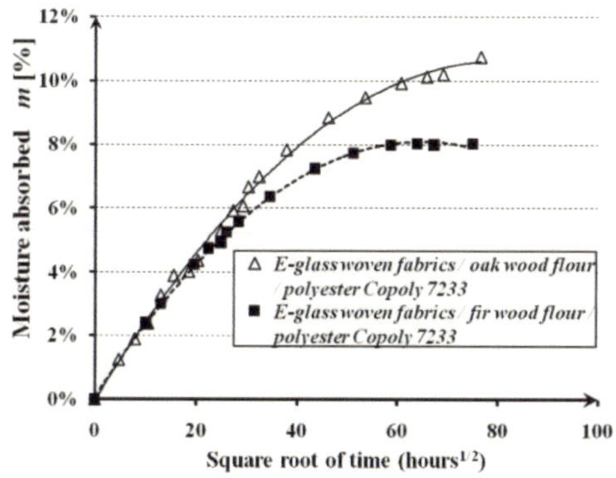

Fig. 10. Data of the absorbed moisture during immersion in water in case of the E-glass woven fabrics / wood flour / polyester

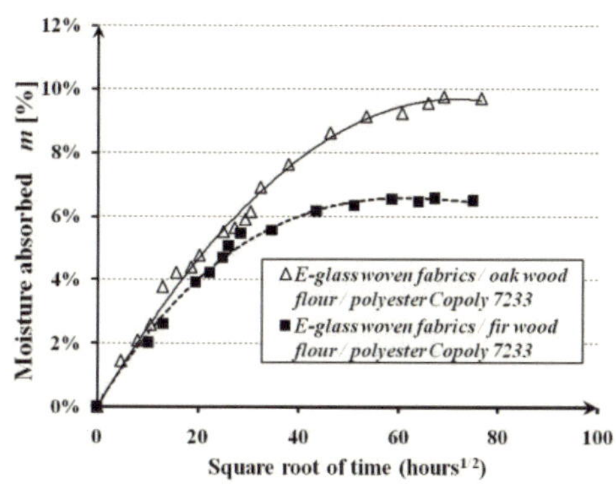

Fig. 11. Data of the absorbed moisture during immersion in seawater in case of the E-glass woven fabrics / wood flour / polyester

It may be easily observed that the two absorption curves recorded in case of the composite material filled with fir wood flour is located below the one recorded in case of the other one composite filled with oak wood flour. The cause may be assigned to resinous nature of the fir wood. Therefore, the greater resin content of the fir wood flour acts as a barrier against the water absorption. The average value of the water content (Fig. 10) was 10.73% while the seawater content (Fig. 11) recorded was 9.72% after immersion during 5853 hours, in case of the composite filled with oak wood flour. In case of the other one composite material filled with fir wood flour, the water content (Fig. 10) was equal to 8.02% while the seawater content was 6.50% after 5612 hours of immersion. Therefore, like the other previous works showed, it was recorded again a smaller quantity of the moisture absorbed during the immersion in seawater than in case of the immersion in water. The salts of the seawater act again like a barrier against the moisture absorption.
There is a small difference between the absorption curves recorded during the first 400-600 hours of immersion. It follows that the diffusivity of the moisture inside the composite material, has approximately the same value in the both cases: water environment and seawater environment.

3.2 Mechanical behaviour in tensile test after immersion in different environments

After approximately 7000 hours of immersion (≈ 10 months) the tensile specimens made of polymer resins reinforced only with glass fibres, were subjected to the tensile test. A photo of these specimens after the tensile test, is shown in the figure 12.

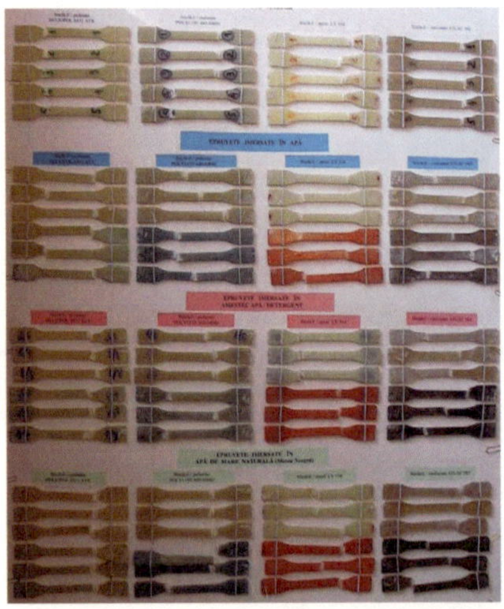

a.

b.

c.

d.

Fig. 12. Tensile Specimens reinforced only with E-glass fibres after flexural test: a. dried specimens; b. specimens after immersion in water; c. specimens after immersion in detergent solution; d. specimens after immersion in seawater

Comparatvely analysing of the experimental results (Fig. 13 - 16) obtained in case of both dried and wet specimens it may observe:

- Tensile strength decreases in case of all composites;
- Decreasing of the tensile strength (40 %) is greater for the specimens made of E-glass / Heliopol 8431 ATX and E-glass / epoxy LY 554 composites after immersion in water than in case of the other two environments (Fig. 10 and 12);

- Conservation of the tensile strength was not very different if all sides of the specimens were coated using the resin of the matrix of the composite;
- Tensile strength of the specimens decreases with 10 - 20 % in case of the immersion in seawater and water / detergent mix (Fig. 10 - 12). The reason could be that moisture content was much smaller in case of these environments.

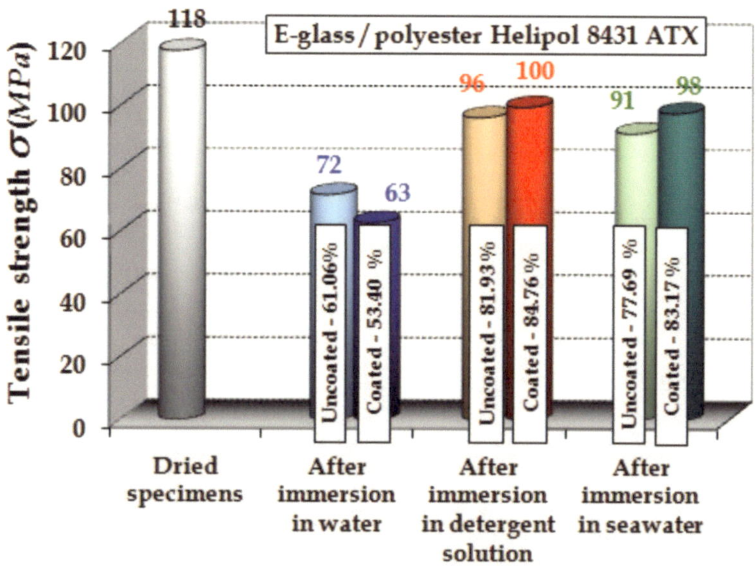

Fig. 13. Changes of the tensile strength in case of E-glass / Heliopol 8431 ATX composite

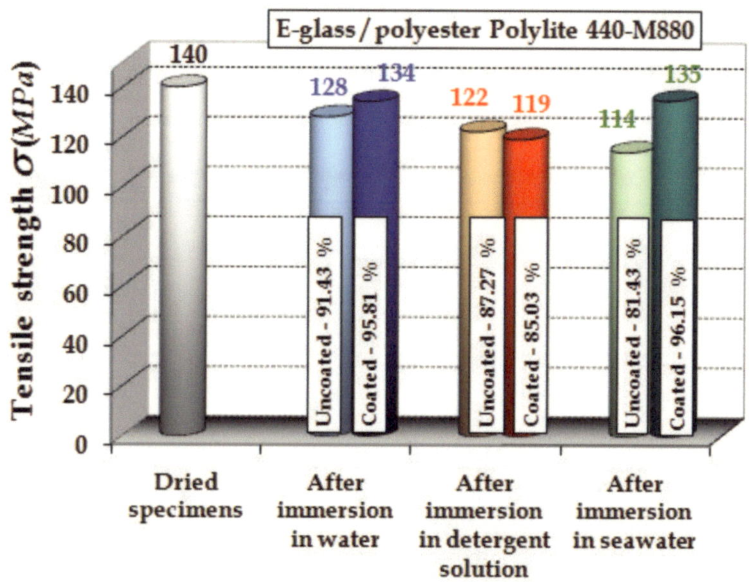

Fig. 14. Changes of the tensile strength in case of E-glass / Polylite 440-M880 composite

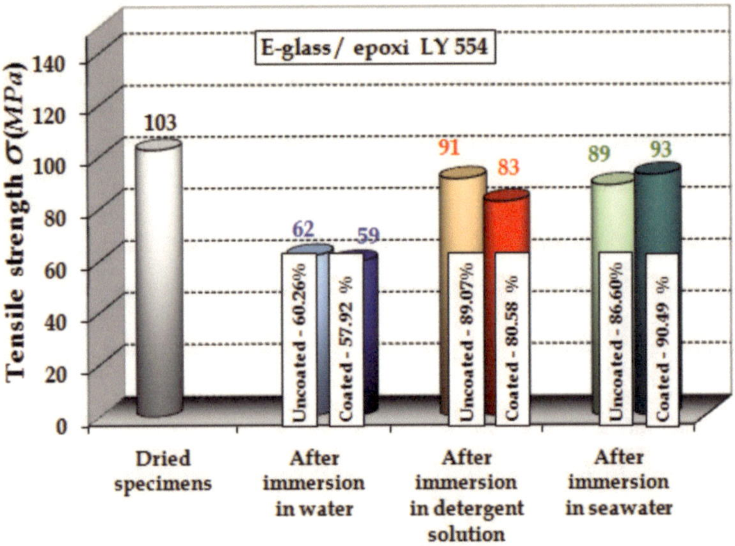

Fig. 15. Changes of the tensile strength in case of E-glass / epoxy LY 554 composite

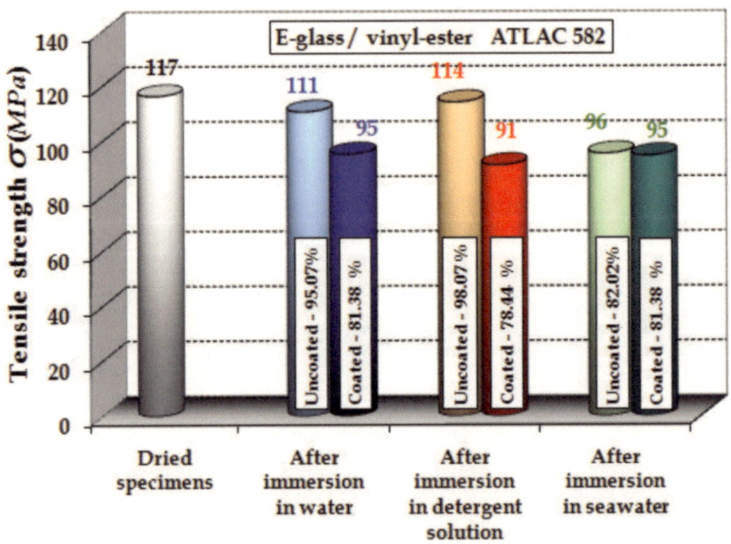

Fig. 16. Changes of the tensile strength in case of E-glass / vinyl-ester ATLAC 582 composite

3.3. Mechanical behaviour in flexural test after immersion in different environments

Then, flexural test by using the three-point method, was considered the immersion in the three kinds of wet environment. The specimens made of polymer resins reinforced with only glass fibres, after they were subjected to the flexural test, are shown in the figure 17. The results obtained in case of the wet specimens were compared with the ones obtained in case of the dried specimens.

Two photos of the flexural specimens filled with both E-glass woven fabrics and wood flour, after immersion in water, are shown in the figures 18 and 19, respectively.
Figures 20 - 23 comparatively show the force - deflection (F-v) curves recorded during the flexural tests, in case of both wet specimens in case of the following composite materials:
- E-glass / polyester Heliopol 8431 ATX (Fig. 20);
- E-glass / polyester Polylite 440-M880 (Fig. 21);
- E-glass / epoxy LY 554 (Fig. 22);
- E-glass / polyester Polylite 440-M880 (Fig. 23).

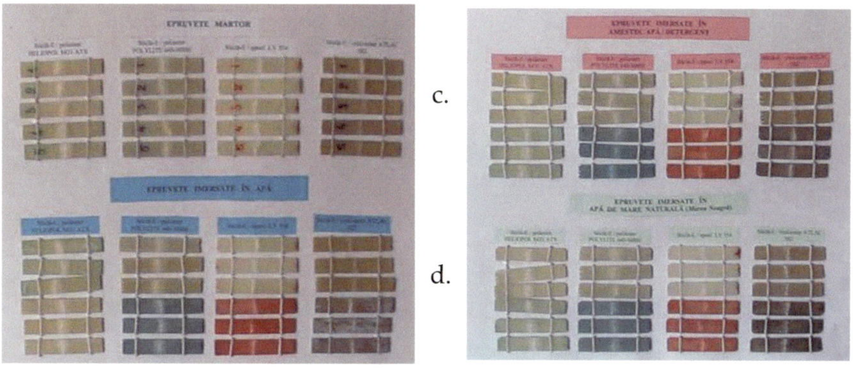
a. b. c. d.

Fig. 17. Flexural specimens reinforced only with E-glass fibres after flexural test: a. dried specimens; b. specimens after immersion in water; c. specimens after immersion in detergent solution; d. specimens after immersion in seawater

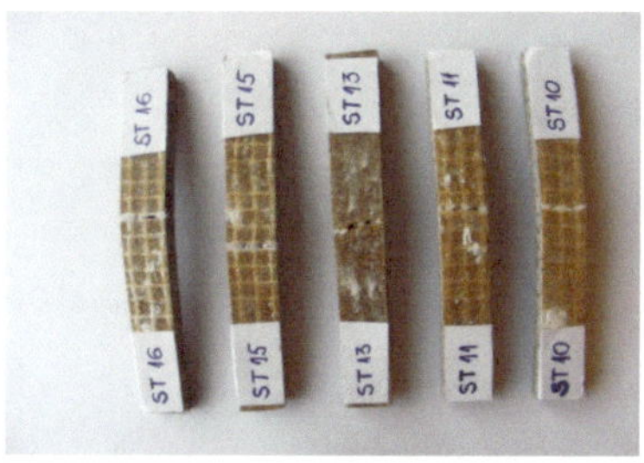

Fig. 18. Flexural specimens made of E-glass EWR145 / oak wood flour / polyester Colpoly 7233 after flexural test

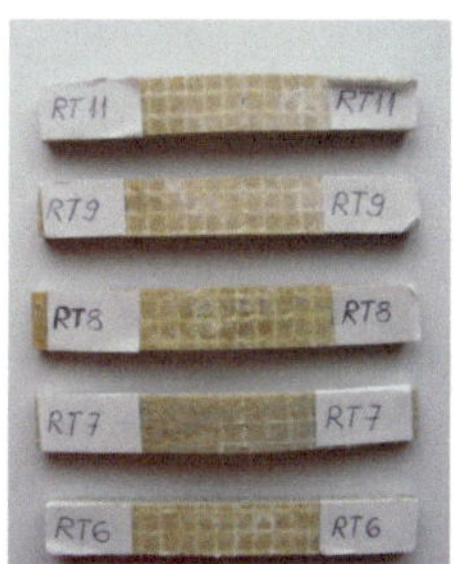

Fig. 19. Flexural specimens made of E-glass EWR145 / fir wood flour / polyester Colpoly 7233 after flexural test

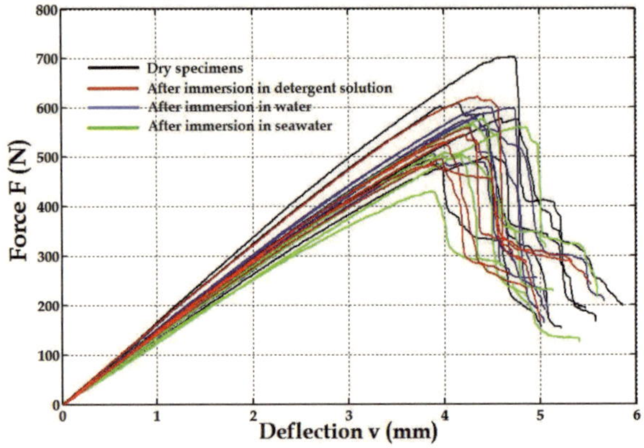

Fig. 20. Curves F-v recorded during the flexural tests in case of E-glass / polyester Heliopol 8431 ATX composite

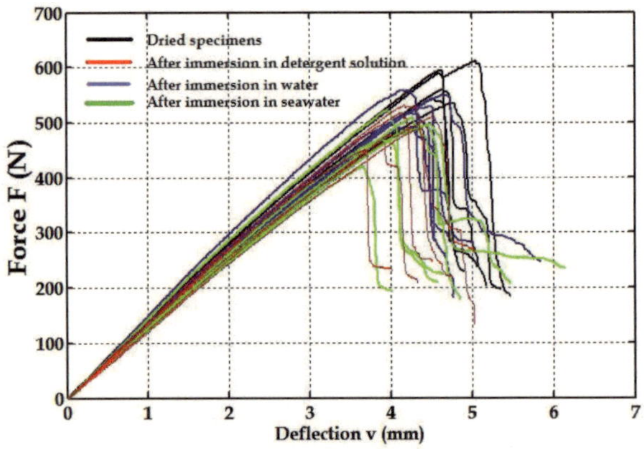

Fig. 21. Curves F-v recorded during the flexural tests in case of E-glass / polyester Polylite 440-M880 composite

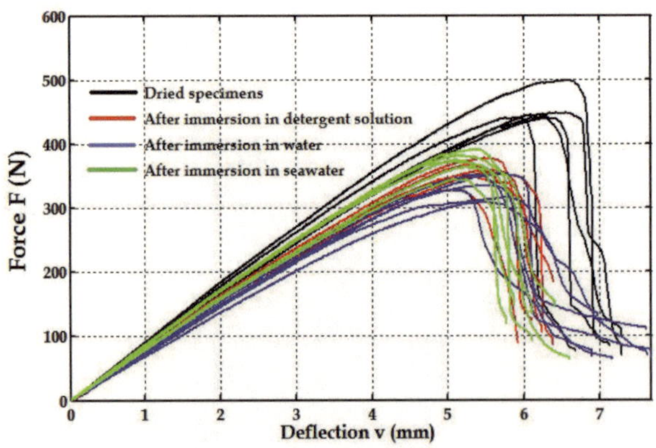

Fig. 22. Curves F-v recorded during the flexural tests in case of E-glass / epoxy LY 554 composite

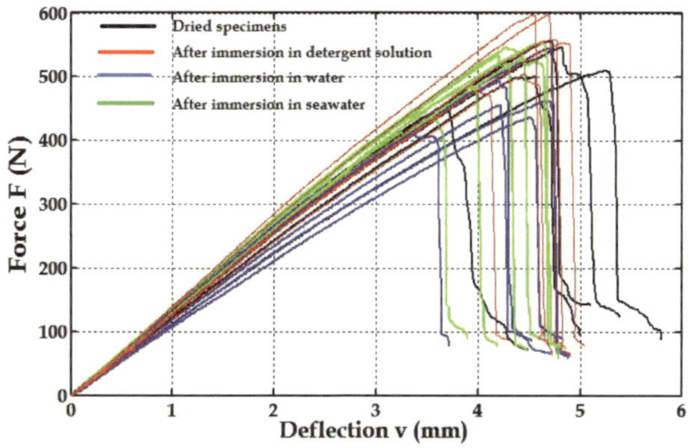

Fig. 23. Curves F-v recorded during the flexural tests in case of E-glass/vinyl-ester ATLAC 582 composite

The flexural modulus was computed on the linear portion of the force-displacement curve. Figures 24 and 25 graphically show the experimental results obtained in case of the glass / polyester composites (E-glass / Heliopol 8431 ATX and E-glass / Polylite 440-M880), figure 26 represents the results in case of E-glass / epoxy LY 554 composite material and figure 27 shows the flexural properties measured in case of the E-glass / vinyl-ester Atlac 582 composite.

Analysing of the results of the experimental research shown in the figures 24 - 27, lead to important remarks that are noted below.

- Effects of the seawater are more pronounced than the action of the water in case of E-glass / polyester composites (E-glass / polyester Heliopol 8431 ATX and E-glass / polyester Polylite 440-M880) as shown in figures 24 and 25.
- Decreasing of the Young's modulus E was ≈ 11 % while the change of the flexural strength was ≈ 12 % in case of the E-glass / polyester Heliopol 8431 ATX composite when the specimens were kept in seawater and detergent solution (Fig. 24). One may observe a good conservation of the flexural characteristics in case of the specimens after 9200 hours of immersion in water.
- Decreasing of the Young's modulus E was ≈ 5 % when the specimens were kept in water and detergent solution while the change was ≈ 10 % when were submerged in seawater in case of the E-glass / Polylite 440-M880 composite (Fig. 25, a).
- A decreasing of the flexural strength was also observed in case of the E-glass / Polylite 440-M880 composite (Fig. 25, b) - about 11%, 23% and 15 % when the specimens were kept in water, seawater and detergent solution, respectively.
- On the other hand, when the E-glass / epoxy LY 554 composite was submerged in water, the decreasing of the Young's modulus was much more pronounced - about 21 % (Fig. 26, a) while the decreasing of the flexural strength was approximately 31 % (Fig. 26, b).
- The decreasing of the flexural strength was about 23%, 26% when the specimens were kept in seawater and detergent solution respectively, in case of the E-glass / epoxy LY 554 composite (Fig. 26, b).
- The decreasing of the Young's modulus E was about 10 %, 15 % when the specimens were kept in seawater and water / detergent mix in case of the E-glass / epoxy LY 554 composite (Fig. 26, a).

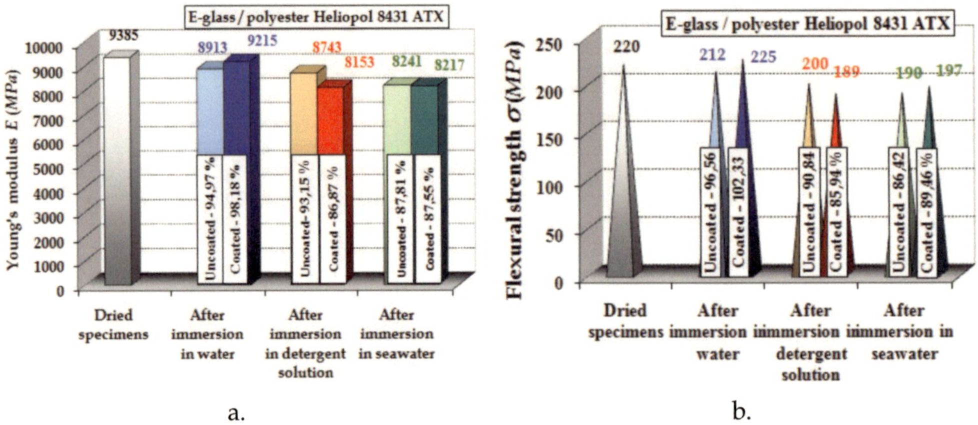

Fig. 24. Experimental results of the flexural test in case of E-glass / Heliopol 8431 ATX composite

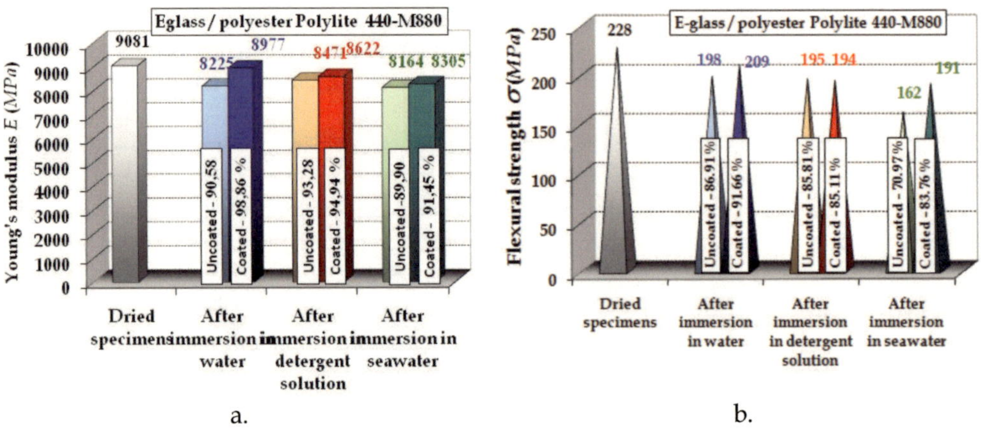

Fig. 25. Experimental results of the flexural test in case of E-glass / Polylite 440-M880 composite

Several researchers also found that water absorption causes degradation of matrix-dominated properties such as interface and in-plane shear strengths, compressive strength and transverse tensile strength (Corum et al., 2001; Pomies et al., 1995; Cerbu, 2007; Takeshige et al., 2007). In (Pomies et al., 1995) E-glass / epoxy and carbon / epoxy composites were studied. Finally, the loss in the mechanical properties has been attributed to the plasticity of the matrix by water and degradation of the fibre/matrix interfacial bond due to moisture swelling of the matrix.

In case of the composite materials reinforced only with glass fibres, tested during our experimental research the above reason could be again the cause of the decreasing of the mechanical characteristics of the composite materials.

Experimental results recorded during bending tests, are graphically drawn in case of the hybride composite materials: E-glass EWR145 / fir wood flour / polyester Colpoly 7233 (Fig. 28 and 29) and E-glass EWR145 / oak wood flour / polyester Colpoly 7233 (Fig. 30 and 31). It may be noted that Young's modulus was computed again, for data points located on the linear portion of the F–v curve.

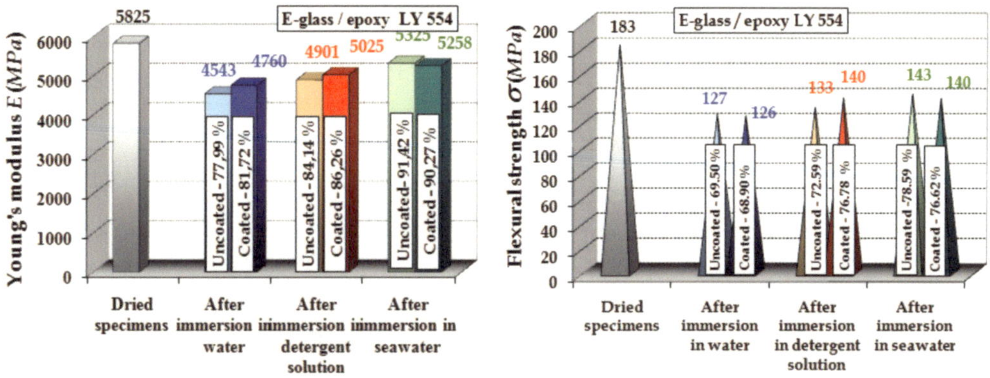

a. b.

Fig. 26. Experimental results of the flexural test in case of E-glass / epoxy LY 554 composite

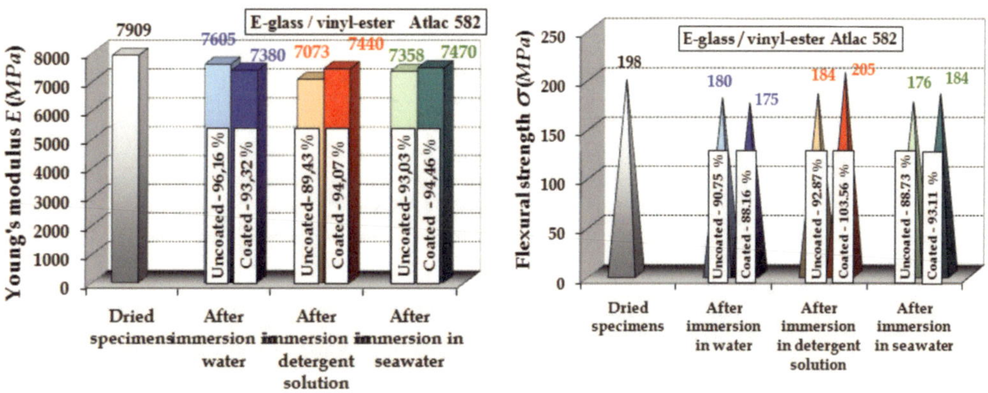

a. b.

Fig. 27. Experimental results of the flexural test in case of E-glass / vinyl-ester Atlac 582 composite

The first of all, it is discussed the changing of the mechanical properties in case of the E-glass / fir wood flour / polyester composite. Young's modulus E (Fig. 28) decreases from 601.1MPa down to 356.2 MPa (with 40.7 %) after 5621 hours of immersion in water while it increases up to 766.0 MPa (with 27.5%) after the same immersion time in seawater. In the same manner, the maximum flexural stress σ_{max} (Fig. 29) decreases from 27.7 MPa down to 16.0 MPa (with 42.2%) after immersion in water and it decreases down to 23.5 MPa (with 15.2%) after immersion in seawater.

In case of the E-glass woven fabric / oak wood flour / polyester composite, it may observe generally speaking, the increasing of both Young's modulus E (Fig. 30) and maximum flexural stress σ_{max} (Fig. 31) after immersion in wet environment. Therefore, this remark confirms once again the well-known property of the oak wood concerning the hardening by aging over the years. In fact, the keeping of the materials completely immersed in water, represents an accelerate process of aging. More exactly, in case of the composite material filled with oak wood flour, Young's modulus E (Fig. 30) increases from 215.0 MPa up to

500.9 MPa (with 132.9%) after 5853 hours of immersion in water while it increases up to 482.0 MPa (with 124.2%) after the same immersion time in seawater. In the same time, the maximum flexural stress σ_{max} (Fig. 31) increases from 21.0 MPa up to 25.0MPa (with 19.05%) after 5853 hours of immersion in water while it decreases down to 17.5MPa (with 16.67%) after the same immersion time in seawater.

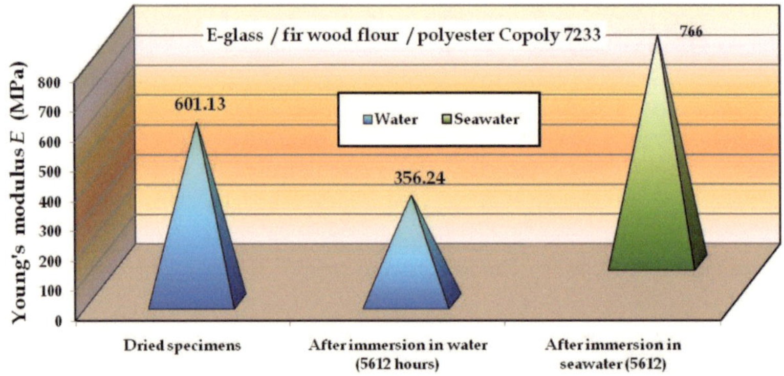

Fig. 28. The effects of water / seawater absorption on *Young*'s modulus E in case of E-glass EWR145 / fir wood flour / polyester Colpoly 7233

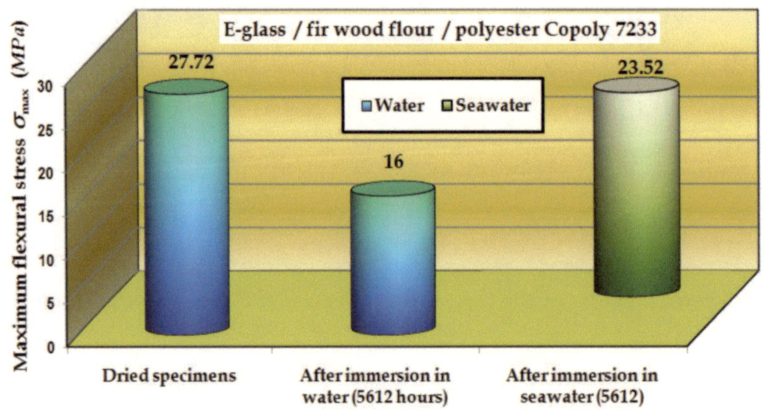

Fig. 29. The effects of water / seawater absorption on on the maximum flexural stress σ_{max} in case of E-glass EWR145 / fir wood flour / polyester Colpoly 7233

In case of the composite material filled with oak wood flour, it was also analysed the effect of the immersion time in water on the changing of the mechanical characteristics. Therefore, Young's modulus increased with 182.28 % after 861 hours of immersion in water while the increasing was only with 132.9%) after 5853 hours of immersion. With other words the increasing of the immersion time leads to a decreasing of the rigidity. The results concerning the changing of the maximum flexural stress σ_{max} show contrary that the maximum flexural stress σ_{max} increases with 7.14 % after 861 hours and it increases with 19.05% after 5853 hours of immersion.

But, the most important remark remains that concerning the values of the maximum deflection v_{max} of the midpoint of the specimens during and after the flexural test. The values recorded for this quantity is shown in the Table 4 in case of the dried specimens made with oak wood flour. It may easily observe that the maximum residual deflection v_{max} after approximately 30 min. after flexural test had finished, was much smaller than the maximum deflection recorded at maximum load and also, than the one recorded at the final test.

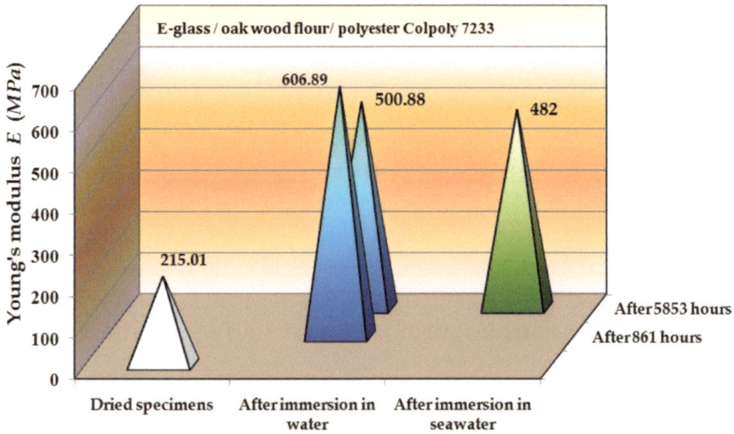

Fig. 30. The effects of water /seawater absorption on Young's modulus E in case of E-glass EWR145 / oak wood flour / polyester Colpoly 7233

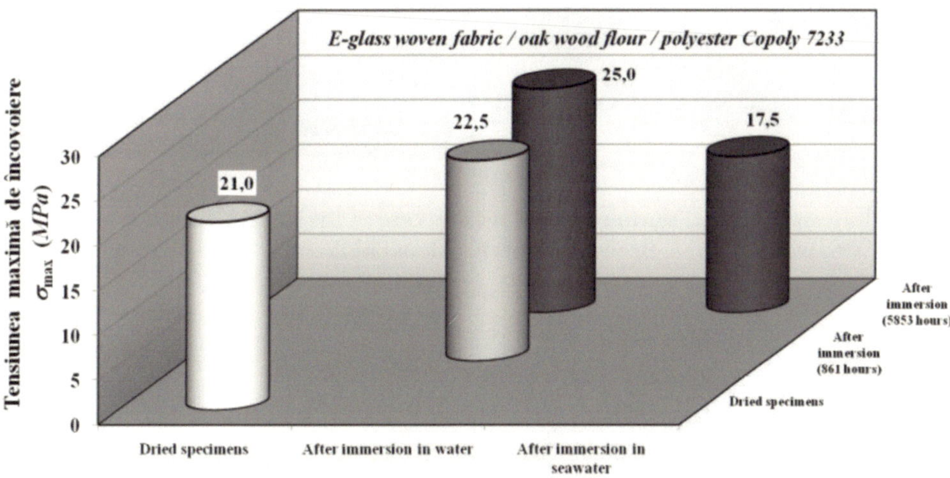

Fig. 31. The effects of water /seawater absorption on on the maximum flexural stress σ_{max} in case of E-glass EWR145 / oak wood flour / polyester Colpoly 7233

The reason of this mechanical behaviour could be assigned to the wood flour used to manufacture the composite specimen because no suchlike observation was recorded in case of E-glass / polyester composite materials tested within this work. Practically, this unexpected mechanical behaviour of the new hybrid composite after the flexural test could be owing to a good combination between the rheological behaviour of wood and the shape memory, property that is assigned to the E-glass fibres.

Specimen No.	1	2	3	4	5
vmax at max. load	22.143	29.513	19.984	39.387	31.504
vmax at final of the flexural test	58.674	54.592	59.396	59.400	59.396
vmax after ≈ 30 minutes after test	7.1	5.2	5.8	4.1	5.7

Table 4. Maximum values of deflection v_{max} in case of the dried specimens made of E-glass EWR145 / oak wood flour / polyester Colpoly 7233 composite material

3.4. Failure mode

Figure 32, a shows a specimen made of the E-glass EWR145 / fir wood flour / polyester Colpoly 7233 composite material, after it was subjected to the flexural test. On the other hand, figure 32, b is a photo of the failure area aquired in case of a flexural specimen made of E-glass EWR145 / oak wood flour / polyester Colpoly 7233 composite material.

It could be observed that only 1-2 layers was partially failured during the testing of the specimens filled with wood flour. Contrary, almost all plies was failured during the flexural test in case of the specimens made of composite materials reinforced only with glass fibres (Fig. 33).

a.

b.

Fig. 32. Failure area occured during flexural tests in case of the specimens additionaly reinforced with wood flour: a. fir wood flour; b. oak wood flour

Fig. 33. Failure area of some specimens reinforced only with E-glass fibres, after flexural test

3.5. Degradation of the composite materials

The figure 34, a shows a photo of the two specimens made of composite material reinforced only with glass fibres, after immersion in water while the figure 34, b is a detailed photo of a damaged area located on the surface of the specimen. It was observed that more specimens analysed had similar brown spots located on the cut edge of the specimens. Since there was no spot before immersion in water, it may assume that the oxidation of the resin could be the cause of the spot appearance. The photos shown in figure 35, acquired by using a metallographic microscope, confirms this opinion.

a. b.

Fig. 34. Photos of the damaged composite materials

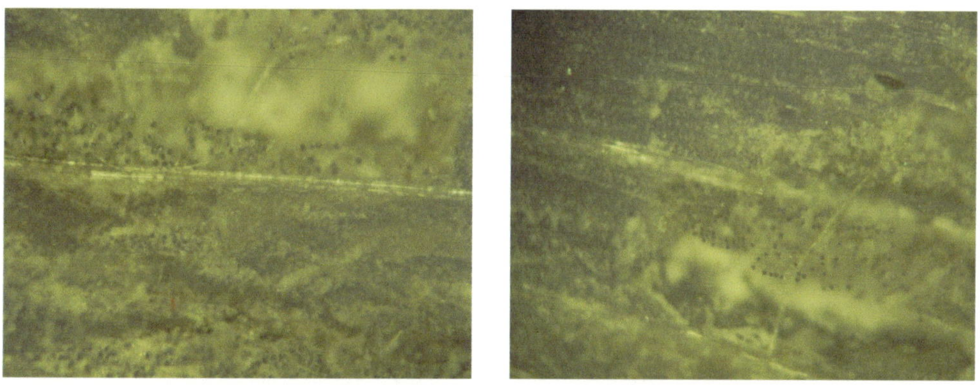

Fig. 35. Specimen photos (zoom 100x) acquired by using a metallographic microscope, after 7197 hours of immersion in water, in case of E-glass/polyester Heliopol 8431ATX composite

Fig. 36. Photo of the specimen surface made of E-glass EWR145 / fir wood flour / polyester Colpoly 7233 after immersion in water (5612 hours)

Concerning the degradation of the surfaces of the specimens made of E-glass EWR145 / fir wood flour / polyester Colpoly 7233, it was observed no spots or colour changes (Fig. 36).
The figure 37 shows two photos of some specimens made of E-glass / oak wood flour / polyester composite material after 5853 hours of immersion in water (Fig. 37,a) and seawater (Fig. 37, b), respectively. It may be easily observed that more specimens analysed had similar black spots located on the cut edges of the specimens, especially in case of the specimens immersed in seawater.

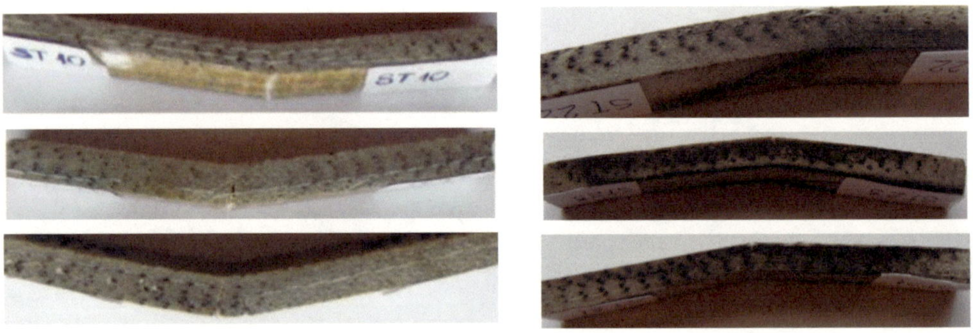

a. b.

Fig. 37. Photos of the edges of the specmens made of E-glass / oak wood flour / polyester after 5853 hours of immersion in: a) water; b) seawater

Since there was no spot before immersion in water, it may assume that the oxidation of the resin could be one of the causes of the spot appearance.
Moreover, it was remarked that in case of the specimens immersed in seawater, the damaged areas located on the specimen edge are larger in case of the composite material filled with oak wood flour than in case of the other one type of composite. The greater content of the tannin in case of oak wood than in case of the fir wood, could be the cause of the greater degradation of the specimens made of the E-glass / oak wood flour / polyester composite material.

4. Conclusions and discussions

We note that the moisture absorption into the composite materials reinforced only with glass fibres, leads to the decreasing of the tensile strength, flexural strength and Young's modulus E. It was observed that the flexural modulus E of these composite materials decreases with 10 - 20 %.
Decreasing of the tensile strength is about 40 % in case of E-glass / epoxy LY 554 composite after ≈ 7000 hours of immersion in water while the decreasing of the flexural strength is 30 % after 9200 hours.
In general, it was found that the E-glass / epoxy composite analysed absorbed more water than seawater or detergent solution. The sodium chloride molecules contained in seawater (as well as sulphate), appear to be limiting the diffusion of water into the matrix material. It was observed that the detergent effect was much more pronounced than the sodium chloride effect in case of E-glass / polyester composite.
The E-glass / polyester composites (E-glass / polyester Heliopol 8431 ATX and E-glass / polyester Polylite 440-M880) are recommended as composite materials in case of water or seawater environment.

Finally, it should remark that the absorption of water leads to the increasing of both Young's modulus E and maximum flexural normal stress σ_{max} in case of the hybrid composite material with oak wood flour. Contrary, the absorption of water leads to the decreasing of the mechanical characteristics (Young's modulus E and strength) in case of the other one composite with fir wood flour. It follows that wood oak flour should be recommended as filler for the parts that works in water environment and that made of the hybrid composite material involved. If the humid environment contains salts fir wood flour should be used as filler because the greater content of the tannins associated to the oak wood leads to the appearance of the large dark stains over the surface of the composite.

5. Acknowledgement

This work was supported by CNCSIS - UEFISCSU of Romania, project number PNII - IDEI 733 / 2008 (CNCSIS - National Council of Scientific Research in Higher Education; UEFISCSU - Executive Unit for Financing Higher Education and Research).

6. References

Adhikary, K. B.; PANG, S.; STAIGER, M. P. (2007). Long-term moisture absorption and thickness swelling behaviour of recycled thermoplastics reinforced with Pinus radiata sawdust. In: Chemical Engineering Journal, doi: 10.1016 /j.cej.2007.11.024.

Adhikary, K. B.; PANG, S.; STAIGER, M. P. (2008). Dimensional stability and mechanical behaviour of wood-plastic composites based on recycled and virgin high-density polyethylene - HDPE. In: Composites Journal (Part B) 39: 807–815.

Cerbu C. (2007). Aspects concerning the degradation of the elastic and mechanical characteristics in bending in case of th glass / polymer composite materials due to the moisture absorption, Materiale Plastice, ISSN 0025 / 5289, 44, nr. 2, 2007, pp.97-102;

Cerbu C. (2005) - *Researches concerning structural optimisation of some pieces made of composite materials, loaded in environmental aggressive conditions.* Doctoral thesis. Transilvania University of Brasov

Cerbu, C., Ciofoaia, V., Curtu I., Vişan, A. (2009). The effects of the immersion time on the mechanical behaviour in case of the composite materials reinforced with E-glass woven fabrics, Materiale Plastice, ISSN 0025 - 5289, 46, nr. 2, 2009, p.201;

Cerbu, C.; Teodorescu, H. (2009). Bending behaviour of the composite materials made by recycling of the CDs and DVDs, In: Proceedings of The World Congress on Engineering WCE 2009, vol. II: p.1753-1756.

Cerbu, C., Curtu, I. (2009). Particularities concerning the mechanical behaviour in wet environment in case of a hybrid composite material with wood flour, Rev. ProLigno, ISSN 1841-4737,vol. 5, Nr. 3, 2009, pp.37,

Cerbu, C., Motoc, D., Ciofoaia, V. (2009). Advantages of the using of the poliester resin to manufacturing of the composite materials based on wood flour, Annals of DAAAM for 2009 & Proceed. of the 20th International DAAAM Symposium "Intelligent Manufacturing & Automation: focus on Theory, Practice & Education", vol. 20, no. 1, 25-28th November 2009, Vienna, Austria, ISSN 1726-9679, ISBN 978-3-901509-70-4, Editor Branko Katalinic, pp.1417-1418.

Corum, J. M., Battiste, R. L., Ruggles, M. B., Ren, W. (2001). Durability - based design criteria for a chopped-glass-fibre automotive structural composite, Composite Science and Technology, no. 61;

Kamdem, D. P.; Jiang, H.; Cui, W.; Freed, J.; Matuana, L. M. (2004). Properties of wood plastic composites made of recycled HDPE and wood flour from CCA-treated wood removed from service. In: Composites Journal (Part A) 35: 347–355.

Klyosov, A. A. (2007). Wood-plastic composites, A John Wiley & Sons, Inc. Hoboken, New Jersey, USA.

Pomies, F., Carlson, L. A., Gillespie, J. W. (1995). Marine environmental effects on polymer matrix composites, Composite Materials: Fatigue and Fracture, Fifth Volume, ASTM STP 1230, R. H. Martin, Ed., American Society for Testing and Materials, Philadelphia;

Springer, G. S. (1984). Environmental Effects on Composite Materials, Vol. 2, Technomic Publishing Inc., Lancaster, PA, ,;

SR EN ISO 527 / 1,2,3. Glass fibre reinforced plastics, Determination of the tensile properties;

SR EN ISO 178 (2001). Glass fibre reinforced plastics, Determination of flexural properties, Three point method, CEN, Bruxelles, 2001;

SR EN ISO 62 (2008). Plastics. Determination of water absorption, European Committee for Standardization, Brussels.

Takeshige, F.; Kawakami, Y.; Hayashi, M.; Ebisu S. (2007). Fatigue behavior of resin composites in aqueous environments, In: Dental Materials 23: 893–899.

20

Simulations of Woven Composite Reinforcement Forming

Philippe Boisse
Université de Lyon, LaMCoS, INSA-Lyon, 69621
France

1. Introduction

Complex preforms can be obtained by forming an initially flat textile composite reinforcement. Then the resin is injected on the preform in LCM processes (Liquid Composite Moulding)(Advani, 1994, Rudd & Long, 1997, Parnas, 2000). These processes and especially the RTM process (Resin Transfert Moulding) can be used to manufacture highly loaded composite part for aeronautical application (e.g. the helicopter frame and the motor blade presented in figures 1 and 2).

Numerical optimization of products and production processes becomes increasingly important in the design phase of composite structures. Numerical simulations of the composite forming processes are an essential part of these optimization tools. They permit to determine the conditions of the feasibility of a process without defect (wrinkling, fracture of yarns, porosities...) but above all, they give the fibre orientations after shaping. This is mainly important because redistribution of the fibres is inevitable when double curved products are considered. The fibre orientations strongly influence the mechanical behaviour of the final part and the permeability of the reinforcement and thus the injection of the resin in the case of a liquid moulding process

The first method used for analysing composite forming processes, and especially draping in woven composite reinforcements, is "kinematic models". The fibre distribution of a woven cloth is predicted on a given geometry based on a pin jointed net assumption (fishnet algorithm). The yarns of the fabric are assumed to be inextensible and the rotation between warp and weft yarns is free. The woven reinforcement is placed progressively from initial lines. Several packages are commercially available. This method is briefly described in section 2. It is fairly efficient for hand draping in classical prepreg fabrics, but the models do not account for load boundary conditions, for possible sliding of the fabric in relation to the tools, and for the mechanical behaviour of the woven reinforcement

For a physical analysis of a composite forming process, the complete model must include all the equations for the mechanics, especially equilibrium, constitutive equations, and boundary conditions. These equations must be solved numerically, with some approximations. Finite Element Analysis of the composite forming process includes the tools modelling, the contact and friction between the different parts, and above all, the mechanical behaviour of the composite during forming. If these models can be numerically costly, problems of computation time are steadily reduced through improved processing capabilities. The main problem for the FE approach therefore lies in the requirement for accurate models of all the significant aspects of the forming process.

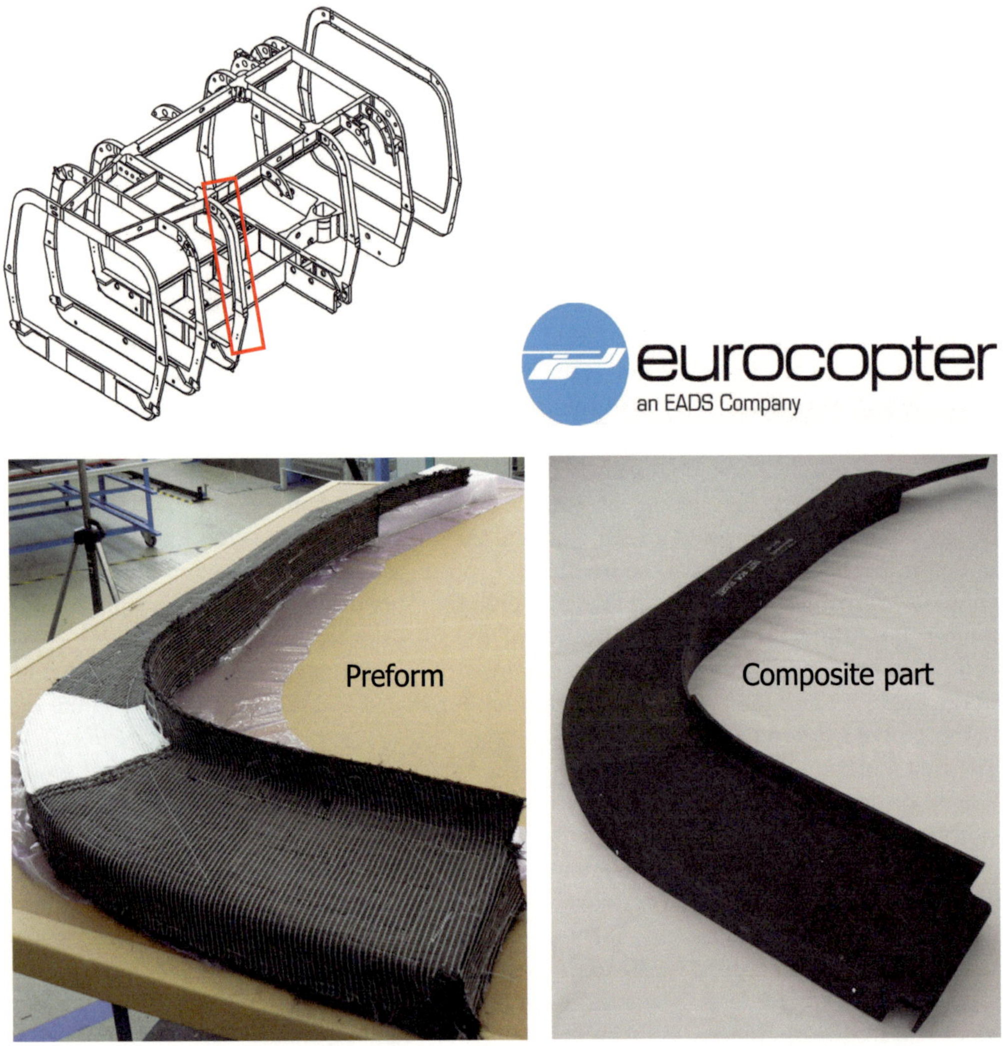

Fig. 1. Preform/RTM parts in NH90 (Dumont et al. 2008) (Courtesy of Eurocopter, EADS Group)

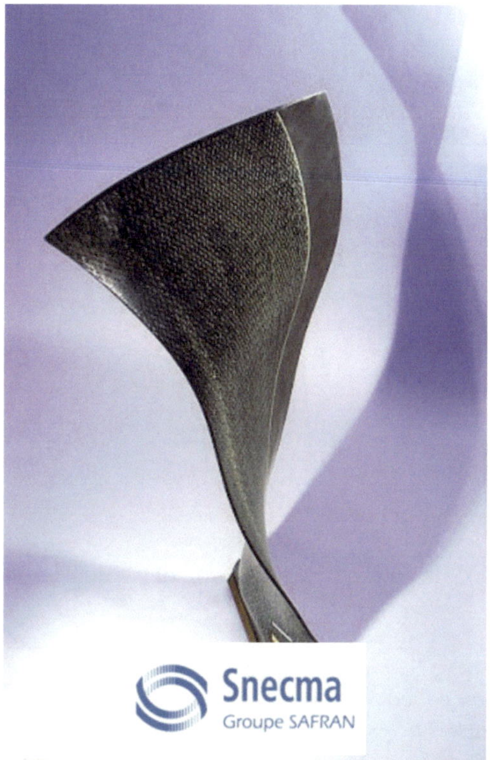

Fig. 2. Plane motor blade (Courtesy of Snecma, Groupe Safran) (De Luycker et al. 2009)

The mechanical behaviour of fabrics is complex due to the intricate interactions of the yarns and fibres. It is a multi-scale problem. The macroscopic behaviour is very dependent on the interactions of yarns at the meso-scale (scale of the woven unit cell) and at the micro-scale (level of the fibres constituting yarns). Despite the many works in the field, there is no widely accepted model that accurately describes all the main aspects of a composite woven reinforcement mechanical behaviour. The approaches to model the forming of textile composite reinforcements belong to two main families that are related to the scale at which the analysis is made. The textile reinforcement is a set of yarns (or fibres). The analysis of the forming can be made considering and modelling each of these yarns (or fibres) and their interactions (contact with friction). In this case the approach is called discrete or mesoscopic. Of course the number of yarns is high and the interactions are complex. On the opposite, the continuous approaches consider a continuous medium juxtaposed with the fabric and the mechanical behaviour of which is equivalent to those of the textile reinforcement. This mechanical behaviour is complex because it concerns large strains and strong anisotropy. Furthermore, it strongly changes during the forming.

The present chapter aims to present continuous and discrete approaches for composite reinforcements forming simulations. First two continuous approaches are described within a membrane assumption. The first one is based on a hyperelastic model and the second on a hypoelastic one. Then simulations of woven fabric forming based on a discrete approach are presented. Finally a semi-discrete approach which can be seen as an intermediate method between continuous and discrete ones is presented. This approach is extended to 3D interlock forming simulations. The advantages and drawback of the different approaches are discussed.

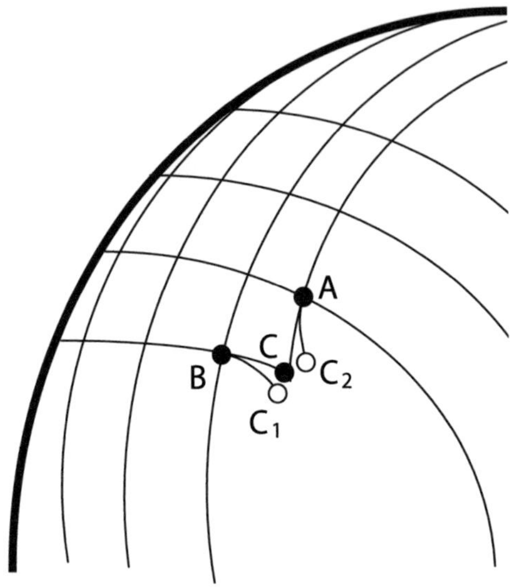

Fig. 3. Fishnet algorithm: Calculation of the position of point C knowing those of A and B

2. Kinematic models

"Kinematic models" also called fishnet algorithms are commonly used in industry for analysing composite forming processes, and especially draping in woven composite reinforcements (Mark, 1956, Van Der Ween, 1991, Long et al, 1994, Borouchaki et al, 2002). Several packages are commercially available. The method is based on the (strong) following assumptions : i/ The yarns are inextensible, ii/ there is no sliding at the intersection between warp and weft yarns, iii/ The rotations between warp and weft yarns are free, iv/ there is no sliding between the fabric and the tool.
As shown on figure 3, the position of the node C can be determined if those of its neighbours A and C are already known. AC and BC have prescribed length. C is defined as the intersection of two geodesics coming from A and B and that intersect in C. This constitutes a small scalar problem, generally non-linear that can be solved very fast. The surface of the tool must be defined analytically of by curved elements. In order to initiate the draping as shown figure 3, it is necessary to position a first node and to fixe two initial draping directions. These directions are the symmetry axes if they exist. The result of the draping depends on these directions.
This method is very fast and fairly efficient for hand draping in classical prepreg fabrics, but the models do not account for load boundary conditions, for possible sliding of the fabric in relation to the tools, and above all for the mechanical behaviour of the woven reinforcement.

3. Mechanical behaviour of textile composite reinforcement

The diameter of each fibre constituting the textile composite reinforcements is very small: 5 to 7 μm for carbon, 5 to 25 μm for glass, 10 to 20 μm for aramid. A yarn is made up of several thousands juxtaposed fibres (usually $3.\ 10^3$ to $96.\ 10^3$). These yarns are woven following standard weaves (plain, satin, twill) or more complex structures such as braiding or ply to ply interlock weaves (Figure 4c). An alternative consists in stitching a ply made of parallel fibres. This leads to the so-called NCF reinforcement (Non Crimp fabric) in which the fibres are not undulated (Figure 4d). The material resulting from this assembly of continuous fibres exhibits a very specific mechanical behaviour since relative motions are possible between the yarns and the fibres. The textile reinforcement pre-forming stage takes advantage of these possible motions. The forming is made on dry reinforcement (i.e. without resin) since it is performed before the injection stage.

A woven fabric is intrinsically a multiscale material and, depending on the specific application of interest, one or more scales of the woven fabric have to be explored.

Three scales can be distinguished (Figure 5). The macroscopic scale refers to the whole component level, with dimensions in the order of some centimetres to several meters (Figure 5a). At the mesoscopic scale, the woven reinforcement is seen as a set of yarns, respectively the warp and the weft (or fill) yarns in case of a woven fabric (Figure 5b) . Consequently, the corresponding working scale is the one of the yarn dimension, typically one to several millimetres. For periodic materials, mesoscopic models consider the smallest elementary

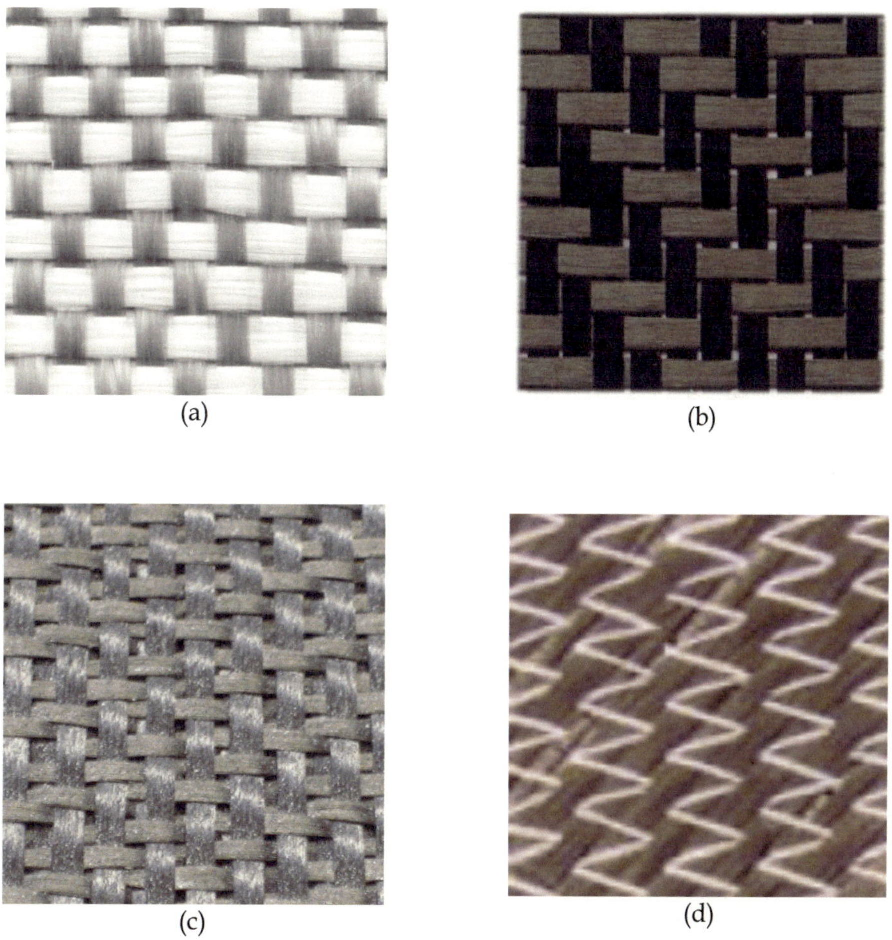

Fig. 4. Textile composite reinforcements (a) plain weave, (b) twill weave (c), interlock, (d) NCF

Macroscopique scale (a) Mesoscopic scale (b) Microscopic scale (c)

Fig. 5. The three scales of the fibre reinforcement

pattern which can represent the whole fabric by several translations. That domain is called the Representative Unit Cell (RUC). Each yarn is made up of thousands of continuous fibres which interact (Figure 5c), and thus the interactions of the reinforcement can be analyzed at the microscopic scale. At the microscopic level, the characteristic dimension is about one to several micrometers. This is the only scale at which the material is actually continuous.
Although the fibrous reinforcement is not strictly continuous because of possible relative sliding between fibres, several mechanical behaviour models have been proposed that consider the textile reinforcement as an anisotropic continuum (Spencer, 2000; Dong et al, 2001; Yu et al, 2002; King et al, 2005, Peng et al, 2005, Ten Thije et al, 2007, Badel et al, 2009) Nevertheless there is no widely accepted model that describes accurately all of the main aspects of fabric mechanical behaviour. Actually, such a model must convey the specificities of the composition of the textile made of yarns and fibres and above all take into account the variation of the properties during the forming. These changes are very large because of the variations of fibre directions and of the local lateral compression of the yarns due to the forming.

4. Experimental texts

The specific mechanical behaviour of textile composite reinforcements has led to the development of specific experimental tests in order to quantify the tensile, in-plane shear and bending behaviour.

4.1 Biaxial tensile behaviour

The tensile behaviour of woven material is specific mainly because of the decrimping of tows when they are stretched. This leads to tensile behaviour non-linearities. The fabric is much softer than the tow for small axial strains. Because of the weaving, the decrimping phenomenon in warp and weft directions are interdependent and the tensile behaviour is biaxial. Some biaxial tests have been developed in order to measure these properties (Kawabata et al, 1973, Buet-Gautier & Boisse, 2001, Carvelli et al, 2008, Willems et al, 2008). Fig. 5. shows a biaxial tensile device using a cross shape specimen (Buet-Gautier & Boisse, 2001). The measurements of tensions in warp and weft directions $T_1(\varepsilon_{11}, \varepsilon_{22})$ and $T_2(\varepsilon_{11}, \varepsilon_{22})$ are shown Figure 5b for different ration between warp and weft strains. It has been experimentally shown that the influence of the shear angle on the tensile behaviour is usually weak and can be neglected (Buet-Gautier & Boisse, 2001).

(a)

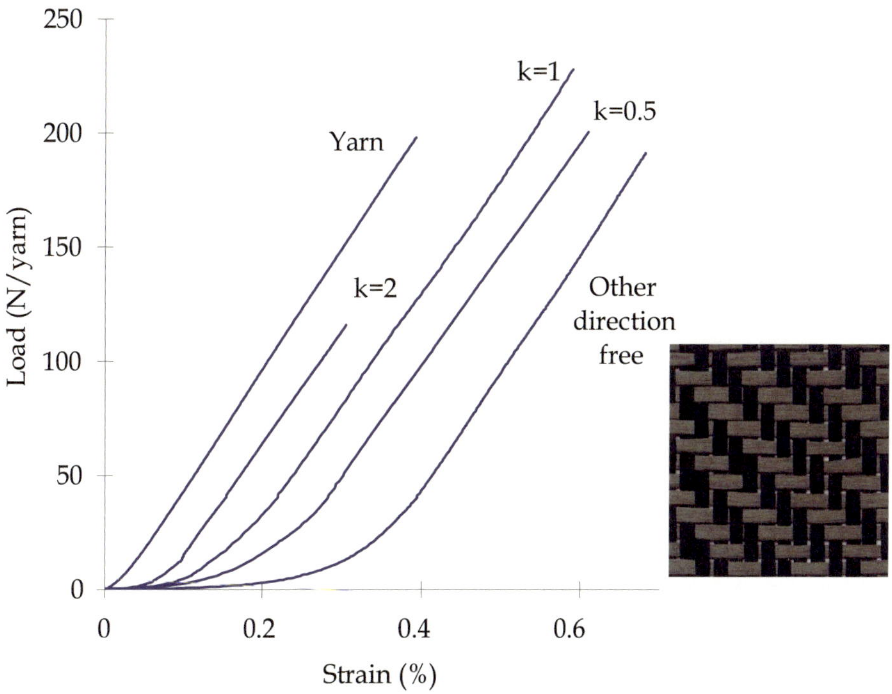

(b)

Fig. 6. (a) Biaxial tensile test on cross-shaped specimen (b) Load versus strain for carbon twill weave. $k = \varepsilon_{warp} / \varepsilon_{weft}$ (Buet-Gautier et al, 2001)

4.2 In-plane shear behaviour

Two experimental tests are used to determine the in-plane shear behaviour of textile composite reinforcements: the picture-frame (Figure 7) and the bias-extension tests (Figure 8). A great literature is dedicated to those tests (Prodromou & Chen, 1997, Rozant et al, 2000, Potter et al, 2002, Lebrun et al, 2003, Sharma et al, 2003, Peng et al, 2004, Harrison et al, 2004, Lomov et al, 2006, Launay et al, 2008, Lomov et al, 2008, Cao et al, 2008) mainly because the in-plane shear is the most dominant deformation mode in woven composite forming when

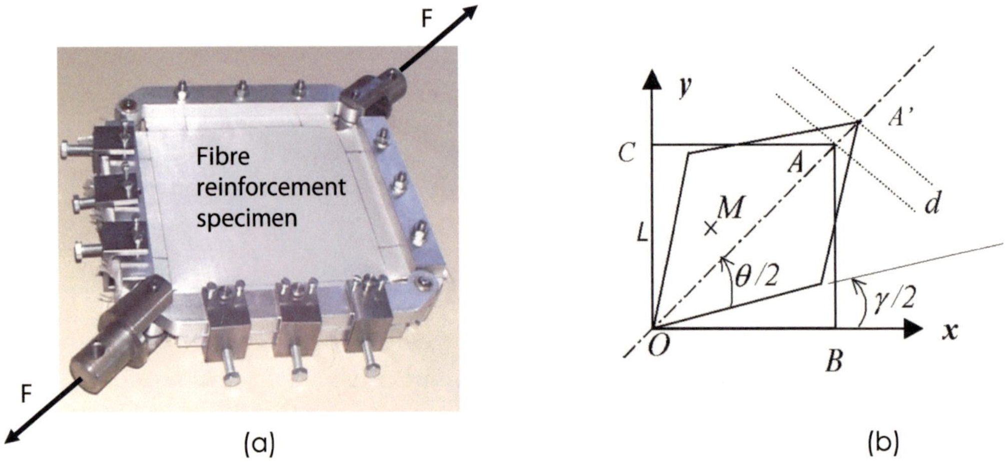

Fig. 7. Picture frame test. (a) Experimental device. (b) Kinematics of the test

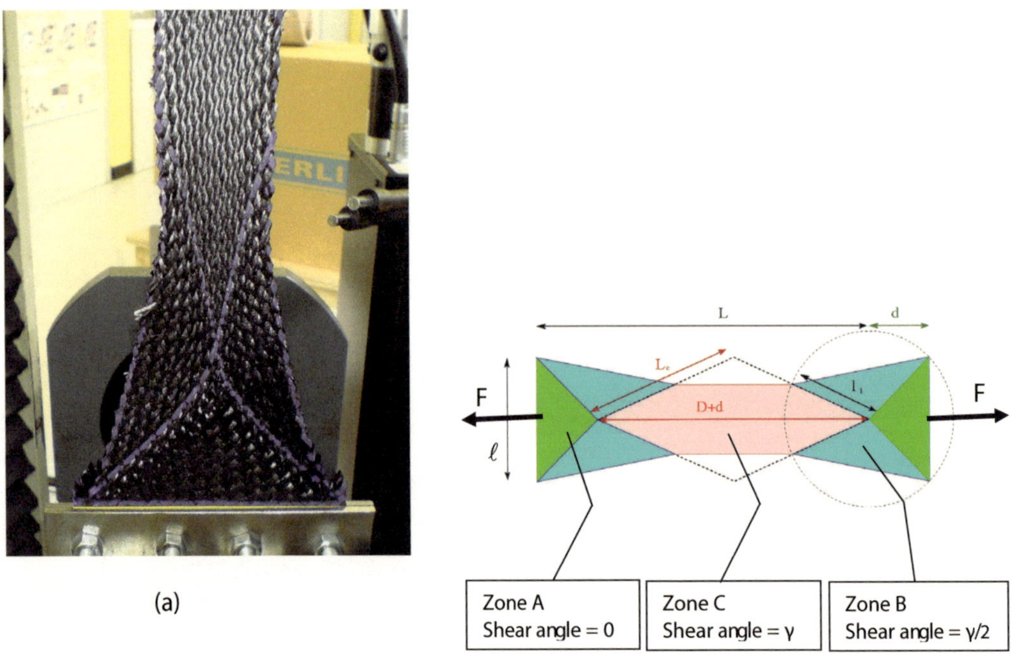

Fig. 8. The bias-extension test and the three zones

the manufactured part is doubly curved. The shear angle can reach 50° (and even more in some cases such as presented in section 5.6). For large values, wrinkling can occur depending on the process parameters and on the material properties. In addition, the two tests are difficult both from the experimental point of view and concerning the interpretation of the results. For these reasons an international benchmark has been launched recently. Eight laboratories of six different countries have performed picture frame and bias-extension tests on the same textile composite reinforcements (Cao et al, 2008).
The picture-frame (or trellis-frame) is made of four hinged bars. The fabric specimen is initially square and the tows are parallel to the bars. Consequently it is theoretically subjected to pure in-plane shear and the shear angle γ is function of the displacement d.

$$\gamma = \frac{\pi}{2} - 2\arccos\left(\frac{1}{\sqrt{2}} + \frac{d}{2L}\right) \tag{1}$$

Neglecting the dissipation due to friction in the hinged bars, the in-plane shear moment on a unit shell M_s is related to the force on the frame F using the equality of the power expressions. S_c is the surface of the unit woven cell in the initial state.

$$M_s(\gamma) = F\frac{S_c}{\sqrt{2}\,L}\left(\cos\frac{\gamma}{2} - \sin\frac{\gamma}{2}\right) \tag{2}$$

The bias-extension test is an alternative to the picture-frame test. It consists in clamping a rectangular specimen of woven fabric with warp and weft directions initially oriented at 45° with respect to the tensile load applied by a tensile machine (Figure 8). The initial length of the specimen L must be larger than twice the width ℓ. The zone C in the centre of the specimen is submitted to a pure shear γ if the yarns are assumed to be inextensible. That is a correct assumption for the type of yarns used as composite reinforcements. This inextensibility imposes that the shear angle in the zone B is half the value in the central region C. The shear angle γ is related to the specimen elongation d in equation (c). The in-plane shear moment is related to the force on the frame F in equation (d).

$$\gamma = \frac{\pi}{2} - 2\arccos\left(\frac{D+d}{\sqrt{2}D}\right) \tag{3}$$

$$M_S(\gamma) = \frac{FDS_C}{\ell(2D-\ell)}\left(\cos\frac{\gamma}{2} - \sin\frac{\gamma}{2}\right) - \frac{\ell}{2D-\ell}M_S\left(\frac{\gamma}{2}\right) \tag{4}$$

A shear curve $M_s(\gamma)$ measured in the case of glass plain weave is presented in Figure 9. In the first part of the curve (i.e. for small shear angles), the in-plane shear stiffness is small. For larger shear angles this rigidity increases and becomes significant. The optical field measurements performed within the tow show that during the first part of the loading, the tows rotate in a rigid body motion (Figure 9b). When the shear angle becomes larger lateral contacts between the yarns occur (Dumont et al, 2003). The tows are progressively compressed and the shear rigidity increases significantly. This increase of shear stiffness leads to wrinkling onset. The corresponding shear angle is called locking angle (in order of 40°- 45° for textile composite reinforcements (Prodromou & Chen, 1997, Cao et al, 2008).

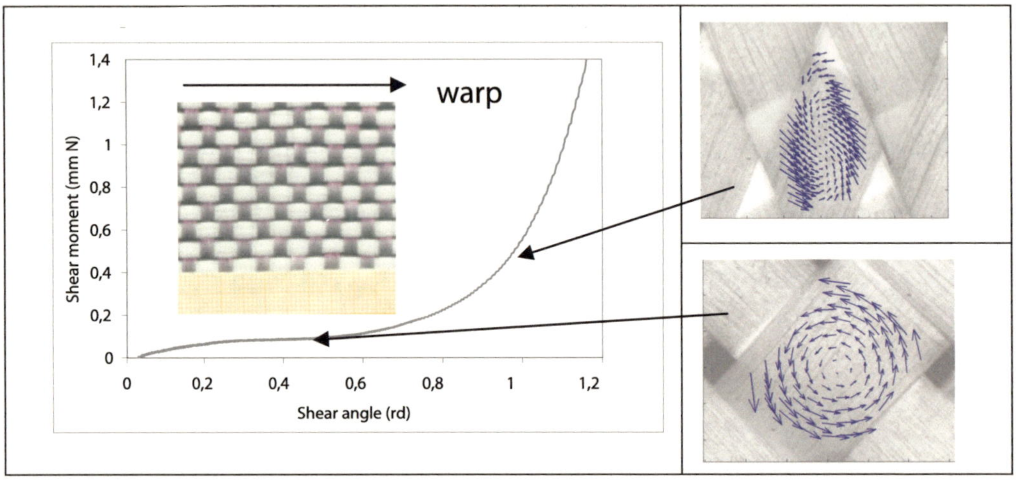

Fig. 9. (a) Shear curve of glass plain weave, (b) relative displacement field inside a yarn

4.3 Bending behaviour

The bending stiffness of the fibrous reinforcement is very low due to the possible motion between fibres. It is often neglected and membrane approaches are often used. Nevertheless this bending stiffness is important to obtain accurate wrinkle shape. Specific tests have been developed for these textile materials, the bending stiffness of which is much smaller than the one given by the plate theory (Kawabata, 1986, Lahey and Heppler 2004, de Bilbao et al, 2010). Figure 10 shows a cantilever bending test (de Bilbao et al, 2010). The own weight of the fabric defines the bending moments for a given length of the specimen. The curvatures are deduced from the measured geometry. These measurements, made for different length of a specimen give the bending moment in function of the curvature. They are made in warp and weft directions.

5. Continuous approach for the simulation of textile composite forming

In the continuous approaches, a woven fabric is seen as a continuous material with a specific mechanical behaviour, including high anisotropy and the ability to exhibit very large shearing and bending deformations. Investigation at the macroscopic level is the most popular for reinforcement forming simulations, as it allows using finite elements codes with standard shell or membrane elements and does not ask the description of the internal textile material structure. Despite the large amount of work in this field (Spencer, 2000; Dong et al, 2001; Yu et al, 2002; King et al, 2005, Peng et al, 2005, Ten Thije et al, 2007, Badel et al, 2009) there is no widely accepted model that accurately describes all aspects of the mechanical behaviour of fabrics. Two continuous approaches are described below.

5.1 Hyperelastic behaviour

The formulation of a hyperelastic behaviour law lies on the proposition of a potential energy from which derives the hyperelastic constitutive model. This potential aims to reproduce the non linear mechanical behaviour of textile composite reinforcements. The proposed potential is a function of the right Cauchy Green and structural tensor invariants defined

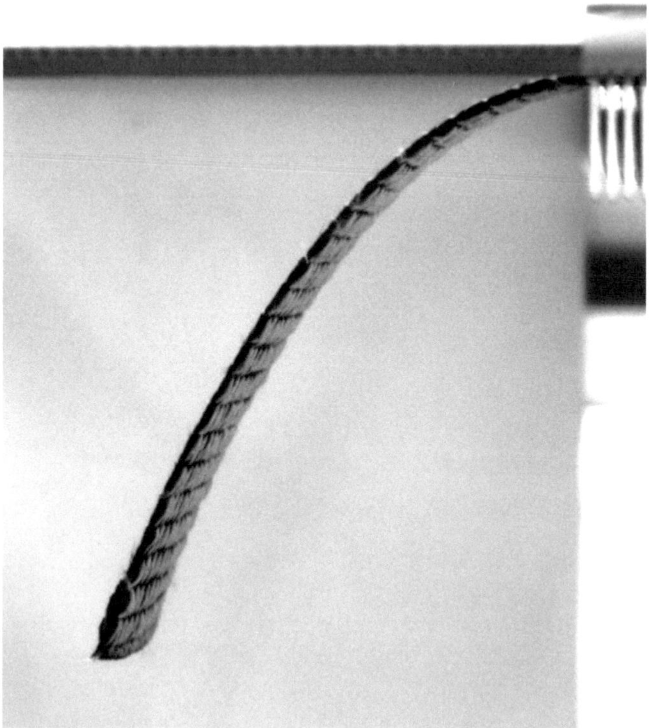

Fig. 10. Cantilever bending test

from the fibre directions. This potential is based on the assumption that tensile and shear strain energies are uncoupled. It is the sum of three terms.

$$W = \bar{W}_1(I_1) + \bar{W}_2(I_2) + \bar{W}_s(I_{12}) \tag{5}$$

This assumption (tensile and shear strain energies are uncoupled) are made for sake of simplicity. The independence of tensile behaviour relatively to in plane shear has been shown experimentally (Buet-Gauthier and Boisse, 2001). The other hypotheses are probably less true, but there are made for sake of simplicity and because there are few data available on some couplings.

The structural tensors $\underline{\underline{L}}_{\alpha\beta}$ are defined from the two unit vectors in the warp and weft directions $\underline{f}_{10}$ and $\underline{f}_{20}$ in the reference configuration C^0 (figure 11):

$$\underline{\underline{L}}_{\alpha\beta} = \underline{f}_{\alpha 0} \otimes \underline{f}_{\beta 0} \tag{6}$$

The two first terms $\overline{W}_1$ and $\overline{W}_2$ are the energies due to the tensions in the yarns. They are function of invariants I_1 and I_2 respectively, themselves depending on the right Cauchy Green strain tensor $\underline{\underline{C}} = \underline{\underline{F}}^T \cdot \underline{\underline{F}}$ and the structural tensors $\underline{\underline{L}}_{\alpha\alpha}$:

$$I_1 = \mathrm{Tr}(\underline{\underline{C}} \cdot \underline{\underline{L}}_{11}) = \lambda_1^2 \qquad I_2 = \mathrm{Tr}(\underline{\underline{C}} \cdot \underline{\underline{L}}_{22}) = \lambda_2^2 \tag{7}$$

λ_α is the deformed length of on initially unit fibre in the direction α.

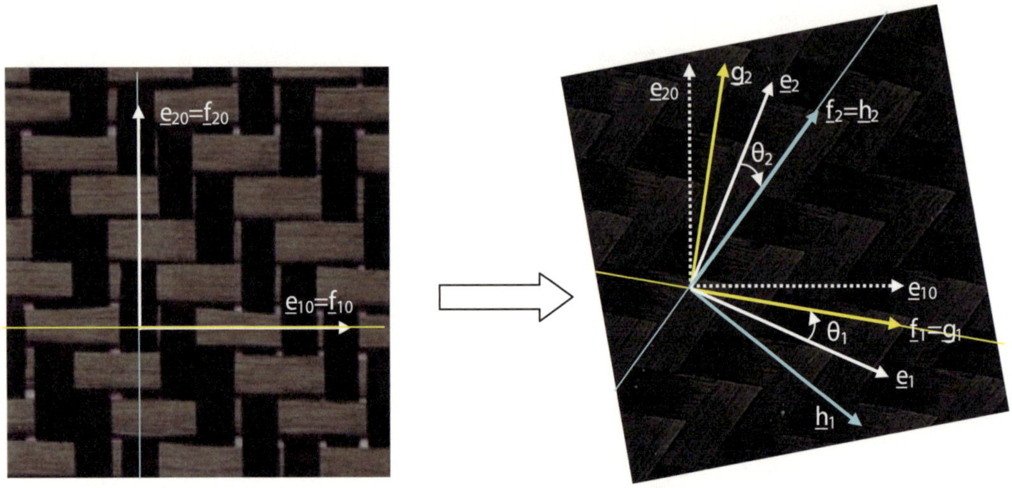

Fig. 11. Fibres axes and GN axes after deformation. Initially these axes are superimposed.

The third term $\overline{W}_s$ in equation (5) is a function of the second mixed invariants of $\underline{\underline{C}}$.

$$I_{12} = \frac{1}{I_1 I_2} Tr(\underline{\underline{C}} \cdot \underline{\underline{L}}_{11} \cdot \underline{\underline{C}} \cdot \underline{\underline{L}}_{22}) = \cos^2\theta \tag{8}$$

The second Piola Kirchhoff stress tensor is derived from this potential (Basar and Weichert, 2000):

$$\underline{\underline{S}} = 2\frac{\partial W}{\partial \underline{\underline{C}}} \tag{9}$$

And in the case of the present potential (5):

$$\begin{aligned}\underline{\underline{S}} = {} & 2\left[\frac{\partial \overline{W}}{\partial I_1} - \frac{I_{12}}{I_1}\frac{\partial \overline{W}}{\partial I_{12}}\right]\underline{\underline{L}}_{11} + 2\left[\frac{\partial \overline{W}}{\partial I_2} - \frac{I_{12}}{I_2}\frac{\partial \overline{W}}{\partial I_{12}}\right]\underline{\underline{L}}_{22} \\ & + 2\left[\sqrt{\frac{I_{12}}{I_1 I_2}}\frac{\partial \overline{W}}{\partial I_{12}}\right]\left(\underline{\underline{L}}_{12} + \underline{\underline{L}}_{21}\right)\end{aligned} \tag{10}$$

In order to define the form of the potential two complementary assumptions are made taking into account the specific woven fabric behaviour and its deformation modes. As assumed above, i/ The tensions in the yarns and the in-plane shear are independent. ii/ The tensions in the warp and weft directions are uncoupled.

The potential has to vanish in a stress free configuration. Polynomial functions of the invariants are considered in the present work. The global form of the proposed potential energy is given by:

$$W\left(\underline{\underline{C}}\right) = \sum_{i=0}^{r} \frac{1}{i+1} A_i \left(I_1^{i+1} - 1\right) + \sum_{j=0}^{s} \frac{1}{j+1} B_j \left(I_2^{j+1} - 1\right) + \sum_{k=1}^{t} \frac{1}{k} C_k I_{12}^k \tag{11}$$

The resulting second Piola Kirchhoff tensor is:

$$\underline{\underline{S}} = 2\left(\sum_{i=0}^{r} A_i I_1^i - \frac{1}{I_1}\sum_{k=1}^{t} C_k I_{12}^k\right)\underline{\underline{L}}_{11} + 2\left(\sum_{j=0}^{s} B_j I_2^j - \frac{1}{I_2}\sum_{k=1}^{t} C_k I_{12}^k\right)\underline{\underline{L}}_{22}$$
$$+2\left(\frac{1}{\sqrt{I_1 I_2}}\sum_{k=1}^{t} C_k I_{12}^{k-1/2}\right)\left(\underline{\underline{L}}_{12} + \underline{\underline{L}}_{21}\right) \tag{12}$$

For strain-free configuration, stresses have to vanish. This condition imposes:

$$\sum_{i=0}^{r} A_i = 0 \quad ; \quad \sum_{j=0}^{s} B_j = 0 \tag{13}$$

To determine the constants A_i, B_j and C_k, three experimental tests are necessary: two tensile tests in the warp and weft directions and one in-plane pure shear test. The details of the calculations to obtain equations **???** to **???** are given in (Aimene et al, 2010). In this paper it is also shown that the form of the potential given above gives correct results concerning the direction of the loads on the boundary of a picture frame while other forms of the potential lead to boundary loads that are not correct for a woven fabric.

The proposed hyperelastic model is implemented in a user routine VUMAT of Abaqus/Explicit and it is applied to membrane elements. The simulation of a hemispherical punch forming process is performed in the case of strongly unbalanced twill (Daniel et al, 2003). The warp rigidity is 50 N/yarn and the weft rigidity is 0.2 N/yarn. The experimental deformed shape are shown figure 12 (a) together with the results of the simulation figure 12 (c). The computed deformed shape (made using the hyperelastic model proposed above) is in correct agreement with the experimental one. Especially the strong difference of the deformation in warp and weft directions is well verified.

5.2 Hypoelastic behaviour

Hypoelastic models have been proposed for material at large strain (Truesdell, 1955, Xiao et al, 1988)

$$\underline{\underline{\sigma}}^{\nabla} = \underline{\underline{\underline{\underline{C}}}} : \underline{\underline{D}} \tag{14}$$

where $\underline{\underline{D}}$ and $\underline{\underline{\underline{\underline{C}}}}$ are the strain rate tensor and the constitutive tensor, respectively. $\underline{\underline{\sigma}}^{\nabla}$, called the objective derivative of the stress tensor, is the time derivative for an observer who is fixed with respect to the material.

$$\underline{\underline{\sigma}}^{\nabla} = \underline{\underline{Q}} \cdot \left(\frac{d}{dt}\left(\underline{\underline{Q}}^T \cdot \underline{\underline{\sigma}} \cdot \underline{\underline{Q}}\right)\right) \cdot \underline{\underline{Q}}^T \tag{15}$$

$\underline{\underline{Q}}$ is the rotation from the initial orthogonal frame to the so-called rotating frame where the objective derivative is made. The most common objective derivatives are those of Green-Naghdi and Jaumann. They use the rotation of the polar decomposition of the deformation gradient tensor $\underline{\underline{F}} = \underline{\underline{R}} \cdot \underline{\underline{U}}$, (standard in Abaqus explicit), and the corotational frame, respectively. These are routinely used for analyses of metals at finite strains (Belytschko et al, 2000). It has been shown that, in the case of a material with one fibre direction the proper objective rotational derivative is based on the rotation of the fibre (Badel et al, 2009)

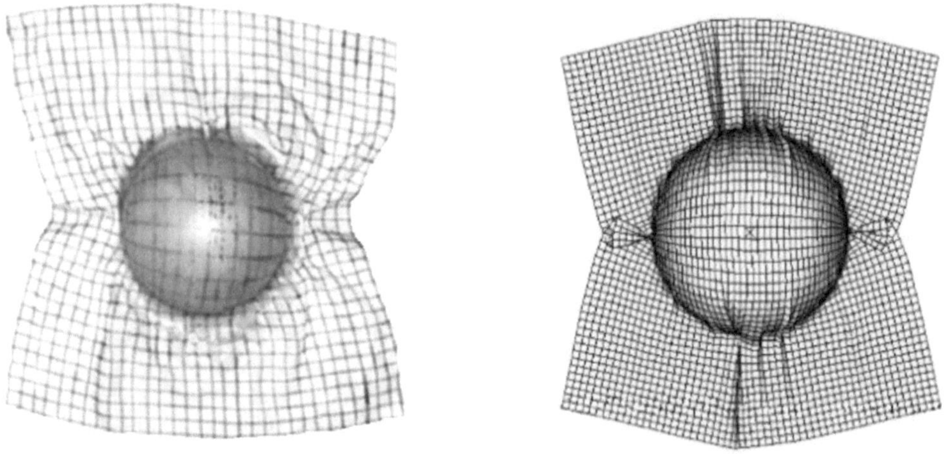

(a) (c)

Fig. 12. Deformed shape in the case of an unbalanced fabric (experimental shape (a); Result of the simulation (c))

A membrane assumption is used. The Green-Naghdi's frame (GN) is the default work basis of ABAQUS/Explicit. Its unit vectors $(\underline{\mathbf{e}}_1,\underline{\mathbf{e}}_2)$ in the current configuration are updated from the initial orientation axes, $(\underline{\mathbf{e}}_1^0,\underline{\mathbf{e}}_2^0)$ using the proper rotation $\underline{\underline{\mathbf{R}}}$:

$$\underline{\mathbf{e}}_1 = \underline{\underline{\mathbf{R}}} \cdot \underline{\mathbf{e}}_1^0 \qquad \underline{\mathbf{e}}_2 = \underline{\underline{\mathbf{R}}} \cdot \underline{\mathbf{e}}_2^0 \tag{16}$$

In the current configuration, the unit vectors in the warp and weft fibre directions are respectively:

$$\underline{\mathbf{f}}_1 = \frac{\underline{\underline{\mathbf{F}}}.\underline{\mathbf{f}}_1^0}{\left\|\underline{\underline{\mathbf{F}}}.\underline{\mathbf{f}}_1^0\right\|} \qquad \underline{\mathbf{f}}_2 = \frac{\underline{\underline{\mathbf{F}}}.\underline{\mathbf{f}}_2^0}{\left\|\underline{\underline{\mathbf{F}}}.\underline{\mathbf{f}}_2^0\right\|} \tag{17}$$

Where $(\underline{\mathbf{e}}_1^0,\underline{\mathbf{e}}_2^0)$ and $(\underline{\mathbf{f}}_1^0,\underline{\mathbf{f}}_2^0)$ are assumed to coincide initially (Figure 3). Two orthonormal frames based on the two fibre directions are defined: $g\left(\underline{\mathbf{g}}_1,\ \underline{\mathbf{g}}_2\right)$ with $\underline{\mathbf{g}}_1 = \underline{\mathbf{f}}_1$, and $h\left(\underline{\mathbf{h}}_1,\underline{\mathbf{h}}_2\right)$ with $\underline{\mathbf{h}}_2 = \underline{\mathbf{f}}_2$ (Figure 11).

The strain increment $\underline{\underline{\mathbf{d\varepsilon}}}$ is given as a code's output in calculation loop from time t^n to time t^{n+1}. (The matrix of the components of this strain increment is given in the GN frame in the case of Abaqus explicit, but it could be any other frame). The components of the strain increment in the two frames g and h are considered (α and β are indexes taking value 1 or 2):

$$\underline{\underline{\mathbf{d\varepsilon}}} = d\varepsilon_{\alpha\beta}^{g}\ \underline{\mathbf{g}}_{\alpha} \otimes \underline{\mathbf{g}}_{\beta} = d\varepsilon_{\alpha\beta}^{h}\ \underline{\mathbf{h}}_{\alpha} \otimes \underline{\mathbf{h}}_{\beta} \tag{18}$$

The fibre stretching strain and the shear strain are calculated for the two frames g and h.

$$d\varepsilon_{11}^{g} = \underline{\mathbf{g}}_1 \cdot \underline{\underline{\mathbf{d\varepsilon}}} \cdot \underline{\mathbf{g}}_1 \qquad d\varepsilon_{12}^{g} = \underline{\mathbf{g}}_1 \cdot \underline{\underline{\mathbf{d\varepsilon}}} \cdot \underline{\mathbf{g}}_2 \tag{19}$$

$$d\varepsilon_{22}^{h} = \underline{\mathbf{h}}_2 \cdot \underline{\underline{\mathbf{d\varepsilon}}} \cdot \underline{\mathbf{h}}_2 \qquad d\varepsilon_{12}^{h} = \underline{\mathbf{h}}_1 \cdot \underline{\underline{\mathbf{d\varepsilon}}} \cdot \underline{\mathbf{h}}_2 \tag{20}$$

From these strain components the axial stress component and shear stress components are calculated in each frame g and h:

$$d\sigma_{11}^{g} = E^{g} d\varepsilon_{11}^{g} \qquad d\sigma_{12}^{g} = G d\varepsilon_{12}^{g} \tag{21}$$

$$d\sigma_{22}^{h} = E^{h} d\varepsilon_{22}^{h} \qquad d\sigma_{12}^{h} = G d\varepsilon_{12}^{h} \tag{22}$$

E^g and E^h are the stiffness in the warp and weft fibre directions respectively and G the in-plane shear stiffness of the fabric (They are not constant, especially G depends strongly on the in plane shear). Following the scheme of Hughes and Winget the stresses are then integrated on the time increment from time t^n to time t^{n+1} (Hughes & Winget, 1980):

$$\left(\sigma_{11}^{g}\right)^{n+1} = \left(\sigma_{11}^{g}\right)^{n} + d\sigma_{11}^{g^{n+1/2}} \qquad \left(\sigma_{12}^{g}\right)^{n+1} = \left(\sigma_{12}^{g}\right)^{n} + d\sigma_{12}^{g^{n+1/2}} \tag{23}$$

$$\left(\sigma_{11}^{h}\right)^{n+1} = \left(\sigma_{11}^{h}\right)^{n} + d\sigma_{11}^{h^{n+1/2}} \qquad \left(\sigma_{12}^{h}\right)^{n+1} = \left(\sigma_{12}^{h}\right)^{n} + d\sigma_{12}^{h^{n+1/2}} \tag{24}$$

The stress at time t^{n+1} in the fabric is the addition of the stresses in the two fibre frames:

$$\underline{\underline{\boldsymbol{\sigma}}}^{n+1} = \left(\underline{\underline{\boldsymbol{\sigma}}}^{g}\right)^{n+1} + \left(\underline{\underline{\boldsymbol{\sigma}}}^{h}\right)^{n+1} \tag{25}$$

For instance, denoting $\underline{\underline{\boldsymbol{\sigma}}} = \sigma_{\alpha\beta}^{e}\, \underline{\mathbf{e}}_{\alpha} \otimes \underline{\mathbf{e}}_{\beta}$ and omitting the superscript n+1 because all the quantities are at time t^{n+1}, the components of the Cauchy stress tensor in the GN frame (that are requested in the Abaqus Explicit code) are:

$$\begin{aligned}\sigma_{\alpha\beta}^{e} = {} & \sigma_{11}^{g}\left(\underline{e}_{\alpha}\cdot\underline{g}_1\right)\left(\underline{e}_{\beta}\cdot\underline{g}_1\right) + \sigma_{22}^{h}\left(\underline{e}_{\alpha}\cdot\underline{h}_2\right)\left(\underline{e}_{\beta}\cdot\underline{h}_2\right) + \sigma_{12}^{g}\left(\underline{e}_{\alpha}\cdot\underline{g}_1\right)\left(\underline{e}_{\beta}\cdot\underline{g}_2\right) \\ & + \sigma_{12}^{h}\left(\underline{e}_{\alpha}\cdot\underline{h}_1\right)\left(\underline{e}_{\beta}\cdot\underline{h}_2\right)\end{aligned} \tag{26}$$

More detail on this approach can be found in (Badel et al, 2009, Khan et al, 2010). This approach is used to simulate the forming double dome shape corresponding to an international benchmark (Khan et al, 2010). An experimental device has been realised in INSA Lyon in order to perform this forming (Figure 13). The woven fabric is a commingled glass/polypropylene plain weave that has been tested in the material benchmark study conducted recently (Cao et al, 2008). The computed and experimental geometries after forming are compared figure 14 and 15. The measured and numerical geometries and shear angles are in good agreement.

6. Discrete approach for the composite reinforcement forming

In discrete modelling (also called meso-modelling in the case of textile material), the modelling does not directly concern the textile material but each fibre bundle. This one is

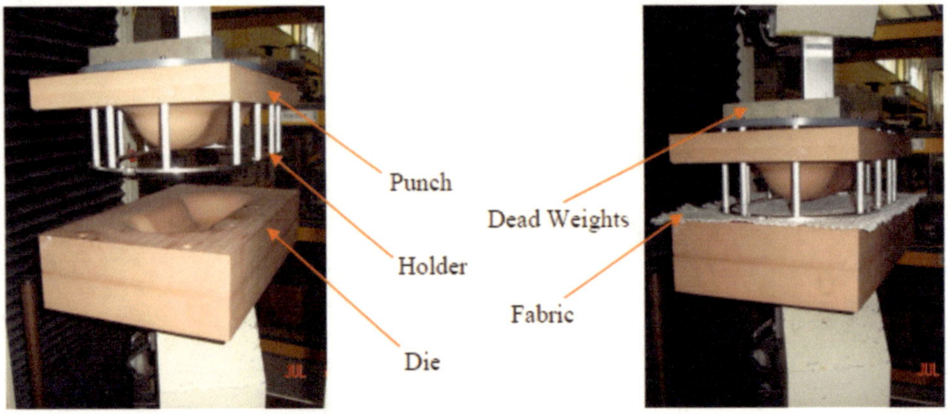

Fig. 13. Double dome forming: experimental device

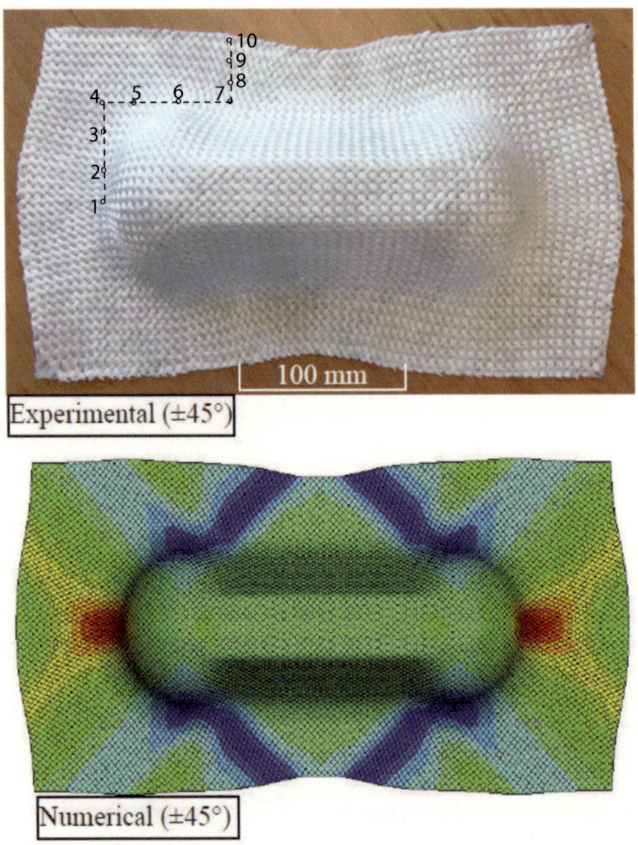

Fig. 14. Experimental and numerical outputs of double dome forming test

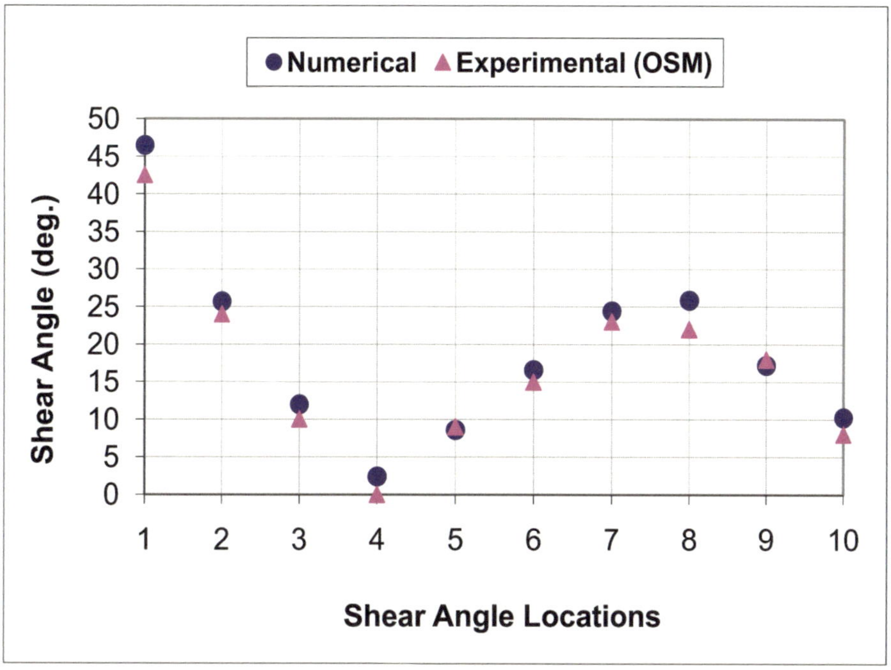

Fig. 15. Comparison of numerical and experimental shear angles at the locations shown in Figure 14.

modelled by elements simple enough to render the computation possible because it concerns the forming of the whole composite reinforcement and the number of yarns and contacts between these yarns is very large. The interactions between warp and weft directions are taken into account explicitly by considering contact behaviour and relative motions between the yarns are possible (Pickett et al, 2005, Duhovic et al, 2006, Ben Boukaber et al 2007).

At the microscopic level, each fibre is satisfactorily described as a beam but this approach is time consuming. The main difficulty is the great number of contacts with friction that have to be taken into account, especially for a woven fabric. For this reason, only very small elements of the fabric have been modelled to date (Durville, 2005, Miao et al, 2008). Nevertheless, this approach is promising because it does not necessitate any assumptions regarding the continuity of the material, the specific mechanical properties resulting at the macroscopic level naturally follow the displacements and deformations of the yarns and it provides an interesting way of taking the weaving operation into account. The fibres constituting the yarns can be modelled directly, but their very large number (3K to 48K per yarn) requires that the computations are made for a number of fibres per yarn significantly smaller than in reality. An alternative possibility is to use a continuous behaviour for each yarn (meso-modelling). This implies that the fibrous nature of the yarn is taken into account in this model especially in order to have rigidities in bending and transverse compression very small in comparison to the tensile stiffness. In any case, a compromise must be found between a fine description (which will be expensive from the computation time point of view) and a model simple enough to compute the entire forming process. Figure 16(b) show

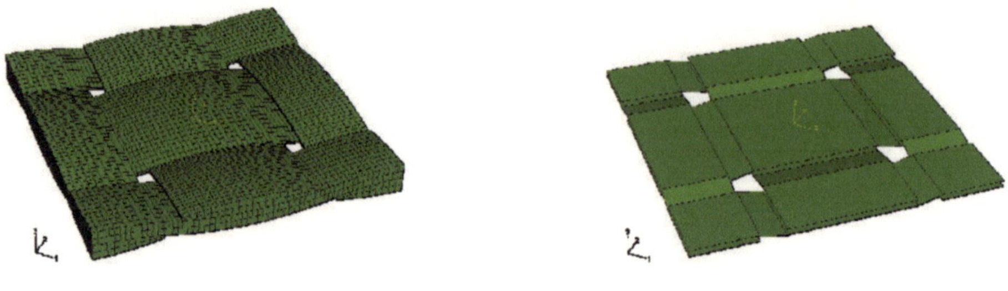

(a) (b)

Fig. 16. Meso-modelling of a unit cell of a plain weave. (a) FE model for the analysis of the behaviour of the unit cell. 47214 Dof. (b) FE model for simulations of the whole composite reinforcement forming. 216 Dof.

the finite element model used for discrete simulations of forming processes (216 dof (degrees of freedom)). It is compared to another FE model of the unit cell used in (Badel et al, 2009)(Figure 16(a)) to analyze the local in plane shear of a plain weave unit cell (47214 dof). It cannot be considered (at least today) to use this last FE model to simulate the forming of a composite reinforcement that is made of several thousands of woven cells. In the simplified unit cell (Figure 16(b)) each yarn is described by few shell elements and the contact friction and possible relative displacement of the yarns are considered. The in-plane mechanical behaviour is the same as the one defined in (Badel et al, 2009). The bending stiffness is independent of the tensile one and very much reduced in comparison to the one given by plate theories.
Two examples are presented in figures 17 and 18 based on a discrete modelling using the unit cell of figure 16(b) (Gatouillat et al, 2010) The first one is a picture frame test for which the wrinkles appear naturally in the simulation when the shear locking angle is reached. It must be noticed that the in-plane shear behaviour of the fabric is not an input data of the analysis and does not need to be known. It results at the macroscopic level of contact and friction between the yarns and lateral compression of the yarns. Figure 18 shows the results of a hemispherical forming simulation. It must be said that this study concerning forming simulation at the meso-scopic scale is beginning at INSA Lyon. If the discrete or mesoscopic modelling is a promising approach because a large part of the mechanical specificity of fabric behaviour is due to yarn and fibre interactions, and following fibre directions is simpler than for continuous models, it must be recognized that the forming simulations made with approaches that permits the relative sliding of the yarns in contact are not many.

7. The semi-discrete finite elements for the composite reinforcement forming

This approach can be seen as intermediate between the continuous and the discrete approaches. The textile composite reinforcement is seen as a set of a discrete number of unit woven cells submitted to membrane loadings (i.e. biaxial tension and in-plane shear) and bending (Figure 19)(Hamila et al, 2009)
In any virtual displacement field $\underline{\eta}$ such as $\underline{\eta} = 0$ on the boundary with prescribed loads, the virtual work theorem relates the internal, exterior and acceleration virtual works:

$$W_{ext}(\underline{\eta}) - W_{int}(\underline{\eta}) = W_{acc}(\underline{\eta}) \tag{27}$$

Fig. 17. Simulation of a picture frame test using the unit cell model of figure 16(b)

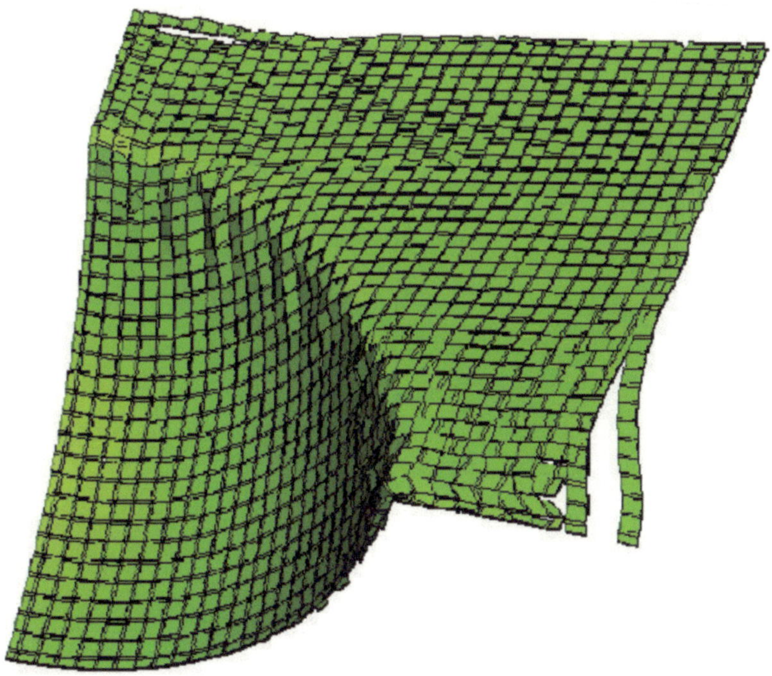

Fig. 18. Simulation of hemispherical forming test using the unit cell model of figure 16(b)

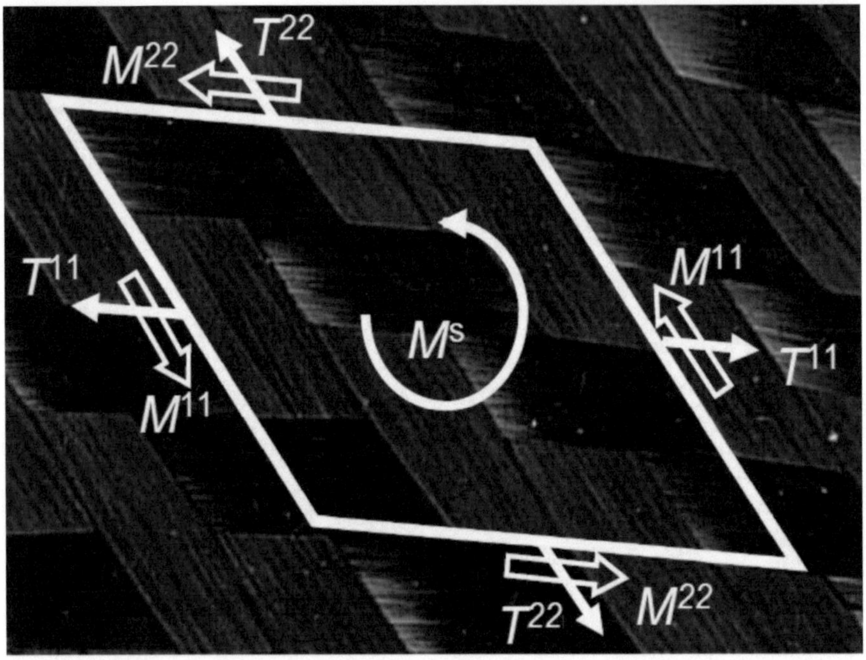

Fig. 19. Unit woven cell submitted to tension, in plane shear and bending

In the case of the woven fabric reinforcement, the internal virtual work is assumed to be separated into:

$$W_{int}(\underline{\eta}) = W_{int}^{t}(\underline{\eta}) + W_{int}^{s}(\underline{\eta}) + W_{int}^{b}(\underline{\eta}) \tag{28}$$

$W_{int}^{t}(\underline{\eta})$, $W_{int}^{s}(\underline{\eta})$, $W_{int}^{b}(\underline{\eta})$ are the internal virtual work of biaxial tension, in plane shear and bending respectively with :

$$W_{int}^{t}(\underline{\eta}) = \sum_{p=1}^{ncell} {}^{p}\varepsilon_{11}(\underline{\eta})\ {}^{p}T^{11}\ {}^{p}L_1 + {}^{p}\varepsilon_{22}(\underline{\eta})\ {}^{p}T^{22}\ {}^{p}L_2 \tag{29}$$

$$W_{int}^{s}(\underline{\eta}) = \sum_{p=1}^{ncell} {}^{p}\gamma(\underline{\eta})\ {}^{p}M^{s} \tag{30}$$

$$W_{int}^{b}(\underline{\eta}) = \sum_{p=1}^{ncell} {}^{p}\chi_{11}(\underline{\eta})\ {}^{p}M^{11}\ {}^{p}L_1 + {}^{p}\chi_{22}(\underline{\eta})\ {}^{p}M^{22}\ {}^{p}L_2 \tag{31}$$

where ncell is the number of woven cell. ${}^{p}A$ means that the quantity A is considered for the woven cell number p. L_1 and L_2 are the length of unit woven cell in warp and weft directions. $\varepsilon_{11}(\underline{\eta})$ and $\varepsilon_{22}(\underline{\eta})$ are the virtual axial strain in the warp and weft directions. $\gamma(\underline{\eta})$ is the virtual angle between warp and weft directions. $\chi_{11}(\underline{\eta})$ and $\chi_{22}(\underline{\eta})$ are the virtual curvatures of warp and weft directions. $\varepsilon_{11}(\underline{\eta})$, $\varepsilon_{22}(\underline{\eta})$, $\gamma(\underline{\eta})$, $\chi_{11}(\underline{\eta})$ and $\chi_{22}(\underline{\eta})$ are function of the gradient of the virtual displacement field. T^{11} and T^{22} are the tensions on the unit woven cell in warp and weft directions. M^{11} and M^{22} are the bending moments on the woven cell respectively in warp and weft directions. M^s is the in-plane shear moment. The loads on an edge of the woven unit cell (presented Figure 19) result in the tensions T^{11} and T^{22} in one hand and shear forces in the other hand. This shear forces on a warp and weft sections have a moment at the centre of the woven unit cell in the direction normal to the unit cell. The component of this moment is called in-plane shear moment M^s. This quantity is conjugated to the in-plane shear angle γ. The internal virtual work of in plane shear is directly given from M^s and the virtual shear angle (Equation 30). In the case of a textile material, the shear angle γ is a significant and clearly defined quantity and it is interesting to express the internal virtual work of in plane shear in function of this quantity.

The mechanical behaviour of the textile reinforcement defines a relation between the loads $T^{\alpha\alpha}$, M^s, $M^{\alpha\alpha}$ and the strain field. The experimental tests specific to textile composite reinforcements that have been presented in Section 4 are used to obtain T^{11}, T^{22}, M^s, $M^{11}$and M^{22} in function of ε_{11}, ε_{22}, γ, χ_{11} and χ_{22}.

An alternative consists in virtual tests i.e. in 3D simulations of the deformation of a unit woven cell submitted to elementary loadings such as biaxial tensions or in plane shear (Badel et al, 2008, Badel et al, 2009)

7.1 Shell finite element made of woven cells

A three node shell finite element have been defined from the simplified form of the principle of virtual works given in equations 28 to 31. The details of its formulation are given in (Hamila et al, 2009). It is summarized below.

A finite element interpolation is introduced within the principle of virtual work. The displacement $\underline{u}$ and virtual displacement $\underline{\eta}$ of any point within an element are in the form:

$$\underline{u} = \mathbf{N}\mathbf{u}^e \text{ and } \underline{\eta} = \mathbf{N}\boldsymbol{\eta}^e \tag{32}$$

$\mathbf{N}$ is the interpolation matrix of the element under consideration and $\mathbf{u}^e$ and $\boldsymbol{\eta}^e$ the single column matrices of its nodal displacements and virtual displacements respectively. Equation 28 leads to:

$$\mathbf{M}\ddot{\mathbf{u}} + \left(\mathbf{F}^t_{int} + \mathbf{F}^s_{int} + \mathbf{F}^b_{int}\right) - \mathbf{F}_{ext} = 0 \tag{33}$$

$\mathbf{M}$ is the mass matrix, $\mathbf{u}$ is the single column matrices of the nodal displacements. $\mathbf{F}^t_{int}, \mathbf{F}^s_{int}, \mathbf{F}^b_{int}$ are the single column matrices of the nodal internal forces respectively for tension, shear and bending.

This dynamic equation can be solved using an explicit scheme (central differences):

$$\ddot{\mathbf{u}}^{i+1} = \mathbf{M}_D^{i\ -1}\left(\mathbf{F}^i_{ext} - \mathbf{F}^{ti}_{int} - \mathbf{F}^{si}_{int} - \mathbf{F}^{bi}_{int}\right) \tag{34}$$

$$\dot{\mathbf{u}}^{i+1/2} = \dot{\mathbf{u}}^{i-1/2} + \frac{1}{2}\left(\Delta t^{i-1} + \Delta t^i\right)\ddot{\mathbf{u}}^{i+1} \tag{35}$$

$$\mathbf{u}^{i+1} = \mathbf{u}^i + \dot{\mathbf{u}}^{i+1/2}\Delta t^i \tag{36}$$

There is no system to solve since $\mathbf{M}_D$ is a diagonal matrix calculated from $\mathbf{M}$ (Zienkiewicz & Taylor, 2000). This explicit scheme requires the time step to be small enough to insure the

stability of the scheme (Belytschko, 1983). It is effective for many dynamic applications and also in material forming. For the sake of numerical efficiency, the speed can be larger than the real one under the condition that the dynamic effects do not modify the solution.
The 3 node shell finite element $M_1M_2M_3$ made up of ncelle woven cells is considered (Figure 20). The vectors $\underline{k}_1=\underline{AM}_2$ and $\underline{k}_2=\underline{BM}_3$ respectively in warp and weft directions are defined. The internal virtual work of tension on the element defines the element nodal tensile internal forces $\mathbf{F}_{int}^{te}$:

$$\sum_{p=1}^{ncelle} {}^{P}\varepsilon_{11}(\underline{\eta})\ {}^{P}T_1\ {}^{P}L_1 + {}^{P}\varepsilon_{22}(\underline{\eta})\ {}^{P}T_2\ {}^{P}L_2 = \boldsymbol{\eta}^{eT}\mathbf{F}_{int}^{te} \tag{37}$$

The internal tensile force components are calculated from the tensions T_1 and T_2:

$$\left(F_{int}^{te}\right)_{ij} = ncelle\left(B_{1ij}T_1\frac{L_1}{\|\underline{k}_1\|^2}+B_{2ij}T_2\frac{L_2}{\|\underline{k}_2\|^2}\right) \tag{38}$$

i is the index of the direction (i=1 to 3), j is the index of the node (j=1 to 3). B_{1ij} and B_{2ij} are strain interpolation components. They are constant over the element because the interpolation functions in equation 32 are linear in the case of the 3 node triangle.
The internal virtual work of in-plane shear on the element defines the element nodal tensile internal forces $\mathbf{F}_{int}^{se}$:

$$\sum_{p=1}^{ncelle} {}^{P}\gamma(\underline{\eta})\ {}^{P}M_s = \boldsymbol{\eta}^{eT}\mathbf{F}_{int}^{se} \tag{39}$$

The internal in-plane shear force components are calculated from the in-plane shear moment:

$$\left(F_{int}^{se}\right)_{ij} = ncelle\ B_{\gamma ij}M^s(\gamma) \tag{40}$$

In order to avoid supplementary degrees of freedom and consequently for numerical efficiency, the bending stiffness is taken into account within an approach without rotational degree of freedom (Onate & Zarate, 2000, Sabourin et Brunet, 2006). In theses approaches the curvatures of the element are computed from the positions and displacements of the nodes of the neighbouring elements (Figure 20). The internal virtual work of bending on the element defines the element nodal bending internal forces $\mathbf{F}_{int}^{be}$:

$$\sum_{p=1}^{ncelle} {}^{P}\chi_{11}(\underline{\eta})\ {}^{P}M_1\ {}^{P}L_1 + {}^{P}\chi_{22}(\underline{\eta})\ {}^{P}M_2\ {}^{P}L_2 = \boldsymbol{\eta}^{eT}\mathbf{F}_{int}^{be} \tag{41}$$

The internal bending force components are calculated from the bending moments M_1 and M_2:

$$\left(F_{int}^{be}\right)_{km} = n_{celle}\left(Bb_{1km}M_1\frac{L_1}{\|\underline{k}_1\|^2}+Bb_{2km}M_2\frac{L_2}{\|\underline{k}_2\|^2}\right) \tag{42}$$

The finite element presented above is used to simulate the hemispherical forming of the very unbalanced fabric presented in section 5.1. The experimental forming by a

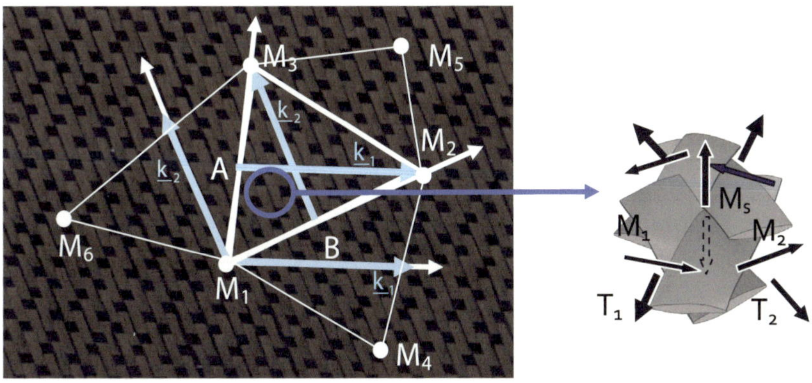

Fig. 20. Three node finite element made of unit woven cells,

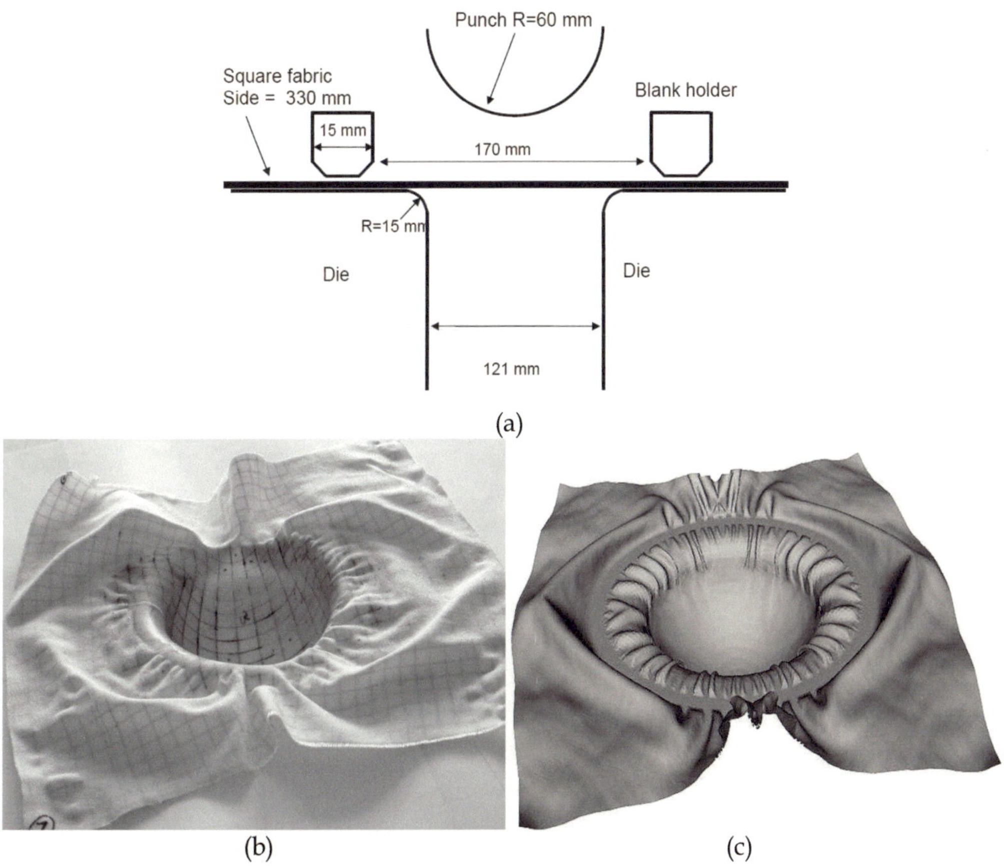

Fig. 21. Forming of an unbalanced textile reinforcement

hemispherical punch has been performed at the University of Nottingham (Daniel et al, 2003). A 6 kg ring was used as blank-holder avoiding reinforcement wrinkling in the hemispherical zone (Figure 21a). The experimental shape obtained after forming is shown in Figure 21b. In warp direction (with the strongest rigidity) large fabric sliding is observed relatively to the die. On the contrary, in the weft direction (weak direction) no edge movement is depicted and the yarns are subjected to large stretch deformations. The computed shape after forming is shown in Figure 21c. It is in good agreement with experiments. Especially the extension ratio at the centre of the hemisphere ($l_{weft}/l_{warp} = 1.8$) is correctly computed. The shape of the many wrinkles in the flat part of the textile is also properly simulated.

7.2 Extension to 3D interlock textile reinforcements

When the thickness of a composite part is large, the use of these laminated composites is restricted by manufacturing problems and their low resistance to delamination cracking. To overcome these difficulties composites with 3D fibre architecture called ply to ply interlock fabric have been proposed (Tong et al, 2002). This material is not fully 3D since there is no third yarn set in the transverse direction but the properties through the thickness are much improved. Above all, the possible delaminations of the 2D laminated composites are overcome. (Figure 22). The semi-discrete approach has been extended to 3D hexahedral finite elements for interlock forming simulations; These elements are made of yarns as shown Figure 23. The simulation of a thick twisted plate made of interlock textile reinforcement is shown figure 24. The formulation of these finite elements for interlock forming simulations is given in (De Luycker et al, 2009).

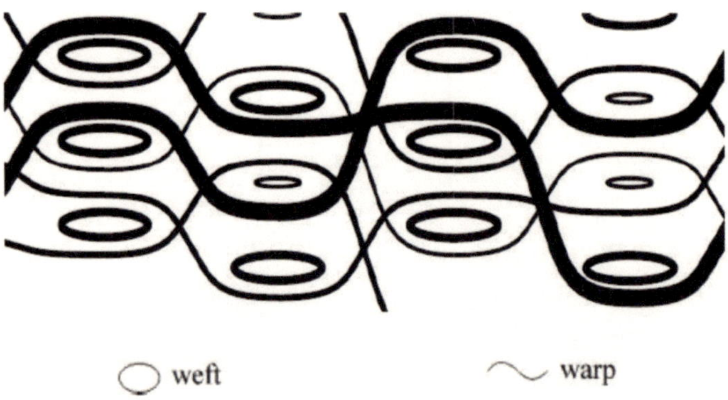

Fig. 22. Example of complex layer interlock weave

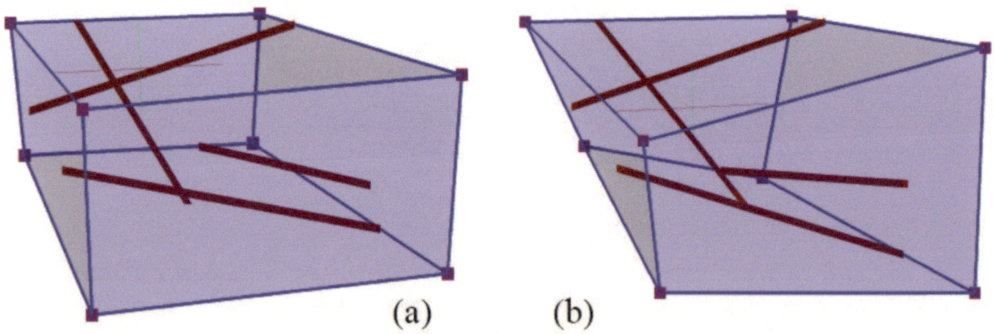

Fig. 23. Height node hexahedral finite element containing yarns - (a) Initial state–
(b) Deformed state.

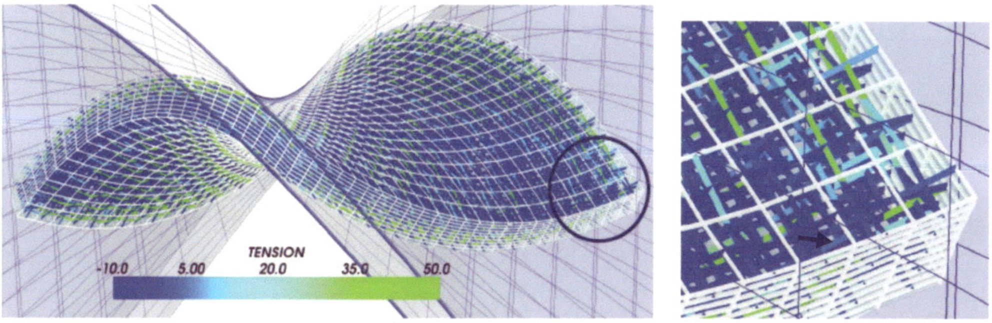

Fig. 24. Forming of a 3D interlock twisted plate

8. Conclusion

Different approaches for composite forming simulations have been presented in this chapter. They are continuous discrete or semi-discrete element. These different approaches are based on the strong multi-scale nature of the textile reinforcements.

The discrete approach is attractive and promising. The very specific mechanical behaviour of the textile material due to the contacts and friction between the yarns and to the change of direction is explicitly taken into account. If some sliding occurs between warp and weft yarns, they can be simulated. This is not possible by the continuous approaches that consider the textile material as a continuum. This is an important point because it can be necessary to prevent such a sliding in a process. Nevertheless, the main difficulty of the discrete approach is the necessary compromise that must be done between the accuracy of the model of the unit woven cell and the total number of degrees of freedom. The modelling of the unit cell must be accurate enough to obtain a correct macroscopic mechanical behaviour, but the number of degrees of freedom of each cell must remain small in order to compute a forming process for which there will be thousand of woven cells. There are a lot of improvements to achieve in the meso-modelling of different textile reinforcements. The continuous increase of the computer power is a strong argument in favour of this approach.

The continuous approach is the most commonly used in composite reinforcement forming today. The main advantage is to use standard shell or membrane finite element. The only mechanical behaviour has to be specified in order to take the very particular behaviour of textile materials into account. Many models exist, but none of them is clearly admitted. The semi-discrete approach aims to avoid the use of stress tensors and directly define the loading on a woven unit cell by the warp and weft tensions and by in-plane shear and bending moments. These quantities are simply defined on a woven unit cell and above all they are directly measured by standard tests on composite reinforcements (biaxial tension, picture frame, bias extension and bending tests).

9. References

Advani SG. Flow and rheology in polymeric composites manufacturing. Elsevier, 1994.

Aimène Y, Vidal-Sallé E, Hagège B, Sidoroff F, Boisse P, A hyperelastic approach for composite reinforcement large deformation analysis, Journal of Composite Materials Vol. 44, No. 1/2010, 5-26

Badel P, Vidal-Sallé E, Maire E, Boisse P (2009) Simulation and tomography analysis of textile composite reinforcement deformation at the mesoscopic scale. Composites Science and Technology 68:2433-2440

Badel P, Gauthier S, Vidal-Salle E, Boisse P. Rate constitutive equations for computational analyses of textile composite reinforcement mechanical behaviour during forming. Composites: Part A 40 (2009) 997–1007

Basar Y, Weichert D (2000) Nonlinear continuum mechanics of solids. Springer, Berlin

Belytschko T. An overview of semidiscretisation and time integration procedures. In: Belytschko T, Hughes TJR, editors, Computation Methods for Transient Analysis. Amsterdam: Elsevier, 1983.

Ben Boukaber B, Haussy G, Ganghoffer JF (2007) Discrete models of woven structures. Macroscopic approach. Composites: Part B 38:498-505

Borouchaki H, Cherouat A. Une nouvelle approche pour le drappage des structures composites. Rev Comp Mat Avanc 2002;32:407–22

Buet-Gautier K., Boisse P. Experimental analysis and modeling of biaxial mechanical behavior of woven composite reinforcements. Experimental Mechanics 2001; 41 (3): 260-269.

Cao J., Akkerman R., Boisse P., Chen J. et al. Characterization of Mechanical Behavior of Woven Fabrics: Experimental Methods and Benchmark Results. Composites Part A 2008; 39: 1037-1053.

Carvelli V., Corazza C., Poggi C. Mechanical modelling of monofilament technical textiles. Computational Materials Science 2008; 42: 679-691.

Dong L, Lekakou C, Bader MG. Processing of composites: simulations of the draping of fabrics with updated material behaviour law. Journal of Composite Materials 2001; 35: 38–163.

Daniel JL, Soulat D, Dumont F, Zouari B, Boisse P, Long AC, Forming simulation of very unbalanced woven composite reinforcements. International Journal of Forming Processes 2003, 6:465-480

de Bilbao E, Soulat D, Hivet G, Gasser A., Experimental Study of Bending Behaviour of Reinforcements, Experimental Mechanics (2010) 50:333–351

De Luycker E, F. Morestin, P. Boisse, D. Marsal, Simulation of 3D interlock composite preforming, Composite Structures, 88, Issue 4, May 2009, Pages 615-623

Duhovic M, Bhattacharyya D (2006) Simulating the deformation mechanisms of knitted fabric composites. Composites: Part A 37:1897-1915

Dumont F, Hivet G, Rotinat R, Launay J, Boisse P, Vacher P. Field measurements for shear tests on woven reinforcements. Mécanique et Industrie, 2003; 4:627–35.

Durville D, 2005, Numerical simulation of entangled materials mechanical properties. Journal of Materials Science 40:5941-5948

Gatouillat S, Vidal-Salle E, Boisse P, Advantages of the meso/macro approach for the simulation of fibre composite reinforcements, Proceedings of the ESAFORM 2010 Conference, Brescia, April 2010, Italy

Hamila N., Boisse P., Sabourin F., Brunet M. A semi-discrete shell finite element for textile composite reinforcement forming simulation. Int J Numerical Methods in Engineering 2009; 79: 1443-1466.

Harrison P., Clifford MJ., Long AC. Shear characterisation of viscous woven textile composites: a comparison between picture frame and bias extension experiments. Compos Sci Tech 2004; 64: 1453-1465.

Hughes TJR, Winget J (1980) Finite rotation effects in numerical integration of rate constitutive equations arising in large deformation analysis. International Journal for Numerical Methods in Engineering 15:1862-1867

Kawabata S., Niwa M., Kawai H. The Finite Deformation Theory of Plain Weave Fabrics Part I: The Biaxial Deformation Theory. Journal of the Textile Institute 1973; 64(1): 21-46.

King MJ, Jearanaisilawong P, Socrate S. A continuum constitutive model for the mechanical behavior of woven fabrics. International Journal of Solids and Structures 2005; 42: 3867–3896.

Kawabata S. The Standardization and Analysis of Hand Evaluation. Osaka: The Textile Machinery Society of Japan, 1986.

M.A. Khan, T. Mabrouki, E. Vidal-Sallé, P. Boisse, Numerical and experimental analyses of woven composite reinforcement forming using a hypoelastic behaviour. Application to the double dome benchmark, Journal of Materials Processing Technology 210 (2010) 378–388

Lahey TJ., Heppler GR. Mechanical Modeling of Fabrics in Bending. ASME Journal of Applied Mechanics 2004; 71: 32-40.

Launay J., Hivet G., Duong AV., Boisse P. Experimental analysis of the influence of tensions on in plane shear behaviour of woven composite reinforcements. Compos Sci Tech 2008; 68: 506-515.

Lebrun G., Bureau MN., Denault J. Evaluation of bias-extension and picture-frame test methods for the measurement of intraply shear properties of PP/glass commingled fabrics. Compos Struct 2003; 61: 341-52

Lomov S., Boisse P., Deluycker E., Morestin F., Vanclooster K., Vandepitte D., Verpoest I., Willems A. Full field strain measurements in textile deformability studies. Composites: Part A 2008; 39: 1232-1244.

Lomov SV., Willems A., Verpoest I., Zhu Y., Barburski M., Stoilova Tz. Picture frame test of woven composite reinforcements with a full-field strain registration. Textile Research Journal 2006; 76 (3): 243-252.

Long A.C., Rudd C.D. (1994), 'A simulation of reinforcement deformation during the production of preform for liquid moulding processes', *I. Mech. E. J. Eng. Manuf.*, 208, 269-278.

Mark C., Taylor H. M. (1956), 'The fitting of woven cloth to surfaces', *Journal of Textile Institute*, 47, 477-488

Miao Y, Zhou E, Wang Y, Cheeseman BA (2008) Mechanics of textile composites: Micro-geometry. Composites Science and Technology 68:1671-1678

Parnas RS. Liquid Composite Molding. Hanser Garner publications, 2000.

Onate E., Zarate F. Rotation-free triangular plate and shell elements. Int J for Num Meth in Eng 2000; 47: 557-603.

Peng XQ., Cao J., Chen J., Xue P., Lussier DS., Liu L. Experimental and numerical analysis on normalization of picture frame tests for composite materials. Compos Sci Tech 2004; 64: 11-21.

Peng X, Cao J. A continuum mechanics-based non-orthogonal constitutive model for woven composite fabrics. Composites Part A 2005; 36: 859–874.

Pickett AK, Creech G, de Luca P (2005) Simplified and Advanced Simulation Methods for Prediction of Fabric Draping. European Journal of Computational Mechanics 14:677-691

Potter K. Bias extension measurements on cross-plied unidirectional prepreg. Composites Part A 2002; 33: 63-73.

Potluri P., Perez Ciurezu DA., Ramgulam RB. Measurement of meso-scale shear deformations for modelling textile composites. Composites Part A 2006; 37: 303-314.

Prodromou AG., Chen J. On the relationship between shear angle and wrinkling of textile composite preforms. Composite Part A 1997; 28A:491-503.

Rozant O., Bourban PE., Manson JAE. Drapability of dry textile fabrics for stampable thermoplastic preforms. Composites: Part A 2000; 31: 1167-1177.

Rudd CD., Long AC. Liquid Molding Technologies. Cambridge: Woodhead Pub. Lim., 1997.

Spencer AJM. Theory of fabric-reinforced viscous fluid. Composites Part A 2000; 31: 1311–1321.

Sabourin F., Brunet M. Detailed formulation of the rotation-free triangular element "S3" for general purpose shell analysis. Engineering computations 2006; 23 (5): 469-502.

Sharma S.B., Sutcliffe M.P.F., Chang S.H. Characterisation of material properties for draping of dry woven composite material. Composites Part A, 2003; 34:1167–1175.

Spencer A.J.M. - Theory of fabric-reinforced viscous fluids - Composites: Part A 31 (2000) 1311–1321

Ten Thije RHW, Akkerman R, Huetink J. Large deformation simulation of anisotropic material using an updated Lagrangian finite element method. Computer methods in applied mechanics and engineering 2007; 196(33–34): 3141–3150.

Tong L, Mouritz AP, Bannister MK. 3D Fibre reinforced polymer composites. Elsevier Science, 2002.

Truesdell C (1955) Hypo-elasticity. J Ration Mech Anal 4:83-133

Van Der Ween F. (1991), 'Algorithms for draping fabrics on doubly curved surfaces', *International Journal of Numerical Method in Engineering,* 31, 1414-1426.

Willems A., Lomov SV., Verpoest I., Vandepitte D. Optical strain fields in shear and tensile testing of textile reinforcements. Composites Science and Technology 2008; 68: 807-819.

Xiao H, Bruhns OT, Meyers A (1998) On objective corotational rates and their defining spin tensors. International Journal of Solids and Structures 35:4001-4014

Yu W.R., Pourboghrata F., Chungb K., Zampalonia M., Kang T. J. - Non-orthogonal constitutive equation for woven fabric reinforced thermoplastic composites - Composites: Part A 33 (2002) 1095–1105

Zienkiewicz OC., Taylor RL. The finite element method, vol. 2: Solid Mechanics. Oxford: Butterworth, Heineman, 2000.